Springer Series in Wood Science

Editor: T. E. Timell

Springer
Berlin
Heidelberg
New York
Barcelona
Budapest
Hong Kong
London
Milan
Paris
Santa Clara
Singapore
Tokyo

Springer Series in Wood Science

Editor: T. E. Timell

M. H. Zimmermann
Xylem Structure and the Ascent of Sap (1983)

J. F. Siau
Transport Processes in Wood (1984)

R. R. Archer
Growth Stresses and Strains in Trees (1986)

W. E. Hillis
Heartwood and Tree Exudates (1987)

S. Carlquist
Comparative Wood Anatomy (1988)

L. W. Roberts/P. B. Gahan/R. Aloni
Vascular Differentiation and Plant Growth Regulators (1988)

C. Skaar
Wood-Water Relations (1988)

J. M. Harris
Spiral Grain and Wave Phenomena in Wood Formation (1989)

B. J. Zobel/J. P. van Buijtenen
Wood Variation (1989)

P. Hakkila
Utilization of Residual Forest Biomass (1989)

J. W. Rowe (Ed.)
Natural Products of Woody Plants (1989)

K.-E. L. Eriksson/R. A. Blanchette/P. Ander
Microbial and Enzymatic Degradation of Wood and Wood Components (1990)

R. A. Blanchette/A. R. Biggs (Eds.)
Defense Mechanisms of Woody Plants Against Fungi (1992)

S. Y. Lin/C. W. Dence (Eds.)
Methods in Lignin Chemistry (1992)

G. Torgovnikov
Dielectric Properties of Wood and Wood-Based Materials (1993)

F. H. Schweingruber
Trees and Wood in Dendrochronology (1993)

P. R. Larson
The Vascular Cambium: Development and Structure (1994)

M.-S. Ilvessalo-Pfäffli
Fiber Atlas: Identification of Papermaking Fibers (1995)

B. J. Zobel/J. B. Jett
Genetics of Wood Production (1995)

C. Mattheck/H. Kubler
Wood – The Internal Optimization of Trees (1995)

T. Higuchi
Biochemistry and Molecular Biology of Wood (1997)

T. Higuchi

Biochemistry and Molecular Biology of Wood

With 200 Figures and 25 Tables

Springer

Prof. Dr. TAKAYOSHI HIGUCHI
Wood Research Institute
Kyoto University (retired)
Momoyamacho, Yosai 22-8
Fushimi-ku, Kyoto 612
Japan

Series Editor:

T. E. TIMELL
State University of New York
College of Environmental Science and Forestry
Syracuse, NY 13210, USA

Cover: Transverse section of *Pinus lambertiana* wood. Courtesy of Dr. Carl de Zeeuw, SUNY College of Environmental Science and Forestry, Syracuse, New York

ISBN-13:978-3-642-64419-1 e-ISBN-13:978-3-642-60469-0
DOI: 10.1007/978-3-642-60469-0

Library of Congress Cataloging-in-Publication Data. Higuchi, Takayoshi, Biochemistry and molecular biology of wood/T. Higuchi, T. Umezawa. p. cm. – (Springer series in wood science) Includes bibliographical references and index. ISBN-13:978-3-642-64419-1 1. Wood – Anatomy. 2. Wood – Chemistry. 3. Plant molecular biology. I. Umezawa, T. (Toshiaki), 1957- . II. Title. III. Series. QK647.H53 1997 582'.0495 – dc20 96-28429

Softcover reprint of the hardcover 1st edition 1997

Cover design: Design & Production GmbH, Heidelberg

Typesetting: Best-set Typesetter Ltd., Hong Kong

SPIN: 10493637 31/3137/SPS – 5 4 3 2 1 0 – Printed on acid-free paper

Preface

Progress in wood chemistry has been related mainly to chemical wood pulping and bleaching and chemical utilization of wood and wood extractives. Methods of wood analysis were developed by Schorger (proximate analysis in 1917) and Dore (summative analysis in 1919), and standard methods based on Schorger's method, e.g., TAPPI standards (Technical Association of the Pulp and Paper Industry), have been widely used for chemical analysis of woods in many countries. Thus it is generally known that wood is composed of about 50% cellulose, 20–35% of lignin, 15–25% of hemicelluloses, and variable amounts of extractives. Chemical characterization and efficient utilization of these wood components have been studied in laboratories of wood chemistry and technology in universities and government institutions.

In the last decade, biochemistry and molecular biology of microorganisms, animals, and plants have greatly progressed. At the same time wood has been recognized as a unique renewable ecomaterial produced by trees using solar energy. In addition, many desirable properties of wood and wood components as biomaterial that affects physiology and psychology in humans have recently attracted attention.

In order to elucidate the properties of wood and wood components produced by trees, characterization of genes encoding enzymes involved in the biosynthesis of wood components, differentiation of the cambium into phloem and xylem, and the mechanisms of the expression of these genes needed to be investigated. Research on the molecular biology of trees and wood has just begun in Europe, USA, Canada, New Zealand, and Japan, among other countries, and rapid progress is being made.

I became interested in wood biochemistry during my stay at Gifu University, where the first Wood Biochemistry Laboratory in Japan was founded in 1949, and at the Wood Research Institute, Kyoto University, through research on the biosynthesis of lignin and formation of heartwood.

I am very much indebted to Dr. Toshiaki Umezawa from the Wood Research Institute at Kyoto University, who contributed to the section on "Lignans" (Sect. 4.3.5). Dr. Umezawa was the first to synthesize the lignans, (–)-secoisolariciresinol, and (–)-matairesinol enzymatically from coniferyl alcohol. I wish to thank Dr. Stewart A. Brown (Trent University, Canada), Dr. Fernand Barnoud (Grenoble University, France), Dr. M. Shimada and Dr. F. Nakatsubo (Kyoto University), Dr. T. Yamasaki (Kagawa University), Dr. H. Ohashi and Dr. M. Tanahashi (Gifu University), and Dr. Y. Nakamura, Dr. H. Kuroda, Dr. H. Kutsuki, Dr. S. Kawai, and Dr. H. Fushiki for their assistance during my studies on lignin biochemistry. I also wish to thank Dr. T. Itoh

(Wood Research Institute, Kyoto University), Dr. D. Shibata and Dr. T. Hibino (Mitsui Plant Biotechnology Research Institute), Dr. H. Fukuda (Tokyo University), Dr. H. Ohashi (Gifu University), Dr. K. Fukazawa (Emeritus Professor at Hokkaido University), Dr. T. Minamikawa (Tokyo Metropolitan University), Dr. T. Hayashi and Dr. J. Sugiyama (Wood Research Institute, Kyoto University), Dr. K. Takabe (Kyoto University), Dr. H. Kuroda (Wood Research Institute, Kyoto University), Dr. R. Funada (Hokkaido University), and Dr. K. Fukushima (Nagoya University) for their support and help with this book. I thank the many scientists and publishers who kindly granted permission for me to use their original illustrations for this book. Finally, I would like to thank Dr. T. E. Timell, the editor of this series, and the staff of Springer-Verlag for their help and encouragement during the preparation of this book.

Kyoto, Japan
Autumn 1996

TAKAYOSHI HIGUCHI

Contents

1 Structure and Functions of Wood

Trees used as wood materials for timber, furniture, and pulp industries are softwoods (conifers), which are gymnosperms, and hardwoods, which are dicotyledonous angiosperms. Bamboos, other grasses, and palms are included in the monocotyledons.

Conifers are usually evergreen, with narrow, needlelike or scalelike leaves, and bear exposed seeds usually in cones, whereas hardwoods are mostly deciduous, have net-veined leaves, and bear seeds enclosed in fruits (Little 1993).

Bamboos are perennial grasses with woody stems or culms that occur mostly in tropical, subtropical, and temperate regions and are abundant in tropical Asia. Bamboos belong to the Bambusoideae, a subfamily of the Gramineae. There are about 750 species of bamboo in about 45 genera.

Abies, *Chamaecyparis*, *Cedrus*, *Cupressus*, *Cryptomeria*, *Juniperus*, *Larix*, *Picea*, *Pinus*, *Podocarpus*, *Pseudotsuga*, *Sciadopitys*, *Taxodium*, *Taxus*, *Tsuga*, *Thuja*, and *Thujopsis* (conifers) and *Acacia*, *Acer*, *Aesculus*, *Alnus*, *Betula*, *Castanea*, *Celtis*, *Eucalyptus*, *Fagus*, *Fraxinus*, *Juglans*, *Magnolia*, *Nothofagus*, *Populus*, *Quercus*, *Salix*, *Tilia*, *Ulmus*, and *Zelkova* (hardwoods) are some of the trees found in temperate and warm-temperate zones that are useful in the wood and pulp industries.

In addition, *Cassia*, *Diospyros*, *Dipterocarpus*, *Hevea*, *Mansonia*, *Ochroma*, *Pentacme*, *Shorea*, *Swietenia*, *Tectona*, *Terminalia*, *Tieghemella*, among others, are important tropical hardwoods used as wood-based material and furniture.

The tree is composed of a crown, a stem, and a root. Leaves of the crown produce sugars by photosynthesis from CO_2 absorbed by stomata and from water transported from the root. The synthesized sugars (mainly in the form of sucrose) are transported via the phloem to the meristematic tissues of the trunk and branches and are partly used as energy sources. They are converted to tree components such as proteins, cellulose, hemicelluloses, lignin, and wood extractives during tree growth.

The stem is a conductive organ for water and photosynthates; it is composed of phloem (bark) and xylem (wood) and supports the crown, along with the roots. The xylem is involved in transportation of water and minerals from the root, while the phloem transports photosynthesized and metabolic products. The root absorbs water and minerals from the soil and fixes the tree trunk in soil (Fig. 1).

The outermost layer of the tree trunk is composed of outer bark and phloem (inner bark), which covers the xylem. Between the phloem and the

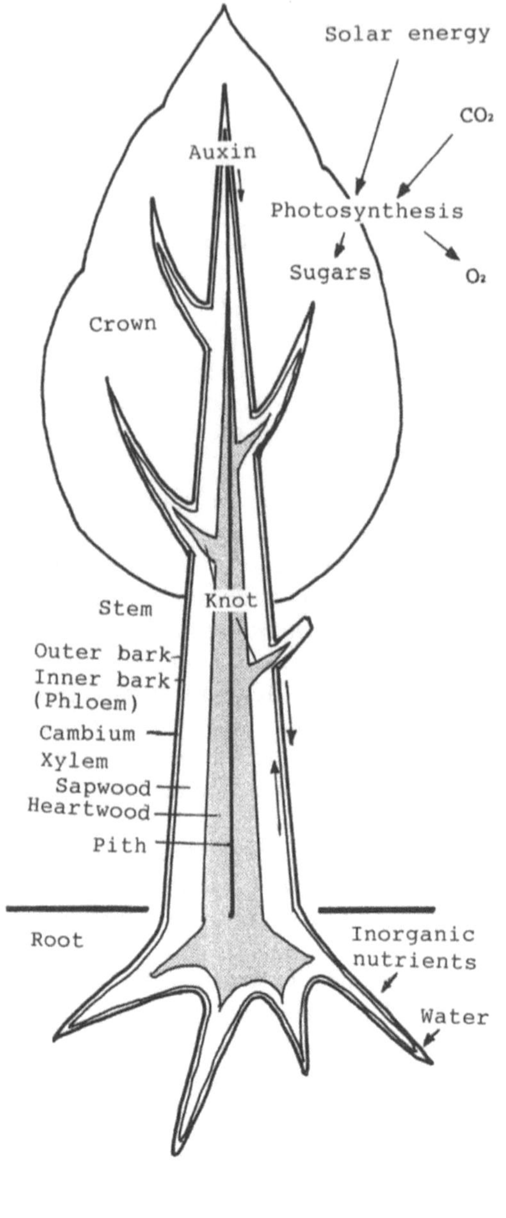

Fig. 1. Organization and functions of a tree (schematic). Water and inorganic nutrients are transported from root to upper tissues via outer sapwood, while photosynthesized sugars and auxins are transported from leaves and meristems to lower tissues and supplied to cambial tissues via phloem. (Courtesy of Dr. M. Fujita)

xylem, a cambial layer is located. Pith is present at the center of the trunk. The shapes and arrangements of the cells in the xylem are different at cross, radial, and tangential surfaces of the cut wood. Microscopic characteristics of tissues on the three surfaces of the cut wood have therefore been used to identify wood species (Saiki 1982; Fig. 2).

The cambium is derived from a procambium located slightly below the growing point of a tree stem. The procambium is arranged as a concentric

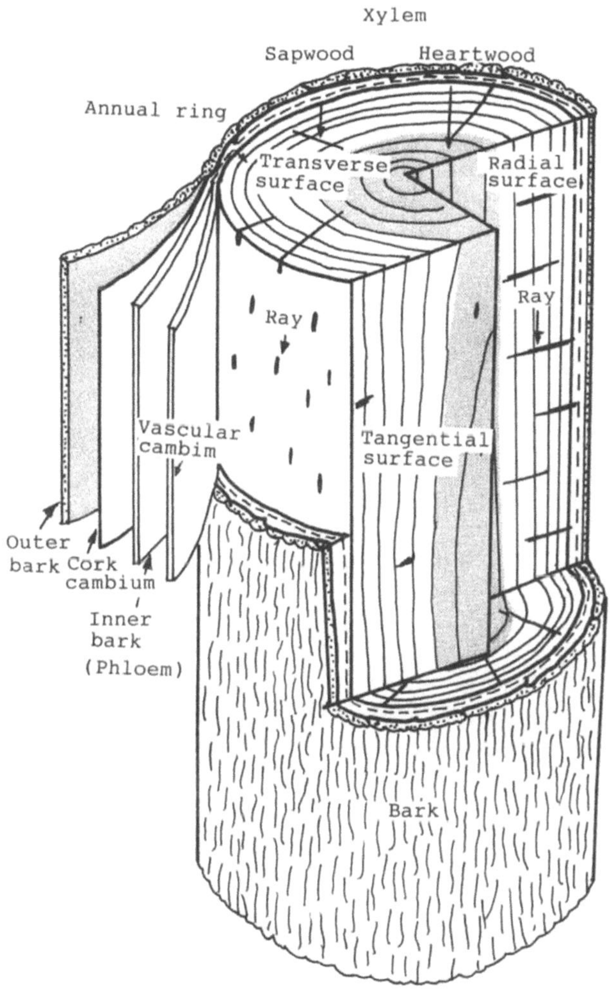

Fig. 2. Planes of wood and their designation: transverse surface or cross section, radial surface, and tangential surface. (Courtesy of Dr. M. Fujita)

circle on the cross surface of the growing tip of the tree and becomes a vascular bundle, a permanent tissue, as the tree grows. The vascular bundle is composed of two different tissues, fascicular phloem and fascicular xylem, and between the two tissues there is a thin layer of meristematic cells, the fascicular cambium.

During tree growth, fascicular cambial cells are divided tangentially to produce xylem tissue on the inner side and phloem tissue on the outer side. In conjunction with this process, interfascicular cambium is formed in the fundamental tissue between the vascular bundles, connecting the neighboring fascicular cambia to form a vascular cambium. Thus the formation of the vascular cambium, which covers pith and primary xylem, is completed as a secondary meristematic tissue (Shimaji 1976a; Fig. 3).

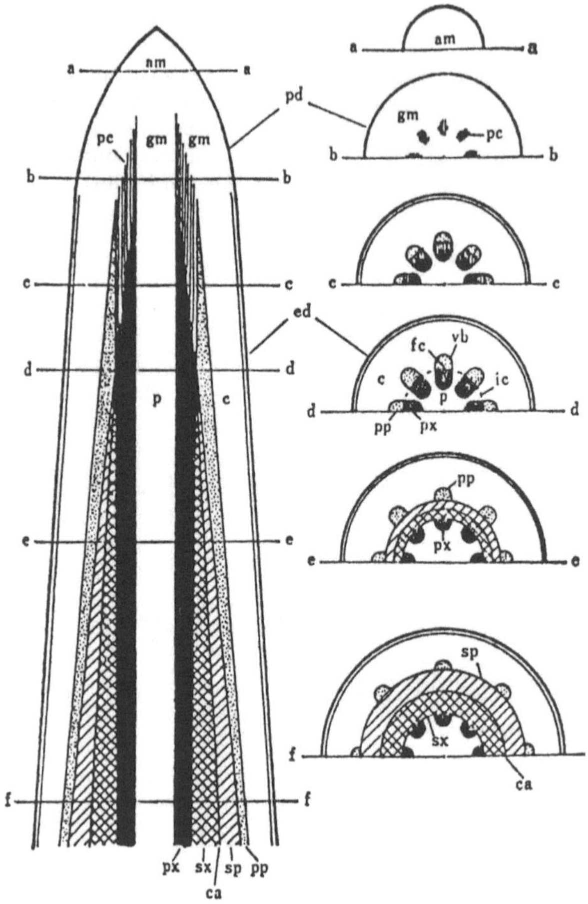

Fig. 3. Schematic model of a developing tree stem. *am* Meristematic tissue of growing point; *pd* dermatogen; *gm* fundamental meristem; *pc* procambium; *ed* epidermis; *p* pith; *c* cortex; *vb* vascular bundle; *fc* fascicular cambium; *ic* interfascicular cambium; *pp* primary phloem; *px* primary xylem; *sp* secondary phloem; *sx* secondary xylem; *ca* cambium. *a*, *b*, *c*, *d*, *e*, *f* denote transverse surfaces of a developing tree stem, respectively. *a* Growing point (apical meristem); *b* tissue consisting of procambium and fundamental meristem; *c* tissue consisting of vascular bundle and ground tissue; *d* tissue consisting of epidermis, cortex, and vascular bundles differentiating to fascicular and interfascicular cambia, and pith; *f* mature xylem tissue consisting of secondary xylem and secondary phloem. (Shimaji 1976a)

The vascular cambium produces xylem tissue at the inner side and phloem tissue at the outer side, followed by enlargement of the circumference of the cambium. The xylem and the phloem newly formed by the division of the cambial cells are called secondary xylem and secondary phloem, respectively.

Figure 4 shows the cross surface of wood of a 3-year-old *Pinus resinosa* and the mode of cell division and growth of the cambial cells.

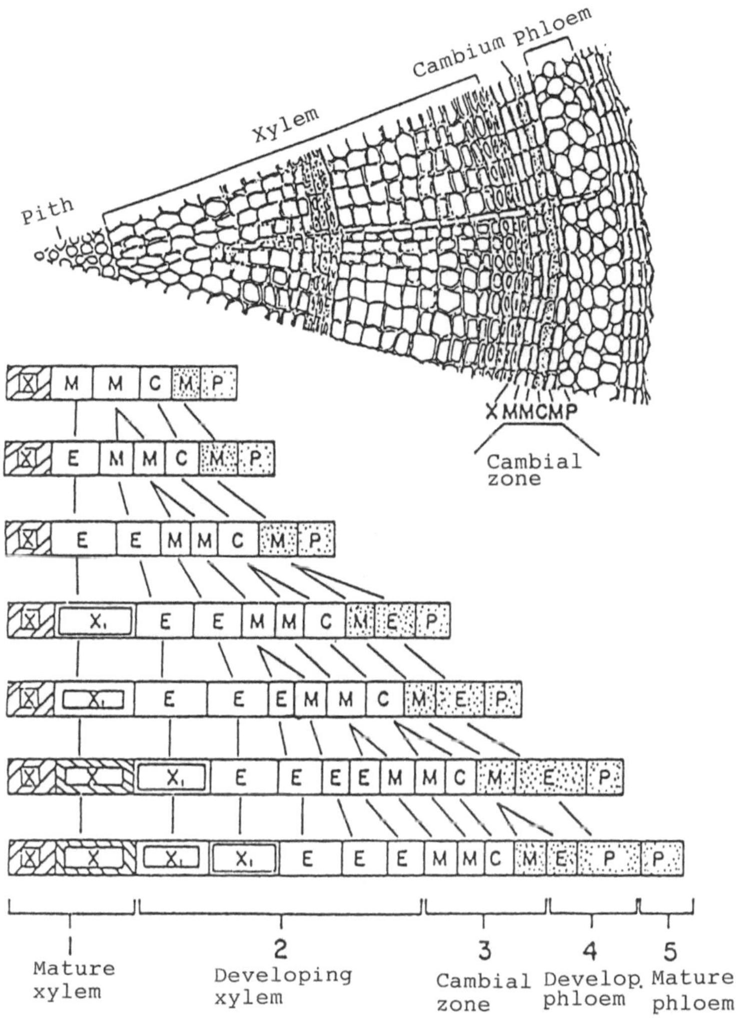

Fig. 4. Transverse surface of wood of a 3-year-old red pine, and the mode of cell growth of the cambial zone. *1* Mature xylem; *2* developing xylem; *3* cambial zone; *4* developing phloem; *5* mature phloem. *C* Cambial initial; *M* xylem mother cell (*left of C*); phloem mother cell (*right of C*); *X* mature xylem cell; *P* mature phloem cell; *E* growing cell

Since cambial cells divide continuously during growth, the secondary xylem continuously increases in thickness, a feature typical of woody plants, especially trees.

Cambial initials are composed of fusiform initials and ray initials. The former divide both in radial and tangential directions. In radial division of xylem and phloem mother cells, one of the divided cells always remains a fusiform initial, while another cell is transformed into a xylem or phloem

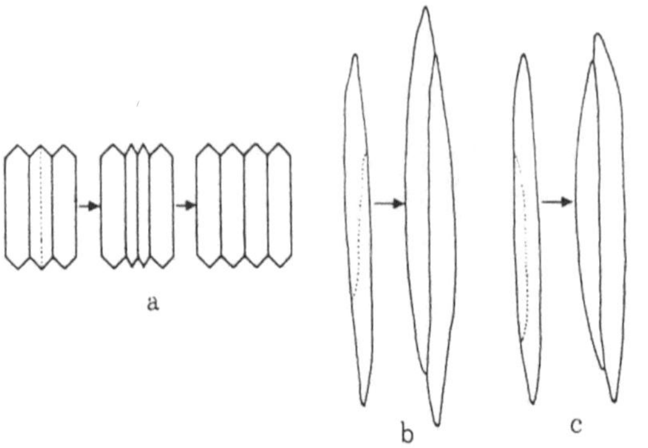

Fig. 5a–c. Pattern of the increase in fusiform initials. **a** Radial division; **b** pseudotransverse division; **c** lateral division. (Shimaji 1976b)

mother cell. The xylem and phloem mother cells divide further to give xylem on the inner side and phloem on the outer side, respectively.

In tangential division, fusiform initials first divide longitudinally, after which the initial cells formed divide radially (radial division) to give xylem tissues (stratification of hardwoods; Fig. 5a).

The risk of harmful mutations from errors occurring during the mitotic cycles of permanent initials that divide hundreds or even thousands of times in long-lived species has been stressed by Gahan (1988). Such a risk would be avoided if the initiating role could possibly pass to one of the nondetermined derivatives, although the molecular mechanism of this has not been elucidated.

In a different type of division the initial cell first divides up and down at the slightly S-shaped surface (pseudotransverse division; Fig. 5b), and the two cells formed grow contiguously in a tangential direction, following which both cells continuously divide (lateral division) in the radial direction as fusiform initials (Fig. 5c).

Enlargement of the cambial circle occurs mainly by an increase in the number of fusiform initials, but ray initials also increase, accompanied by the enlargement of the cambium. Bailey (1923) found that the diameters of fusiform initials (16 μm) and ray initials (14 μm) of a 1-year-old *Pinus strobus* increased to 42 and 17 μm, respectively, in a 60-year-old tree, while the number of fusiform initials ($n = 720$) and ray initials ($n = 70$) in the cambium of the 1-year-old tree increased to 23 000 and 8800, respectively, in the 60-year-old tree, indicating that the increase in the circumference of the cambium is largely due to the increase in the number of initials, especially fusiform initials.

The increase in ray initials occurs by a process of repeated tangential divisions of short fusiform initials formed by pseudolateral (Fig. 6a,b) and lateral divisions (Fig. 6c) of fusiform initials. In this process, short fusiform initials are rapidly contracted to form simple ray initials. Furthermore, short fusiform initials are generally divided at cross surfaces to give ray initials (Shimaji 1976b).

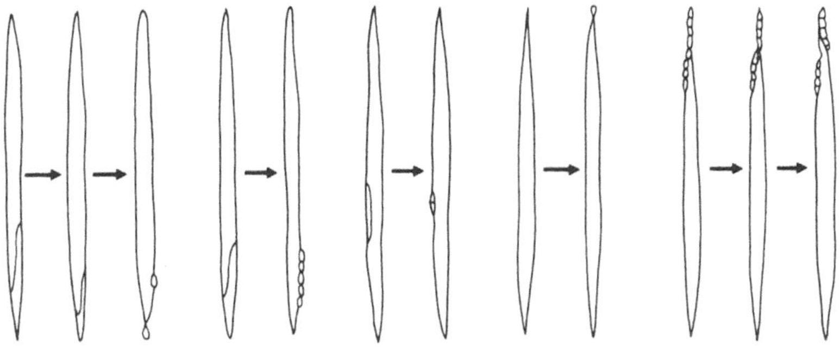

Fig. 6a–e. Pattern of the increase in ray initials. **a,b** By pseudolateral division; **c** by lateral division; **d** from a tip of fusiform initial; **e** by intrusive growth of a fusiform initial. (Shimaji 1976b)

In addition, the tip of a long fusiform initial can be divided to form ray initials (Fig. 6d), and by intrusive growth of a fusiform initial a group of ray initials is divided into two groups of initials that increase by division (Fig. 6e).

Newly formed constituent cells in the ray initials groups increase by repeated lateral divisions. The ray initials and fusiform initials are interwoven and contribute to the strength of the whole cambial fabric and also to that of the vascular tissues with their horizontal and vertical systems. The relative ratio between the two types of initials is therefore roughly maintained throughout the life of the tree.

The production of new rays by transverse divisions of the fusiform initials and then splitting and uniting of preexisting rays is considered to arise in response to complex axial and radial flows of developmental signals such as the polar auxin flow (Catesson 1994).

In temperate and warm-temperate zones, cambial activity has a certain periodicity, and active cell divisions and growth occur in the spring to summer seasons. Xylem produced in spring and early summer is called earlywood and is composed of relatively large xylem cells with thin walls. Xylem produced from late summer to autumn is called latewood. These cells are relatively small, and the cell walls are thicker than those of earlywood. Xylem composed of earlywood and latewood in temperate and warm-temperate zones therefore appears as annual rings of concentric circles on the cross-section of a stem.

Cell-length variations with season, age, and systematic position influence wood properties. Occasionally, false annual rings are formed in defoliated trees by disease or insect attack.

Wood produced in tropical rain forests generally has no annual rings, but sometimes contains false growth rings produced under different growth conditions in wet and dry seasons.

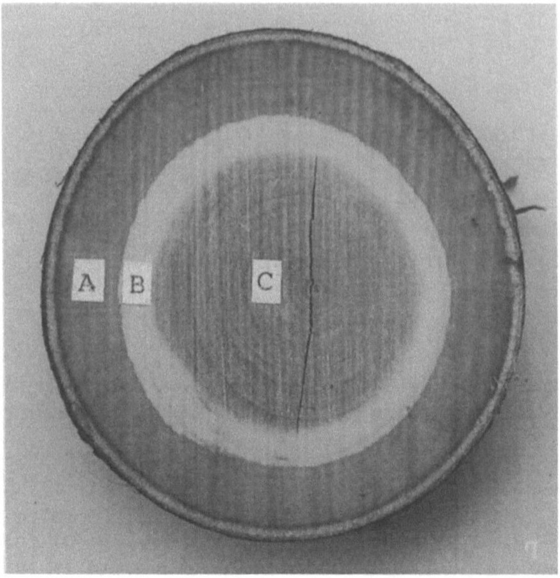

Fig. 7. A cross section of wood of *Prunus sargentii*. *A* Sapwood; *B* transition wood; *C* heartwood. (Courtesy of Dr. S. Ishida)

After a variable number of years, the central portion of the stem is inactivated; it becomes filled with resinous materials and species-characteristic phenolics and develops a dark color. This part is called heartwood; all cells in the heartwood are dead. However, in some species of *Picea*, *Abies*, *Tsuga*, and *Populus*, the heartwood is not dark and is difficult to identify. The xylem outside the heartwood contains living cells (mainly ray parenchyma cells) and is called sapwood. The water content in sapwood is generally higher than in heartwood. Heartwood generally contains a considerable amount of extractives. In the heartwood of some hardwoods such as white oak, tyloses develop in the vessels. The innermost sapwood is gradually transformed into heartwood during growth, and this xylem tissue is called intermediate or transition wood. Transition wood is generally pale in color and contains extractives in the parenchyma cells and tyloses in the vessels (Higuchi 1992; Fig. 7).

1.1 Microscopic Structure

1.1.1 Softwoods (Conifers)

In softwoods, about 95% of the constituent cells are long, fibrous tracheids which are arranged longitudinally in the stem (Table 1).

Tracheids serve to transport water from the roots to growing points and leaves; they also have a mechanical function in supporting the tree. Earlywood

Table 1. Volume percentages of cell elements in Japanese gymnosperm woods. (Sudo 1976)

Species	Tracheids	Parenchyma	Rays
Pinus densiflora	95.87		3.43
P. thunbergii	97.03		1.89
Picea jezoensis	95.22		4.25
Abies sachalinensis	95.80		4.20
A. firma	93.86	0.29	5.81
Thuja standishii	97.44		2.56
Thujopsis dolabrata	96.61	0.81	3.21
Chamaecyparis obtusa	97.09	0.58	2.33
C. pisifera	96.52	0.39	3.09
Pseudotsuga japonica	94.71		4.73
Cryptomeria japonica	97.20	0.80	2.00
Larix leptolepis	95.16		4.58
Sciadopitys verticillata	98.61		1.39
Taxus cuspidata	96.98		3.02
Ginkgo biloba	92.74	0.25	7.01

Fig. 8. Scanning electron micrograph of the wood of *Picea ezoensis*. Tracheid size changes abruptly from early wood (*ew*) to latewood (*lw*). Ray (*r*) on the radial surface, and resin canals (*rc*) on the transverse surface. (Courtesy of Dr. H. Saiki)

tracheids are mainly involved in water transport, while latewood tracheids have a mechanical function. Pits, spiral thickening, and spiral checking are present in the tracheids. Pits are passages for water between neighboring cells. Bordered pits occur between tracheids, while simple pits occur between parenchyma cells. Half-bordered pits connect a parenchyma cell with a tracheid (Saiki 1982; Fig. 8).

Rays, which consist of ray tracheids and ray parenchyma cells, play a role in transportation and storage of water and organic nutrients in the horizontal direction of wood tissues. In addition, resin cell canals and epithelial cells, which excrete resin, are present in some conifers.

1.1.2 Hardwoods (Dicot Angiosperms)

The cells of hardwoods are more differentiated than those in conifers. Instead of the tracheids in conifers, vessels serve to transport water, and wood fibers assume the mechanical function of supporting the tree. Volume percentages of the constituent cells are significantly different among species (Table 2).

Vessels are arranged in a longitudinal direction, and the contacting wall ends between neighboring vessels disappear to form perforations through which water is transported smoothly. Many bordered pits are located on the side of vessels to transport water between vessels and other cells in a horizontal direction. Vessels, especially those of ring-porous wood, are filled with tyloses, and tyloses are present even in sapwood vessels. The diameter and distribution patterns of vessels on the cross surface of woods are different in different tree species and are used to identify hardwoods (Saiki 1982; Fig. 9).

Wood fibers consist of fiber tracheids and libriform fibers. The former are shorter and have thicker cell walls and fewer bordered pits than the tracheids, while the latter resemble fiber tracheids, except that the pits are fewer in number and are simple.

Table 2. Volume percentages of cell elements in Japanese hardwoods. (Sudo 1976)

Species	Vessels	Fiber	Parenchyma	Ray
Fraxinus lanuginosa	4.74	83.23	6.17	5.86
F. spaethiana	11.67	65.09	10.08	13.16
Fagus japonica	49.10	33.79	4.58	12.53
F. crenata	41.22	32.09	9.23	17.46
Quercus crispula	12.64	65.54	6.78	15.04
Betula maximowicziana	18.27	71.81	1.56	8.36
B. grossa	16.51	67.45	3.47	12.57
Populus maximowiczii	36.63	59.54	0.17	3.66
Pterocarya rhoifolia	10.43	82.02	2.92	4.63
Castanea crenata	21.74	58.67	13.35	6.24
Acer mono	14.16	66.86	3.49	15.47
Paulownia tomentosa	17.85	41.21	36.88	4.06
Tilia maximowicziana	31.99	56.49	6.87	4.65
Zelkova serrata	14.31	58.48	16.73	10.48
Magnolia obovata	30.85	59.03	0.55	9.57

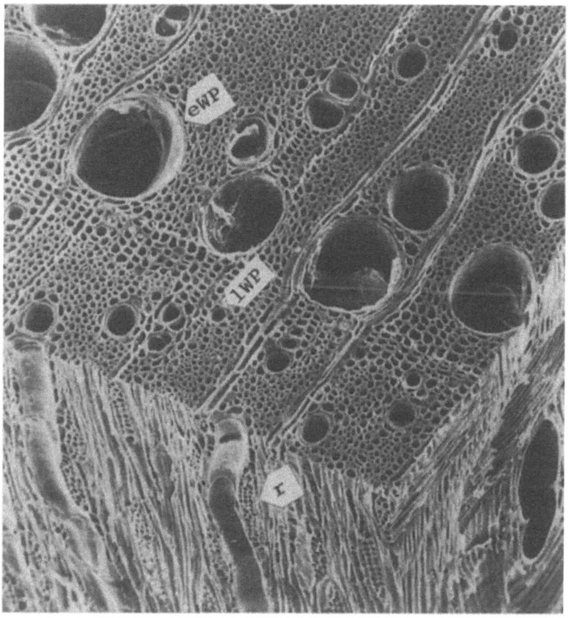

Fig. 9. Scanning electron micrograph of the wood of *Fraxinus lanuginosa*. Vessel pores of earlywood (*eWP*) are large and concentrated in the earlywood. Latewood pores (*lWP*) are much smaller; *r* ray. (Courtesy of Dr. H. Saiki)

Hardwood rays consist of only ray parenchyma, and the ray is divided into two types by the pattern in which the parenchyma cells are arranged. Ray parenchyma consists of procumbent cells, which are brick-like cells oriented radially, upright cells, which are brick-like cells arranged longitudinally, and square cells (Harada 1968; Fig. 10).

These ray parenchymal cells occur in uniseriate rays, multiseriate rays, and aggregate rays depending on the aggregation patterns.

1.1.3 Bamboos

The bamboo culm is characterized by nodes occurring periodically along its length. At the node is a solid cross wall, called the diaphragm; the internodes are usually hollow. The culm can be up to 36 m long and 25 cm in diameter at the base (as in *Phyllostachys pubescens*), depending on the species. The outermost part of the culm is composed of a single layer of epidermis, and the inside is covered by a layer of sclerenchyma cells. The body of the culm wall consists of vascular bundles embedded in ground tissue made up of parenchyma cells. The vascular bundles are composed of vessels, sieve tubes with companion cells, and fibers.

Bamboo culms have no cambium and there are no radial cell elements, such as rays, in the internodes. Bamboo tissue is therefore quite different from

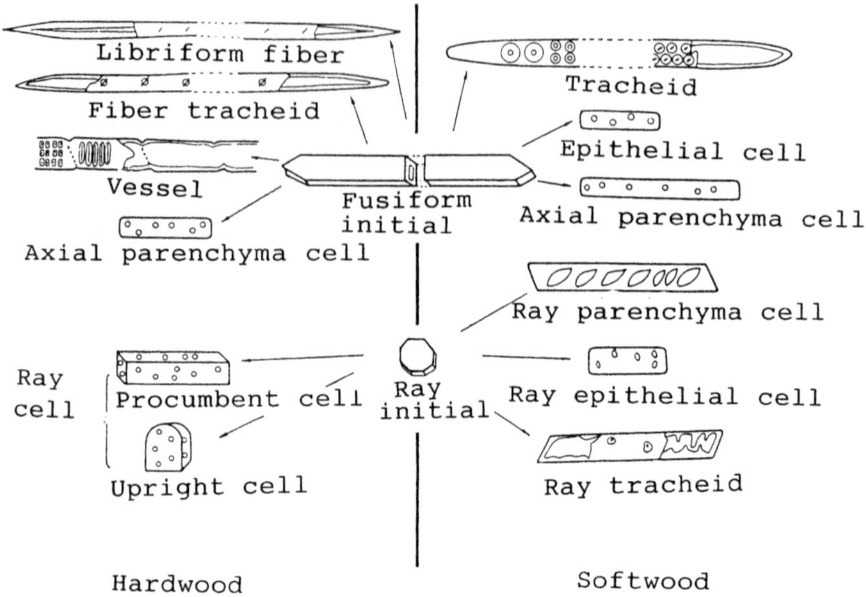

Fig. 10. Various cells derived from cambial initials (fusiform initials and ray initials) in softwoods and hardwoods. Cells of softwood: tracheid, epithelial cell, axial parenchyma cell, ray parenchyma cell, ray epithelial cell, and ray tracheid. Cells of hardwood: libriform fiber, fiber tracheid, vessel element, axial parenchyma cell, procumbent cell (ray parenchyma cell), and upright cell (ray epithelial cell). (Harada 1968)

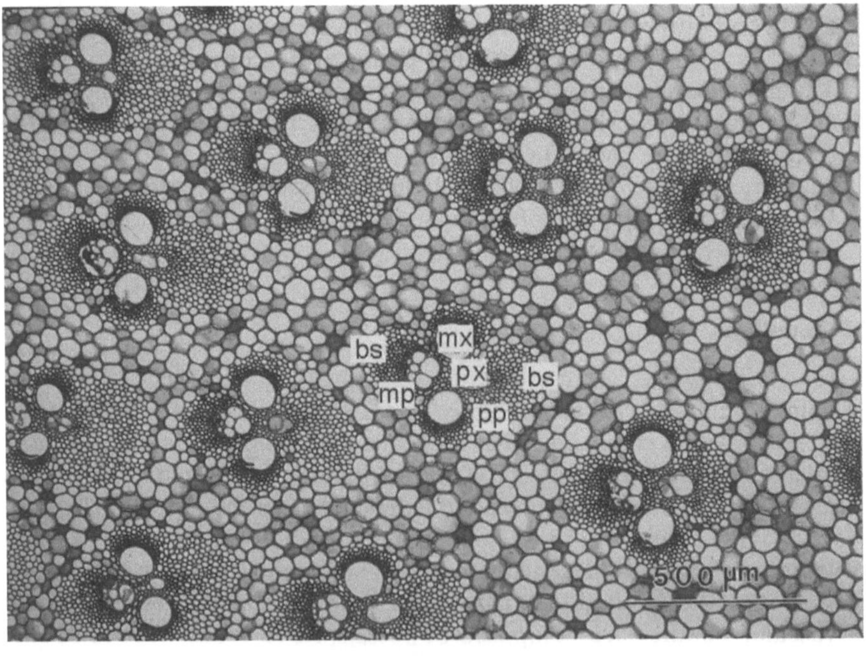

Fig. 11. Vascular bundle of bamboo culm (*Phyllostachys aurea f. albo-variegata*) transverse section. Numerous vascular bundles are scattered in a ground tissue composed of parenchyma cells. *mx* Metaxylem; *px* protoxylem; *mp* metaphloem; *pp* protophloem; *bs* bundle sheath. (Courtesy of Dr. T. Itho)

wood. The culm consists of about 50% parenchyma, 40% fibers, and 10% conducting cells (vessels and sieve tubes) on average. The vascular bundles in the culm consist of xylem, with two large metaxylem vessels (40–120 μm) and one or two protoxylem elements, and of phloem, with thin-walled, unlignified sieve tubes connected to companion cells (Higuchi 1989; Fig. 11).

1.2 Ultrastructure of Wood Cell Walls

Electron microscopy observations show that wood cell walls are composed of an intercellular layer, a primary wall, and a secondary wall. The primary cell wall is a thin layer produced by cell division and the subsequent growth of xylem mother cells, while the secondary wall is a thick layer deposited inside the primary wall. It consists of an outer layer (S_1), a middle layer (S_2), and an inner layer (S_3), each with a different orientation of cellulose microfibrils (Harada and Côté 1985; Fig. 12).

The ultrastructure of most fibers in bamboos is characterized by thick, lamellated secondary walls. This lamellation consists of alternating broad and

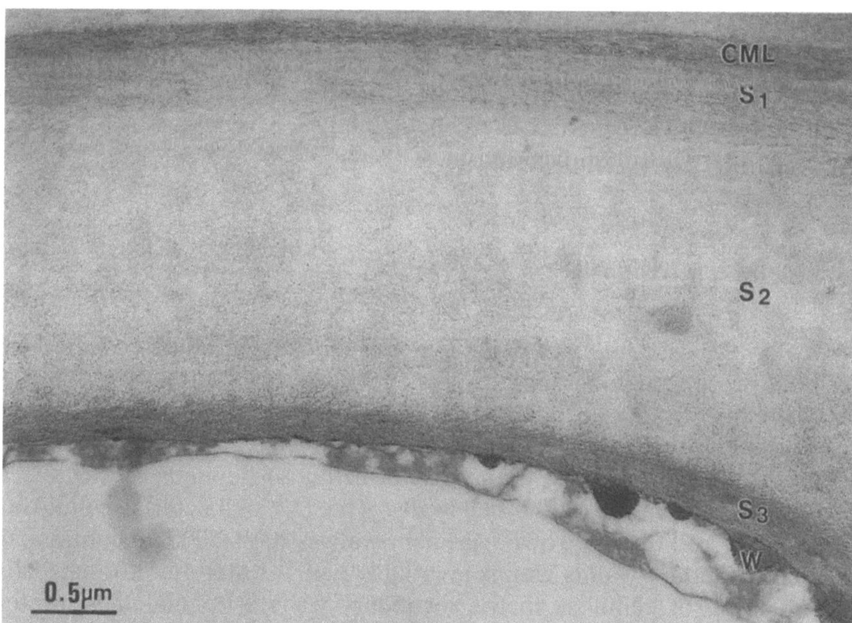

Fig. 12. Transmission electron micrograph of ultrathin transverse section from *Cryptomeria japonica* showing the compound middle lamella (*CML*) containing primary wall, different layers of the secondary wall (S_1,S_2,S_3), and the warty layer (*W*) with cytoplasm in a tracheid. (Courtesy of Dr. Takabe)

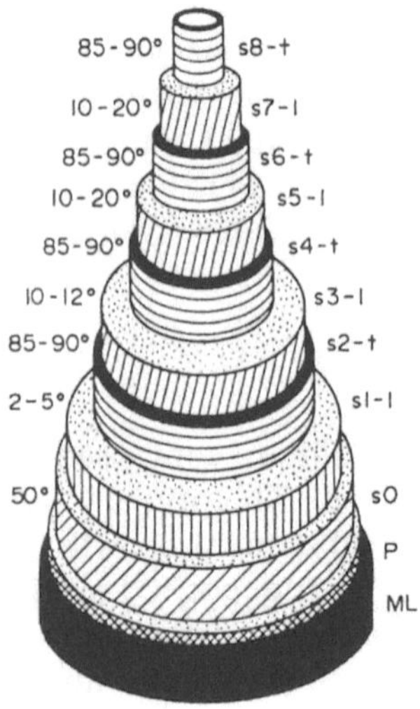

Fig. 13. Polylamellated wall structure of a bamboo culm fiber. *ML* Middle lamella; *P* primary wall; *s* secondary wall; *l* and *t* denote longitudinal and transverse orientation of microfibrils. (Parameswaran and Liese 1981)

narrow lamellae with differing fibrillar orientation. In the broad lamellae, the microfibrils are oriented at a small angle to the fiber axis, but in the narrow ones they are mostly horizontal (Parameswaran and Liese 1981; Fig. 13).

The narrow lamellae are more highly lignified than the broad ones. A typical tertiary wall is not present, but warts cover the innermost layer in some taxa. The polylamellated wall structure of the fibers, especially in the outer part of the culm, results in an extremely high tensile strength, which is used to advantage in structural applications.

1.2.1 Cellulose Microfibrils

Wood cell walls are composed of laminated cellulose microfibrils covered with hemicelluloses and lignin–hemicellulose matrices (Ruel et al. 1978; Fig. 14). Native cellulose is a linear polysaccharide in which 2000–26500 molecules of D-glucose are connected with β-1,4-linkages (Table 3).

The degree of polymerization (DP) of cellulose depends on the species and the growing stages of the cell walls. The DP of bacterial cellulose is relatively low, and that in growing primary walls is 2000–6000. In contrast, the DP of the secondary wall is almost invariably high (10000–15000), suggesting that the length of cellulose in the secondary walls is regulated accurately. Wood cellulose has a DP of about 10000 (Goring and Timell 1962).

The DP of valonia microfibrils is exceptionally high (26500). The glucose molecule is 0.5 nm in length, and the length of valonia cellulose is accordingly about 13.3 μm. However, the actual length of valonia microfibrils is known to be far greater than the length of the calculated glucose chains. There must be

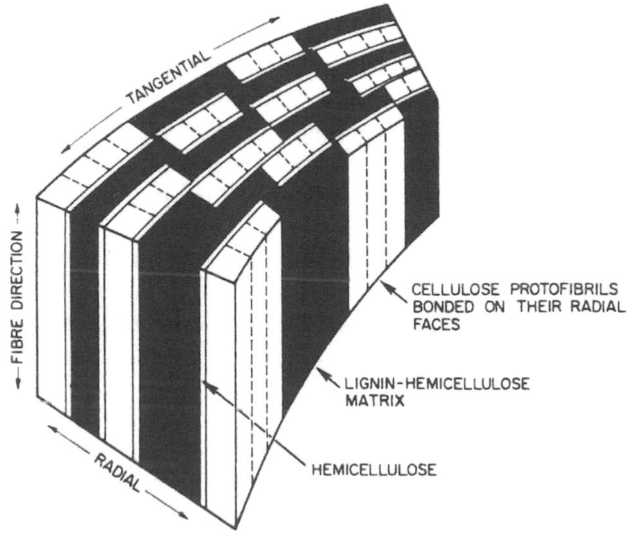

Fig. 14. Localization and relationship of cellulose, hemicelluloses, and lignin in the S_2 layer. (Ruel et al. 1978)

Table 3. Degree of polymerization (DP) of cellulose from different organisms. (Itoh 1994a)

Species	Degree of polymerization
Acetobacter xylinum	2000–3700, 5700
Gossypium hirsutum	
Primary walls	2000–6000
Secondary walls	13000–15000
Acanthamoeba castellani	2000–6000
Valonia	26500
Seed hairs	10000
Phloem fibers	10000
Angiosperm wood	10000
Gymnosperm wood	10000

several discontinuous parts of the molecular chains in the microfibril, corresponding to the initiation and termination of the synthesis of cellulose molecules. It is likely that the ends of glucan chains of the cellulose constitute a disordered, amorphous part in the microfibril (Itoh 1985).

Electron microscopy observation and X-ray diffraction of cellulose microfibrils of many plants have shown that the width of the crystalline regions of the microfibrils is different in different species. In a study of the difference in size of crystalline regions of microfibrils by species, Frey-Wyssling (1954) suggested that microfibrils are made up of a number of crystallites, each of which is separated by a paracrystalline region. He proposed an elementary fibril as a fundamental crystalline unit of native cellulose which is composed of dozens of cellulose molecules connected by hydrogen bonds with a cross surface of about 30 × 70 Å. He suggested that four elementary fibrils constitute a microfibril.

Kolpak and Blackwell (1976) supported Frey-Wyssling's proposal, and observed small elementary fibrils 3.5 nm in width in *Acetobacter*, primary wall of higher plants, and fractured parts of large microfibrils. They suggested that the difference in the size of microfibrils depends on the number of elementary fibrils (3.5 nm in width) common to all biological species and the packing mode in the microfibrils.

Preston and Cronshaw (1958) considered the microfibril to have a single core of cellulose crystllite surrounded by a paracrystalline region. They proposed that microfibrils are not a bundle composed of elementary fibrils, but single crystals of different sizes which are specific for each plant species (Fig. 15).

Recent studies (Heyn 1966, Harada and Goto 1982, Revol 1985) have shown that native cellulose is composed of microfibrils of different widths in different species (Table 4). The structure corresponding to an elementary (3.5 nm) or subelementary fibril (1.0–2.0 nm) is only found in the cambium,

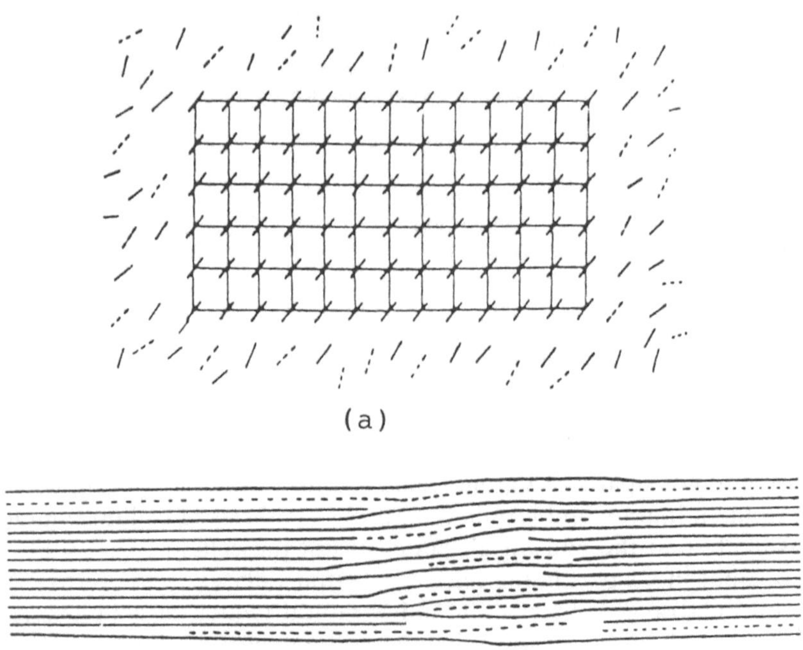

Fig. 15a,b. Structure of cellulose microfibrils. **a** Transverse section; central part shows crystalline region (100 × 50 Å). **b** Longitudinal section; *solid* and *dotted lines* show glucan and hemicellulose chains, respectively; *central irregular part* amorphous region. (Preston 1974)

Table 4. Microfibril widths of cellulose in different organisms. (Itoh 1994a)

Organism	Microfibril width (nm)
Wood cambium	1.5
Primary wall of plants	2–2.5
Secondary wall of wood	5–10
Quince slime (*Cydonia vulgaris*)	3–5
Cotton and ramie	5–6
Acetobacter xylinum	7.5
Acetobacter xylinum	1.6–5.8
Tunica	12–20
Valonia	20
Boergesenia and *Chaetomorpha*	30

primary walls, fractured ends, microfibrils dispersed by ultrasonic or grinding treatments, stretched and bent parts of microfibrils, homogenized cotton linters, and in valonia dispersed by ultrasound.

There is currently no evidence of the existence of elementary fibrils, and it seems that the elementary fibrils hitherto observed are an artifact formed by physical and mechanical disruption of cellulose (Itoh 1985).

Sugiyama et al. (1985a,b) studied the crystalline lattice of valonia cellulose microfibrils by electron microscopy. The lattice showed that the interplanar space is the same as that observed by X-ray diffraction. The structure corresponding to the elementary fibril was not found in the image of the crystal, and the valonia microfibril was constructed by a single fundamental crystalline unit of 30 nm in width.

Microfibrils of valonia and some algae (*Boergesenia* and *Chaetomorpha*) are 20 and 30 nm in width, respectively, while the width of microfibrils in a wood cell wall such as *Pinus densiflora* is reported to be 5 nm (Harada 1968). It is likely that the size of the microfibrils depends on the size and mode of arrangements of the cellulose-synthesizing complexes, as will be discussed below.

Cellulose fibers are aggregate materials of cellulose molecules, which are linear polymers composed of β-1,4-linkages of D-glucose. In cellulose molecules aligned in parallel, hydrogen bondings are formed between hydroxyl groups of neighboring glucose residues to form crystalline regions, while in randomly oriented cellulose molecules and the ends of microfibrils hydrogen bondings are scarce, resulting in amorphous regions (Preston 1971; Fig. 15).

Although the width of microfibrils is specific for each plant species, cellulose molecules are present between crystalline and amorphous regions without any definite boundary, forming "fringe micelles." In amorphous regions of randomly oriented cellulose molecules and at the ends of cellulose molecules and in their periferal region, cellulose molecules are physically linked to surrounding hemicelluloses, while the hemicellulose molecules are known to be linked chemically via ether or ester linkages to lignin, forming lignin–hemicellulose complexes (Higuchi 1990).

Raman microprobe spectra from the secondary wall of earlywood tissue from *Picea mariana* have revealed that the aromatic rings of the phenylpropane structural units of lignin molecules in intermicrofibril spaces are parallel to the plane of the cell wall surfaces (Atalla and Agarwal 1985) and form a composite wood structure.

1.2.2 Orientation of Microfibrils

Cell walls of tracheids (conifers) and fibers and vessels (hardwoods) are composed of an intercellular layer (M), a primary wall (P), and secondary layers composed of an outer layer (S_1), an intermediate layer (S_2), and an inner layer (S_3). Differences in secondary layers are ascribed to the orientation of the cellulose microfibrils in respective layers. The respective layers can be identified by polarization microscope observations of cell walls. Details of the microfibril orientations can be established by electron microscopy of replicas of longitudinal surfaces of the secondary walls (Table 5).

The width of intercellular layers is 0.2–1.0 µm. Early in cell wall formation, the intercellular layer is composed mostly of pectin, but in mature cell walls it consists largely of lignin. Primary walls are only 0.1–0.2 µm in width. Early in cell wall growth, pectin, hemicelluloses, cellulose, and protein are the major constituents. Later, this layer becomes highly lignified. The combined intercellular layer and primary wall is sometimes referred to as the compound intercellular layer or compound middle lamella.

The S_1 layer, which is 0.2–0.3 µm in width, is composed of three to four lamellae. The microfibrils in the lamellae circle to the right or left against the

Table 5. Microfibril angle of the secondary walls in tracheids; all values are in degrees. (Saiki 1970)

Wood species	Earlywood			Latewood		
	S_1	S_2	S_3	S_1	S_2	S_3
Pinus densiflora (45 years)						
Radial wall	71	26.5	83	77	8.0	79
Tangential wall	68	20	83	75	5.83	83
Cryptomeria japonica (55 years)						
Radial wall	68	14	81	83	3.5	85
Tangential wall	64	3.5	85	80	3.0	82

Table 6. The distribution of lignin in black spruce tracheids as determined by UV spectroscopy. (Fergus et al. 1969)

Wood	Morphological region[a]	Tissue volume (%)	Lignin	
			Total (%)	Conc. (%)
Earlywood	S	87	72	23
	ML	9	16	50
	MLcc	4	12	85
Latewood	S	94	82	22
	ML	4	10	60
	MLcc	2	8	100

[a] S, Secondary wall; ML, middle lamella; MLcc, corner of middle lamella.

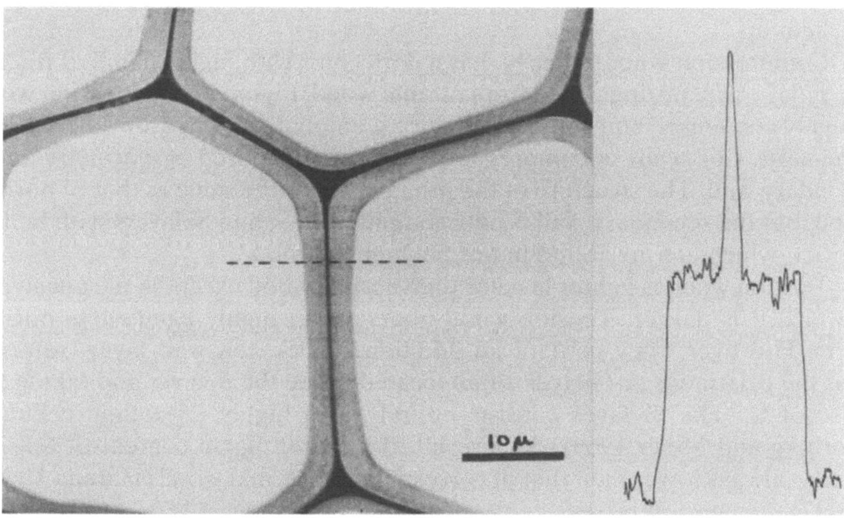

Fig. 16. Ultraviolet photomicrograph, taken at 240 nm, of the cell walls of black spruce earlywood tracheids. The densitometer tracing was taken along the *dotted line*. (Courtesy of Dr. D.A.I. Goring)

direction of cell length. The inclination of the crossed microfibrils is 60–80°. The S_2 layer is the main wall layer. Its width is 1 μm in earlywood, and 5 μm in latewood of conifer tracheids, which are sometimes composed of 150 lamellae. Their inclination in the direction of cell length is 5–30° in earlywood, and 5–10° in latewood. The S_3 layer, which is 0.1 μm in width, is composed of several lamellae. Microfibrils circle in the right or left direction with an inclination of 70–90° (Côté 1967).

In addition, a so-called warty layer is an integral part of S_3, forming an innermost wall layer in some softwoods and a few hardwoods.

The lignin concentration in wood cell walls as determined by ultraviolet (UV) micrography is highest in the compound middle lamella (50–100%). In the secondary wall, the concentration is 20–25%. However, 70–80% of the total lignin in wood cell walls occurs in the secondary wall, because about 90% of the volume of a wood cell wall is occupied by the secondary wall (Fergus et al. 1969, Fergus and Goring 1970; Table 6; Fig. 16).

1.2.3 Cell Walls of Reaction Woods

Reaction woods are abnormal tissues often associated with eccentric growth of a leaning stem. In conifers such tissue is called compression wood because it often appears on the underside of a tree stem, where it exerts a compressive stress to right the stem. In hardwoods it is termed tension wood and is formed on the upper side of a leaning stem, where it exerts a tensile stress that pulls the stem up.

Compression wood is heavy, has a dark color with highly lignified tracheids, and is easily distinguished from normal wood. Lignin in compression wood is highly condensed and contains significant amounts of *p*-hydroxyphenyl propane units. Cell walls of compression wood are composed of a primary and a secondary wall. The structure of the primary wall is the same as that of normal wood, but the secondary wall is only composed of S_1 and S_2 layers with helical cavities, which are not found in normal conifer wood.

Tension wood is lighter in color than normal wood except in the eucalypts, in which it is darker. Tension wood fibers are as highly lignified as normal fibers. However, they contain an additional thick cell wall layer, referred to as the gelatinous or G layer, often located inside the S_2 layer and taking the place of S_3. The G layer consists entirely of a highly crystalline cellulose (Norberg and Meier 1966). As a result, the overall lignin content of tension wood is always lower than that of corresponding normal wood (Harada 1976b; Fig. 17).

Microfibrils of the poplar gelatinous layer were found by electron microscopy to have a paracrystalline region surrounding a crystalline core (Goto et al. 1978).

Compression wood contains much more galactan (essentially in the S_1 and outer part of the S_2) and less galactoglucomannan and cellulose than tracheids of normal wood (Côté et al., 1968, Timell 1982, 1986). In addition, a slightly acidic 1,3-linked β-D-glucan, called laricinan (Hoffmann and Timell 1970, 1972), is also present in small amounts in compression wood. It is absent in normal wood.

A highly complex, acidic galactan occurs in tension wood of some genera, while it is absent in others. The cellulose content is always higher than in normal wood, while the content of xylan is the same or slightly lower (Timell 1969, Meier 1985).

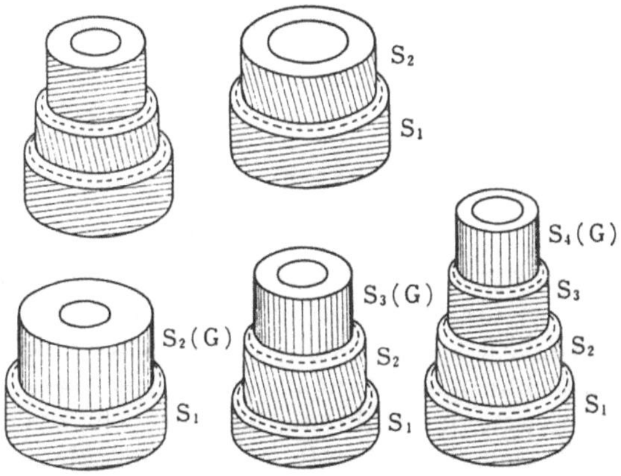

Fig. 17. Cell wall structures of reaction woods. *Above* Normal tracheid (*left*), compression wood tracheid (*right*). *Below* Three types of fiber walls of tension woods. *G* Gelatinous layer. (Harada 1976b)

Smolinski et al. (1972) and Phelps et al. (1974, 1977) reported that morphactin, which inhibits tropistic phenomena in herbaceous plants, induced compression wood. Yamaguchi et al. (1980, 1983) found that compression wood was formed by treatment with morphactin and 1-*N*-naphthylphthalamic acid, inhibitors of indole acetic acid (IAA) transport, in the upper segment of the treated stem, but no abnormal tissues were produced in the lower segment. They concluded that compression wood was induced by a high ratio of endogenous IAA to sugar in the differentiating xylem tissue by IAA transport inhibitor.

Compression wood is formed in the Ginkgoales, Coniferales, and Taxales, but not in the Cycadales and Gnetales. However, the compression wood of *Ginkgo biloba* differs from that in most of the gymnosperms in the more angular outline of its tracheids, their thinner walls, and their lack of helical cavities.

Timell (1978, 1986) suggested that compression wood has probably existed since the Devonian period, but that helical cavities probably developed relatively late in the evolution of compression wood.

Recently, Yoshizawa et al. (1993) found that *Buxus microphylla* (a hardwood) exhibits a pronounced growth on the underside of inclined stems. Formation of tension wood (gelatinous fibers) was not observed. Xylem formed on the underside showed interesting features resembling the compression wood formed in gymnosperms. The reaction wood tracheids and vessels showed an excessive lignification in their secondary walls, but lacked both helical cavities and an S_3 layer, features that are almost the same as those of primitive gymnosperms. These results indicate that *B. microphylla* is able to form compression wood and that the genus *Buxus* acquired a different mechanism in the formation of conducting elements in phylogenic evolution.

Details on compression wood can be found in *Compression Wood in Gymnosperms* by Timell (1986).

1.3 Ultrastructure and Functions of Wood Cells

Cell activities such as cell division, growth, and deposition of cell wall components are ascribed to the metabolic activities of cell organelles. Plant cells are made up of organelles such as chloroplasts and cell walls. Vacuoles are found in animal cells, but not cell walls and chloroplasts (Takabe 1994). The characteristics of plant cells are due to the activities of these organelles. The main structures and functions of plant cell organelles are outlined below in relation to wood formation.

1.3.1 Nucleus

Plant cells (eukaryotes) have a nucleus in the center. Most of the deoxyribonucleic acid (DNA) in the cell is present in the nucleus, and genes in the DNA

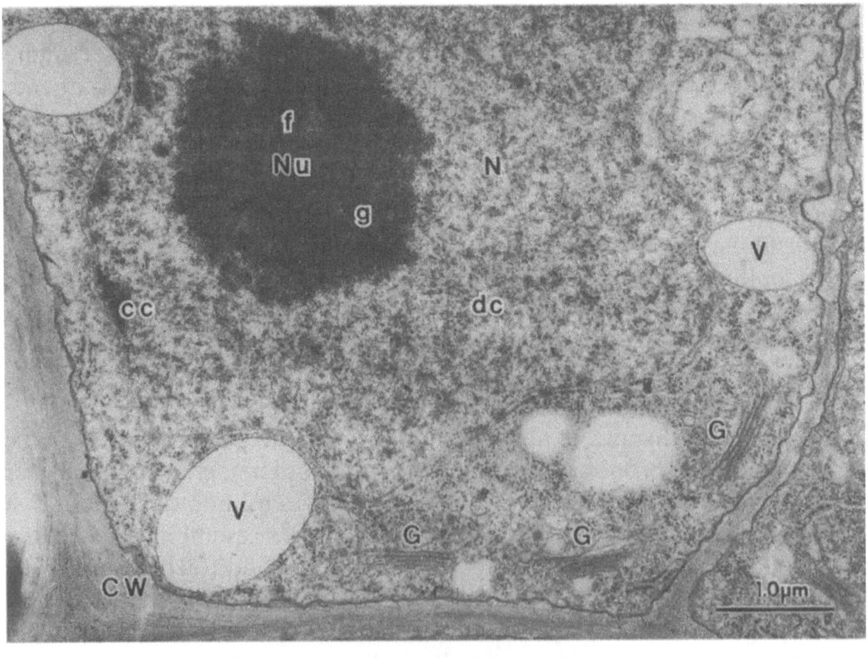

Fig. 18. Electron micrograph of xylem cell of *Morus bombycis* (eukaryote). *N* Nucleus; *Nu* nucleolus; *cc* concentrated chromatin; *dc* diffuse chromatin; *f* fibrous part; *g* granular part; *G* Golgi apparatus; *V* vacuole; *CW* cell wall. (Courtesy of Dr. K. Takabe)

control the synthesis of cell proteins and enzymes. The nucleus is therefore considered to be the control center of the cell or a center of cell functions. The nucleus is encased in a double layer of nuclear membrane and contains a nucleolus in the center (Fig. 18).

The shape of nuclei in cambial cells changes depending on the physiological condition of the cell. Cambial nuclei are generally globular and large. In conifer parenchyma cells, the shape of nuclei changes to oval during cell growth. Cell and nuclear volumes increase with age, at least in nonstoried cambia. The increase in the genome size probably serves to maintain a balance between DNA and cell volume. In hardwood parenchyma cells, nuclei are several times smaller than those of conifers, and the shape is scarcely changed during cell growth, although the shape is different among species (Higuchi et al. 1964, Fukazawa and Higuchi 1965).

Mellerowicz et al. (1989) found that, in balsam fir, all interphase nuclei are in the G_1 phase at the end of the growing season (September) and that they remain at a 2C DNA level until after December, when normal DNA synthesis (S phase) is resumed. Similarly, in shoot apical meristems, the nuclei remain either in a noncycling G_1 phase or an S phase during dormancy (Tepfer et al. 1981, Cottignies 1983). Nuclear volume and DNA content are therefore maximal in early spring. During the growing season, one third of the 2C level DNA

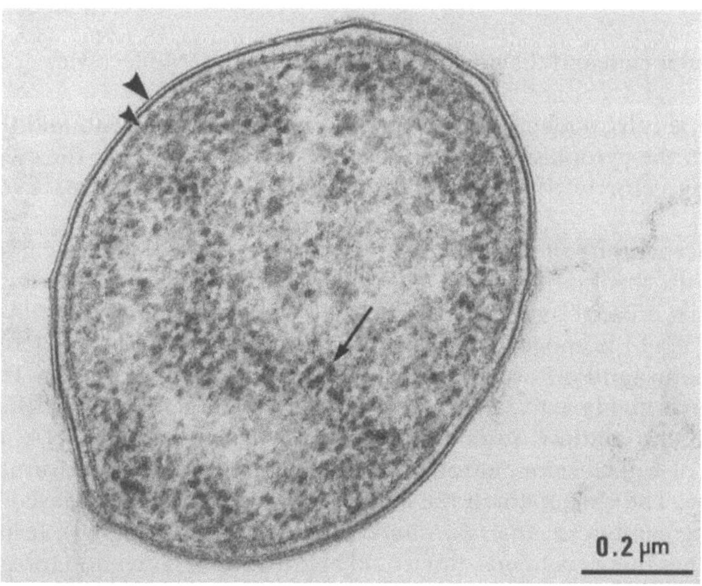

Fig. 19. Electron micrograph of *Acetobacter xylinum* cell (prokaryote). The cell is covered with an outer membrane (*large arrowhead*), and an inner membrane (*small arrowhead*) of lipopolysaccharides. No nuclei exist, and no organelles with membrane structure. Many ribosomes (*arrow*) are distributed in the cytoplasm. (Courtesy of Dr. Oogiya)

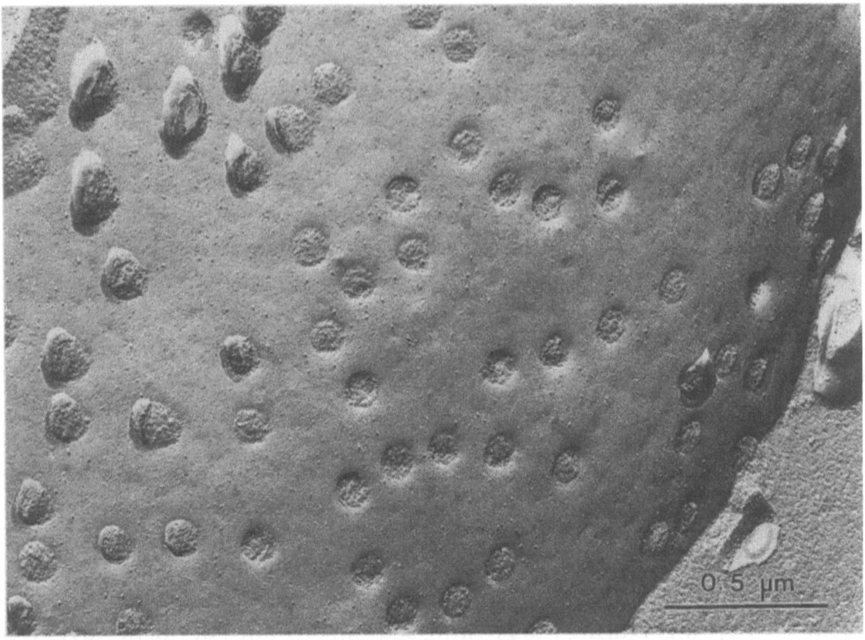

Fig. 20. Electron micrograph of the nuclear membrane (freeze fracture) of *Morus bombycis*. Many nuclear pores can be seen. (Courtesy of Dr. K. Takabe)

(extra DNA) disappears and the nuclear volume decreases (Mellerowicz et al. 1992b).

In prokaryotic cells, such as bacteria cells, no nucleus is present, and the DNA is folded in the cytoplasm. Many ribosomes are distributed in the cytoplasm of bacteria cells, but no organelles with membranous structures are present (Fig. 19).

Electron micrographs of plant cells in metaphase show that the caryoplasm is filled with chromatin, and in the central part a nucleolus with high electron density is present as a black particle. In this stage the chromatin is dispersed, but it is not homogeneously distributed. Concentrated parts (concentrated chromatin) and diffused parts (diffuse chromatin) are present. The diffuse chromatin is highly active in synthesis of ribonucleic acid (RNA), while the concentrated chromatin is somewhat less active in synthesis of RNA. In the interkinesis of cell division, chromatin is concentrated and transformed into chromosome. The chromatin of the interphase nuclei is less condensed in old than in young cambia in *Abies balsamea* (Mellerowicz et al. 1989). It has been suggested that this decondensation could be a means of activation for the greater number of genes necessary to maintain larger cells. One to five nucleoli per nucleus with an extensive granular zone, which indicates a high synthetic activity, were found in old cambial cells.

The nucleolus is composed of a fibrous and a granular part and produces ribosomes. Ribosome chromatins located in the fibrous parts play a role in transcription of ribosomal RNA (rRNA) and synthesis of precursor of ribosomes, while processing and maturation of ribosome precursors occur in the granular part to produce ribosome subunits.

The nuclear envelope is composed of inner and outer membranes. The inner membrane is smooth membrane, while the outer has ribosomes attached at the cytoplasm side and is partly connected to rough endoplasmic reticulum (ER). Between the inner and the outer membranes there is a perinuclear space (about 20 nm in width), which is connected to the lumen of rough ER. Many nuclear pores, 60–100 nm in diameter, are located on the nuclear membrane, and the exchange of metabolic substances between caryoplasm and cytoplasm is conducted through nuclear pores. (Fig. 20).

Messenger RNA (mRNA), transfer RNA (tRNA), and ribosome subunits are translocated from caryoplasm to cytoplasm, while nuclear proteins, enzymes, and ATP etc. are translocated from the cytoplasm to the nucleus.

1.3.2 Ribosomes

Ribosomes 25–30 nm in diameter with 80 S (Svedberg units) and composed of two different sizes of subunits, with 60 S and 40 S, are distributed in the cytoplasm. A ribosome is a complex body with rRNA and proteins. Large subunits are composed of 28 S, 5.8 S, and 5 S rRNA, and about 40 kinds of proteins, while small subunits are composed of 18 S rRNA and about 30 kinds of proteins. Ribosomes are distributed in the cytoplasm and on the membranes of rough ER, and these are often connected by mRNA to form moniliform polysomes.

Ribosomes are involved in protein synthesis in cooperation with tRNA. Polysomes in the cytoplasm are involved in the synthesis of proteins in the cell, while ribosomes on the rough ER take part in the synthesis of excreting proteins and acummulating proteins in vacuoles (Watanabe 1990a). On the other hand, ribosomes in mitochondria and chloroplasts in plant cells are 70 S and are composed of 50 S and 30 S subunits, similar to those in prokaryotic cells.

1.3.3 Endoplasmic Reticulum

In 1945 Porter and coworkers observed fibrous network structures that were attached to granules in the cytoplasm of cultured animal tissues and named them the endoplasmic reticulum (ER). Later, the organelle has generally been observed in ultrathin sections of various cells by electron microscopy. The ER is composed of several layers of cisternae and of tubes 50–100 nm in diameter connected in a netlike fashion on the surface of the cisternae.

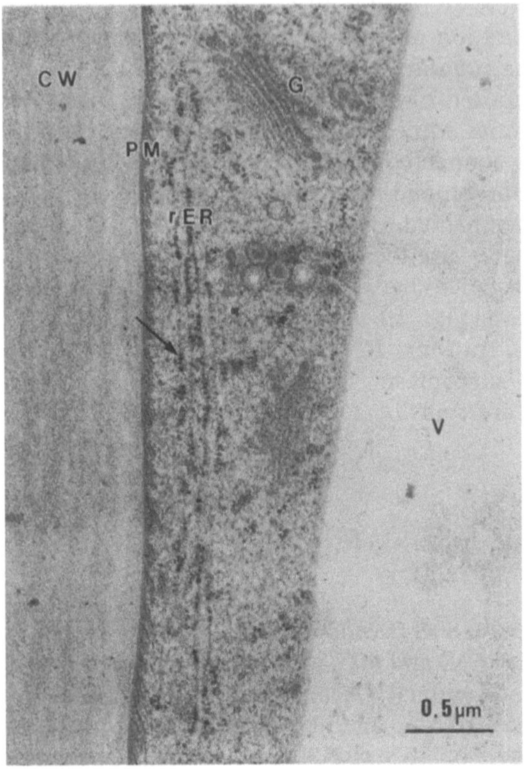

Fig. 21. Electron micrograph of rough endoplasmic reticulum (*rER*). Two layers of cisternae are arranged parallel to the cell wall. Many ribosomes are attached to cisternae from the cytoplasm side (*arrow*). *G* Golgi apparatus; *V* vacuole; *PM* cell membrane; *CW* cell wall. (Courtesy of K. Sano)

In preparation of a cell homogenate, the ER is disintegrated to form a microsome fraction which can be purified by centrifugation. Biochemical studies on the microsome have significantly contributed toward elucidating the metabolic and biosynthetic functions of the ER.

The ER can be divided into rough and smooth ER. The former has ribosomes on the surface of its membrane, while the latter lacks ribosomes.

In rough ER, several layers of cisternae are arranged in parallel, and ribosomes are attached on the surface of the cisternae membrane on the cytoplasm side. The ER membrane contains a specific protein that is associated with a 60S subunit. In many cases, ribosomes are present as a spiral polysome connected to mRNA (Figs. 21, 22).

Polypeptides newly synthesized by ribosomes are translocated to the ER lumen through pores on the ER membrane. The protein is glycosylated in the ER lumen and Golgi apparatus. Glycosylated proteins are stored in the ER lumen or vacuoles via the Golgi apparatus.

The ER in the cells of woody tissues was first investigated in 1969 by Porter and Machado (1969, also Wardrop 1965, Cronshow 1965). They found many ER at the beginning of cell division and noted that the vesicles derived from them were connected to the ER to form network structures from which cell

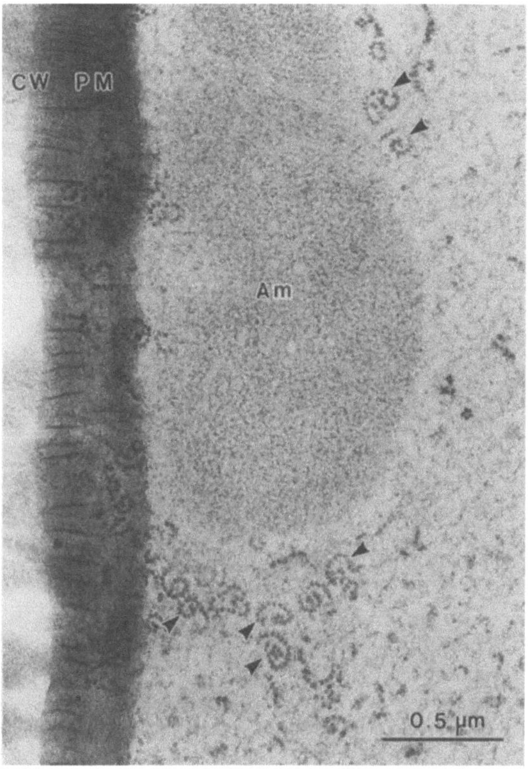

Fig. 22. Electron micrograph of rough endoplasmic reticulum. Ribosomes attached to cisternae (section parallel to cisternae) can be seen. Many ribosomes are arranged to form polysomes (*arrowheads*). *Am* Amyloplast; *PM* plasma membrane; *CW* cell wall. (Courtesy of K. Sano)

plates were formed. They suggested that the ER contains an enzyme system catalyzing the synthesis of cell wall polysaccharides and their precursors. Variously shaped ER were observed along cell walls, especially in cells forming the S_2 layer, and several ER were connected to neighboring cells to penetrate cell membranes. During cell wall formation, many rough ER are present in wood cells. In particular, during primary cell wall formation a rough ER develops to synthesize wall proteins and enzymes involved in cell wall formation.

Smooth ER are branch or netlike structures of tubes of about 100 nm in diameter. As ribosomes are attached to a part of the smooth ER membrane, the strict classification of rough and smooth ER is difficult. The number of smooth ER increases during secondary wall formation in wood cells, and it has been suggested that smooth ER is involved in lignification (Takable et al. 1989).

1.3.4 Golgi Apparatus

The Golgi apparatus was first observed in animal cells by Golgi in 1898. In wood tissues, Mühlethaler (1965) and Cronshow (1965) studied it in detail by

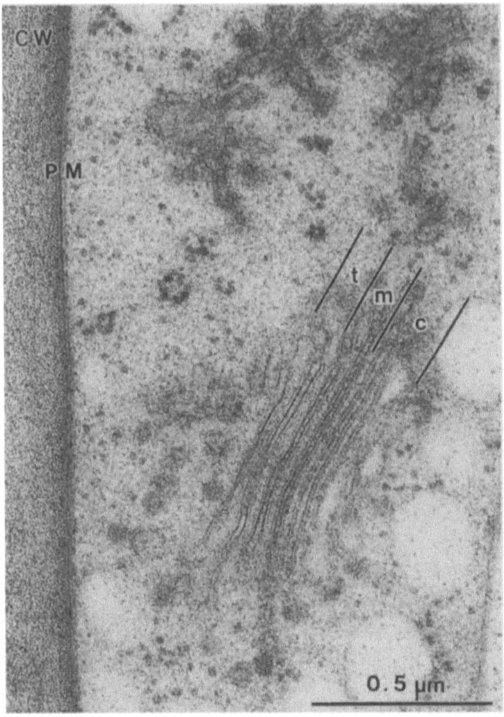

Fig. 23. Electron micrograph of Golgi apparatus of parenchyma cells of *Morus bombycis*. Layered structure of disk cisternae are observed (cisternae are cut perpendicularly). Cisternae are classified into *cis*, *medial*, and *trans* faces. Cisternae on the *cis* face are irregular, but become straight on the *trans* face. *CW* Cell wall; *PM* plasma membrane; *t trans* face; *m medial* face; *c cis* face of Golgi apparatus. (Courtesy of Dr. K. Takabe)

electron microscopy. The Golgi apparatus is a disk, about 1 μm in diameter, which is composed of four to seven layers of curved cisternae. Many Golgi vesicles and fine tubes are present around the apparatus.

It seems that the Golgi apparatus has a polarity, because the membrane structure of the cisternae, the staining properties of the apparatus, and substances in the cisternae gradually change in a particular direction. Cisternae are classified into *cis*, *medial*, and *trans* faces. The cisternae of the *cis* face have irregular forms, the membranes are weakly stained, and the substances in the cisternae are scarce.

Such cisternae become gradually thinner on the *trans* face, staining of the membrane becomes stronger, and substances appear in the cisternae. Cisternae on the *trans* face are constricted and partly come off to form vesicles (Takabe 1994; Fig. 23). Near the cisternae of the *trans* face, netlike structures of fine tubes sometimes occur as *trans*-Golgi networks associated with coated vesicles.

In animal cells, many vesicles originate from rough ER face on the Golgi apparatus, and the vesicles migrate to the *cis* face of the Golgi apparatus and

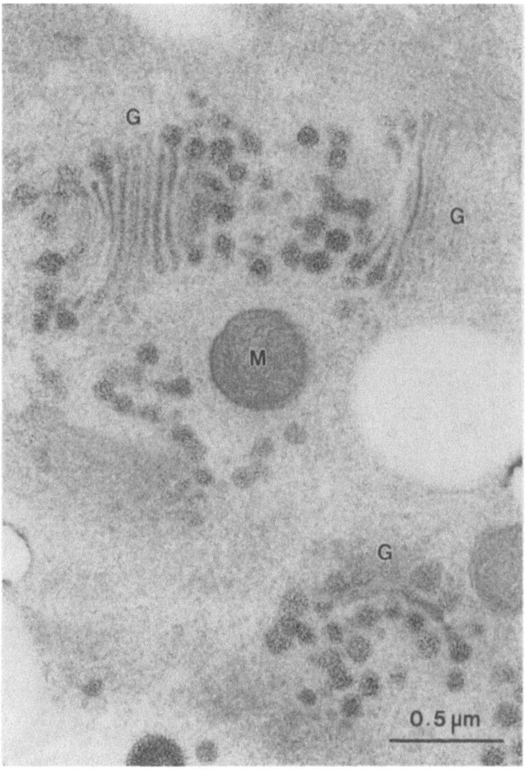

Fig. 24. Golgi apparatus in a tracheid (*Cryptomeria japonica*) with polysaccharides selectively stained. Cisternae and Golgi vesicles at *trans* face are well stained. *G* Golgi apparatus; *M* mitochondria. (Courtesy of K. Sano)

fuse to form *cis* face cisternae, while in plant cells transition vesicles are rarely observed.

In plant cells, polysaccharides are selectively stained, the Golgi apparatus is seen to be strongly stained, and the staining intensity increases from the *cis* face to the *trans* face (Fig. 24). Immunoelectron microscopy has shown that hemicellulose antibodies are combined with the Golgi apparatus, suggesting that the Golgi apparatus is involved in synthesis of hemicelluloses in plant cells (Moore and Staehelin 1988, Northcote et al. 1989). The synthesized polysaccharides are translocated to the cell walls by fusion of the Golgi vesicles with the cell membranes. Tritium-labeled phenylalanine fed to plant cells is incorporated into the Golgi apparatus, indicating that the Golgi apparatus is also involved in the synthesis and transportation of lignin precursors (Picket-Heaps 1968, Takabe et al. 1985; Fig. 25).

Recent immunoelectron microscopy using hemicellulose antibodies has shown that a xyloglucan is located in the *trans* face cisternae of the Golgi apparatus and in a *trans*-Golgi network, while polygaraturonan and rhamnogaracturonan are located in the *cis* and *medial* face cisternae. These results suggest that the function of the Golgi apparatus in the synthesis of polysaccharides is different for different polysaccharides (Zhang and Staehelin 1992).

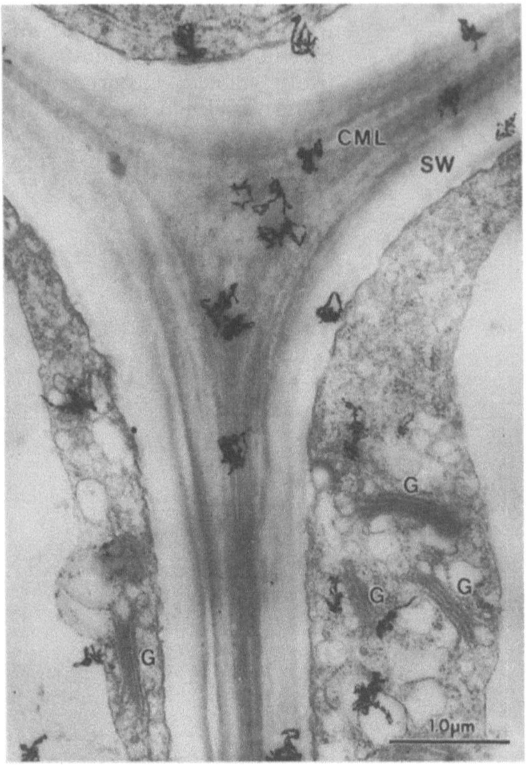

Fig. 25. Electron micrograph-autoradiograph of *Cryptomeria japonica* tracheids fed with L-phenylalanine-^{3}H. ^{3}H of L-phenylalanine is incorporated into the Golgi apparatus and the compound middle lamella of lignifying cells. *G* Golgi apparatus; *CML* compound middle lamella; *SW* secondary wall. (Courtesy of Dr. K. Takabe)

In animal cells, glycosylproteins synthesized in the rough ER are translocated to the Golgi apparatus. The glycosylprotein is partly processed and substituted with new sugars and is then excreted from the cells. In contrast, in plant cells extensin (a glycosylprotein) in primary walls is synthesized by the rough ER, processed in the Golgi apparatus, and then transported to the cell wall.

1.3.5 Mitochondria

Mitochondria are granulous particles in cells. The size, number, and distribution patterns are variable. Mitochondria are stained with Janus green and easily observed by light microscopy. Phase-contrast micrographs show that mitochondria are always moving and changing shape. Electron micrography shows the sectional structures of round or oval mitochondria (Fig. 26).

Mitochondria are sacks covered with an outer membrane layer. The inner membrane is bent to the inner side to form cristae. These membranes, about 4 nm thick, can be stained with osmium tetraoxide. The matrix covered with

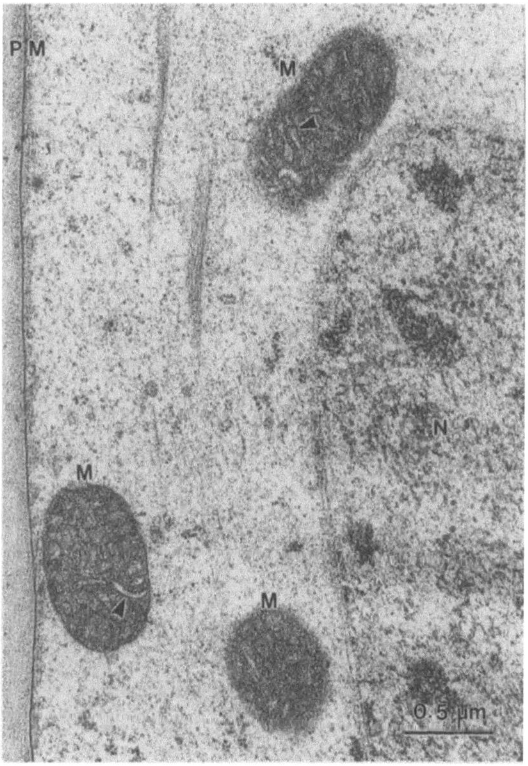

Fig. 26. Electron micrograph of mitochondria in *Cryptomeria japonica* tracheids. Round or oval mitochondria with cristae (*arrowheads*) are observed. *M* Mitochondria; *N* nucleus; *PM* plasma membrane. (Courtesy of K. Sano)

the membrane has a high electron density and is difficult to stain with osmium tetraoxide. Cristae generally occur perpendicular to the long axis, but sometimes parallel to the axis. ATP synthetase is embedded in the inner membrane of mitochondria, while the part covered with the inner membrane is filled with matrix substances.

The function of mitochondria is to produce ATP (energy source) by the degradation of carbohydrates, lipids, and proteins for the synthetic metabolism. Accordingly, various enzymes such as pyruvate oxidase, fatty acid oxidases, and enzymes in the tricarboxylic acid (TCA) cycle are located in the matrix. In the cytoplasm, pyruvic acid and fatty acids are formed by glycolysis of carbohydrates, hydrolysis and deamination of proteins, and hydrolysis and oxidation of fats. These acids are selectively incorporated into the matrix of mitochondria and oxidized to produce acetylcoenzyme A (acetyl-CoA), which is metabolized via the TCA cycle to form ATP, (reduced) nicotinamide adenine dinucleotide (NADH), (reduced) flavin adenine dinucleotide ($FADH_2$) etc. Electrons released by oxidation of substrates are transported to the electron transfer system in the inner membrane of mitochondria, and along with transfer of electrons via the electron transfer system hydrogen ions are pumped out from the matrix. The hydrogen ions are used to synthesize

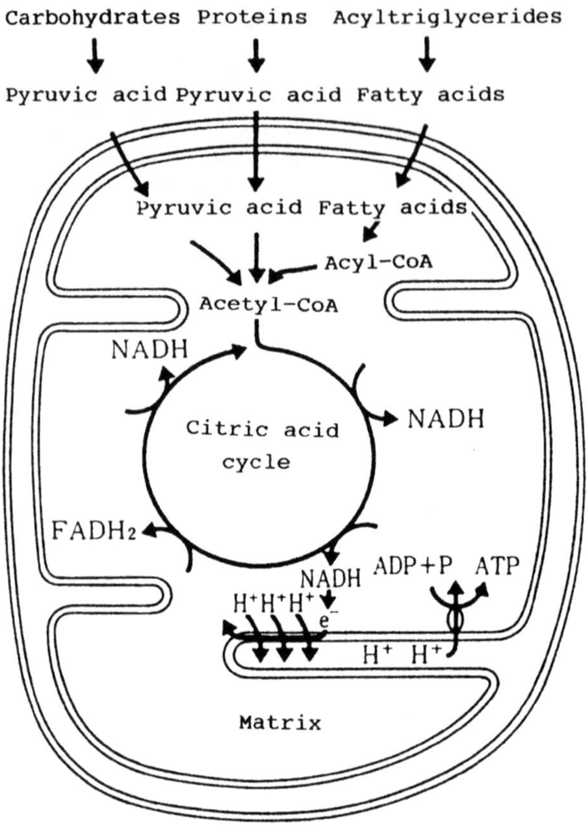

Fig. 27. Pathways for ATP synthesis in mitochondria

ATP in association with oxidative phosphorylation of ADP in the matrix (Fig. 27).

In the matrix of mitochondria, DNA, tRNA, ribosomes, and many enzymes are located. The DNA in mitochondria has a circular structure, both ends of which are connected by covalent bonds, and differs from nuclear DNA. Mitochondrial ribosomes are smaller than those in the cytoplasm. They are similar to those of prokaryotic cells and undergo the same reactions with antibiotics.

Most of the proteins and enzymes required by the mitochondria are synthesized in the cytoplasm and transported to the mitochondria, but some are encoded by mitochondrial DNA and synthesized within the mitochondria (Watanabe 1990b).

1.3.6 Plastids

Several plastids are present in cells. Proplastids, the precursors of plastids, are present in young growing cells. Inner membrane systems of proplastids are formed during differentiation and develop into chloroplasts containing chlorophyll, while proplastids in plants grown in the dark develop into etioplasts lacking chlorophyll. Some proplastids develop into amyloplasts, whose function is to store starch. Plastids contain circular DNA 70 S ribosomes (which are involved in the synthesis of ribosome proteins), and several enzymes used in photosynthesis and they synthesize ATP as in mitochondria, using light energy. Plastids are specific plant cell organelles which store starch, fat, and protein and provide energy using light energy.

1.3.6.1 Amyloplasts

Amyloplasts are present both in dormant cambial cells and in lignifying cells. Amyloplasts are ellipsoids 0.5–1 μm wide and 2–5 μm long and store starch granules. Like mitochondria, amyloplasts are covered with an outer and an inner membrane, the latter of which is incompletely developed. In many cases, rough ER covers the periphery of amyloplasts (Fig. 28).

Starch granules are stored in amyloplasts in dividing and cell wall-forming cells, but the starch granules disappear after completion of cell wall formation,

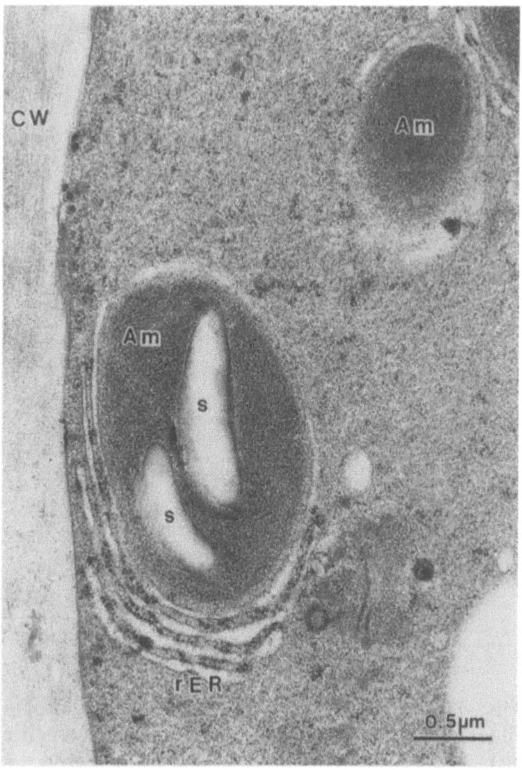

Fig. 28. Amyloplasts in *Cryptomeria japonica* tracheids. Amyloplasts containing starch granules are observed. Rough endoplasmic reticulum (*rER*) surrounds the amyloplasts. *Am* Amyloplast; *s* starch granule; *CW* cell wall. (Courtesy of K. Sano)

indicating an efficient conversion of the starch granules to energy and to cell wall components.

1.3.6.2 Chloroplasts

Chloroplasts are covered with a double-layered membrane composed of outer and inner membranes. In the stroma in the inner membrane, many ribosomes and substances with high electron density are present. Inner membranes develop to form cisternae named thylakoids. The thylakoids are stacked on top of one another and form grana, which are connected with lamellae (Fig. 29). Chlorophyll and carotenoids are present in the thylakoid membrane.

The main function of chloroplast is the synthesis of ATP and (reduced) nicotinamide adenine dinucleotide phosphate (NADPH) using energy from sunlight and conversion of CO_2 to carbohydrates using ATP and NADPH. The former reaction is called the light reaction and is conducted by thylakoid

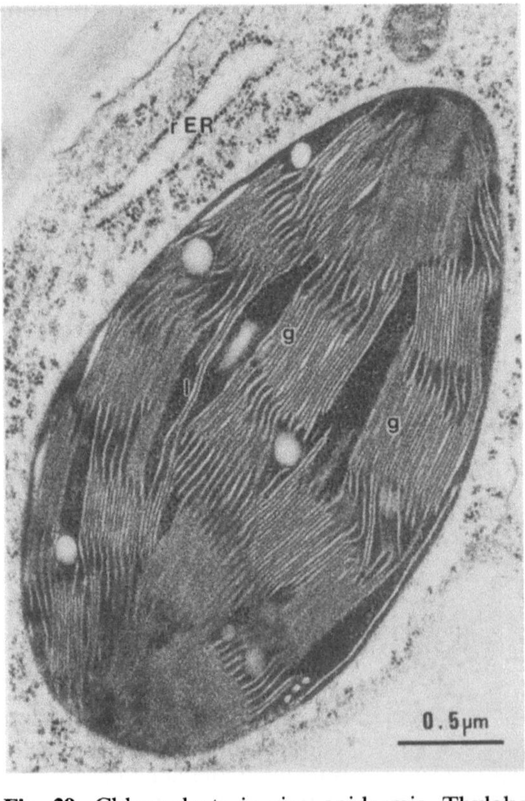

Fig. 29. Chloroplasts in rice epidermis. Thylakoids occur in multilayers, and form grana, which are connected with lamellae. *g* Grana; *l* lamellae; *rER* rough endoplasmic reticulum. (Courtesy of Dr. K. Takabe)

membranes, while the latter reaction is called the dark reaction and is conducted by stroma and cytoplasm (Watanabe 1990c).

1.3.7 Vacuoles

Vacuoles occupy 70–90% of the volume of elongated plant cells and are separated from the cytoplasm by a vacuole membrane called the tonoplast. In earlier studies, vacuoles were thought to be an organelle for storing secondary metabolites and waste materials and for regulating the osmotic pressure of cells. Heartwood phenolics, such as flavonoid and antocyanins, sugars, and alkaloids are present in the vacuoles of ray parenchyma cells in wood.

However, recent studies have shown that vacuolation of cells is an essential physiological event in cell differentiation and that proteases, nucleases, and wall-degrading enzymes hydrolyzing tonoplasts increase in the vacuoles at a late stage of cell wall differentiation (Fukuda et al. 1993).

1.3.8 Cytoskeletons

In eukaryotic cells, complex protein fiber networks are present in the cytoplasm. For example, microtubules and microfilaments which function as bone and muscle are present in eukaryotic, but not in prokaryotic cells. These are therefore called cytoskeletons (Itoh 1994b).

1.3.8.1 Microtubules

The cytoplasm was earlier thought to be transparent, but when glutaraldehyde was used as a tissue fixative in electron microscopy, microtubules were discovered as small hollow tubes in the cytoplasm. Microtubules were observed for the first time in plant cells by Ledbetter and Porter (1963). They consist of 13 fibrils composed of globular polypeptide, α-tubulin, β-tubulin, and other proteins, known as microtubulin-associated proteins. The average diameter and length are about 24 nm and several micrometers, respectively. Transmission electron microscopy of their cross section showed that the electron-translucent core of the microtubules is about 10 nm in diameter and the wall is about 7 nm thick.

Microtubules are elements constituting cilia of protozoa, while in plant cells they are found at various stages of cell growth. They are classified as preprophase microtubules, spindle microtubules, phragmoplast microtubules, and cortical microtubules, depending on their functions (Hogetsu 1989).

Preprophase microtubules appear at the site of cell plate formation and may function to regulate the location of the cell plate before cell divisions. Spindle microtubules serve to pull apart chromosomes and pull them to the poles, while phragmoplast microtubules play a role in aggregating Golgi vesicles in the equatorial plane as tracks during cell plate formation. Cortical microtubules are thought to regulate the direction of the regular deposition of cellulose microfibrils in the formation of cell walls. They appear as a layer of bands at certain intervals immediately inside the cytoplasmic membrane at the sites of future wall thickening.

Treatment of growing cell walls with a microtubule-depolymerizing agent, such as colchicine and taxol, a microtubule-stabilizing agent, causes the formation of unusual secondary wall thickening, suggesting that microtubules determine the wall pattern by determining the position and orientation of secondary walls (Fukuda et al. 1993). Recently, microtubules have been observed under the fluorescence microscope using the fluorescent antibody method. By this method the distribution and arrangements of microtubules can be seen in whole cells. The arrangements of the microtubules in tracheids of secondary thickenings of *Pinus taeda* and in growing fiber cells of cotton are shown in Figs. 30 and 31 (Itoh 1994b).

30 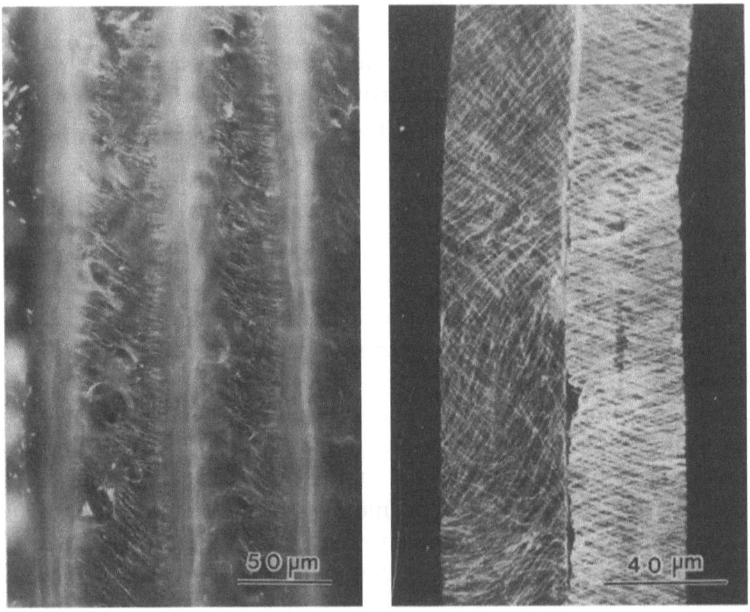 31

Fig. 30. Microtubules in secondary thickenings in tracheids of *Pinus taeda*. Immunofluorecent micrograph. (Courtesy of Dr. T. Itoh)

Fig. 31. Microtubules in cotton fiber cells at the beginning of secondary wall formation. Immunofluorecent micrograph. (Courtesy of Dr. T. Itoh)

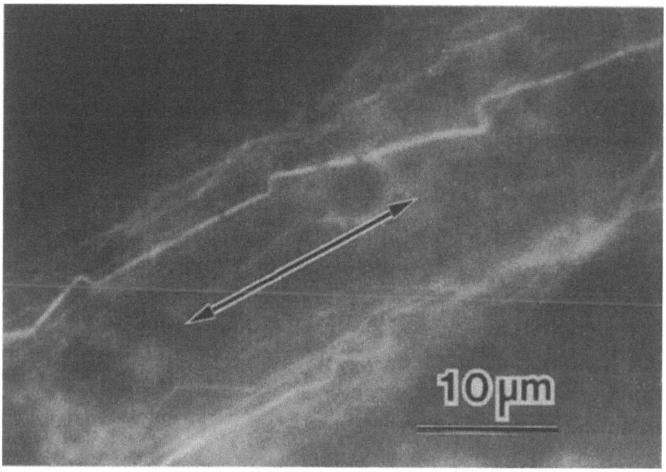

Fig. 32. Immunofluorecent micrograph of microfilaments in cotton fiber cells at the beginning of secondary wall formation. (Courtesy of Dr. T. Itoh)

1.3.8.2 Microfilaments

Microfilaments are composed of actin and are known as actin filaments. They are thinner than microtubules, with a diameter of 7 nm. A globular actin molecule is composed of double-helical strands, which are twisted around one another with a periodicity of 70 nm. In plant cells, the actin filaments can be shown by various immunofluorescence staining procedures with actin filament-binding compounds or with their antibodies (Fig. 32).

Fungal metabolite phalloidin, a stabilizer of actin filaments, and cytochalasin, a inhibitor of self-assembly of actin, have been used to demonstrate and elucidate the role of microfilaments in plant cells. The nucleus at the center of cells is moved near the cell membrane by cytochalasin treatment, suggesting that microfilaments function to keep the nucleus in the center of the cell and regulate the direction of the cell plates formed during cell divisions (Sonobe 1989).

In xylem differentiation of tracheary elements, actin filaments regularly appear in patches, and secondary walls are formed in the the same direction as the microtubules. Increases in tubulin content are responsible for increases in numbers of microtubules during differentiation, and α- and β-subunits of tubulin appear in a coordinated manner. In summary, microfilaments appear at the site of microtubules and are involved in cell wall formation, nuclear division, and phragmoplast formation. There are at least three actin genes in *Arabidopsis*, eight in soybean, ten in tomato, and six in maize. The actin proteins encoded by these genes are 90–92% homologous with each other and approximately 86% homologous with animal actins. It seems likely that these actins evolved from the cytoplasmic actin in protozoa and are highly conserved in evolution (Fukuda et al. 1993).

1.3.9 Cell Membrane

In plant cells, the terms plasma membranes and plasmalemma are used synonymously. Electron microscopy of plant cells shows that the cytoplasm is surrounded by a three-layered membrane composed of two dark layers with a light layer between them. The three layers comprise a unit membrane. Many organelles, such as the ER and the Golgi apparatus, are covered by a unit membrane. These membranes are composed of lipid and protein; they have the structure of a lipid bilayer 8–10 nm in width and contain protein with a diameter of 7.5 nm. Water freely penetrates the membrane; ions, sugars, and amino acids cannot penetrate it, but are transported by granules of the inner membrane proteins which function as membrane transport proteins.

Macromolecules such as proteins, polynucleotides, and polysaccharides cannot penetrate the membrane. These molecules are transported into the cytoplasm by two different mechanisms, namely exocytosis and endocytosis. In exocytosis, vesicles containing macromolecules covered with unit membranes are fused to the cell membrane, while in endocytosis vesicles are removed from the cell membrane and enter the cytoplasm. During cell wall formation, precursors of matrix components, and probably also cellulose synthases (rosettes), are transported from the inside to the outside of the cells as vesicles (exocytosis). Figure 33 shows endocytosis in cells in secondary wall formation of a sea alga, *Boergesenia forbesii.*

The lipid bilayer of cell membranes is fluid, while membrane proteins function as a part of membrane and move on the membrane. It has been suggested that crystalline cellulose micrifibrils are deposited at the surface of cell membranes by mobile cellulose synthase complexes (Brown and Montezinos 1976).

Protoplasts isolated by polysaccharidase treatments are covered only with plasma membranes and can only live in the hypertonic solutions, whose os-

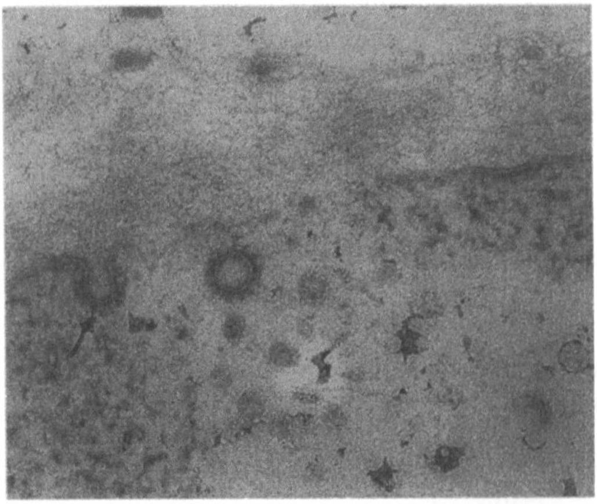

Fig. 33. Endocytosis in cells of *Boergesenia forbesii* during secondary wall formation. *Arrow* Endocytosis. (Courtesy of Dr. T. Itoh)

motic pressures are controlled by sorbitol or mannitol. This is due to the turgor pressure of the cells. Intact plant cells are covered with firm cell walls, and turgor pressure caused by water absorption by cells during cell growth functions as a driving force in cell growth.

In differentiation and development of cells, primary and secondary walls are deposited, and cell wall components are selectively excreted to the outside through the cell membranes.

The inside of cell membranes can only be observed by the freeze–fracture technique. Many protein particles are present on membranes, but the distribution pattern of the particles on inner and outer surfaces of cell membranes are different, suggesting nonsymmetrical functions of cell membranes. In living cells, the membranes are tightly pressed onto the cell walls by turgor pressure. Electron microscopy of freeze–fractured cells shows many molds of microfibril located in the innermost layers of the cell walls. In addition, at the end of the molds aggregates of granules are sometimes observed (Fig. 34). These granules are called terminal complexes (TCs). It has been suggested that they are enzymes that synthesize cellulose.

In animal cells, there are layers rich in glycosylproteins and oligosaccharides with side chains of glycosyllipids at the outside of the cell membranes. These oligosaccharides play important roles in identifying each cell, while in plant cells the outside of cell membranes are covered with thick cell walls composed of polysaccharides.

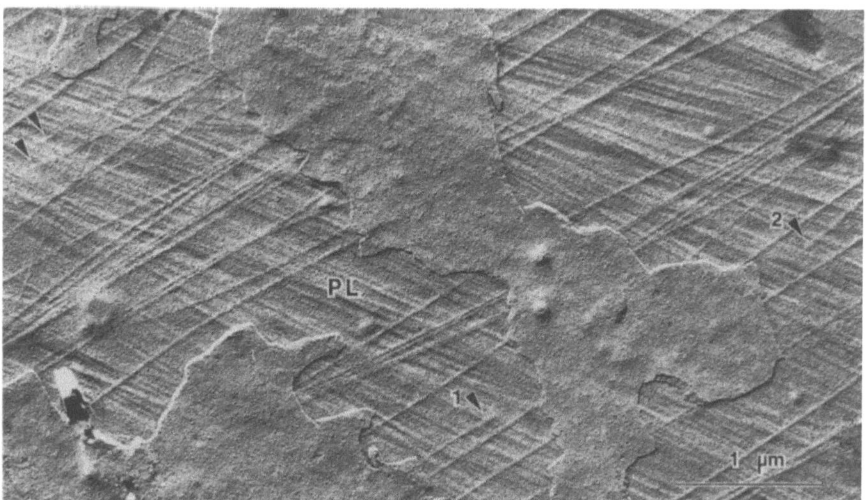

Fig. 34. Electron micrograph of freeze-fractured cells of valonia during secondary wall formation. Regularly oriented microfibrils are imprinted as templates on the plasma membrane (*PL*). *Arrowheads* indicate terminal complexes (TCs). TCs at *arrowheads 1* and *2* are thought to be moving in opposite directions. (Courtesy of Dr. T. Itoh)

1.3.10 Cell Wall

Cell walls are specific to plant cells, and wood cells in particular are covered with highly differentiated cell walls, while microbial cells have less differentiated cell walls with different chemical components. It has been suggested that primitive organisms first appeared on earth about 3.5×10^9 years ago as a result of a complicated chemical evolution and gradually evolved to primitive bacteria and the blue-green algae (prokaryotes). These organisms subsequently further developed into eukaryotes, in which chromosomes are covered with a nuclear membrane, and cells enlarged and the organelles increased. Primitive eukaryotes gradually evolved into plants and animals.

Some green algae developed into freshwater and land plants such as mosses, ferns, and seed plants. The base sequence analysis of 5S rRNA suggests that mosses and flowering plants successively derived from sea lettuce, spirogyra, and stonewort (Kimura and Osawa 1989).

Plant cells are covered with relatively firm cellulosic cell walls. In the development of land plants, cellulose cell walls were strengthened with lignin, an aromatic polymer, which enabled upright growth against gravity and offered protection against environmental stresses such as wind and rain.

Higher plant cell walls contain a hemicellulose–protein matrix, which is encrusted by lignin and in which cellulose microfibrils are embedded. The

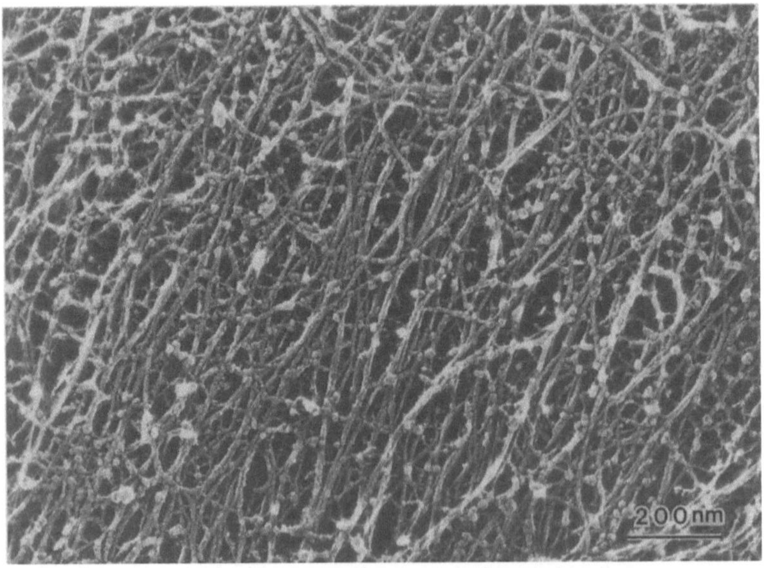

Fig. 35. Electron micrograph of freeze-fractured replica of the cell wall of suspension-cultured poplar cells. Bridge structures are observed between cellulose microfibrils. (Courtesy of Dr. T. Itoh)

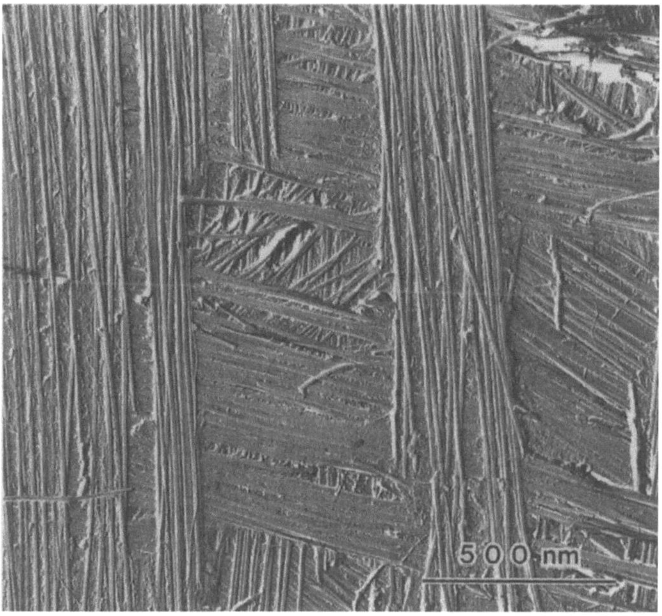

Fig. 36. Electron micrograph of freeze-fractured replica of a secondary wall of valonia. Microfibrils in each lamella cross at almost right angles. (Courtesy of Dr. T. Itoh)

matrix substance generally consists of protein (10%), pectin (35%), and hemicelluloses in primary cell walls. There are no proteins and no pectin present in secondary walls. The structure of the cell walls is similar to that of carbon fibers reinforced with resins or reinforced concrete (Fig. 35).

The composition and the amount of matrix substances are different in different plant species. Cell walls of cotton fibers are composed of only cellulose. In lower plants such as algae, some species such as valonia have cell walls with large amounts of cellulose, while the cell walls of other algal species contain little or no cellulose.

Cell walls are divided into primary and secondary walls in developmental processes. Primary wall is deposited during cell growth and enlargement, and the secondary wall is formed after completion of cell growth. In these walls, lamellae of cellulose microfibrils are laid down upon one another, and the degree of overlay depends on the kind of cell. For example, cell walls of parenchyma cells are formed by sheets of microfibrils with lateral windings, and cell walls of epidermis, cortex, and collenchyma cells are formed by many overlaid sheets of microfibrils with different orientations. In cell walls of algae, microfibrils in each lamella are often crossed almost at right angles (Fig. 36).

In cell walls of epidermal, collenchyma, algal, and stone cells, the orientation of the microfibrils gradually changes from inside to outside in a clockwise direction. Such types of cell walls are called helicoidal. Wood cell walls are thought to be helicoidal cell walls. Microfibrils in the lamellae of each cell wall cross each other or gradually change their orientation and the substances that they deposit, resulting in a strong structure as in plywood.

Cell walls support the weight of a plant and build up a structure to transport water to the top of trees 100 m in height. Cell walls also function as a barricade to prevent microbial and insect attacks and play a role in the circulation of liquids and in the transportation of water. They are also involved in cell communication through plasmodesmata and determine cell growth and cell shapes through regulation of cell wall looseness (Brett and Waldron 1990). Such functions are related to the complexity and diversity of cell wall structures and their components.

2 Primary Metabolism in Woody Plants

Wood is composed of 40–50% cellulose, 15–30% hemicelluloses, and 20–35% lignin, with extractives as minor constituents. These compounds are synthesized in the processes of growth and differentiation of cambial cells. Therefore, it is important to understand the metabolism of the cambium, the mechanism of xylem and phloem differentiation, and the synthesis of wood components by cambial and xylem tissues.

It has been shown that the approximate composition of primary cell walls in trees is the following: 25% cellulose, 20% galactoxyglucan, 35% pectin, 10% hemicelluloses (arabinogalactan), and 10% protein, largely based on hydroxyproline (Simson and Timell 1978).

2.1 Primary and Secondary Metabolism

Plants convert sunlight energy to chemical energy (ATP, NADPH etc.) by the mediation of chlorophyll in chloroplasts and synthesize sugars and starch from CO_2 by using ATP and NADPH. These carbohydrates are stored and used for differentiation and formation of plant tissues. Energy metabolism and metabolism of lipids, amino acids, protein, and nucleic acids are a common feature of animals, plants, and microorganisms. These biological reactions are essential to maintain life in living organisms and are known as primary metabolisms.

In contrast, the types of metabolism involved in the synthesis of lignin, tannins, flavonoids, suberin, alkaloids etc., which are not directly related to maintaining life, are known as secondary metabolisms. Metabolism involves synthesis and degradation of particular compounds in a given organism. The products formed by secondary metabolism are called secondary metabolites. Secondary metabolites play a role in reinforcement of tissue and tree body (e.g., cellulose, lignin, suberin), protection against insects, disease etc. (e.g., alkaloids, tannins, terpenoids), and growth regulation (plant hormones).

Secondary metabolism is closely related to primary metabolism. Raw materials and energy required for the synthesis of secondary metabolites are supplied by primary metabolism. Secondary metabolites have been elucidated by organic chemists studying natural products. More recent progress in biochemistry has revealed biosynthetic pathways of secondary metabolites and enzymes involved, as noted in Chapter 4.

2.2 Photosynthesis

Higher plants with chloroplasts absorb light energy by mediation of chlorophyll and synthesize sugars from CO_2 and H_2O, while releasing O_2. The reaction is known as photosynthesis and can be summarized by the following formula:

$$CO_2 + 2H_2{}^{18}O \rightarrow C(H_2O) + (H_2O) + {}^{18}O_2$$

By tracer experiments with $H_2{}^{18}O$, Ruben and Kamen (Kamen 1963) demonstrated that O_2 released in photosynthesis is formed by photolysis of H_2O. Photosynthesis is divided into a light reaction and a dark reaction. In the light reaction, water molecules are decomposed by the energy of sunlight absorbed by chlorophyll molecules, followed by the reduction of $NADP^+$ to NADPH and the conversion of ADP to ATP by mediation of ATPase. NADPH and ATP formed by the light reaction are used as a reducing reagent and energy source to produce sugars from CO_2 in the dark reaction (Sato Ki 1992).

The light reaction takes place on the thylakoid membrane of the chloroplast. The thylakoid membrane is composed of a lipid bilayer, in which photosystem I (PS I), photosystem II (PS II), cytochrome b6/f complex, and ATP synthase complex are embedded.

PS I contains a chlorophyll–protein complex (P700), which has a maximum absorption at 700 nm and is directly involved in photochemical reactions as an electron transfer entity, and light-harvesting chlorophyll, chlorophyll a, a small amount of chlorophyll b, and carotenoids. Generally about 300 molecules of light-harvesting chlorophyll are associated with a P700 protein. A part of the light-harvesting chlorophyll is linked to the reaction center protein, but most of it is present as a chlorophyll–protein complex (LHCP), which efficiently catches photons and transfers them to P700. PS II contains as a reaction center a chlorophyll–protein complex (P680) which has an absorption maximum at 680 nm, chlorophyll a and b as light-harvesting chlorophyll, and associated carotenoids. As in PS I, photons collected by the light-harvesting chlorophyll are finally transferred to P680 (Fig. 37).

P700, excited by light, transfers electrons to another linked chlorophyll a (A_0) and is oxidized itself. Oxidized P700 is reduced by plastcyanin, while reduced A_0 rapidly reduces A_1 (vitamin K_1). The electron is then transferred from A_1 to $NADP^+$ via the bound iron-sulfur protein acceptors of PS I (F_x), (F_A and F_B may operate in parallel), ferredoxin (Fd), and ferredoxin-$NADP^+$ reductase, successively (Fig. 38).

The excited P680 in PS II transfers electrons to pheophytin, and the reduced pheophytin rapidly transfers electrons to plastoquinone (PQ) via two kinds of linked quinones (Q_A, Q_B). The component that transfers electrons to P680 was recently identified as a tyrosine residue of the protein in the reaction center. On the Mn-linked protein in the reaction center (Mn cluster), water is

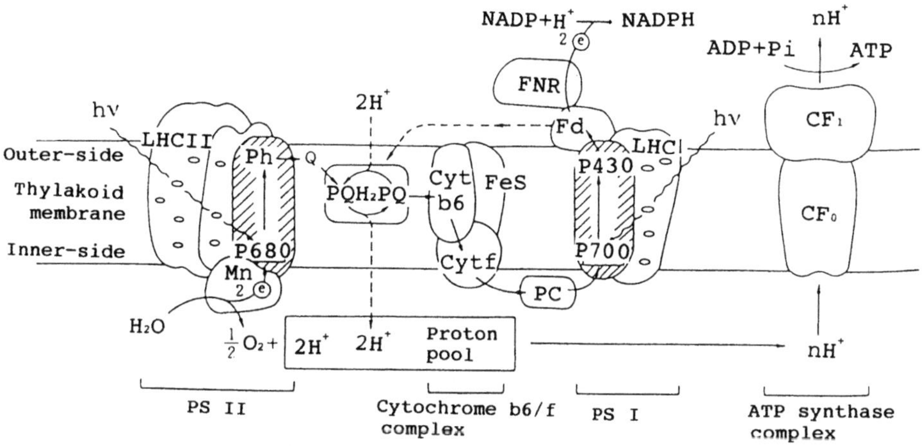

Fig. 37. Organization and function of protein complexes in thylakoid membrane of higher plant chloroplasts. *-e-* Routes for electron transport from water to NADP; *dotted line* from *Fd* (ferredoxin): circulated electron transfer route; *hv* light energy transfer route; *hatched area* reaction center proteins of PSI and PSII complexes, respectively; *LHCI and LHCII* light harvesting-chlorophyll-protein complexes I and II

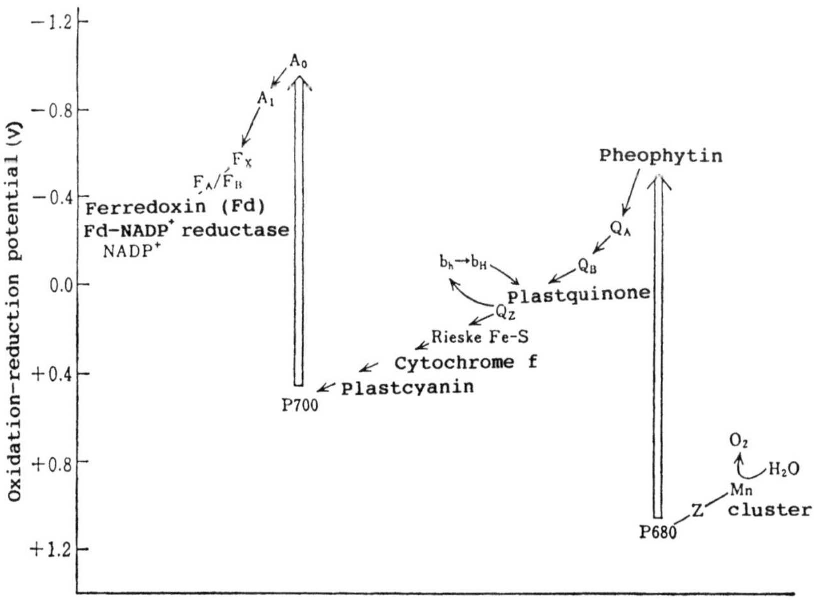

Fig. 38. Electron transfer system in photosythesis

oxidized and the electron is transferred to the tyrosine residue. Accompanied by water oxidation, O_2 is generated and two H^+ are released and stored in the lumen of the thylakoid (Sato Ka 1992).

The ATP synthase complex located in the thylakoid membrane has an intermembrane channel for proton transport and synthesizes ATP by using chemical energy generated by different concentrations of protons between inner and outer membranes.

Light energy absorbed via series of reactions by chlorophyll molecules is stored as NADPH and ATP and used for CO_2 fixation in stroma (dark reaction). Synthetic reactions of sugars from CO_2 and water in the dark reaction were elucidated by feeding experiments with ^{14}C-labeled compounds. The enzymes involved in each reaction were characterized. Calvin and Benson (1948, 1949) found that, when *Chlorella* and *Scenedesmus* cultured in the presence of $^{14}CO_2$ for a short period of illumination are extracted with hot methanol, hexose phosphates, triose phosphates and 3-phosphogylceric acid (3-PGA) are labeled. After 5s of illumination of the culture, about 70% of the incorporated ^{14}C was located at the carboxyl group of 3-PGA. The incorporation reaction of CO_2 into 3-PGA has since been investigated by many investigators.

It was finally found that ribulose-1,5-bisphosphate is the initial acceptor of CO_2. Ribulose-1,5-bisphosphate is carboxylated by the mediation of ribulose-1,5-bisphosphate carboxylase/oxygenase (Rubisco), and the labile α-keto acid intermediate formed is hydrolyzed, giving two molecules of 3-PGA (Fig. 39). Rubisco catalyzes the covalent attachment of CO_2 to ribulose 1,5-bisphosphate and the simultaneous hydrolysis of the six-carbon α-keto acid intermediate to form two molecules of 3-PGA, of which one bears the carbon introduced from CO_2. Rubisco is a large enzyme protein with a molecular mass of 550kDa, located in the stroma of the chloroplast. It can account for 50% of the total soluble protein in chloroplasts. This enzyme protein is the most abundant single protein in the biosphere.

Rubisco is made up of two types of subunits designated large (L) and small (S). There are eight L subunits each with a molecular mass of 51–58kDa, and eight S subunits with a molecular mass of 12–18kDa, forming a structure of L_8S_8. The L subunit contains the active site and is encoded in the chloroplast genome: its mRNA is present as part of the chloroplast RNA. This subunit is synthesized as a precursor protein with a polypeptide leader sequence that directs the S subunit to be transported across the chloroplast envelope. Inside the chloroplast, the two subunits are assembled with the assistance of an assembly or L subunit-binding protein (Jensen 1990). The activity exhibited by Rubisco is modulated in the plant leaf by a number of factors, most notably its response to light intensity.

A series of reactions to form sugars from CO_2 and the enzymes involved in these reactions have been characterized; the complete reaction was named the Calvin or C_3 cycle (Fig. 40). The 3-PGA formed is reduced to glyceraldehyde-3-phosphate (GAP) by glyceraldehyde-3-phosphate dehydro-

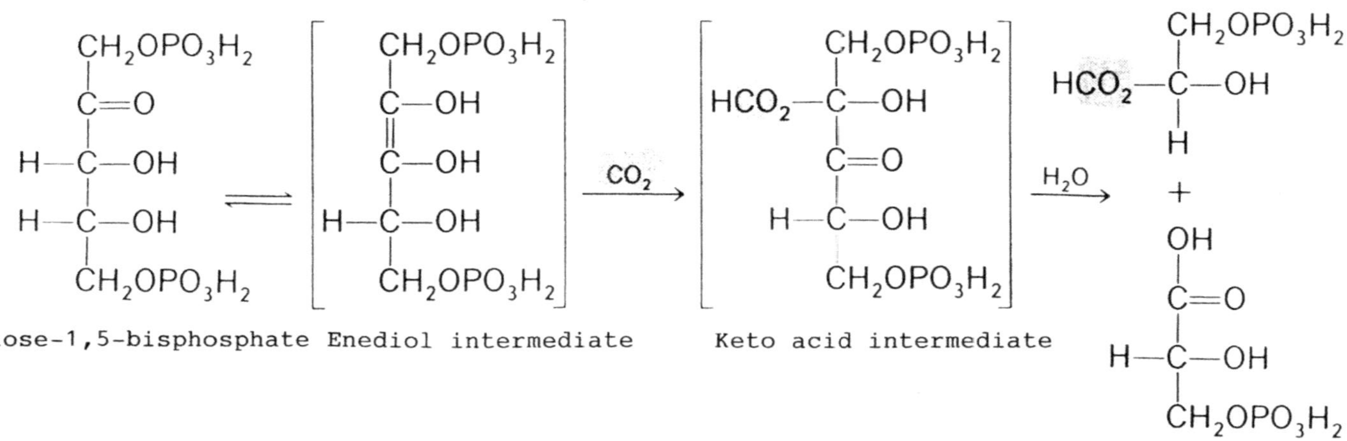

Ribulose-1,5-bisphosphate Enediol intermediate Keto acid intermediate

3-Phosphoglyceric acid

Fig. 39. Incorporation of CO_2 into 3-phosphoglyceric acid (3-PGA) via ribulose-1,5-bisphosphate in the C_3 pathway of photosynthesis

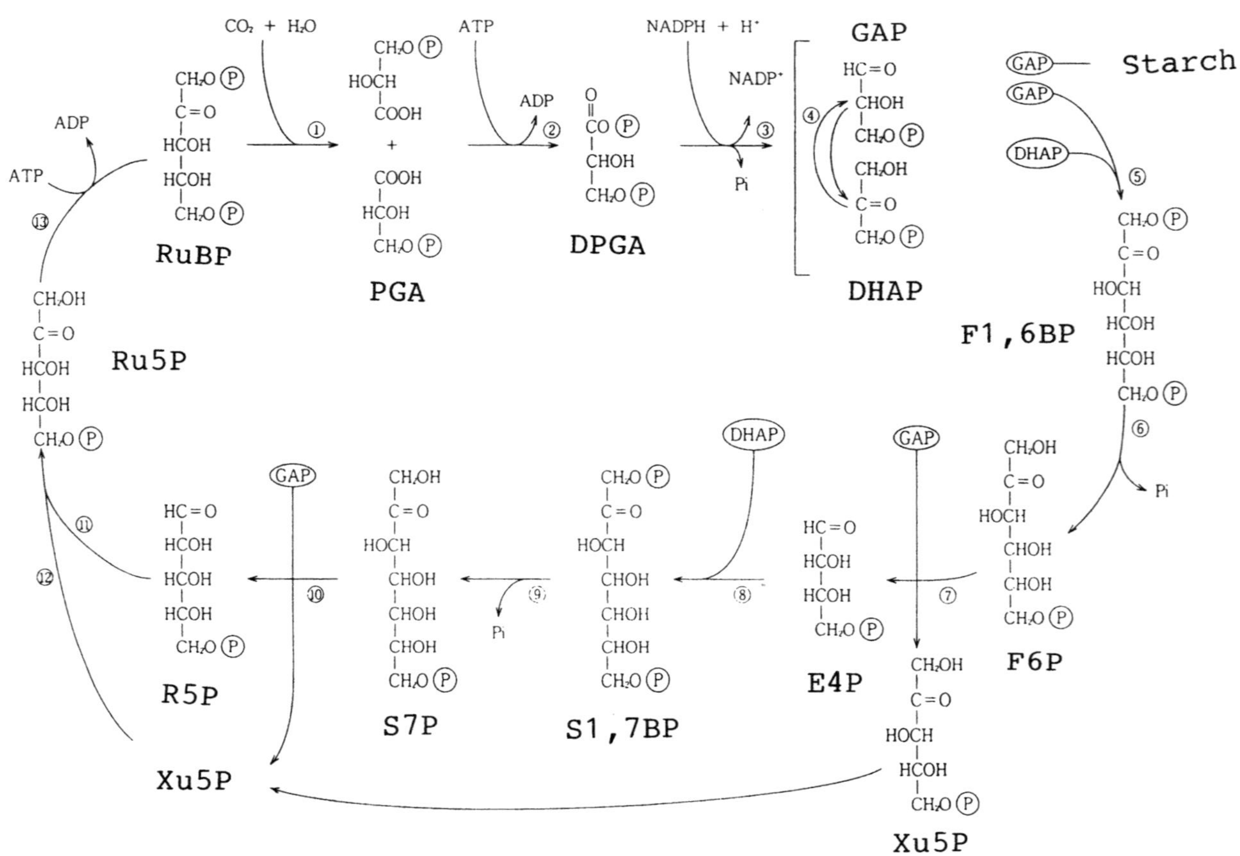

Fig. 40. The Calvin cycle in photosynthesis. *RuBP* Ribulose-1,5-bisphosphate; *PGA* 3-phosphoglycerate; *DPGA* 1,3-bisphosphoglycerate; *DHAP* dihydroxyacetone phosphate; *GAP* glyceraldehyde-3-phosphate; *F1,6BP* fructose-1,6-bisphosphate; *Ru5P* ribulose-5-phosphate; *R5P* ribose-5-phosphate; *Xu5P* xylulose-5-phosphate; *S7P* sedoheptulose-7-phosphate; *S1,7BP* sedoheptulose-1,7-bisphosphate; *E4P* erythrose-4-phosphate; *F6P* fructose-6-phosphate. Enzymes involved in respective reactions: *1* ribulose-1,5-bisphosphate carboxylase (Rubisco); *2* phosphoglycerate kinase; *3* glyceraldehyde-3-phosphate dehydrogenase; *4* triosephosphate isomerase; *5* aldolase; *6* fructose-1,6-bisphosphatase; *7* transketolase; *8* aldolase; *9* sedoheptulose-1,7-bisphosphatase; *10* transketolase; *11* ribose-5-phosphate isomerase; *12* ribulose-5-phosphate epimerase; *13* phosphoribulokinase

genase in the presence of NADPH and ATP obtained by the light reaction. Glyceraldehyde-3-phosphate is finally converted to sucrose and starch via fructose-1,6-bisphosphate. In higher plants, the final products of photosynthesis are sucrose, starch, and fructans.

Plants that produce 3-PGA as the first product in the dark reaction are the most common and are called C_3 plants (Sato Ka 1992). Most photosynthetic organisms assimilate atmospheric CO_2 directly into 3-PGA by mediation of Rubisco, but certain families of flowering plants (angiosperms) produced C_4 dicarboxylic acids (C_4-DCA) such as oxaloacetic acid, malic acid, and aspartic acid as the first product in the dark reaction. These plants are called C_4 plants.

In C_4 plants, phosphoenolpyruvic acid (PEP) is the acceptor of CO_2. In the leaves of sugar cane (a C_4 plant), about 90% of CO_2 is first used to form malic acid and aspartic acid, after which hexose phosphates are formed from these acids via 3-PGA. The direct incorporation of CO_2 into PGA only accounts for 10% of the total. The first CO_2 incorporation into C_4 DCA takes place at the chlorophyll of mesophyll cells, and the synthesized malic acid or aspartic acid is transported to bundle sheath cells via plasmodesmata. Malic acid is oxidatively decarboxylated to CO_2 and NADPH, after which the CO_2 enters the Calvin cycle in the chloroplast of the bundle sheath cells, with NADPH used for reduction. Aspartic acid releases CO_2 via oxaloacetic acid, and the CO_2 enters the Calvin cycle (Fig. 41).

By such compartmentalization, the CO_2 formed by photorespiration is reused for photosynthesis in C_4 plants, and a high rate of photosynthesis takes place even in leaves with closed stomata. C_4 plants include monocots such as Gramineae and some dicots of tropical origin and contain pyruvate orthophosphate dikinase (PPDK), which is not present in C_3 plants.

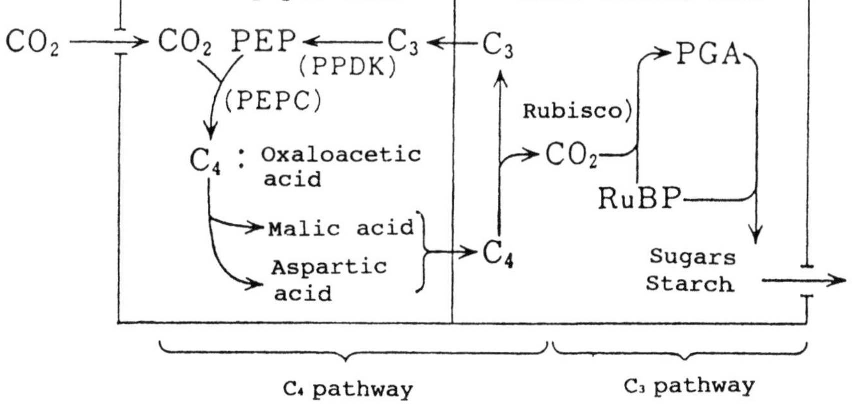

Fig. 41. Pathway of carbon dioxide fixation in leaves of a C_4 plant showing the spatial separation of fixation of carbon dioxide by phosphoenol pyruvate (*PEP*) carboxylase in the mesophyll cells prior to fixation in the Calvin cycle in the bundle sheath cells. *PGA* 3-Phosphoglycerate; *RuBP* ribulose-1,5-bisphosphate; *PPDK* pyruvate: orthophosphate dikinase; *Rubisco* ribulose-1,5-bisphosphate carboxylase

Succulent plants such as *Opunta ficus-indica* fix CO_2 via a special C_4-DCA pathway. Formation of acids occurs at night and decarboxylation of the acids in the daytime. This type of acid metabolism is known as crassulacean acid metabolism (CAM). CAM plants open stomata at night and fix CO_2, which is converted to oxaloacetic acid by PEP carboxylase in the cytoplasm. Oxaloacetic acid is converted to malic acid by malate dehydrogenase with NADPH and is stored in vacuoles. The malic acid stored at night is decarboxylated in the daytime, and the CO_2 enters the Calvin cycle. Malic enzyme–NADP or malate dehydrogenase–NAD and PEP carboxylase are involved in decarboxylation.

2.3 Photorespiration

C_3 plants, which fix CO_2 via ribulose-1,5-bisphosphate, take up O_2 and release CO_2 upon illumination at high O_2 pressure. This process is called photorespiration. In photorespiration, Rubisco, which plays a central role in CO_2 fixation in the Calvin cycle, acts as an oxygenase in chloroplast as follows (Imahori and Yamakawa 1990):

$$\text{Ribulose-1,5-bisphosphate} + CO_2 \xrightarrow{\text{Photosynthesis}} \text{2,3-PGA}$$

$$\text{Ribulose-1,5-bisphosphate} + O_2 \xrightarrow{\text{Photorespiration}} \text{3-PGA} + \text{glycolate-2-phosphate}$$

The reactions involving CO_2 and O_2 are catalyzed by the same part of the Rubisco enzyme. The reactions are competitive, and both substrates inhibit the counterpart reaction. When leaves or isolated chloroplast are illuminated in the presence of $^{18}O_2$, ^{18}O is incorporated into the carboxyl group of the glycolate within a few seconds. Mutants of *Arabidopsis thaliana* lacking a specific phosphatase for glycolate-2-phosphate assimilate glycolate-2-phosphate, but do not form any glycolate, showing that glycolate-2-phosphate is the precursor of glycolate. Glycolate-2-phosphate is converted to 3-PGA in a sequence of reactions that involves cooperation between several subcellular compartments: the chloroplast, the mitochondria, and the peroxisome (Leegood 1993). In this process, O_2 uptake and CO_2 release (photorespiration) occur (Fig. 42). C_4 plants, which incorporate CO_2 into PEP, exhibit no photorespiration, and the efficiency of photosynthesis is high.

2.4 Glycolysis

Organisms store energy required for synthesis, differentiation, and growth as high-energy compounds, such as ATP, creatine phosphate, enolphosphate and

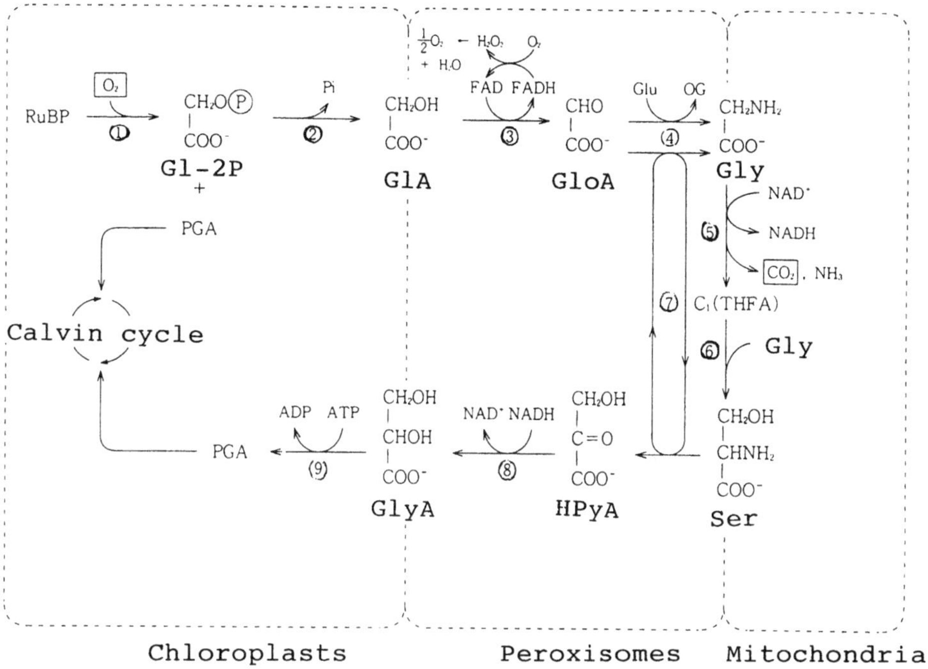

Fig. 42. The photorespiratory pathway and its subcellular location. Rubisco in the chloroplast can both oxygenate and carboxylate RuBP. Oxygenation leads to the generation of glycolate 2-phosphate from which glycerate 3-phosphate is regenerated in the photorespiratory pathway. During this process, one quarter of the carbon in glycolate is lost as carbon dioxide, together with half the nitrogen, as NH_3 in glycine. *Gl-2P* Glycolate-2-phosphate; *GlA* glycolic acid; *GloA* glyoxylic acid; *Gly* glycine; *Ser* serine; *HPyA* hydroxypyruvic acid; *GlyA* glyceric acid. Enzymes involved: *1* RuBP carboxylase/oxygenase (Rubisco); *2* phosphoglycolate phosphatase; *3* glycolate oxidase; *4* glycine aminotransferase; *5* glycine synthase; *6* serine hydroxymethyltransferase; *7* serine glyoxylate transaminase; *8* hydroxypyruvate reductase; *9* glycerate kinase

acylmercaptide. ATP is the most common high-energy compound and releases 13.6 kcal/mol by hydrolysis to ADP and orthophosphoric acid (Pi) under physiological conditions. Various compounds are synthesized by using this energy. Glycolysis is a series of reactions in which sugars are converted anaerobically to pyruvic acid and ATP is produced. The reaction is common in microorganisms, plants, and animals. Starch produced by photosynthesis is converted to glucose-1-phosphate by α-glucan phosphorylase. Glucose-1-phosphate is converted to fructose-6-phosphate via glucose-6-phosphate by phosphoglucomutase and glucosephosphate isomerase. Fructose-6-phosphate is then converted to fructose-1,6-bisphosphate by phosphofructokinase with ATP. Fructose-1,6-bisphosphate is degraded to two molecules of triosephosphate (dihydroxyacetone phosphate and glyceraldehyde-3-phosphate) by aldolase.

Glyceraldehyde-3-phosphate is converted to pyruvic acid via several reactions, as described in Fig. 43.

The characteristics of the enzyme involved in glycolysis are as follows:

1. Aldolases (fructose-1,6-bisphosphate; D-glyceraldehyde 3-phosphate lyase, EC 4.1.2.13), which catalyze the aldol cleavage of fructose-1,6-bisphosphate, yielding dihydroxyacetone phosphate and GAP, are classi-

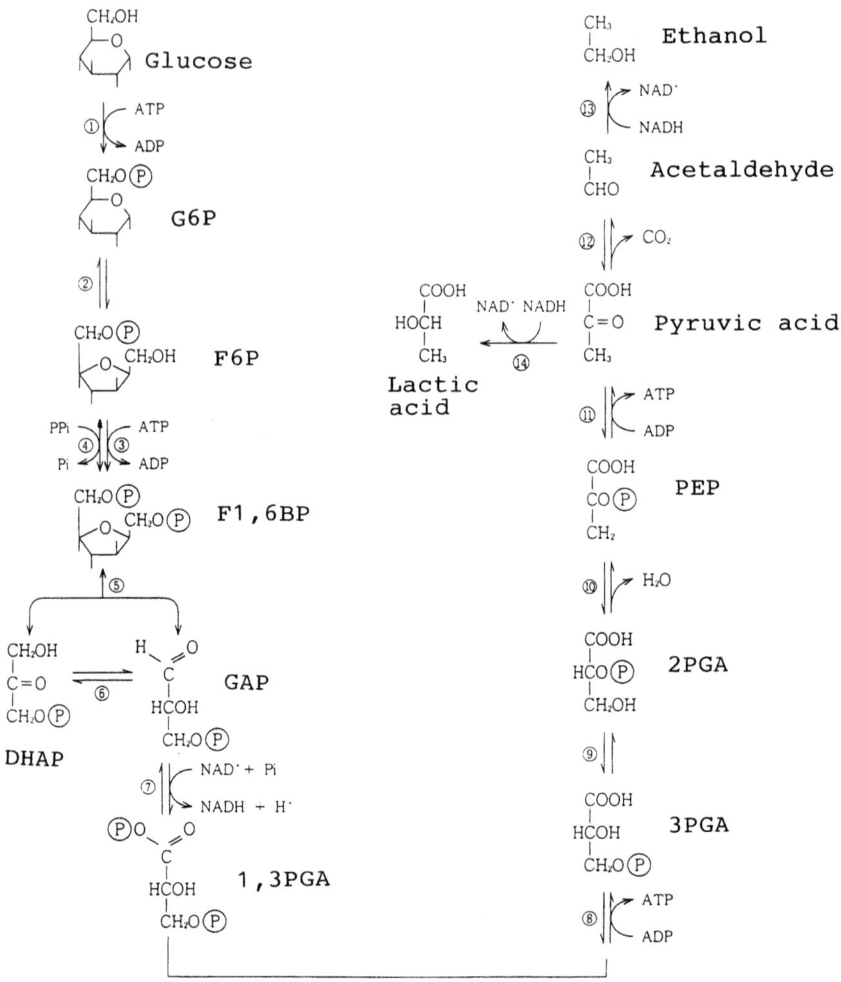

Fig. 43. Glycolysis (Embden-Meyerhof pathway) in higher plants. *1* Hexokinase; *2* phosphoglucoseisomerase; *3* phosphofructokinase; *4* pyrophosphate:fructose 6-phosphate phosphotransferase; *5* aldolase; *6* triosephosphate isomerase; *7* glyceraldehyde 3-phosphate dehydrogenase; *8* phosphoglycerokinase; *9* phosphoglyceromutase; *10* enolase; *11* pyruvate kinase; *12* pyruvate decarboxylase; *13* alcohol dehydrogenase; *14* lactate dehydrogenase

fied into classes I and II. Class I aldolases are relatively large homotetramers (M_r, approximately 160kDa), have an essential sulfhydryl group, and have a Schiff base reaction mechanism. Class II aldolases are smaller (M_r, approximately 65kDa), monomeric proteins that require potassium and divalent cations and do not have a Schiff base mechanism. Animal aldolases are class I, while class II aldolases occur in bacteria and fungi (Miernyk 1990). Higher plants, beginning with ferns and mosses, have class I aldolases. The amino acid sequence of *Zea mays* root aldolase has been determined from cDNA clones (Kelley and Tolan 1986).

2. Triose phosphate isomerases (D-glyceraldehyde 3-phosphate ketol isomerase, EC 5.3.1.1), which catalyze the reversible isomerization of glyceraldehyde-3-phosphate and dihydroxyacetone phosphate, are homodimeric proteins with each subunit having a molecular mass of 27kDa. Complete primary sequences, from either direct amino acid sequencing or nucleotide sequencing of cDNA, have been reported for triose phosphate isomerases from bacteria, fungi, invertebrates, and animals.

   The physicochemical and catalytic properties of plant enzymes are very similar to those of enzymes from other sources. Nucleotide sequences of the small multigene family of maize triosephosphate isomerases showed a high degree of homology with the previously reported animal and microbial sequences (Miernyk 1990).

3. Glyceraldehyde-3-phosphate dehydrogenase (EC 1.2.1.12), which catalyzes the only oxidative reaction that occurs in glycolysis, has been purified and characterized from several plant seeds (pea, green gram, castor oil) and leaves (spinach, white mustard, tobacco). In all cases, the properties of the plant enzyme are similar to those of other organisms. The amino acid sequence for tobacco leaf cytoplasmic NAD^+ – glyceraldehyde-3-phosphate dehydrogenase has been derived from the nucleotide sequence of a cloned cDNA. This sequence has a higher homogeny with *Escherichia coli* or animal enzymes (Miernyk 1990).

4. Phosphoglycerate kinase (EC 2.7.2.3) is a key enzyme in the generation of ATP during glycolysis. The enzyme has been purified and characterized from various plants. The catalytic and physicochemical properties of phosphoglycerate kinase from plant and cyanobacterial sources are very similar to those from yeast or mammals. The molecular mass of purified plant PGA kinase is 46–48kDa (Miernyk 1990).

5. Phosphoglycerate mutase (EC 5.4.2.1), which catalyzes an intermolecular phosphoryl group transfer, has been purified from wheat, rice germ, and developing *Ricinus communis* endosperm. In each instance the enzyme is monomeric, and the subunits have a molecular mass of 62–64kDa. Plant enzymes seem to be metalloenzymes containing 1 mol of tightly bound divalent cation per mol protein (Miernyk 1990).

6. Enolase (2-phospho-D-glycerate hydrolyase, EC 4.2.1.11) is a metalloenzyme that catalyzes the freely reversible interconversion between 2-PGA and phosphoenolpyruvate. The enzyme has been purified and character-

ized from potato tubers, spinach leaves, and developing *Ricinus communis* endosperm. The physicochemical properties of *Ricinus* cytoplasmic protein are similar to those of yeast enzyme (Miernyk 1990).

7. Pyruvate kinase (EC 2.7.1.40) catalyzes the transfer of phosphate from phosphoenolpyruvate to ADP, yielding ATP and pyruvate. Pyruvate kinase has been purified and extensively characterized from several mammalian sources. However, relatively little has been reported for plant sources. All animal, fungal, and higher plant pyruvate kinase have an ordered, sequential kinetics mechanism.

   In most instances, the glycolytic enzymes from plant cells are similar to mammalian and yeast enzymes (Miernyk 1990).

In the reaction process of glycolysis, two molecules of ATP and NADPH are formed by oxidation of glyceraldehyde-3-phosphate to 1,3-bisphosphoglyceric acid and dehydration of 2-PGA to PEP. The reaction pathway is also called the Embden–Meyerhof pathway (Conn and Stumpf 1972).

Glycolysis, the anaerobic metabolism of sugars, yields pyruvic acid, but in muscle contraction in animals and in lactic acid fermentation pyruvic acid is reduced to lactic acid. In yeast fermentation, pyruvic acid is decarboxylated to acetaldehyde, which is reduced to ethanol (ethanol fermentation).

Ethanol, ethanol dehydrogenase, and pyruvate decarboxylase activities occur in the cambium of several tree species (Kimmeres and Stringer 1988). The presence of enzymes involved in fermentation and ethanol indicates that cambial respiration might be oxygen limited. Intercellular spaces are scarce in the cambium itself, but they exist in the adjoining tissues, where they are most probably continuous with the general air space network of the tree (Catesson 1994).

2.5 Tricarboxylic Acid Cycle

Pyruvic acid formed by glycolysis is decomposed to CO_2 and H_2O via acetyl-CoA in the tricarboxylic acid (TCA) cycle. This metabolic pathway is also known as the citric acid cycle or Krebs cycle and was characterized by Krebs in 1937 (Conn and Stumpf 1972; Fig. 44).

In this cycle, acetyl-CoA (C_2) derived from pyruvic acid (C_3) is combined with oxaloacetic acid (C_4) to yield citric acid (C_6), which is successively converted to oxaloacetic acid (C_4) via *cis*-aconitic acid (C_6), isocitric acid (C_6), 2-oxoglutaric acid (C_5), succinyl-CoA (C_4), succinic acid (C_4), fumaric acid (C_4), and malic acid (C_4). The cycle comprises the following six reactions:

1. Oxidative decarboxylation of pyruvic acid to acetyl-CoA

$$\text{Pyruvic acid} + NAD^+ + CoA \xrightarrow{\text{Pyruvate dehydrogenase}} \text{acetyl-CoA} + CO_2 + NADH + H^+$$

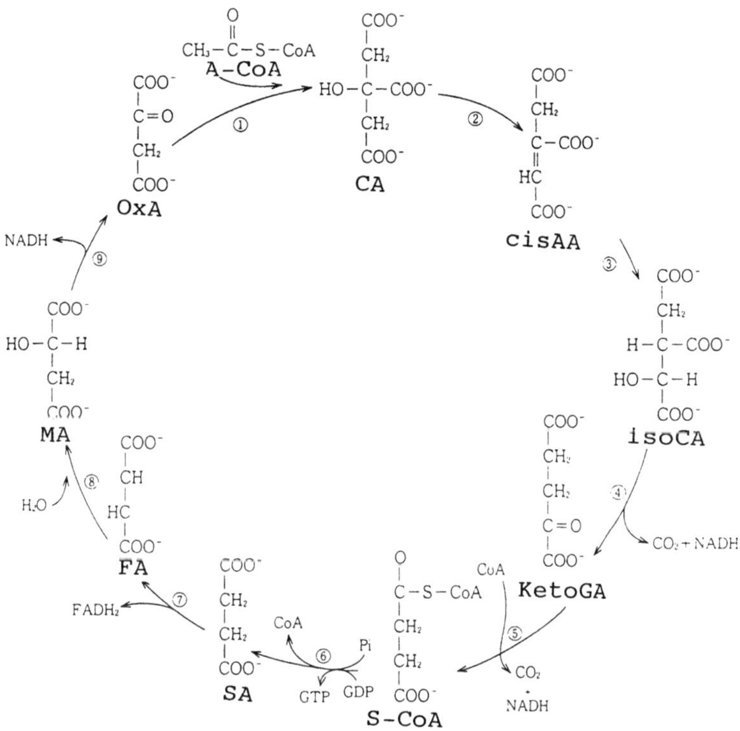

Fig. 44. The tricarboxylic acid (TCA) cycle in higher plants. *1* Citrate synthase; *2,3* aconitase; *4* isocitrate dehydrogenase; *5* 2-oxoglutarate dehydrogenase; *6* succinyl-CoA synthetase; *7* succinate dehydrogenase; *8* fumarase; *9* malate dehydrogenase

2. Condensation of acetyl-CoA and oxaloacetate to yield citric acid

$$\text{Acetyl-CoA} + \text{oxaloacetate} + H_2O \xrightleftharpoons{\text{Citrate synthase}} \text{citric acid} + \text{CoA}$$

3. Oxidative decarboxylation of citric acid (via isocitric acid) to 2-oxoglutaric acid

$$\text{Isocitric acid} + NAD^+ \left(NADP^+\right) \xrightarrow{\text{Isocitrate dehydrogenase}} \text{2-oxoglutaric acid} + CO_2 + NADH \left(NADPH\right) + H^+$$

4. Oxidative decarboxylation of 2-oxoglutaric acid to succinyl-CoA

$$\text{2-Oxoglutaric acid} + NAD^+ + \text{CoA} \xrightarrow{\text{2-Oxoglutarate dehydrogenase}} \text{succinyl-CoA} + CO_2 + NADH + H^+$$

5. Dehydrogenation of succinic acid to fumaric acid

$$\text{Succinic acid} + \text{FAD enzyme} \xrightarrow{\text{Succinate dehydrogenase}} \text{fumaric acid} + \text{FADH}_2 \text{ enzyme}$$

6. Dehydrogenation of malic acid to oxaloacetic acid

$$\text{Malic acid} + \text{NAD}^+ \xrightarrow{\text{Malate dehydrogenase}} \text{oxaloacetic acid} + \text{NADH} + \text{H}^+$$

By these six reactions, pyruvic acid is completely decomposed to CO_2 and H_2O, and three NADH, one NADPH, and one $FADH_2$ are produced, corresponding to 15 ATP molecules.

$$\text{Pyruvic acid} + 3\text{H}_2\text{O} + 3\text{NAD}^+ + \text{NADP}^+ + \text{FAD} \rightarrow 3\text{CO}_2 + 3\text{NADH} + \text{NADPH} + \text{FADH}_2 + 4\text{H}^+$$

Characteristics of the enzymes involved in the tricarboxylic acid cycle are as follows:

1. Pyruvate dehydrogenase (EC 1.2.2.2) is a large multienzyme complex that catalyzes the oxidation of pyruvate to yield acetyl-CoA and CO_2. The complex contains pyruvate decarboxylase, dihydrolipoamide acetyltransferase, and dihydrolipoamide dehydrogenase with their respective prosthetic groups, thiamine pyrophosphate, lipoic acid, and FAD. Pyruvate dehydrogenase catalyzes three steps in a coupled sequence in which the intermediates remain enzyme bound. Initial decarboxylation of the pyruvate gives CO_2 and the α-hydroxyethyl derivative of the thiazole ring of thiamine pyrophosphate. The hydroxyethyl group is dehydrogenated, and the resulting acetyl group is transferred to the lipoic acid. This acetyl group is then transferred to the thiol group of CoA. The dihydrolipoic acid on the acetyltransferase is oxidized by dihydrolipoamide dehydrogenase, with the resulting $FADH_2$ being oxidized by NAD^+. Plant cells contain two pyruvate dehydrogenase complexes, one in the mitochondria and the other in the plastids (Rees 1990).
2. Citrate synthase (EC 4.1.3.7) catalyzes an aldol condensation between the methyl group of acetyl-CoA and the carbonyl group of oxaloacetate. The condensation gives citroyl-CoA as the intermediate, and the thioester bond is immediately hydrolyzed to yield citrate and CoA. In plants, citrate synthase is confined to the mitochondria, except for glyoxosomes in glycogenic tissues.
3. Aconitase (EC 4.2.1.3) catalyzes the reversible interconversion of citrate and isocitrate via the enzyme-bound intermediate *cis*-aconitate. Dehydration is followed by hydration, and the net effect is an interchange of H and OH.
4. Isocitrate dehydrogenases (EC 1.1.1.41–42) catalyze two reactions, the oxidation of isociatrate to enzyme-bound oxalosuccinate, which is then decarboxylated to give CO_2 and 2-oxoglutarate. Plants contain both NAD- and NADP-linked enzymes. The former is confined to the mitochondria, and the latter is found in the cytosol and the chloroplast.

5. 2-Oxoglutarate dehydrogenase (EC 1.2.4.2) catalyzes oxidation of 2-oxoglutarate to succinyl-CoA. The enzyme is constituted as a complex comparable to pyruvate dehydrogenase.
6. Succinate dehydrogenase (EC 1.3.99.1) catalyzes the oxidation of succinate to fumarate. This enzyme is a component both of the citric acid cycle and, as complex II, of the respiratory electron transport chain. The mammalian enzyme was shown to have a large subunit with a molecular mass of 70 kDa that contains covalently bound FAD and two binuclear 2Fe–4S iron-sulfur clusters.
7. Fumarase (EC 4.2.1.2) catalyzes the *trans* addition of H and OH to the double bond of fumarate. The addition is stereospecific in that the OH is added to only one side of the double bond so that only L-malate is formed.
8. Malate dehydrogenases (EC 1.1.1.38–40) catalyze the oxidation of malate to oxaloacetate. Isozymes of NAD-linked malate dehydrogenase are found in the mitochondria, cytosol, and glyoxosomes in plants.

The complete operation of the cycle involves the coordination and sequential interaction of eight enzymes, seven of which are in the mitochondrial matrix with the eight having access to it. Specific interactions between purified enzymes of the cycle have been demonstrated in vitro. A number of the matrix enzymes have been shown to bind preferentially to the inner membrane. Light sonication of rat liver mitochondria produces a sedimentable fraction, in which the membranes had been disrupted but enzymes remained bound to the inner membrane (Rees 1990).

2.6 Pentose Phosphate Cycle

In addition to metabolism by glycolysis and the TCA cycle, glucose is also metabolized via the aerobic pentose phosphate cycle (Fig. 45). This cycle is also called the Warburg–Dickens–Horecker cycle. Glucose-6-phosphate formed from glucose by the mediation of hexosekinase is oxidized to gluconic acid-6-phosphate by glucose-6-phosphate dehydrogenase, which is then oxidatively decarboxylated to ribulose-5-phosphate and CO_2.

$$\text{Glucose-6-phosphate} + NADP^+ \xrightarrow{\text{Glucose-6-phosphate dehydrogenase}} \text{6-phosphogluconolactone} + NADPH + H^+$$

$$\text{6-Phosphogluconolactone} + H_2O \xrightarrow{\text{6-Phosphogluconolactonase}} \text{gluconic acid-6-phosphate}$$

$$\text{Gluconic acid-6-phosphate} + NADP^+ \xrightarrow{\text{Phosphogluconate dehydrogenase}} \text{ribulose-5-phosphate} + CO_2 + NADPH + H^+$$

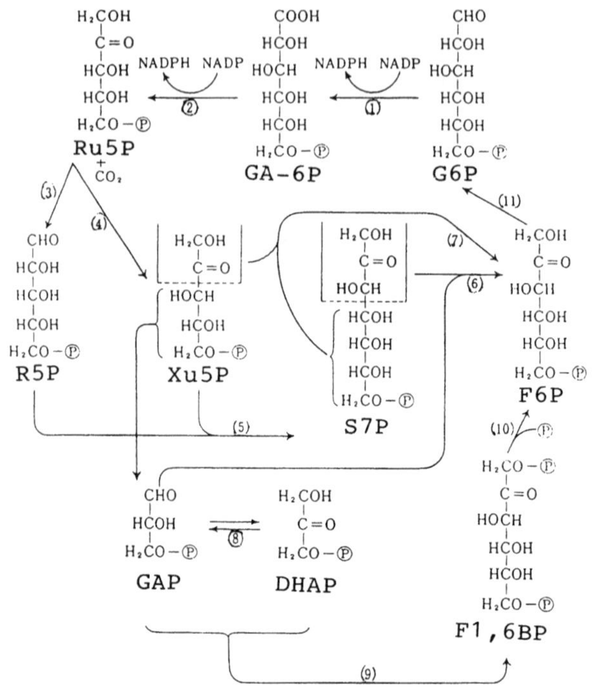

Fig. 45. Pentose phosphate cycle in higher plants. *P* Orthophosphate; *1* glucose 6-phosphate dehydrogenase and 6-phosphogluconolactonase; *2* 6-phosphogluconate dehydrogenase; *3* ribose phosphate isomerase; *4* ribulose phosphate 3-epimerase; *5* transketolase; *6* transaldolase; *7* transketolase; *8* triose phosphate isomerase; *9* aldolase; *10* hexose diphosphatase; *11* hexosephosphate isomerase

The ribulose-5-phosphate thus formed is converted to either xylulose-5-phosphate by epimerase or ribose-5-phosphate by isomerase. Xylulose-5-phosphate and ribose-5-phosphate are converted to sedoheptulose-7-phosphate and glyceraldehyde-3-phosphate by mediation of transketolase; transketolase transfers a two-carbon unit, whereas transaldolase transfers a three-carbon unit from the upper part of the sugar. Sedoheptulose-7-phosphate and glyceraldehyde-3-phosphate are converted to erythrose-4-phosphate and fructose-6-phosphate by transaldolase. Erythrose-4-phosphate and xylulose-5-phosphate are converted to fructose-6-phosphate and glyceraldehyde-3-phosphate by transketolase. In addition, two molecules of glyceraldehyde-3-phosphate are converted to fructose-1,6-bisphosphate by mediation of aldolase (Harrison 1960).

Accordingly, one molecule of glucose is oxidized to $6CO_2$ by circulating of the reaction cycle six times as follows:

$$\text{6-Glucose-6-phosphate} + 12NADP^+ + 6H_2O \rightarrow$$
$$\text{6 ribulose-5-phosphate} + 6CO_2 + 12NADPH + 12H^+$$

$$\text{6 Ribulose-5-phosphate} \rightarrow$$
$$\text{5-glucose-6-phosphate} + \text{orthophosphate (Pi)}$$

$$\text{Glucose-6-phosphate} + 12\text{NADP}^+ + 6\text{H}_2\text{O} \rightarrow 6\text{CO}_2 + \text{Pi} + 12\text{NADPH} + 12\text{H}^+$$

Characteristics of the enzymes involved in the pentose phosphate cycle are as follows:

1. Glucose-6-phosphate dehydrogenase (D-glucose-6-phosphate; NADP oxidoreductase, EC 1.1.1.49), which catalyzes the oxidation of glucose-6-phosphate to δ-glucono-1,5-lactone-6-phosphate, is a key enzyme in the pentose phosphate pathway. The molecular masses of the subunits of microbial glucose-6-phosphate dehydrogenases are 50–60 kDa, while those of mammalian enzymes are somewhat larger (58–67 kDa). Two isozymes of glucose-6-phosphate dehydrogenase, one located in the cytoplasm and the other in the plastid, are found in plants. The pea cytoplasmic isozyme is a 244-kDa homotetramer with subunits of 60 kDa (Fickenscher and Scheibe 1986). Like the cytoplasmic form, the plastid isozyme is a homotetramer (M_r, 224 kDa), but the molecular mass of the subunit is slightly smaller (56 kDa). The enzymes are absolutely specific for glucose-6-phosphate and NADPH (Miernyk 1990).
2. 6-Phosphogluconate dehydrogenase (6-phosphogluconate; NADP oxidoreductase, EC 1.1.1.44), which catalyzes the irreversible oxidative decarboxylation of 6-phosphogluconate, yielding ribulose-5-phosphate plus CO_2, is specific for $NADP^+$ and requires divalent cations. 3-Oxo-6-phosphogluconate is believed to be a reaction intermediate. The enzyme from mammalian and fungal sources is a homodimer with a molecular mass of 100–110 kDa, and the molecular mass of the subunit is around 51 kDa. The enzyme has been purified from several plants. Its physicochemical and catalytic properties are similar to those of mammalian and fungal enzymes (Miernyk 1990).
3. Transaldolase (EC 2.2.1.2) catalyzes the freely reversible interconversion between sedoheptutulose-7-phosphate plus glyceraldehyde-3-phosphate, and erythrose-4-phosphate plus fructose-6-phosphate. Transaldolase mediates a base-catalyzed aldol cleavage reaction, in which a Schiff base intermediate is formed. Transaldolase from mammalian and some fungal sources is a dimer with a molecular mass of between 65 and 75 kDa, and the molecular mass of the subunit is around 33 kDa. In other fungi and bacteria, the enzyme is a monomeric protein with native and subunit molecular masses of 50–65 kDa. However, the plant enzyme has not been purified (Miernyk 1990).
4. Transketolase (EC 2.2.1.1) is a glycolaldehyde transferase, reversibly transferring the two-carbon ketol moiety from sedoheptulose-7-phosphate and glyceraldehyde-3-phosphate to give ribose-5-phosphate plus xylulose-5-phosphate. Thiamine pyrophosphate and divalent cations are tightly bound cofactors required for catalytic activity. Both the mammalian and yeast enzymes are homodimers with a molecular mass of 140 kDa, and the molecular mass of the subunit is 69 kDa (Paoletti and Aldinucci 1986). Both

cytoplasmic and plastid isozyme of transketolase have been found in plant cells. The molecular mass of spinach and wheat leaves transketolase is 150kDa with a subunit molecular mass of 37.6kDa.

5. Pentose phosphate isomerase (D-ribose ketol isomerase, EC 5.3.1.6) catalyzes the interconversion between ribose-5-phosphate and ribulose-5-phosphate. The enzyme has been reported to exist in dimeric, trimeric, and tetrameric structures, with molecular masses of 20–58kDa (Noltmann 1972). Ribose-5-phosphate isomerase has been purified from several plant sources, and the physicochemical and catalytic properties are generally similar to those from non-plant sources. The enzymes from alfalfa shoots and spinach chloroplast are homodimers with a subunit molecular mass of 26kDa and a native molecular mass of 53kDa.
6. Pentose phosphate epimerase (D-ribulose-5-phosphate 3-epimerase, EC 5.1.3.1) catalyzes the interconversion between ribulose-5-phosphate and xylulose-5-phosphate. The reaction mechanism is thought to proceed through an enediol intermediate. Both yeast and mammalian enzymes have a native molecular mass of 45kDa. The mammalian enzyme is a homodimer made up of 23kDa subunits. The plant enzyme has not been characterized.

The pentose phosphate pathway operates within plant cells. There are clearly two discrete pathways, one in the cytoplasm and the other within the plastid. Distinct isozymes are responsible for the activities in the two compartments (Miernyk 1990).

In the pentose phosphate cycle, $NADP^+$ is used as an oxidation reagent, and the NADPH formed is used as a reducing reagent in the synthesis of fatty acids, shikimic acid, etc. In addition, ribose-5-phosphate required for nucleotide synthesis and erythrose-4-phosphate required for shikimic acid synthesis are formed as intermediates.

For every one molecule of NADH or NADPH oxidized via respiration, three molecules of ATP are formed; 12 molecules of NADPH correspond to 36 molecules of ATP.

2.7 Glyoxylate Cycle

Seeds containing fat as a storage product rapidly convert the fat to sugars upon germination. In these tissues, fat is hydrolyzed to fatty acids and glycerol by mediation of lipase, and the fatty acid formed is converted to acetyl-CoA by β-oxidation. Acetyl-CoA is combined with oxaloacetic acid to yield citric acid by mediation of citrate synthase as usual. Citric acid is converted to isocitric acid, but isocitric acid is converted to glyoxylic acid and succinic acid by isocitrate lyase in the germinating seeds. The glyoxylic acid thus formed is combined with acetyl-CoA to give malic acid as follows (Conn and Stumpf 1972; Fig. 46):

Citrate synthase

$$\text{Acetyl-CoA} + \text{oxaloacetic acid} + H_2O \rightarrow \text{citric acid} + \text{CoA}$$

Isocitrate lyase

$$\text{Isocitric acid} \rightarrow \text{glyoxylic acid} + \text{succinic acid}$$

Malate synthase

$$\text{Acetyl-CoA} + \text{glyoxylic acid} + H_2O \rightarrow \text{malic acid} + \text{CoA}$$

This metabolic pathway, when combined with the TCA cycle, is known as the glyoxylate cycle. The enzymes involved in the glyoxylate cycle are located in the glyoxosome in the endosparm or in cotyledons of oily or fatty seeds during the early stages of growth after germination. This metabolic

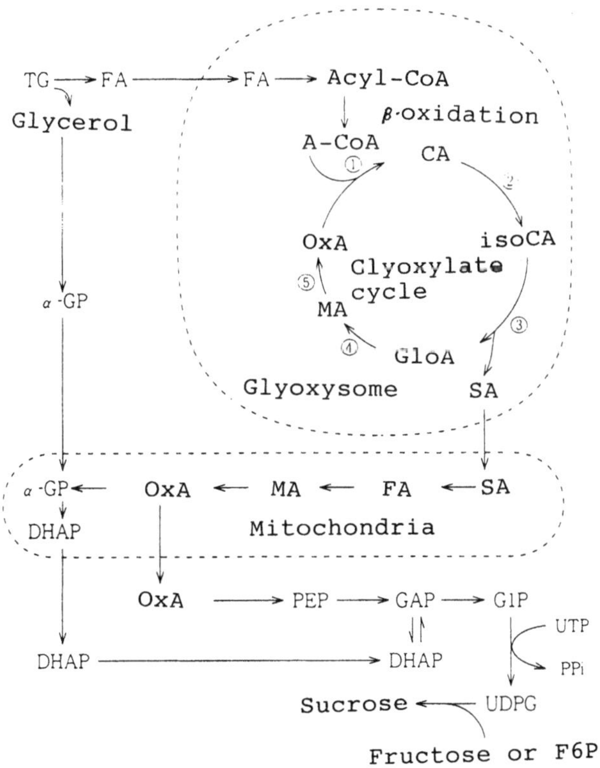

Fig. 46. Formation of sucrose from fatty acids via the glyoxylate cycle. Triglycerides (*TG*) are hydrolyzed to fatty acids (*FA*) and glycerol by esterases. Both TG and FA are converted to sucrose. Enzymes involved in glyoxylate cycle: *1* citrate synthase; *2* aconitase; *3* isocitrate-lyase; *4* malate synthase; *5* malate dehydrogenase; *DHAP* dihydroxyactone phosphate; *α-GP* α-glycerophosphate; *PEP* phosphoenol pyruvate; *GAP* glyceroaldehyde 3-phosphate; *UDPG* uridine diphosphate glucose; *G1P* glucose-1-phosphate; *UTP* uridine triphosphate; *PPi* pyrophosphate

activity serves to provide developing shoots with energy in the form of sugar, at the expense of the fat stored in the seed, until the first green leaves appear and photosynthesis begins. It was demonstrated that 75% of the lipid carbon is converted to sucrose in the endosperm of germinating castor bean.

In this cycle, one molecule of succinic acid is formed from two molecules of acetyl-CoA. Succinic acid is converted to sugars via PEP and the reverse reaction of glycolysis. Accordingly, malate synthase and isocitrate lyase play central roles in the glyoxylate cycle. Both enzymes are formed upon germination of fat seeds, but they are not found in starch seeds. A remarkable amount of fat is stored in seeds and ray parenchyma in trees, especially in conifers, as well as other species. It seems that the glyoxylate cycle is involved in the production of energy for wood formation and synthesis of wood extractives in these trees.

2.8 Nitrogen Metabolism

2.8.1 Nitrogen Fixation

The first step in biological nitrogen fixation is the reduction of N_2 to form ammonia. Nitrogen-fixing bacteria – *Azotobacter*, *Clostridium*, and *Rhizobium* (legume plants) and *Frankia* and *Actinomycete* (nonlegume plants) – are able to fix atmospheric nitrogen as organic nitrogen. These bacteria have nitrogen-fixing enzymes, composed of two proteins, namely, MoFe protein, containing iron, sulfur, and molybdenum, and Fe protein, containing iron and sulfur. In addition to the two proteins, a source of low-potential electrons and ATP is required for nitrogen fixation.

MoFe protein (M_r, about 220 kDa) and Fe protein (M_r, about 64 kDa) have been purified and characterized from a variety of bacterial sources, including the bacteroides of soybean and other legume nodules (Winter and Burris 1976, Orme-Johnson 1977). The properties of the enzyme from nodules are quite similar to those of the complex isolated from *Azotobacter* or other free-living microorganisms.

Fe protein, which is composed of two identical subunits (M_r, 30 kDa) with a single (4Fe–4S) iron sulfur cluster transfers electrons from ferredoxin to MoFe protein. MoFe protein is composed of two subunits (α, 50 kDa; β, 60 kDa) and contains four (4Fe–4S) clusters and two (6Fe–Mo–6S) clusters. Two molecules of ATP combined with the reduced Fe protein are hydrolyzed to ADP and orthophosphate (Pi) during the process of one electron transfer from Fe protein to MoFe protein. N_2 is reduced to ammonia in three steps of two electron reductions on the MoFe protein (Smith and Gallon 1993; Fig. 47):

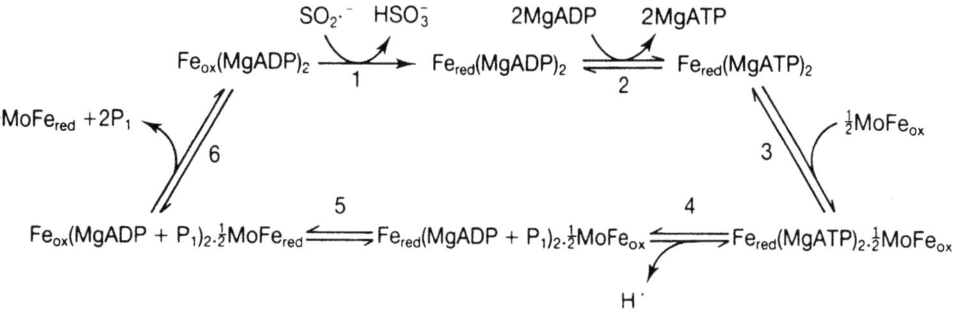

Fig. 47. Catalytic cycle for reduction of the Fe-protein and MoFe-protein of nitrogenase. Fe_{ox} and Fe_{red} refer to oxidized and reduced forms of Fe-protein of nitrogenase, respectively. *1/2MoFe* shows one independently functioning half of the tetrameric $\alpha_2\beta_2$ structure of the MoFe-protein of nitrogenase. Each half contains one *FeMoco* (iron-molybdenum cofactor) substrate binding site and one Fe-protein binding site. The MoFe-protein also exists in oxidized and reduced states. *1* Reduction of the Fe-protein by dithionite; *2* ADP:ATP exchange; *3* association with the MoFe-protein; *4* ATP cleavage; *5* electron transfer; *6* dissociation of the Fe-protein:MoFe-protein complex. (Smith and Gallon 1993)

$$16ATP + 8e\text{-}\left(\text{ferredoxin}\right) + 8H^+ + N_2 + 8H_2O \xrightarrow{\text{Nitrogenase}} 2NH_3 + H_2 + 16ADP + 16Pi$$

In addition to N_2, the nitrogenase complex reduces a variety of other substrates, including H_3O^+, C_2H_2, N_3, CN, and CH_3CN. These alternative substrates compete with N_2 for activated electrons and function as inhibitors of N_2 reduction. One of the alternative reactions, the ATP- and reductant-dependent evolution of H_2, is characteristic of the purified nitrogenase complex from all sources.

Nitrogenases are rapidly inactivated by O_2 and require a protecting mechanism. In the root nodules of legumes, leghemoglobin, which constitutes about 20–30% of the soluble protein of the nodule cytosol, facilitates the diffusion of O_2 into the bacteroides to sustain the O_2 flux essential for the aerobic metabolism of the endophyte, while maintaining the low partial pressure of free O_2 necessary to protect the nitrogenase complex from inactivation (Schubert 1982).

The genes associated with nitrogen fixation are designated *nif*. The *nif* gene in *Klebsiella pneumoniae* has been well elucidated. All the *nif* genes are clustered into a single regulon containing 23280 base pairs. This entire region has been sequenced. It consists of 20 genes encoding Fe protein, α,β-subunits of MoFe protein, synthesis of FeMo pyruvate:flavodoxin oxidoreductase etc., which are probably organized into seven operons (Smith and Gallon 1993).

NH_3 formed by nitrogen fixation and in decomposition processes of animal and plant proteins is oxidized to NO_2 by *Nitromonas* and *Nitrococcus* in the soil:

$$NH_3 + 3/2O_2 \rightarrow NO_2^- + H_2O + H^+$$

NO_2 is then oxidized to nitrate by *Nitrobacter*:

$$NO_2^- + 1/2\,O_2 \rightarrow NO_3$$

Non-nitrogen-fixing plants absorb NO_3 or NH_3 from roots, but generally absorb nitrate ions from root hairs. The nitrate absorbed is reduced to NH_3 by nitrate reductase (NR) and nitrite reductase (NiR) in the root or the shoot, depending upon the plant species. In the reduction of nitrate to ammonia, two enzyme systems, NR and NiR, located in both root and leaf tissue, operate in sequence (Lea 1993).

NR is NADH dependent and consists of two identical subunits of 110–115 kDa, each containing the prosthetic groups FAD, cytochrome b557, and a molybdenum that are involved in electron transport within the protein:

$$NO_3^- + NADH + H^+ \xrightarrow{\text{Nitrate reductase}} NO_2^- + H_2O + NAD^+$$

NR has been cloned from various plants, and it has been shown that it is composed of FAD-binding, heme-binding, and molybdenum-binding domains. Genomic sequences of plant NR showed the presence of three introns and four exons. NiR reduces NO_2^- to ammonia. The enzyme is ferredoxin dependent and comprises one subunit of 60–64 kDa, which contains a siroheme prosthetic group and 4Fe-4S cluster at its active site.

It has been suggested that, for the use of reduced ferredoxin as an electron donor, NiR requires a second subunit with a molecular mass of 24 kDa. NiR cDNA has been cloned and characterized from spinach and maize (Lea 1993).

$$NO_2^- + 6e^- + 7H^+ \xrightarrow{\text{Nitrite reductase}} NH_3 + 2H_2O$$

2.8.2 Amino Acid Formation

NH_3 formed by the above reactions is mainly converted to organic nitrogen compounds by the following three reactions.

$$\text{L-Glutamic acid} + NH_3 + ATP \xrightarrow{\text{Glutamine synthase}} \text{L-glutamine} + ADP + Pi + H_2O \quad (1)$$

It is evident that glutamine synthase (GS), which is contained in chloroplasts in considerable amounts, is the sole port of entry of ammonia into amino acids in higher plants. GS catalyzes the ATP-dependent conversion of glutamate to glutamine (Lea 1993).

Nitrogen-fixing root nodules have an extremely high GS activity. The GS activity increases severalfold during the nodulation of many legume species. This increase occurs at about the same time that nitrogenase is expressed in the rhizobium bacteroid and leghemoglobin is produced in the plant cytoplasm.

Higher plant GS is an octameric protein with a molecular mass of 350–400 kDa. Nodule GS was isolated from *Phaseolus vulgaris*, and the isozyme pattern was characterized in detail. It has been established that GS in *P. vulgaris* is encoded by a small multigene family.

$$\text{2-Oxoglutaric acid} + \text{L-glutamine} + 2\text{NADH} + 2\text{H}^+ \left(\text{reduced ferredoxin}\right) \xrightarrow{\text{Glutamate synthase}} 2\ \text{L-glutamic acid} + 2\text{NAD}^+ \left(\text{oxidized ferredoxin}\right) + H_2O \quad (2)$$

Glutamate synthase is responsible for the transfer of the amide group of glutamine to 2-oxoglutarate to yield two molecules of glutamate. Two different forms of glutamate synthase are present in higher plants: one utilizes reduced ferredoxin as a reductant and the other utilizes NADH.

Ferredoxin-dependent glutamate synthase is located in the chloroplasts in high concentrations. The enzyme is an iron sulfur flavoprotein with a single polypetide of 140–160 kDa. NADH-dependent glutamate synthase is present in low levels, but appears to play a major role in nitrogen-fixing root nodules. The enzyme from root nodules is a monomer of 200 kDa.

The gene for ferredoxin-dependent glutamate synthase has recently been cloned from maize (Lea 1993).

$$\text{2-Oxoglutamic acid} + NH_3 + \text{NAD(P)H}^+ \xrightarrow{\text{Glutamate dehydrogenase}} \text{L-glutamic acid} + \text{NAD(P)}^+ + H_2O \quad (3)$$

It was originally thought that ammonia is incorporated into the organic form by the reductive amination of 2-oxopglutarate catalyzed by glutamate dehydrogenase. This enzyme uses ammonia as a substrate, but the affinity for glutamic acid is about 100-fold greater than for ammonia. Thus it is now thought that the enzyme functions by glutamate breakdown to yield 2-oxoglutarate.

Glutamate dehydrogenase and glutamine synthase are widely distributed in plants.

Many investigations have been conducted on the analysis of amino acids in tree sap. The most common amino acids are aspartic acid and glutamic acid and their amides, asparagine and glutamine. However, in some species, arginine and citrulline, and in other species (e.g., *Acer*, *Aesculus*, *Sassafras*) allantoin, allantoic acid, and urea are found as major nitrogen compounds (Stewart 1966).

2.8.3 Transamination

By transamination from glutamine or glutamic acid to 2-oxo acids by transaminase, various amino acids are formed. Transaminase requires pyridoxal phosphate as a cofactor:

$$\text{L-Glutamic acid} + \text{phenylpyruvic acid} \xrightarrow[\text{Pyriodoxal phosphate}]{\text{Transaminase}} \text{2-oxoglutaric acid} + \text{L-phenylalanine}$$

$$\text{L-Glutamic acid} + \text{oxaloacetic acid} \xrightarrow[\text{Pyriodoxal phosphate}]{\text{Transaminase}} \text{2-oxoglutaric acid} + \text{L-aspartic acid}$$

$$\text{L-Glutamic acid} + \text{pyruvic acid} \xrightarrow[\text{Pyriodoxal phosphate}]{\text{Transaminase}} \text{2-oxoglutaric acid} + \text{L-alanine}$$

Transaminase is widely distributed in plants, and alanine, phenylalanine, glycine, leucine, γ-aminolactic acid, lysine, methionine, proline, serine, valine, threonine, typtophane, tyrosine etc., which are all formed by transaminase, are found in trees.

Barnes (1962, 1963a,b) found that the amino acid composition of tree saps changes with the season and that the amount of organic nitrogen is lowest in summer and winter and reaches a maximum just before the beginning of growth in spring and before dormancy. It seems that the maximum amounts of nitrogen before dormancy are due to the formation of proteins in roots for storage of amino acids required for initial growth in spring.

In *Salix alba*, aspartic acid, glutamic acid, asparagine, and glutamine are present in all seasons, but tyrosine, leucine, and proline are found in small amounts only in spring. In contrast, alanine is found in large amounts in spring and autumn.

Gardner and Leaf (1960) found that radioactive carbon and nitrogen fed to the roots of *Alnus glutinosa* are rapidly incorporated into amino acids, especially aspartic acid, glutamic acid, and citrulline. These amino acids might be synthesized in roots, leaves, growing buds and cambial tissues and transported to meristematic tissues.

Considerable amounts of citrulline were found in the tree saps of species of *Betula*, *Alnus*, *Nothofagus*, *Annona*, *Diopyros*, and *Chamaecyparis*, among others (Bollard 1957, Barnes 1963a). In these species, citrulline amounted to more than 50% of the total amino acids in the tree sap.

From saplings of *Cryptomeria japonica*, 23 kinds of amino acids, including asparagine and glutamine, were detected. Citrulline, which amounted to two thirds of the total nitrogen compounds, and relatively large amounts of proline, glutamic acid, aspartic acid, and γ-aminolactic acid were found in the tree sap.

L-Citrulline (carbamyl-^{14}C) was transformed into arginosuccinic acid, arginine, and urea, suggesting the involvement of the ornithine cycle in *Cryptomeria* (Mori 1977). L-Citrulline is a major compound transferred from root to stem; it is stored as a storage product and is used rapidly when nitrogen nutrition decreases. ^{14}C-Ornithine administered to saplings of *C. japonica* was effectively incorporated into soluble proteins in October, indicating that protein synthesis occurs mainly in autumn to be stored in winter. In spring, the protein is hydrolyzed to amino acids for growth. It has been found that the amounts of arginine and amides are unusually large in pine species.

3 Genome Organization, Protein Synthesis, and Gene Expression

Many excellent and comprehensive books on molecular biology and gene organization have been published (Watson et al. 1977, 1992, Lea and Leegood 1993, Alberts et al. 1994). The basic principles that are required for wood biochemistry are outlined in this chapter.

3.1 Genome Organization

3.1.1 DNA and Genes

Molecular biology has shown that life phenomena appear by the expression of various information involved in cells. The information is conveyed from parent to descendants, and called genetic information.

Genetic information is weaved into a high polymer DNA throughout lower organisms such as bacteria, and higher organisms such as the higher plants and animals. The information weaved in DNA is more diverse and complex in higher organisms. DNA itself is a linear polymer where four kinds of organic bases (nucleotides) are connected in various order. The mode of connection of nucleotides is called the base sequence. The base sequence itself is the entity of gene information, which is similar to the number sequence of data and programs of the computer (Shibata 1994).

When information is derived from genes the process is called gene expression. The gene information in DNA is transcribed to RNA, and proteins are synthesized according to the information transcribed in the RNA. (Fig. 48).

By mediation of the synthesized proteins (enzymes), various chemical reactions proceed in the cell, and cell constituents are synthesized. In other words, the gene information is ultimately expressed as proteins. The synthesis of RNA from DNA is called transcription, and the synthesis of proteins from RNA is called translation.

As noted later proteins are not synthesized from transfer RNA (tRNA) and ribosomal RNA (rRNA) which bear special functions other than protein synthesis. These RNAs are also derived from gene expressions of DNAs. In some viruses RNA bears the genetic information instead of DNA. In this case the genetic information is transcribed to other RNA from which proteins are synthesized.

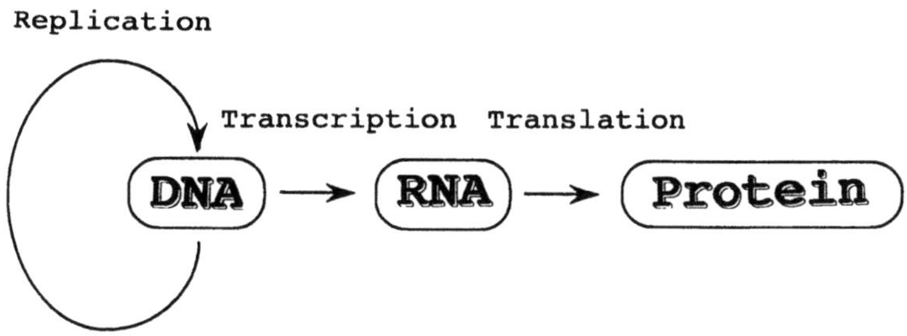

Fig. 48. Expression of gene information

The elucidation of the mechanism of the gene expression is the most important subject in modern biology, and until now the emphasis has been on the establishment of fundamental mechanisms (Watson et al. 1977, 1992).

3.1.2 Genes and Chromosomes

Not all base sequences of DNA are involved in gene information, but special base sequences are involved in special information. This special base sequence in genomic DNA is called gene and is scattered in DNA molecules. Higher organisms, which are more diverse, bear more genes. The total DNA of one organism is called the genome, and the size of genome is shown as the number of bases contained in the base sequence. Generally the number of bases is expressed as base pair (bp), and 1000 bp is expressed as 1 kbp, or 1 kb, and 1000 kbp as 1 Mbp or 1 Mb.

The size of the genome differs considerably among organisms; *Escherichia* 4.8 Mbp, *Drosophila* 165 Mbp, and humans 2900 Mbp. In higher plants the size is 145 Mbp in *Arabidopsis thaliana*, 4900 Mbp in barley, 15 300 Mbp in onion, and 24 700 in tulip. Large differences occur among plant species (Table 7).

As the number of functional genes in plants is supposed to be not much different among plant species, large differences of the genome size are due to the presence of varying amounts of noncoding repetitive DNA in plant genome. The amount of such noncoding DNA are different even in related plant species, and such DNA regions are mainly composed of reiterated sequences.

The reiterated sequence consists of a repeated occurrence of special base sequences on chromosomes. There are several reiterated sequences, and the degree of repetition ranges from 100 to 100 000.

In eukaryotic organisms, a long DNA molecule is connected to a basic protein, histone, in the nucleus to make a nucleosome. In each nucleosome about 145 bp of DNA are coiled, and the nucleosome aggregates to form chromatin fibers. The chromosome is the aggregate body (solenoid structure) of chromatin fibers (Fig. 49).

Table 7. Differences in genome size of different organisms. (Shibata 1994)

Species	Genome size (Mbp)
Escherichia coli	4.8
Yeast	13.5
Myxomycete	70
Nematode	80
Drosophila	165
Dactylethridae	3000
Chicken	1200
Mouse	3000
Human	2900
Arabidopsis thaliana	145
Barley	4900
Onion	15300
Tulip	24700

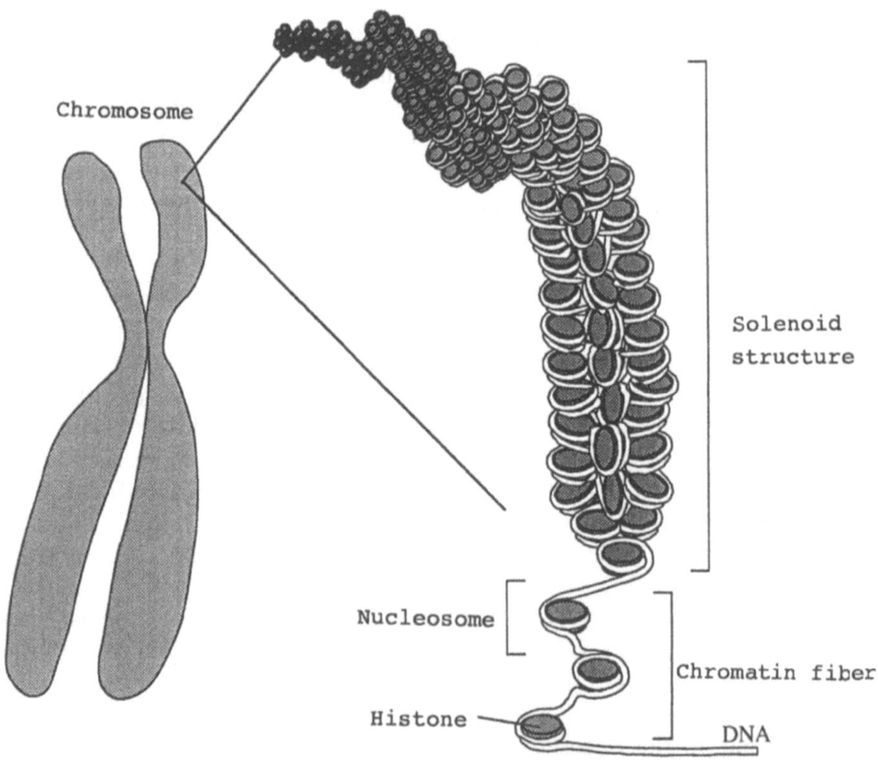

Fig. 49. DNA in nucleus and higher structure of chromosome. DNA constitutes nucleosome with histone. Nucleosome aggregates to form higher structure of chromosome

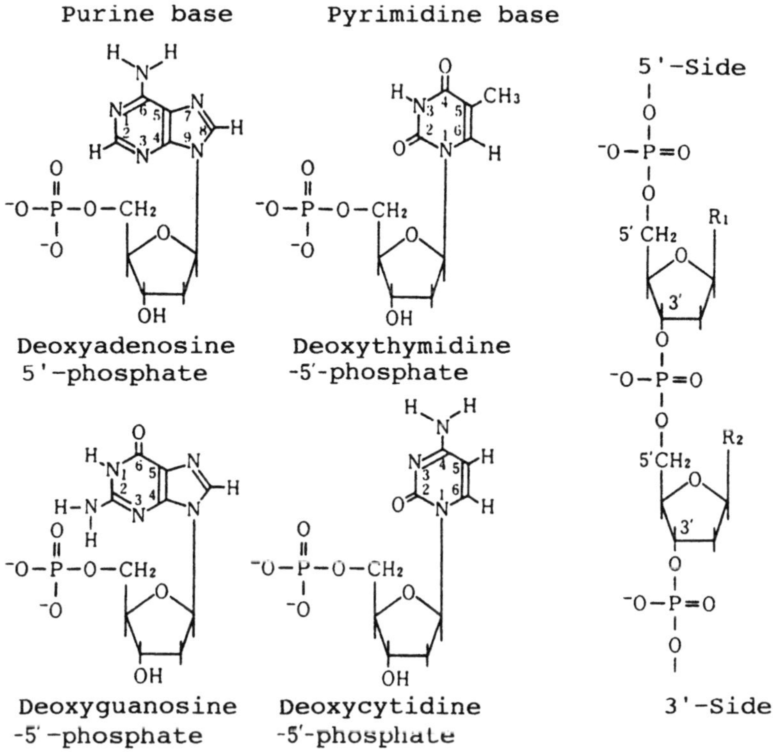

Fig. 50. Part of the linear structure of DNA

This highly complicated structure seems to be responsible for gene expression.

3.1.3 Structure and Duplication of DNA

DNA (deoxyribonucleic acid) is a linear polymer connected by 3′–5′ phosphodiester linkages of deoxyribose residues of four different nucleotides (Watson et al. 1977; Fig. 50). Pyrimidine and purine bases are connected to deoxyriboses. A set of pentose and base is called a nucleoside, and the phosphate ester of the nucleoside is called a nucleotide. The pyrimidine bases are thymine or cytosine, while the purine bases are adenine or guanine. The nucleosides are thymidine (T), cytidine (C), adenine (A) and guanosine (G). In other words DNA is a chain of T, C, A, and G in various combinations, and the order of the bases in the chain is the base sequence.

To indicate the direction of the base sequence, the side of the 5′ hydroxyl group of deoxyribose is named 5′-side, while the side of 3′ hydroxyl group is

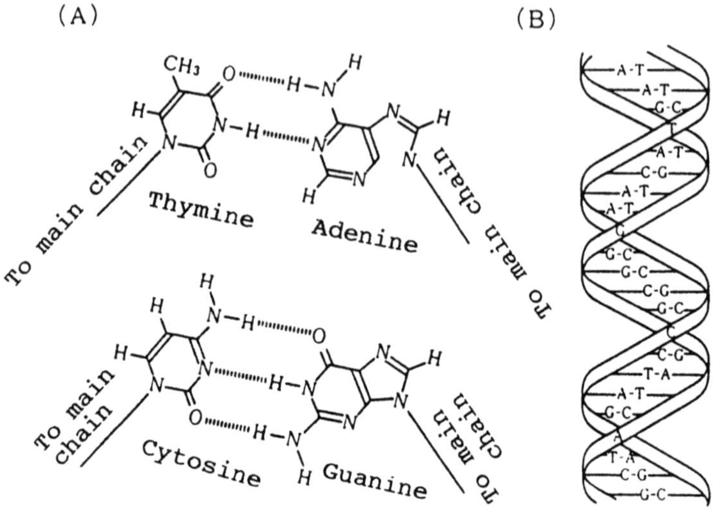

Fig. 51. A Hydrogen bonds between DNA bases **B** helix structure of two chains of nucleotides. *Dotted line* between O and H, and H and N indicate hydrogen bonds

named 3′-side. In general, the hydroxyl group at the 5′-end is phosphorylated, while the hydroxyl group at the 3′-end is not. DNA is an interwoven structure of two chains of nucleotides (Fig. 51).

The directions of the two chains are reversed, and the nucleotides in one side of the chain are always connected to the nucleotides in the other side by hydrogen bonds. The hydrogen bondings are only possible between T and A, or C and G. Therefore, when the base sequence of one side of the chain is determined, the base sequence of the other chain is determined automatically. This relationship of the DNA is named complementation, and the respective chains are called complementary chains. Such complementation is very important for the ability of DNA to carry the gene information. The two chains of DNA are composed of a helix structure like a spiral staircase. The double chain of DNA may then unravel to form two single chains and complementary chains may then be synthesized from the single chains. By such synthesis two sets of the same DNA chains are formed (Fig. 52).

The two sets of DNA chains formed are distributed to daughter cells through cell division, and the genetic information is conveyed accurately from cell to cell. In such a way DNA is synthesized from complementary DNA. The process of DNA synthesis is called replication of DNA, and starts from a specific site (starting point of replication). Two DNA chains are partly untied, and complementary chains are synthesized from nucleotides in cooperation with DNA polymerase and several other factors. The DNA chain which is the original from which a complementary chain is formed is called the template.

DNA is a long linear molecule, and one molecule of DNA is folded into one chromosome in higher organisms. The ends of the DNA molecules are

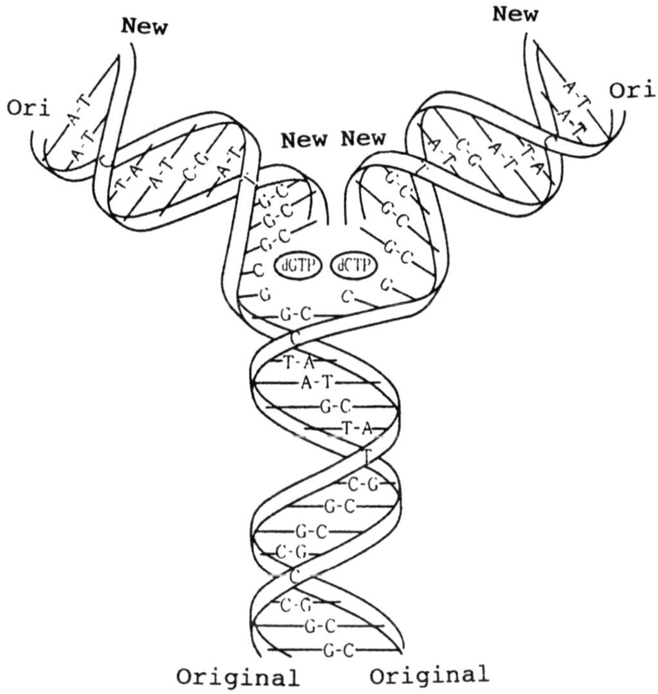

Fig. 52. Replication of DNA

susceptible to splitting by exonucleases, resulting in smaller molecules. To prevent such shortening of the DNA a special base sequence called the telomere exists at the end of DNA molecules. A mechanism to identify such a sequence and keep the DNA at a certain length is present in the cell.

The telomere does not exist in bacteria, and in many cases bacterial DNA is circular. Plasmids, which multiply in bacterial cells, are also composed of circular DNA. In addition, in chloroplasts of higher plants there is a special circular DNA which is different from the nuclear DNA. Based on the mechanism of gene expression, such chloroplast DNA is thought to be similar to the DNA in prokaryotes.

One-chain DNA occurs in some viruses and phages, and in these organisms two complementary DNAs are synthesized once by DNA duplication, after which one DNA chain is synthesized. DNA bases are sometimes chemically modified. For example cytosine is often methylated in higher plants, and such methylation may be involved in inactivation of gene expression.

3.1.4 RNA Synthesis

DNA contains deoxyribose as its sugar residue, but the sugar residue of RNA (ribonucleic acid) is ribose (Fig. 53). The pyrimidine bases of RNA are uracil

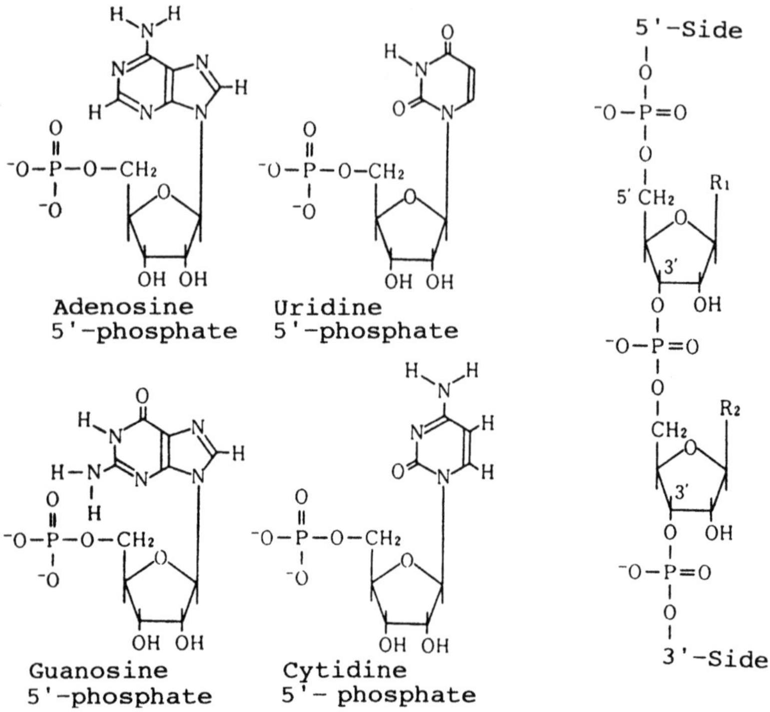

Fig. 53. Part of the linear structure of RNA

or cytosine, while its purine bases are adenine or guanine. Thymine, a pyrimidine base of DNA, is not present in RNA. Nucleosides composed of these bases and ribose are named uridine (U), cytidine (C), adenosine (A) and guanosine (G), respectively. Hydrogen bonds are formed between uracil and adenine residues, and between cytosine and guanine residues of the respective nucleosides. Therefore, a complementary chain is formed between RNA and DNA.

Nucleoside components of RNA are described as U, C, A, and G for convenience. Since C, G, and A are the same symbols as those in DNA one should be careful to indicate whether the base sequence occurs in DNA or RNA.

RNA is synthesized from a part of the base sequence of DNA by forming a complementary base sequence; in other words, RNA is synthesized based on a template of DNA. RNA polymerase identifies a certain point along the base sequence of DNA and starts the synthesis of RNA. The enzyme catalyzes the incorporation and connection of ribonucleoside-3-phosphates complementary to the base sequence of one chain of double-stranded DNA, via phosphodiester linkages of the 3′ hydroxyl group of one nucleotide to the 5′ hydroxyl

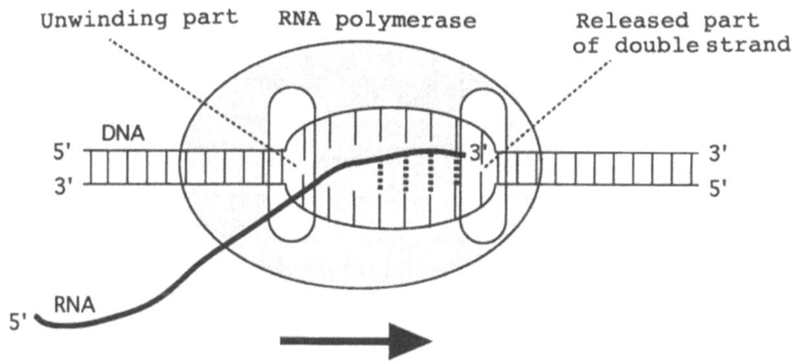

Fig. 54. Synthesis of RNA from double-stranded DNA. RNA polymerases untie double strands of DNA and elongate the RNA chain by incorporation of ribonucleoside 3-phosphate. The untied DNA is returned to double strands

group of the next nucleotide. In the course of this reaction, the RNA chain is extended from the 5′-side to the 3′-side (Fig. 54).

There is a particular base sequence in DNA which stops RNA synthesis, and RNA polymerase leaves the DNA when the enzyme reaches that point. Thus, the one chain of RNA is synthesized from one of the two chains of DNA.

The synthesis of the RNA chain starts from this special point along the DNA and progresses in one direction. The length of the DNA chain up to where RNA synthesis starts is called upstream of the gene, while the part of the DNA chain along which RNA synthesis progresses is called downstream of the gene.

Proteins are synthesized according to the base sequences of messenger RNA (mRNA), and not according to the sequences of some other RNAs (transfer RNA and ribosomal RNA). However, both transfer RNA (tRNA) and ribosomal RNA (rRNA) play important roles in protein synthesis.

3.1.4.1 Structure and Function of mRNA

mRNA bears genetic information for the synthesis of proteins and structures of mRNAs of prokaryotes, such as bacteria, and eukaryotes, such as higher plants, are different.

A single chain of mRNA often carries the genetic information for several proteins in prokaryotes, but for only one protein in eukaryotes, except in a few cases. In addition, 7-methylguanosine is linked to the 5′-end of mRNA of eukaryotes. This chemically modified part is called the cap structure (Fig. 55),

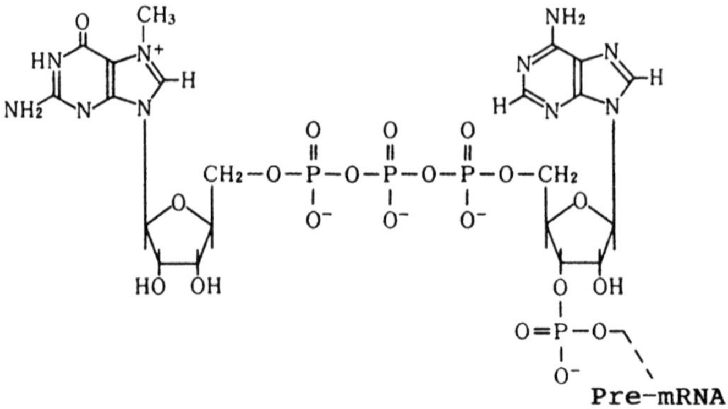

Fig. 55. Cap structure at the 5′ end of messenger RNA of eukaryotes

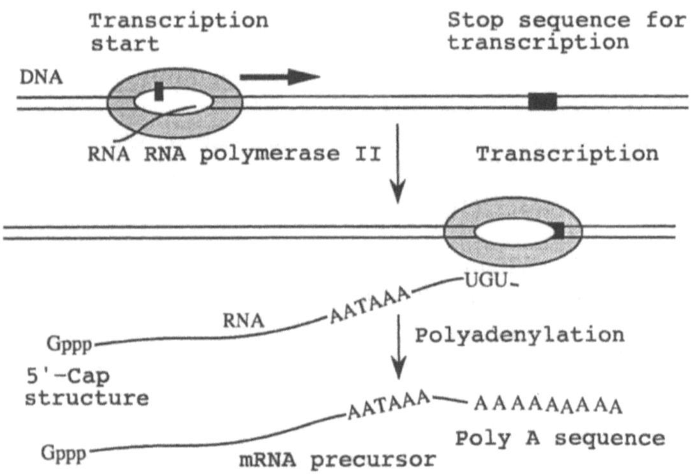

Fig. 56. Transcription of mRNA and addition of adenosine molecules to the newly formed 3′ end (polyadenylation). When RNA polymerase II reaches the stop sequence for transcription and RNA synthesis is completed, part of the sequence is cleaved and adenosines are added to the 3′ end of the RNA chain

and is identified by ribosomes during the translation of mRNA. In addition to the protein-coding sequences (translation region), nontranslation regions exist at the 5′- and 3′-sides as excess base sequences on the mRNA. In eukaryotes, after transcription of the DNA, the part adjacent to the 3′-side of the RNA is cut off, and several tens to several hundreds of molecules of adenosine are added to the newly formed 3′-end (polyadenylation) (Fig. 56).

Such a polyadenosine sequence is not found in the corresponding gene. The sequence AATAAA marks the upstream end of the polyadenylation site, whereas the downstream end is marked by a sequence involving UGU. Thus these sequences constitute a system for identifying the region which undergoes polyadenylation. However, a few genes, such as the histone gene, are known not to accept polyadenylation.

3.1.4.2 Structure and Function of tRNA

tRNA transfers the corresponding amino acid for the codon in protein synthesis. tRNA is a short RNA composed of about 80 nucleotides, but forms a special structure like a cloverleaf by intramolecular hydrogen bonding (Fig. 57).

The base in the nucleotides in t-RNA e.g., N,N-dimethylguanine, dihydrouridine, isopentenyladenine and 4-thiouridine, is often (about 10%) chemically modified. The modification effects the conformation of the t-RNA and complementary base pairing of anticodon, helping the identification of codon by the tRNA. tRNA is partly composed of an anticodon structure which connects in a complementary way to that of the corresponding codon. Amino acids corresponding to respective tRNAs are linked to the 3′-end. tRNA linked to an amino acid is called aminoacyl-tRNA. This reaction is catalyzed by aminoacyl-tRNA synthetase.

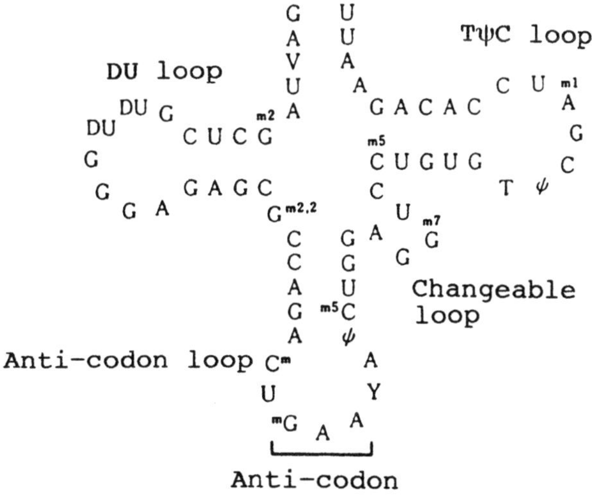

Fig. 57. An example of the structure of tRNA of eukaryotes

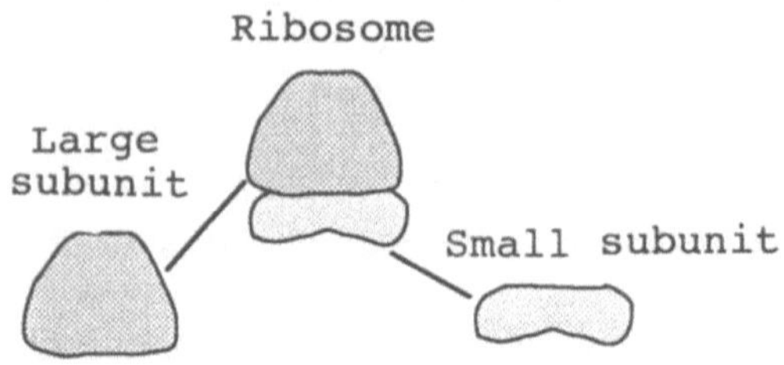

		Large subunit	Ribosome	Small subunit
Prokaryotes	Size	50S	70S	30S
	RNA	23S 5S		16S
	Proteins	34		21
Eukaryotes	Size	60S	80S	40S
	RNA	28S 5.8S 5S		18S
	Proteins	45-50		30-35

Fig. 58. Ribosomes of prokaryotes and eukaryotes

3.1.4.3 rRNA

The synthesis of proteins based on the codon of mRNA is conducted by the ribosomes which are complex bodies of protein and nucleic acid. RNA in ribosomes is called ribosomal RNA (rRNA). A ribosome is composed of two subunits (Fig. 58).

In prokaryotes the size of the ribosome is 70 S, composed of 50 S and 30 S subunits. The former contains 5 S and 23 S rRNA and 34 proteins, while the latter contains 16 S rRNA and 21 proteins. In eukaryotes the size of ribosome is 80 S, composed of 60 S and 40 S subunits. The former contains 18 S rRNA and 30–35 proteins, while the latter has 5 S, 5.8 S and 28 S rRNA and 45–50 proteins.

In higher plants, in addition to the eukaryotic type of rRNA in the chromosomes, a prokaryotic type of rRNA occurs in the chloroplasts. The function of rRNA is not fully elucidated but a cooperation with mRNA has been suggested.

There are many copies of rRNAs in the chromosomes, and rRNA is always synthesized. Therefore, rRNA amounts to about 97% of the total RNA, while tRNA and mRNA amount to only 2% and 1%, respectively.

3.1.4.4 Splicing of RNA

In eukaryotes seemingly useless base sequences frequently exist in mRNA which has just been transcribed from the genes. These sequences are removed when the transcribed mRNA is transferred from nucleus to cytoplasm and are called introns, whereas other base sequences which are used for protein synthesis are called exons. The removal of introns is termed splicing. Numbers and positions of introns on DNA are different among genes, but in the same kinds of gene they are frequently common among species. In contrast, the base sequences and lengths of introns are considerably different even in the same kind of gene.

The intron of higher plants is generally shorter than that of animals. The longest one is several thousands of base pairs, while in animals introns with several tens of thousands of base pairs are found. Introns exist not only in the base sequences of translating regions for protein synthesis but also in the base sequences of nontranslating regions. Between introns and exons common base sequences – the GU sequence at the borders of the 5′-side and the AG sequence at the 3′-side of introns – exist. In yeast-type mRNA the base sequence at the border of the introns is more strictly determined. For splicing of introns, protein–RNA complexes (snRNPs) called spliceosomes are required (Fig. 59).

Introns are found not only in mRNA derived from nuclear genes but also in mRNA derived from mitochondria and chloroplasts. Introns also exist in tRNA and rRNA. Different splicing mechanisms are involved in these different RNAs.

3.1.4.5 RNA Editing

Generally, genetic information is expressed by the exact transcription of the base sequence of DNA to that of RNA. However, it was recently found that some of the bases of transcripted RNA are substituted by other bases. This is known as RNA editing and was first observed in mitochondrial mRNA of trypanosomes. RNA editing is also found in higher plants.

3.1.5 mRNA and the Genetic Code

The base sequences of gene DNA are transcribed to those of mRNA, and the transcribed information of mRNA is expressed as proteins through translation.

Protein molecules have a complex, folded structure and form the higher order structure that determines the functions of the proteins. The higher order structure is determined by the mode of linkages (amino acid sequences) of the

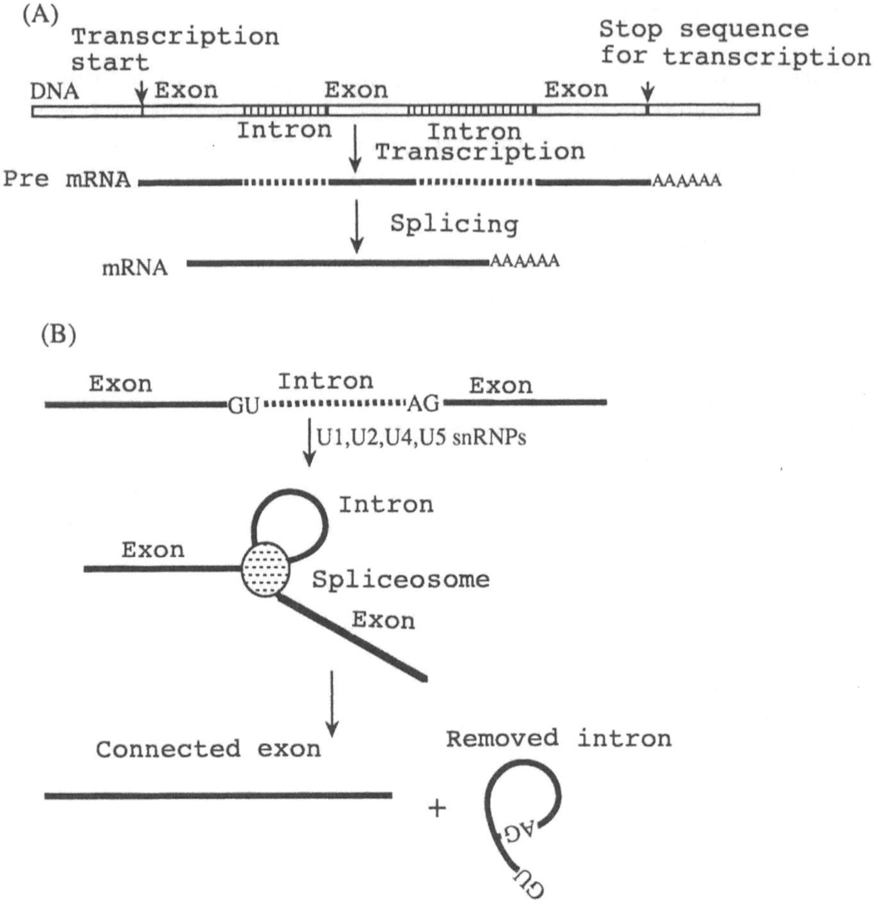

Fig. 59. A,B Introns and splicing. **A** The intron sequence in DNA is transcribed on RNA, and then cut off. **B** The intron on the RNA is cut off by means of a spliceosome

20 kinds of amino acids that constitute proteins. A line of base sequences, which corresponds to a line of amino acid sequences, exists on mRNA. Therefore, the genetic information corresponds to the amino acid sequences.

Four kinds of bases specify 20 kinds of amino acids, and the arrangement of three bases specify one amino acid. Such a set of the base sequence is called a codon, and the table of codon, and corresponding amino acids is called the genetic code (Table 8).

There are 64 possible arrangements of three kinds of bases, and all except three of these correspond to particular amino acids, so that most of the amino acids correspond to more than one codon. The three codons that do not specify amino acids are termination (stop) signals in protein synthesis and are called termination codons.

Table 8. The genetic code. (Watson et al. 1992)

First position	Second position				Third position
	U	C	A	G	
U	Phe	Ser	Tyr	Cys	U
	Phe	Ser	Tyr	Cys	C
	Leu	Ser	Stop	Stop	A
	Leu	Ser	Stop	Trp	G
C	Leu	Pro	His	Arg	U
	Leu	Pro	His	Arg	C
	Leu	Pro	Gln	Arg	A
	Leu	Pro	Gln	Arg	G
A	Ile	Thr	Asn	Ser	U
	Ile	Thr	Asn	Ser	C
	Ile	Thr	Lys	Arg	A
	Met	Thr	Lys	Arg	G
G	Val	Ala	Asp	Gly	U
	Val	Ala	Asp	Gly	C
	Val	Ala	Glu	Gly	A
	Val	Ala	Glu	Gly	G

Given the position of the bases in a codon, the corresponding amino acid is found. The codon (5′) AUG (3′) on mRNA specifies methionine, while CAU specifies histidine. UAA, UAG, and UGA are termination signals. AUG is part of the initiation signal and codes for internal methionine as well.

The amino acids are linked together to give proteins according to the base sequences of the mRNA transcribed from the genes. This process is called protein synthesis. The functions of the proteins are determined by the amino acid sequences, and amino acid sequences are determined by the arrangements of the codons in the mRNA.

3.2 Protein Synthesis

Protein synthesis is initiated by connection of the initiator protein to mRNA, and is followed by connection of a subunit of the ribosome. When a complex of mRNA and ribosome is formed, the tRNA connected to methionine (met-tRNA) is incorporated into the complex. In prokaryotes but not eukaryotes, the methionine is formylated. However, the tRNA that carries methionine to the initiation codon is different from the tRNA that carries methionine to other parts. A new tRNA that carries an amino acid corresponding to the next codon is then transferred adjacently to the methionine. Thus, two amino acids are arranged at the ribosome, and a peptide linkage is formed between them

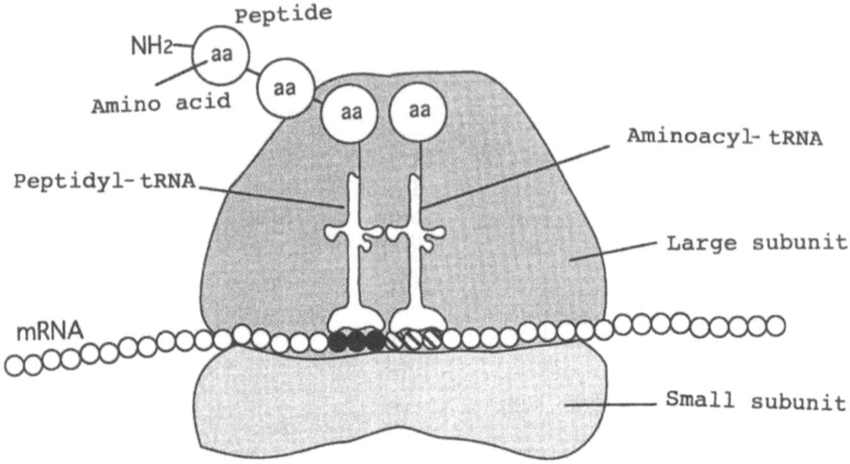

Fig. 60. Organization of a ribosome. *aa* Amino acid

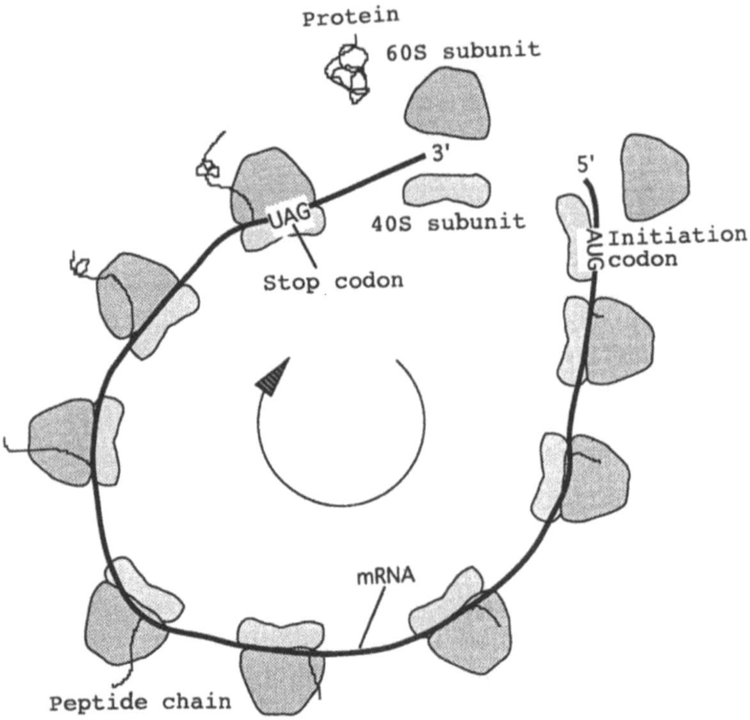

Fig. 61. Protein synthesis based on the codon of mRNA. Two subunits are connected to mRNA and move along the 3′ side on the mRNA, extending the polypeptide chain. When the subunit reaches the stop codon, the polypeptide (protein) formed is released, and the ribosome leaves the mRNA

by mediation of peptidyltransferase. The first amino acid is then pushed out from the complex enabling the next one to enter. The site where the new amino acid enters is called the "A" site (the loading site of aminoacyl-tRNA), while the loaded site of previous amino acid is called the "P" site (the loading site of peptidyl-tRNA; Fig. 60).

Thus the amino acids are connected step by step from the N-end to the C-end direction, resulting in an elongated peptide chain. When the ribosome complex reaches the termination codon the ribosome is released from the mRNA. In this way the amino acids are connected in a linear chain to afford a molecule of protein (Fig. 61).

Several AUG sequences exist in mRNA, but only a special AUG sequence is identified as the initiation (start) codon. In prokaryotes, Shine-Degarno sequences (S-D sequences) are found near the initiation codon which is identified by prokaryote-type ribosomes. An S-D sequences (identified by Shine and Degarno) is composed of 6–7 bases, 5′-AGGAGGU-. It is complementary paired to 16S rRNA by which AUG codon is arranged to initiate translation. In eukaryotes the S-D sequence does not exist but the cap structure of the 5′-end is identified by the ribosomes, and the nearest AUG to the 5′-end is used as the initiation codon. The base sequence around the initiation codon contains specific bases that seem to be required for exact translation.

3.3 Gene Expression

Genes are scattered along the base sequences of the long DNA molecule in the chromosomes. The generation of phenotypes of organisms is ascribed to the transcription of many genes to RNA, but the transcription of each gene is controlled in various ways. The fundamental mechanism of gene expression is common to lower and higher organisms, but the mechanism is somewhat different in prokaryotes and eukaryotes.

3.3.1 Fundamental Structure of the Gene

The transcription of RNA is initiated from a certain point (initiation point of transcription) of the gene by mediation of RNA polymerase. The gene consists of the transcription region and the regulation region for transcription (Fig. 62).

The transcription region contains exons and introns. The gene for transcription of mRNA contains the regions for translation to proteins (coding regions) and the nontranslation regions located upstream and downstream of the coding regions.

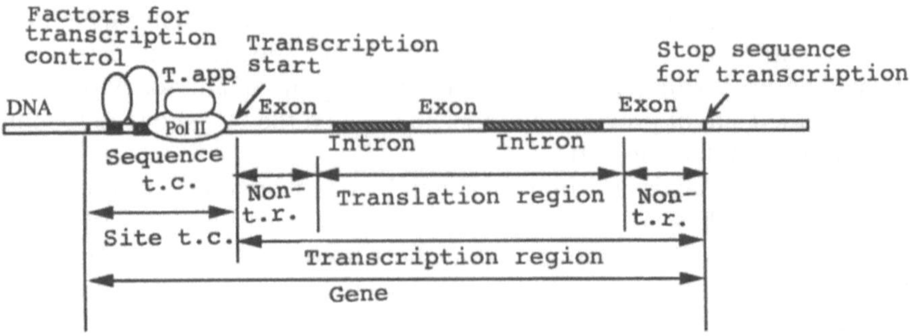

Fig. 62. Basic organization of a gene of a eukaryote. Part of the transcription control sequence is sometimes contained in the gene. The intron is sometimes located in nontranslation regions. *T. app* Transcription apparatus; *sequence t.c.* sequences for transcription control; *non-t.r.* nontranslating region; *site t.c.* transcription control region

In prokaryotes, at about 10 bases above the initiation point of transcription (–10 region), and at 35 bases above (–35 region), a base sequence for promoting of transcription (promoter) is present. RNA polymerase attaches to these regions, and the transcription is initiated.

In eukaryotes, in the upper region of the initiation point for transcription there is a proximal sequence (the TATA, CAAT or AGGA box) responsible for correct initiation of transcription, and a distal sequence responsible for regulating and modulating the expression of individual genes further upstream. These sequences for regulation of transcription are called the promoter regions.

The proximal promoter contains the region which becomes linked to the transcription apparatus and which enables the initiation of transcription by RNA polymerase. Special protein factors (*trans*-acting factors) are connected to the promoter sequences. These factors regulate transcription in cooperation with the transcription apparatus.

The base sequence of the transcription region of the same kind of gene is homologous among species. In the promoter regions, the sequence of the promoter of the same kind of gene is homologous within a species, but the sequences are considerably heterogeneous in other regions.

3.3.2 RNA Polymerase and Transcription Apparatus

In prokaryotes the transcription of all RNAs (mRNA, rRNA and tRNA) is mediated by a RNA polymerase. RNA polymerase of *Escherichia coli* consists of 5 subunits (α, β, β′, σ, and ω). An enzyme that consists of all subunits is called the holoenzyme of RNA polymerase, while an enzyme without σ (the sigma factor) is called the core enzyme of RNA polymerase.

The holoenzyme is strongly connected to the special part (promoter sequence) of genes by the action of the sigma factor. By this connection, DNA chains are partly loosened, and synthesis of RNA is initiated. When the synthesis of several bases has occurred the sigma factor is released from the holoenzyme, and the resultant core enzyme catalyzes the extension of the RNA chain up to the termination sequence of transcription. The released sigma factor is used for other core enzymes. There are several sigma factors, and a variety of gene expressions is regulated in prokaryotes by combination with the promoters of the respective genes.

In eukaryotes, three RNA polymerases (Pol I, Pol II, and Pol III) exist, and the transcription of RNA is mediated by one of the three enzymes. The transcription is regulated differently by different systems.

tRNA is synthesized by the mediation of Pol I, and rRNA and some other RNAs are synthesized by the mediation of Pol III. mRNA, which is involved in protein synthesis, is synthesized by the mediation of Pol II.

In transcription by Pol II and Pol III, only one RNA is involved as a transcription unit. In the case of Pol I, however, three rRNAs are synthesized as a transcription unit as in the transcription of rRNA genes by prokaryotes, and the three rRNAs are then separated later.

The base sequence specifying the initiation of transcription of Pol III is found upstream of the initiation point of the transcription gene of mRNA. The transcription apparatus is connected to this region and the base sequence known as the TATA box exists here.

The transcription apparatus is composed of several proteins, and four protein factors (TFIIA, TFIIB, TFIIC, and TFIID) connected to the TATA sequence. TFIID identifies TATA and connects to the sequence, and then TFIIB is connected. Subsequently, Pol III is connected and this is followed by connection of TFIIE and TFIIF to initiate transcription (Fig. 63).

The amino acid sequence of TFIID is well conserved in yeast, animals and plants, and the transcription apparatus is well maintained with little change during the evolution of higher organisms from lower organisms.

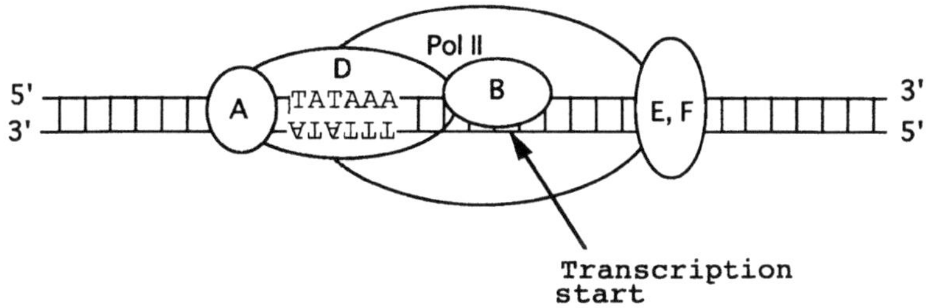

Fig. 63. Connection of transcription apparatus with TATA sequence in proximal promoter

3.3.3 Base Sequence of the Transcription Promoter

The transcription promoter determines the initiation of RNA polymerase in cooperation with the transcription apparatus (Fig. 62). The base sequence connected by the promoter is known as the base sequence of the transcription promoter, or *cis* element. In the gene of higher organisms, the *cis* element exists not only upstream of the initiation point of transcription but also within the transcription region. Even in the former case the *cis* element is sometimes present far from the initiation point. Several *cis* elements are known among genes, but in the same kind of gene in related organisms *cis* elements are similar. In addition, among genes with a common expression pattern a common *cis* element generally exists.

3.3.4 The Factor of Transcriptional Control

The factor of transcriptional control is connected to the base sequence of the transcription promoter and exerts transcriptional control. This factor is also called the *trans*-acting factor. The *trans*-acting factor is generally

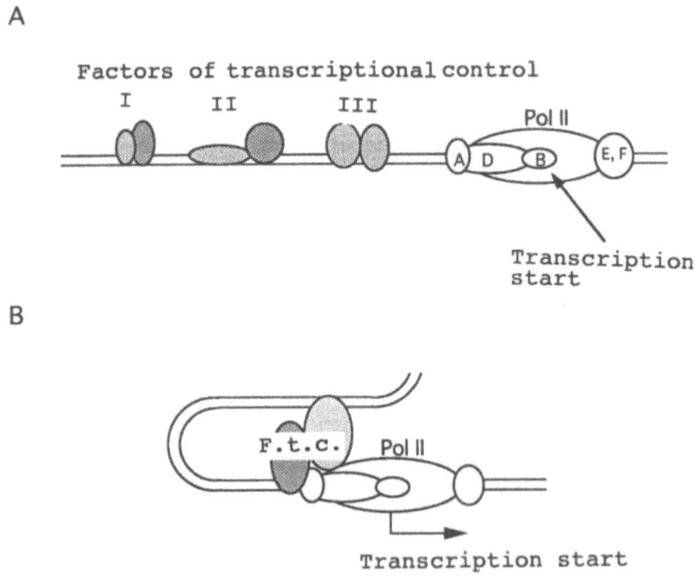

Fig. 64. A,B Transcription control factors located in transcription control sequence. **A** Several transcription control factors (*F.t.c.*) connected to transcription control sequence. **B** Interactive reactions in RNA polymease and transcription control factor

composed of two identical protein subunits. The protein subunit consists of two regions. The first region identifies the transcription control sequence and connects to the sequence, and the second region interacts with the transcription apparatus.

In addition to the DNA-connected factor, a protein factor that interacts with the DNA-connected factor is known as the transcriptional control factor. Transcriptional control of genes is conducted by the interaction of these several factors (Fig. 64).

The factors that promote or inhibit transcription are called the enhancer or repressor, respectively. The base sequences to which the enhancer and the repressor connect are called the enhancer sequence and the repressor sequence, respectively.

3.3.5 Regulation of Gene Expression

Factors of transcriptional control are composed of proteins. The syntheses of the factors of transcriptional control seem to be regulated by their respective genes. Accordingly, the expression of general genes could be regulated by genes situated higher in the hierarchy as shown in Fig. 65.

Several factors of transcriptional control are involved in the expression of one gene, and it seems that different factors are involved in the gene expression at different growing stages and in different tissues. Phenotypic appearance is the result of complex interactions of genes and responses to various environmental stimuli and the mechanism of these interactions remains to be elucidated.

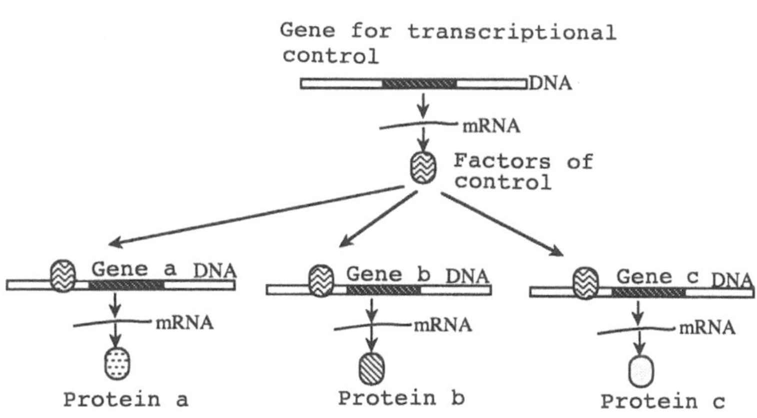

Fig. 65. A hierarchy structure in gene expression

3.4 Gene Expression in Plants

Gene structure and regulation mechanisms of gene expression in plants are similar to those of animals and yeast. Plant proteins of the transcription apparatus and regulators of transcription consist of amino acid sequences very similar to those of animals and yeast, indicating that a fundamental framework of gene expression is involved in these organisms (Sugiura 1992).

The expression of specific plant genes increases or decreases in response to environmental factors, and an inherent developmental program. Environmental factors affecting plant gene expression include interactions with pathogens and abiotic factors such as light, temperature, anoxia, drought, excess and deficiency of nutrients.

Plant life intimately depends on light as in photosynthesis, morphogenesis, regulation of flowering, germination, and bud formation, for example. This situation suggests that light is intimately involved in the regulation of gene expression in plants.

It is known that chloroplasts contain a unique system of gene expression, and that their DNA is regenerated independently of the nucleus. Gene expression of the chloroplast DNA is fundamentally the same as that of the prokaryotes.

3.4.1 Regulation of Gene Expression by Light

It is known that phytochrome exists in plants as a light receptor. Phytochrome is composed of a pigment moiety and a protein. Many of the phytochrome-

```
                              Box II
ATCCAAAAGCTTGGACAGGAACA [AATGTTACCCATACA] TAAAAGAT
        Box III
ATTT [GTGAAGTAACAGTC] ACAAAATTCCATGAGGCCAACATACTAC

AATTGAATTTTCATGGATACAATTCTTACAAAATAAAAATATCGACAT
                  Box III
AACCACCATCACA [CATTTACACT] CTTCACATGAAAAGATAAGATCA
    Box II
 [GTGAGGTAATAT] CCACATGGCACTGTCCTATTGGTGGCTTATGATA
                                      Box II
AGGCTAGCACACAAAATTTCAAATCTT [GTGTGGTTAATATG] GCTGC
            Box III
AAACTTT [ATCATTTTCACT] ATCTAACAAGATTGGTACTAGGCAGTA

GCTAAGTACCACAATATTAAGACCATAATATTGGAAATAGATAAATAA

AAACATTATATATAGCAAGTTTTAGCAGAAGCTTTGCA
                                       Transcription
                                       start
```

Fig. 66. Control of gene expression by phytochrome. Base sequence in transcription control region and the transcription control sequence (*boxes*) for photosynthesis, both of which are essential for regulation by the phytochrome

regulated responses are induced by red (r) light and, with certain exceptions, reversed by far-red (fr) light. The phytochrome polypeptide is encoded by a minimum of three distinct genes in angiosperms, often designated *phyA*, *phyB* and *phyC*. The prosthetic group associated with the protein is composed of an open chain of four tetrapyrrole rings.

The Pr form of phytochrome has an absorption maximum at 660 nm and absorption of red light converts it to the Pfr form with an absorption maximum at 730 nm. The Pfr absorption of far-red light converts the Pfr form to Pr form. The interconversion between Pr and Pfr allows phytochrome to initiate many light-mediated responses, including modification of gene expression. Some of these responses, such as the positive regulation of the gene for the chlorophyll a/b-binding protein (*cab*) and the gene for the ribulose bisphosphate carboxylase small subunit (*rbcS*), have been shown to require de novo protein synthesis. These genes exist in the nucleus, and several transcriptional control sequences, which contain the sequence to amplify transcription by phytochrome, are found in the transcriptional control region (Fig. 66).

A gene to code the promoter connected to the phytochrome-dependent base sequence of transcriptional control has been isolated, and the mechanism of the transcriptional control by light has been investigated on the molecular level.

3.4.2 Gene Expression in Chloroplasts

By cooperative effort of many research laboratories a large volume of gene sequence data has accumulated for the chloroplast-encoded gene ribulose-1,5-bisphosphate carboxylase/oxygenase (*rbcL*). The data are for seed plants but include most major lineages of the angiosperms (Clegg 1993).

The chloroplast DNAs from *Nicotiana tabacum* (155844 bp), *Oryza sativa* (134525 bp) and *Pinus thunbergii* (119707 bp) have been sequenced by Sugiura (1986, 1989a, 1993).

The number of copies of chloroplasts is about 20 in meristematic tissue cells and 10000 in green leaves. Chloroplasts have a circular DNA with 140 to 190 kb, and more than 100 genes are found in the DNA (Fig. 67; Table 9). Among these are genes for the function of chloroplasts, the regulation of their expression, and the genes to code rRNA and tRNA responsible for protein synthesis. In other words, the transcription and translation of such genes are carried out in these organelles. However, not all genetic information for the chloroplast function is contained in the chloroplast DNA. Chloroplasts require the function of proteins coded by genes in the nuclear DNA. It has been suggested that these nuclear genes were originally present in the chloroplast but were gradually transferred from the chloroplast to the nucleus in the process of plant evolution (Sugiura 1989b).

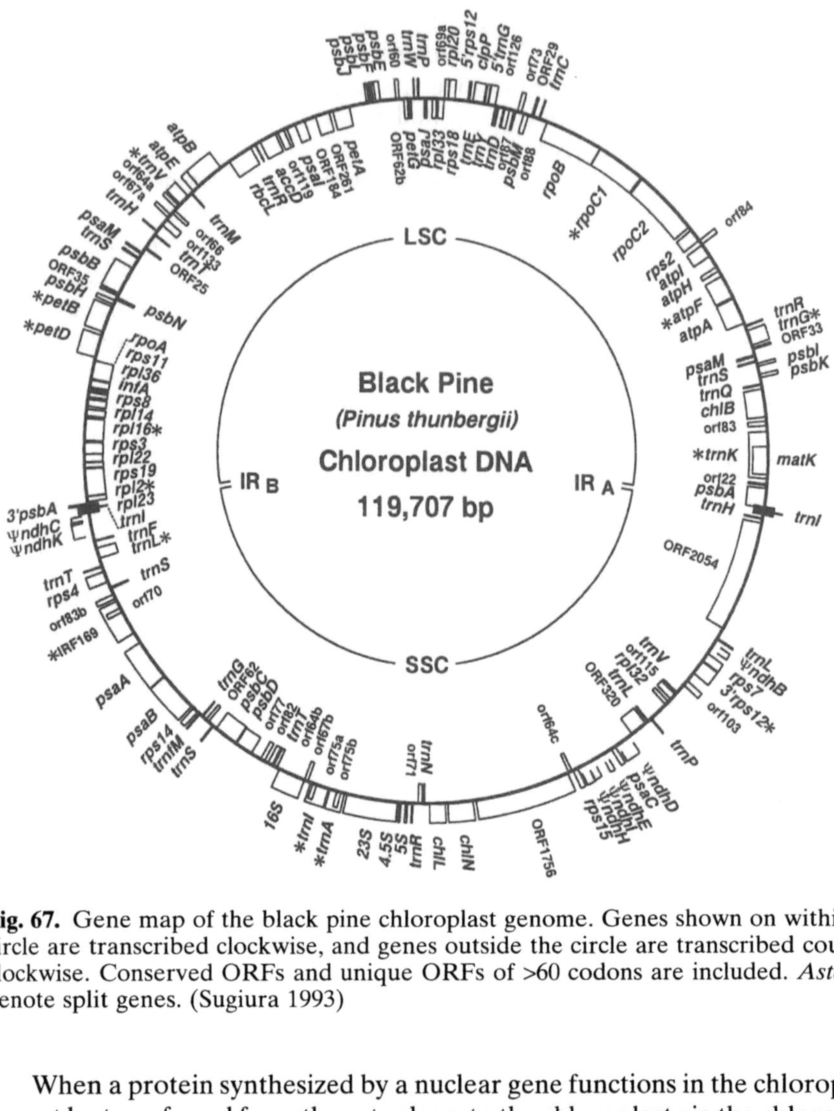

Fig. 67. Gene map of the black pine chloroplast genome. Genes shown on within the circle are transcribed clockwise, and genes outside the circle are transcribed counter-clockwise. Conserved ORFs and unique ORFs of >60 codons are included. *Asterisks* denote split genes. (Sugiura 1993)

When a protein synthesized by a nuclear gene functions in the chloroplast, it must be transferred from the cytoplasm to the chloroplast via the chloroplast membranes. In fact, an amino acid sequence required for the membrane penetration exists at the N-end of the protein and is called the transit peptide.

Nuclear genes required for maintenance of the chloroplasts, and the chloroplast genes are expressed cooperatively. Ribulose-1,5-bisphosphate carboxylase/oxygenase (Rubisco) is composed of a large and a small subunit. The large subunit is encoded by the chloroplast gene and the small subunit by the nuclear gene. Since the ratio of both subunits in chloroplasts is always constant, the nucleus and the chloroplast seem to cooperate to control the gene expression.

Table 9. Gene content of the black pine chloroplast genome. (Sugiura 1993)

RNA genes
Ribosomal RNA genes
23SrDNA, 16SrDNA, 5SrDNA, 4.5SrDNA

Transfer RNA genes
trnA(UGC)*, *trnC*(GCA), *trnD*(GUC), *trnE*(UUC), *trnF*(GAA), *trnG*(GCC), *trnG*(UCC), *trnH*(GUG), *trnI*(CAU)#, *trnI*(GAU)*, *trnK*(UUU)*, *trnL*(CAA), *trnL*(UAA)*, *trnL*(UAG), *trnM*(CAU), *trnM*(CAU), *trnN*(GUU), *trnP*(GGG), *trnP*(UGG), *trnQ*(UUG), *trnR*(ACG), *trnR*(CCG), *trnR*(UCU), *trnS*(GGA), *trnS*(GCU)#, *trnS*(UGA), *trnT*(GGU)#, *trnT*(UGU), *trnV*(GAC), *trnV*(UAC)*, *trnW*(CCA), *trnY*(GUA), ψ*5'trnG*(UCC)

Polypeptide genes
Ribosomal protein genes
*rps2, rps3, rps4, rps7, rps8, rps11, rps12*** ##, rps14, rps15, rps18, rps19, rpl2, rpl14, rpl16*, rpl20, rpl22, rpl23, rpl32, rpl33, rp136*

Transcription/translation component genes
rpoA, rpoB, rpoC1, rpoC2, infA, matK*

Photosynthetic apparatus genes
rbcL, psaA, psaB, psaC, psaI, psaJ, psaM#, psbH, psbI, psbJ, psbK, psbL, psbM, psbN, ψ*3'psbA, petA, petB, petD, petG, atpA, atpB, atpE, atpF*, atpH, atpI,*

Chlorophyll biosynthesis genes
chlB, chlL, chlN,

NADH dehydrogenase genes
ψ*ndhB*, ψ*ndhC*, ψ*ndhD*, ψ*ndhE*, ψ*ndhH*, ψ*ndhI*, ψ*ndhK*,

Other protein genes
accD, clpP

Conserved ORFs
ORF29, ORF62b***, ORF33, ORF35, ORF62, IRF169##, ORF184 ORF261, ORF320, ORF1756, ORF2054

IRF, intron-containing reading frame.
*Gene containing intron; #two copies; **divided gene; ***homologous to tobacco ORF31; ##gene containing two introns.

Chloroplast DNA is different from nuclear DNA, and the gene expression is of the prokaryote type. Among about 100 genes present in the chloroplast genome, some are known to be transcribed individually, but in many cases several genes with related functions form a cluster to be cotranscribed. It has also been found that genes that have no functional relations are sometimes cotranscribed.

Upstream of the initiation point of the transcription of the chloroplast gene "–35" and "–10" there are regions which function as promoters as in prokaryotes; however, the cap structure located at the 5′-end of the RNA of eukaryotes does not occur in the transcribed RNA of chloroplasts. Thus, the gene expression in chloroplasts is of the prokaryote type, but sometimes introns characteristic of eukaryotes are present.

The prokaryote type rRNA is present in the chloroplast genome, and the translation to proteins is performed by the prokaryote type ribosomes. In addition, genes of one-third of the 60 different ribosomal proteins that constitute ribosomes exist in the chloroplast genome, and the proteins are homologous to the ribosomal proteins in prokaryotes. tRNA located in chloroplasts is also of the prokaryote type and similar to that of *Escherichia coli*. However, many tRNA genes contain introns as in eukaryotes.

3.4.3 Gene Expression in Mitochondria

A DNA independent of that of the nucleus exists in the mitochondria, and expresses genetic information. However, the genetic information is not enough to support the function of the mitochondria. Most proteins are coded in the nuclear genes and some tRNA is therefore transported from the cytoplasm to support these organelles.

DNA of mammalian mitochondria exists in a circular form and has about 16 kbp. The size and constitution of mammalian mitochondria are the same among species, whereas the size of the mitochondrial DNA in plants is very different among species (140–570 kbp). In higher plants mitochondrial genomes are composed of several molecules, which are reconstituted intra- and intermoleculary via repeated base sequences and homologous exchanges, whereas the structure of the mitochondrial DNA of green algae and mosses is relatively simple.

4 Biosynthesis of Wood Components

4.1 Chemical Composition of Wood

Chemical constituents of wood are classified into major components consisting of cell wall materials, and minor components extractable with solvents (wood extractives). The former, components of cell walls and intercellular layers, amount to 90–95% of oven-dry wood. Extractives occur mainly, albeit not exclusively, in the cell lumen and special tissues such as resin cells, and are extracted with water or organic solvents. The amount of minor components is very different in different species, and some compounds are limited to certain species only. Tropical woods often contain large amounts of extractives.

Wood components are difficult to isolate quantitatively as definite chemical compounds, and are often determined as the total of cell wall polysaccharides, lignin and extractives. Polysaccharides composed of cellulose and hemicelluloses are called holocelluloses (Fig. 68).

Examples of the chemical compositions of Japanese and American softwoods, hardwoods, and bamboos are shown in Tables 10–12.

4.1.1 Cellulose

About 50% of wood consists of cellulose, which is a linear polymer connected by 1,4-glucosidic linkage of β-D-glucose. In addition, hydrogen bonding occurs between hydroxyl groups of neighboring glucose residues of cellulose molecules to form microfibrils, about 70% of which exist as crystalline regions. Cellulose occurs in both softwoods and hardwoods and is more resistant toward acids than are hemicelluloses.

4.1.2 Hemicelluloses

Noncellulosic polysaccharides amount to 20–30% of wood and are mostly soluble in aqueous alkali. Hemicelluloses are classified into pentosans which give D-xylose, L-arabinose, and L-rhamnose by hydrolysis, and hexsosans which give D-mannose, D-galactose, and D-glucose by hydrolysis. Hemicelluloses generally occur as heteropolymers such as glucomannan, galactoglucoman-

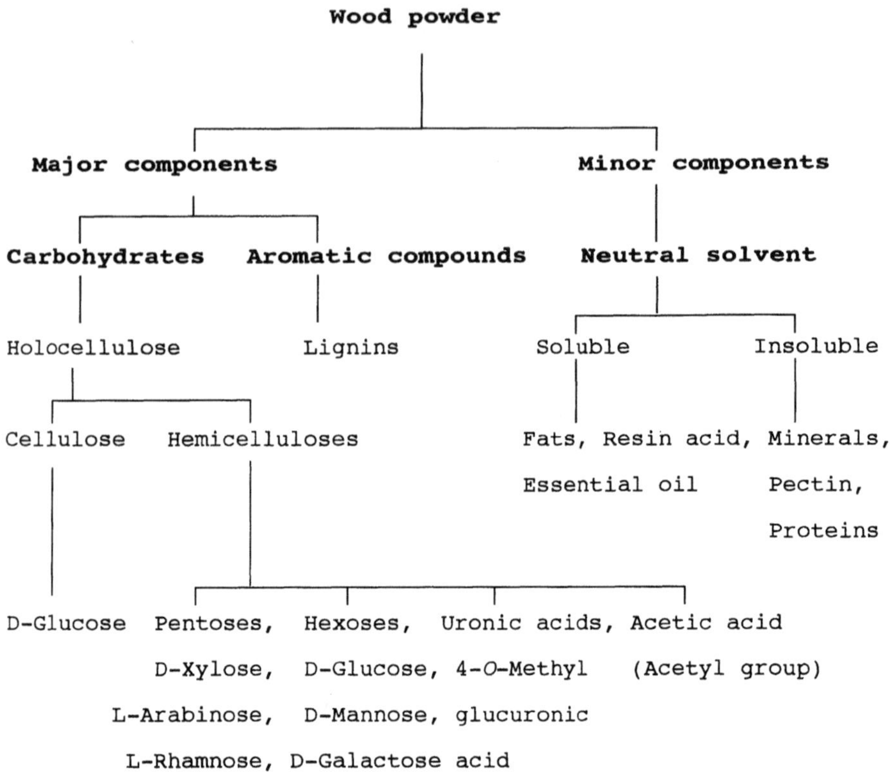

Fig. 68. Constitution of wood components

nan, arabinogalactan and glucuronoxylan, and are sometimes partly acetylated. As homopolymers, galactan (conifer compression wood), arabinan (bark) and β-1,3-glucan (callose in bark, laricinan in compression wood) have been characterized.

The main hemicelluloses of softwoods are *O*-acetylgalactoglucomannan, arabino-4-*O*-methylglucuronoxylan and, in larches, arabinogalactan, while hardwood hemicelluloses are *O*-acetyl-4-*O*-methylglucuronoxylan and glucomannan. The hemicellulose in bamboo is mainly an *O*-acetyl-4-*O*-methylglucuronoarabinoxylan which is different from that of gymnosperms but similar to the hardwood hemicellulose with respect to its degree of branching and the molecular properties (Maekawa 1974).

4.1.3 Lignin

About 20–30% of wood consists of lignin, which is a group of complex aromatic polymers connected by carbon–carbon and alkyl–aryl ether linkages of guaiacylpropane, syringylpropane, and *p*-hydroxyphenylpropane. Protolignin (lignin in situ) is insoluble in organic solvents.

Normal softwood lignin is mainly composed of guaiacylpropane, but compression wood lignin consists of guaiacylpropane and significant amounts of *p*-hydroxyphenylpropane units. Hardwood lignin is composed of guaiacyl-syringylpropanes, in which the ratio of the two phenylpropanes is different in different species. Grass lignin is composed of guaiacyl-syringyl-*p*-hydroxyphenylpropanes, esterified and partly etherified by *p*-coumaric acid.

4.1.4 Aliphatic Compounds

Hydrocarbons, alcohols, fats, waxes and other substances occur as wood extractives. Fats are present mainly in parenchyma tissues, more in sapwood than in heartwood, and are used as a reserve energy source in tree growth.

4.1.5 Sugars

D-glucose, D-fructose, sucrose, and other sugars, such as raffinose and stachyose (Magel and Höll 1993), occur in sapwood, while L-arabinose, D-xylose, L-rhamnose, D-glucose and D-mannose are often present in heartwood. In trees of *Celtis* large amounts of oligosaccharides containing galactose, such as raffinose, stachyose, and verbascose are found together with sucrose (Stewart 1966).

4.1.6 Aromatic Compounds

Phenols, stilbenes, flavonoids, tannins, quinones, tropolones, lignans and other substances occur mainly in heartwood and in bark.

4.1.7 Terpenes

Monoterpenes (such as α-pinene, camphene and cineole), sesquiterpenes, diterpenes (such as resin acids), triterpenes, and polyterpenes (such as rubber) are known as terpenoids. They are especially abundant in softwoods.

4.1.8 Biosynthetic Pathways

The chemical components considered above are derived from intermediary secondary metabolites of growing xylem and phloem, and synthesized by the following biosynthetic pathways:

Table 10. Chemical composition of major Japanese softwoods and hardwoods. (Migita 1968)

Species	Ash	Extractives soluble in				Cellulose	α-Cellulose	Pentosan	Mannan	Galactan	Lignin	Proteins
		Cold water	Hot water	1% NaOH	Alcohol-benzene							
Pinus densiflora	0.1–0.4	0.5–0.9	1.6–3.6	10.7–24.5	1.9–3.9	48.6–58.3	31.0–41.3	9.9–12.9	3.5–10.1	0.3–0.8	24.9–31.6	0.8
P. thunbergii	0.1–0.4	0.2–2.0	2.3–4.3	14.2–24.3	1.8–2.8	55.3–58.1	37.1–41.5	10.3–12.9	3.2–7.9	0.3–0.6	27.2–30.8	0.7
Cryptomeria japonica	0.3–0.8	1.0–2.8	1.3–3.0	13.2–22.7	1.3–5.0	49.0–56.6	33.1–41.9	9.6–12.0	3.6–7.8	0.3–1.3	28.0–34.8	1.5
Chmaecyparis obtusa	0.2–0.9	1.1–3.5	2.1–4.8	11.7–24.3	1.2–4.1	50.8–58.1	39.6–44.7	6.1–11.5	5.9–8.7	0.3–0.5	25.7–32.2	0.7
Abies firma	0.1–0.6	2.1	1.0–3.0	11.5–11.7	1.5	49.0–54.3	38.2	8.5–10.9	7.9	1.3	29.4–32.2	0.9
A. sachalinensis	0.4–0.8	0.7–2.7	1.3–4.2	7.4–14.6	2.6–4.5	50.5–59.4	31.7–41.0	5.1–11.6	3.9–7.2	0.2–0.3	21.8–33.0	0.5–1.0
Larix leptolepis (kaempferi)	0.2–0.5	2.1–11.2	3.9–20.1	10.1–28.3	1.8–5.5	47.2–58.3	30.6–38.9	4.8–11.8	4.1–7.8	0.6–9.0	20.3–31.8	0.3–0.5
Picea ezoensis	0.3–0.6	1.1–3.9	1.8–5.5	8.3–15.0	1.3–4.0	49.5–60.0	33.9–48.8	5.3–12.3	4.7–7.8	0.2–1.4	23.0–32.5	0.3–1.3
Fagus crenata	0.2–1.0	0.6–3.0	1.5–3.6	15.7–19.8	0.6–3.8	51.9–61.2	40.7–45.9	21.3–26.2	0.0	0.1–1.6	8.3–24.2	0.4–1.2
Quercus serrata	0.1–0.6	2.1–5.0	3.3–8.0	14.9–24.3	0.6–1.0	50.4–62.0	37.1–43.5	18.3–24.0	0.0	0.4–1.8	20.5–22.8	0.7–1.0
Betula maximowicziana	0.1–0.7	0.7–1.9	1.5–7.5	12.9–28.0	0.5–7.3	50.7–61.3	36.6–48.6	17.5–29.1	0.0	0.3–0.7	17.1–24.4	0.8–1.1
Tilia japonica	0.2–0.6	2.0–2.3	2.9–4.1	21.1	6.0	55.2–61.5	33.1–42.3	20.4–20.7	0.0	0.1	23.5–26.9	0.7–1.0
Acer mono	0.1–0.7	1.2–3.8	2.2–5.2	16.7–25.8	1.4–3.0	50.9–61.0	37.1–48.9	18.7–25.7	0.0	0.3–1.5	19.5–27.7	0.6–0.7
Poplus maximowiczii	0.2–2.0	1.2–2.2	2.9–4.5	19.5–24.6	1.8–3.6	54.8–66.4	33.8–46.8	16.3–23.6	0.0	0.1–1.1	18.7–25.2	0.4–0.8

Analytical results of α-cellulose, pentosan, mannan, galactan, lignin and proteins are percentages for extractive-free wood powder of the same softwoods and hardwoods.

Table 11. Chemical composition of American softwoods and hardwoods in percent of oven-dried weight. (Migita 1968)

Species	Ash	Extractives soluble in		Acetyl	α-Cellulose	Hemi-celluloses	Lignin
		Hot water	Ethanol				
Pseudotsuga menziesii	0.29	–	–	0.61	57.2	14.1	28.4
Tsuga heterophylla	0.49	–	–	1.21	51.6	15.5	30.4
Pinus taeda	0.28	–	–	1.10	55.0	15.3	29.5
P. ponderosa	0.19	2.12	7.04	0.68	51.9	13.6	24.4
P. elliottii	0.25	0.80	2.75	1.00	53.9	13.8	26.1
Picea mariana	0.36	–	–	1.12	51.5	17.4	28.0
Thuja plicata	0.27	–	–	0.53	52.7	14.7	32.5
Quercus falcata	0.16	–	–	3.30	45.7	23.2	25.2
Q. lyrata	0.41	–	–	2.81	42.4	24.8	27.6

1. Nucleoside diphosphate sugar pathway: cellulose, hemicelluloses, glycosides etc.
2. Shikimate–cinnamate pathway: lignin, lignans, hydrolyzable tannins etc.
3. Acetate–cinnamate pathway: flavonoids, stilbenes, condensed tannins etc.
4. Mevalonate pathway: terpenes, steroids etc.
5. Mixed pathways of 3 and 4: some prenylflavonoids, quinones and stilbenes.

4.2 Biosynthesis of Cell Wall Polysaccharides

4.2.1 Localization of Polysaccharides in Wood Cell Walls

Localization of polysaccharides is different in different layers of the wood cell walls. Meier and Wilkie (1959) and Meier (1964) investigated polysaccharide composition in different cell wall layers of pine tracheids, spruce tracheids, and birch fibers. Young tracheids or fibers in radial sections through the outermost part of the xylem bordering the cambial zone were separated by using a micromanipulator under the polarizing microscope into four layers: (M+P), (M+P+S_1), (M+P+S_1+S_2 outer part), and (M+P+S_1+S_2+S_3). The separated layers were hydrolyzed with sulfuric acid, and from the amounts of the monosaccha-

Table 12. Chemical composition of bamboo culms

Species	Ash	Extractives soluble in			Lignin	Pentosans	Holo-cellulose
		Hot water	1% NAOH	Alcohol-benzene			
Bambusa arundinacea[a]	3.26	5.25	19.35	0.22	30.09	19.62	57.56
B. blumeana[b]	4.8	4.30	39.50	3.10	20.40	19.00	67.40
B. vulgaris[b]	2.8	5.10	27.90	4.10	26.90	21.10	66.50
Dendrocalamus strictus[a]	2.10	5.93	15.00	0.25	32.20	19.56	68.80
Gigantochloa levis[b]	5.30	4.40	28.30	3.20	24.20	18.80	62.90
Melocanna baccifera[a]	1.87	6.48	18.97	1.43	24.13	15.13	62.25
Oxytenanthera nigrociliata[a]	1.25	3.39	17.11	0.24	27.09	17.41	66.72
Schizostacyum lumampao[b]	9.70	4.30	31.40	5.00	20.40	20.60	60.60
Phyllostachys pubescence[c]	1.31	19.96	32.19	4.63	26.06	27.70	49.12[d]
P. nigra Munro var. *Henonis*[c]	2.00	21.47	34.03	3.35	23.82	24.13	42.31[d]
P. reticulata[c]	1.94	23.24	36.17	3.37	25.33	24.54	40.97[d]

[a] Singh and Guha (1981).
[b] Tamolang et al. (1980).
[c] Higuchi (1987).
[d] The cellulose content was determined by a modification of Jenkin's $NAClO_2$ method, and calculated on a pentosan-free basis.

rides in the hydrolyzates, the percentages of the different polysaccharides with known chemical structures were calculated. The composition of the different cell wall layers was finally computed (Meier 1985; Table 13).

There is a marked difference between birch and the two conifers in the arabinan content of the (M+P) layer: in birch it is only half of that in pine. For the xylan content, the reverse is true. In birch, the cellulose content is highest in the inner part of the S_2 and in the S_3 layer. In the S_1 and the outer part of the S_2 layer, the content of 4-*O*-methylglucuronoxylan is very high. The composition of the different cell wall layers is similar in spruce and pine tracheids. The content of glucomannan increases from the outer parts of the cell walls to the inner parts. The content of arabino-4-*O*-methylglucuronoxylan is very high in

Table 13. Polysaccharide composition of the different cell wall layers of *Betula verrcosa* fibers and of tracheids from *Pinus sylvestris* and *Picea abies*. (Meier 1985)

Polysaccharides	Cell wall layers			
	(M + P)[a] (%)	S_1 (%)	S_2 outer (%)	S_2 outer + S_3 (%)
Betula verrucosa Ehrh.				
Galactan	16.9	1.2	0.7	0.0
Cellulose	41.4	49.8	48.0	60.0
Glucomannan	3.1	2.8	2.1	5.1
Arabinan	13.4	1.9	1.5	0.0
O-Acetyl-4-*O*-methyl glucuronoxylan	25.2	44.1	47.7	35.1
Pinus sylvestris L.				
Galactan	16.4	8.0	0.0	0.0
Cellulose	33.4	55.2	64.3	63.6
O-Acetylglucomannan	7.9	18.1	24.4	23.7
Arabinan	29.3	1.1	0.8	0.0
Arabino-4-*O*-methyl-glucuronoxylan	13.0	17.6	10.7	12.7
Picea abies (L.) Karst.				
Galactan	20.1	5.2	1.6	3.2
Cellulose	35.5	61.5	66.5	47.5
O-Acetylglucomannan	7.7	16.9	24.6	27.2
Arabinan	29.4	0.6	0.0	2.4
Arabino-4-*O*-methyl-glucuronoxylan	7.8	15.7	7.4	19.4

M, Middle lamella; P, primary wall; S_1, S_2, and S_3, the outer, middle, and inner layer of the secondary wall, respectively.
[a] Contains also pectic acid, which has not been taken into account.

the S_3 layer. It seems that practically no galactan and arabinan occur in the secondary wall layers of pine and spruce tracheids.

4.2.2 Metabolism of Sugar Nucleotides

It has been shown that cellulose and β-1,3-glucan are synthesized on the plasma membrane by mediation of enzymes and are directly deposited onto the cell walls. Other polysaccharides are synthesized by the Golgi apparatus in the cytoplasm, and are transported in the Golgi vesicles, which fuse with the plasma membrane, depositing the polysaccharides onto the growing cell wall.

In the synthesis of polysaccharides in the Golgi apparatus, sugar nucleotides in the cytoplasm are linked to specific carriers, and penetrate Golgi membranes. The sugar nucleotides transferred are converted to polysaccharides by sugar transferases associated with the inner surface of the Golgi cisternae. The synthesis of hemicelluloses, pectin, and glycoproteins are completed at the lumen. However, immunoelectron micrographs have suggested that different polysaccharides, such as hemicelluloses and pectin are synthesized in different Golgi vesicles.

Electron micrographs of cell membranes have indicated the occurrence of "terminal complexes" (TCs) linked to the end of cellulose microfibrils located at the outer layer of the plasma membrane, while specially arranged TCs, "rosettes" are located at the inner layers of the plasma membrane. These rosettes (TCs) are suggested to be cellulose-synthesizing enzyme complexes in higher green plants (Giddings et al. 1980).

UDP-glucose incorporated as a precursor into the cells is converted to cellulose by the mediation of the TCs and is transported to the cell wall. Cellulose molecules form microfibrils by hydrogen bonding between the cellulose chains. Sugar nucleotides are sugar donors involved in biosynthesis of oligo- and polysaccharides. The chemical structure of a typical sugar nucleotide, UDP-glucose is shown in Fig. 69.

Glycosyltransferases, which are involved in polysaccharide synthesis, catalyze transfer of sugar residues of nucleotide sugars to the acceptor using energy released by hydrolysis of the high energy phosphate linkage between sugar and phosphate in the nucleotide.

In plants, a number of kinases exist that phosphorylate free monosaccharides to form the monosaccharide-1-phosphate. As shown in Fig. 70 (Hayashi 1994), glucose incorporated into cells is phosphorylated to glucose-6-phosphate, which is converted to glucose-1-phosphate by phosphoglucomutase.

Glucose-1-phosphate reacts with nucleoside-3-phosphate to give nucleotide glucose which is converted to several sugar nucleotides by mediation of dehydrogenases, carboxylyases, and epimerases. Sugars linked to the phosphate group of uridine diphosphate (UDP), guanosine diphosphate (GDP), and adenosine diphosphate (ADP) are known as sugar donors in higher plants. However, ADP-glucose is not involved in biosynthesis of wood polysaccharides but in starch biosynthesis. Sugar derivatives of nucleoside-

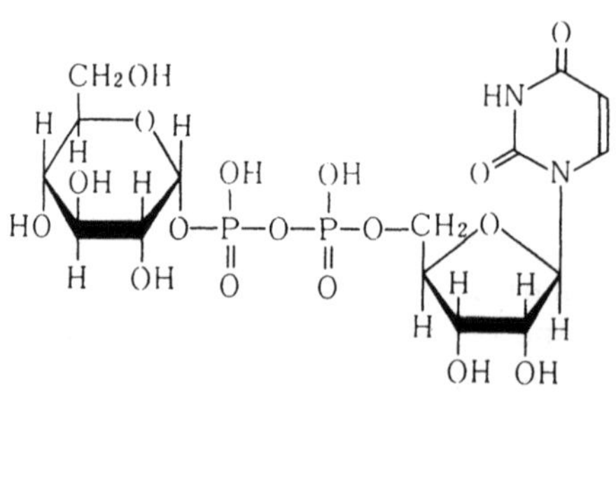

Fig. 69. Chemical structure of UDP-glucose

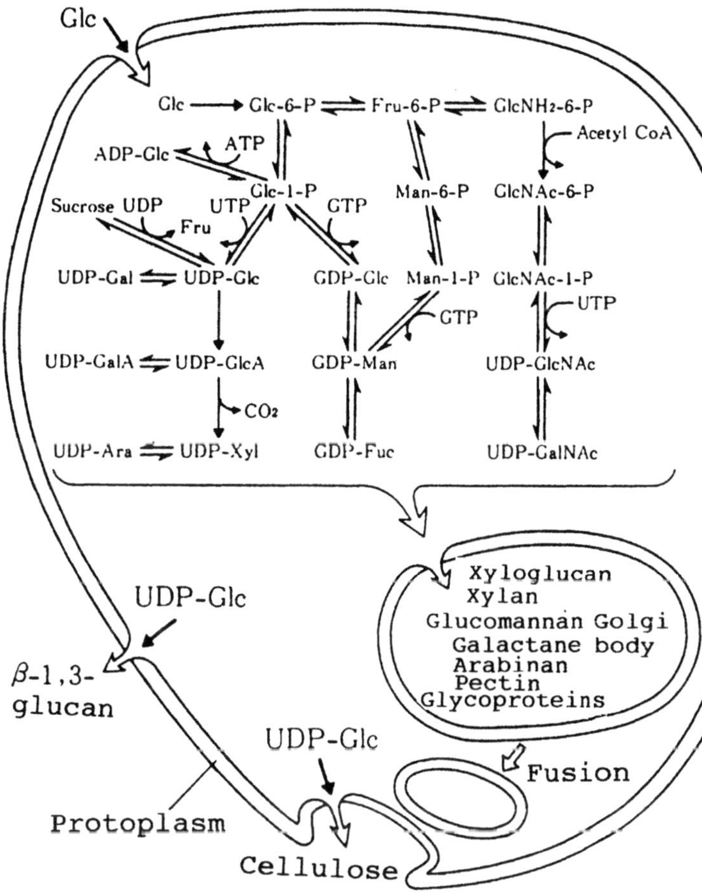

Fig. 70. Biosynthesis of sugar nucleotides and polysaccharides in a cell. (Hayashi 1994)

1-phosphate, nucleoside diphosphate oligosaccharides, and sugar derivatives of nucleosides other than uridine, guanosine and adenosine are found in plant cells, but their physiological roles are not known at present.

Biosynthesis of the monosaccharide units in polysaccharides occurs mostly by isomerization of sugar residues of UDP-sugars, and is regulated in a complex fashion. UDP-glucose has been found to be an allosteric inhibitor of the activity of the pyrophosphorylase at physiological concentrations, so if UDP-glucose is not quickly used for polysaccharide synthesis in developing tissues the flux of carbon through the pyrophosphorylase may be restricted (Smith 1993).

Biosynthetic reactions of UDP-xylose via UDP-glucuronic acid from UDP-glucose are irreversible, and UDP-glucose dehydrogenase and UDP-glucuronate carboxylase are inhibited by UDP-xylose, the final product, whereas UDP-glucose activates UDP-glucuronate carboxylyase.

The properties of the enzymes associated with the major UDP-sugars are outlined in the following Sections.

4.2.2.1 UDP-Glucose Pyrophosphorylase (EC 2.7.7.9)

$$\text{UTP} + \text{Glucose-1-phosphate} \leftrightarrows \text{UDP-Glucose} + \text{Pyrophosphate}\ (\text{PPi})$$

The enzyme is also called glucose-1-phosphate uridyltransferase. For activation of the enzyme, metal ions such as Mg^{2+} are required. The enzyme reaction proceeds as an ordered reaction. The connection order of enzyme to substrate is determined, and the reaction equilibrium is about 1.0.

UDP-glucose occupies a central position in sugar nucleotide metabolism. It serves as a precursor in sucrose synthesis and of sugar nucleotides required for the synthesis of cellulose and hemicelluloses.

4.2.2.2 Sucrose Synthase (EC 2.4.1.13)

$$\text{UDP-Glucose} + \text{Fructose} \leftrightarrows \text{Sucrose} + \text{UDP}$$

The enzyme catalyzes sucrose synthesis by transfer of a glucose residue of UDP-glucose to fructose. The reaction equilibrium is about 1.0. Sucrose synthase catalyzes a reaction that is freely reversible and hence very sensitive to fluctuations in substrate supply. The enzyme is a tetramer of identical subunits of about 90 kDa (Delmer 1972, Su and Preiss 1978).

In most higher plants, sucrose is the major source of carbon translocated to nonphotosynthetic tissues. The enzyme is frequently found in sink tissues where elevated concentrations of sucrose are available. It seems that the enzyme is involved in the cleavage of sucrose to afford UDP-glucose for glucan synthesis in tissues where the concentration of sucrose is elevated.

Recently, the importance of sucrose synthase for synthesis of starch has been demonstrated in maize using the *sh* mutation (Chourey 1981, Chourey et al. 1986, McCarty et al. 1986). The loss of 95% of sucrose synthase activity in endosperm cells is associated with a marked starch reduction in the *sh* mutant. The product of the *Sh* locus has been designated as SS1, and this form of sucrose synthase is highly expressed in endosperm cells at the time of starch synthesis.

A second locus, *Sus*, codes for an additional sucrose synthase gene (enzyme form SS2), which is about 80% homologous to the *Sh* gene at the cDNA level, but the expression pattern is different. When only one gene is expressed only one isozyme is formed. When both are expressed in the same cell type, all five predicted mixed isozyme patterns can be detected, indicating that the products of the two genes can combine to form a functional enzyme.

Delmer et al. (1991) have identified an 84-KDa plasma membrane-localized polypeptide which could be photo-labeled with ^{32}P-UDP-glucose from cotton. The polypeptide was tightly associated with the plasma membrane and was identified as a form of sucrose synthase. The molecular weight is essentially identical to that of the catalytic subunit of the Mg^{2+}-dependent cellulose synthase of *Acetobacter xylinum* (Lin et al. 1990). Cotton fibers do not synthesize starch but synthesize very thick cellulosic secondary cell walls using UDP-glucose as substrate. The results suggest that this form of sucrose synthase may have a critical function in cellulose synthesis.

Delmer and her coworkers (Amor et al. 1995) further found that at least half of the total sucrose synthase of developing cotton fibers (*Gossypium hirsutum*) is tightly associated with the plasma membrane, suggesting that this form of sucrose synthase serves to channel carbon directly from sucrose to cellulose and callose synthases in the plasma membrane. They showed that carbon from sucrose can be converted at high rates to both cellulose and callose by using detached and permeabilized cotton fibers. Synthesis of cellulose or callose was favored by addition of EDTA or calcium and cellobiose, respectively. Immunolocalization studies showed that sucrose synthase can be localized at the fiber surface in a complex consistent with the deposition of cellulose or callose.

4.2.2.3 UDP-Glucose Dehydrogenase (EC 1.1.1.22)

$$\text{UDP-Glucose} + 2\text{NAD}^+ + H_2O \rightarrow \text{UDP-Glucuronic acid} + 2\text{NADH} + 2H^+$$

By mediation of the enzyme 2 mol of NAD^+ are reduced per 1 mol of UDP-glucose, and 1 mol of UDP-glucuronic acid is formed. The reaction is irreversible, and inhibited competitively by its product, UDP-glucuronic acid, and allosterically by UDP-xylose, a subsequent reaction product. UDP-glucuronic acid formed by this reaction is the precursor of UDP-galacturonic acid, UDP-xylose and UDP-arabinose, which are all used in the synthesis of pectin and hemicelluloses.

4.2.2.4 UDP-Glucuronate Carboxylase (EC 4.1.1.35)

$$\text{UDP-Glucuronate} + \text{NAD}^+ \rightarrow \text{UDP-Xylose} + CO_2 + \text{NADH} + H^+$$

The enzyme catalyzes the decarboxylation of UDP-glucuronate to give UDP-xylose. The enzyme protein is firmly linked to NAD, and catalyzes deprotonation from the fourth hydroxyl group of the glucuronic acid residue to give UDP-4-oxoglucuronate as an intermediate. The linkage between the fifth and sixth carbons of the intermediate is loosened resulting in decarboxylation. The enzyme is affected by feedback inhibition by the product, UDP-xylose but activated by UDP-glucose allosterically.

4.2.2.5 UDP-Glucose 4-Epimerase (EC 5.1.3.2)

UDP-Glucose ⇆ UDP-Galactose

The enzyme catalyzes epimerization of the hydroxyl group of the asymmetric carbon 4 of the glucose residue of UDP-glucose to give UDP-galactose. The reaction equilibrium is about 1.0. UDP-galactose is the precursor in the synthesis of galactose-containing polysaccharides in primary cell walls.

4.2.2.6 UDP-Glucuronate 4-Epimerase (EC 5.1.3.6)

UDP-Glucuronate ⇆ UDP-Galacturonate

The enzyme catalyzes epimerization of the hydroxyl group at carbon 4 of a glucuronic acid residue of UDP-glucuronate to give UDP-galacturonate. The equilibrium constant is about 1.0.

4.2.2.7 UDP-Arabinose 4-Epimerase (EC 5.1.3.5)

UDP-Arabinose ⇆ UDP-Xylose

The enzyme catalyzes epimerization of the hydroxyl group at the asymmetric carbon 4 of an arabinose residue of UDP-arabinose to give UDP-xylose. The affinity of the enzyme for UDP-arabinose is much higher than for UDP-xylose, and the enzyme is called as UDP-arabinose 4-epimerase and not UDP-xylose 4-epimerase.

UDP-xylose is synthesized from α-D-xylose-1-phosphate and UTP by xylose-1-phosphate uridyltransferase (EC 2.7.7.11), and also by UDP-arabinose 4-epimerase from UDP-arabinose. However, the most important pathway to form UDP-xylose is thought to be irreversible decarboxylation of UDP-glucuronate by the mediation of UDP-glucuronate carboxylyase.

UDP-xylose is found in microorganisms and animals as well as plants, but its concentration is very low in cells. This is because UDP-xylose regulates the UDP-sugar reaction system as a feedback inhibitor, but also because turnover of UDP-xylose in the synthesis of xylose-containing polysaccharides in cells is very high.

UDP-glucuronate carboxylase bound to the Golgi membrane of soybean was activated by solubilization of Golgi membranes with digitonin, suggesting that the enzyme is located inside of the membrane (Hayashi et al. 1988). The apparent affinity estimated in conversion of UDP-xylose to xyloglucan from UDP-glucuronate administered to Golgi membranes was higher than that of UDP-glucuronate carboxylyase for UDP-glucuronate (Dalessandro and Northcote 1977a,b). The affinity is at the same level as that of xyloglucan synthetase for UDP-xylose, indicating that the turnover of UDP-xylose formed by the Golgi apparatus is very high.

The activity of the soluble UDP-glucuronate carboxylyase in plants is higher in tissues of the secondary wall formation, suggesting that the soluble enzyme is mainly involved in the synthesis of xylan but not xyloglucan. The UDP-xylose formed passes directly to the Golgi membranes via a carrier and is used for xylan synthesis.

Xyloglucan and xylan are major hemicelluloses in primary and secondary walls in plants, respectively. It seems that the growth and differentiation of plants are regulated by the change of metabolism of UDP-xylose (Hayashi 1994).

It is likely that wood polysaccharides are synthesized by polymerization of random repeating units, but recent studies using specific hydrolases have indicated that xyloglucan and glucuronoxylan are synthesized with definite repeating units. The existence of such repeating units suggests that the primary formation of the corresponding oligosaccharide units occurs before polymerization of the hemicelluloses.

It has been suggested that the synthesis of such polysaccharides occurs via dolichol phosphate sugars and dolichol-2-phosphate origosaccharides as intermediates. In cellulose biosynthesis, Khan and Colvin (1961) reported the existence of a lipid intermediate, cellobiose polyprenol linked by phosphoric acid. However, further detailed studies showed that such an intermediate is not present in plants and *Acetobacter xylinum*. Later, however, an intermediate, α-1,5-arabinosyl-oligosaccharide isoprenol linked by phosphoric acid, has been observed and suggested to be an intermediate in the formation of an arabinosyl oligosaccharide residue in extensin (Hayashi and Maclachlan 1984).

4.2.2.8 The *myo*-Inositol Oxidation Pathway

Loewus et al. (1973) found that *myo*-inositol was converted to arabinose, xylose and galacturonic acid residues of plant cell walls. They demonstrated that glucose-6-phosphate is transferred by glucose-6-phosphate cycloaldolase (*myo*-inositol 1-phosphate synthase) into 1L-*myo*-inositol 1-phosphate, which is dephosphorylated to *myo*-inositol by a phosphatase. *myo*-Inositol is then converted to glucuronic acid by mediation of an oxygenase (Fig. 71). In addition, glucuronic acid was phosphorylated to glucuronic acid-1-phosphate by a glucuronokinase. The latter compound could be converted into UDP-glucuronic acid by UTP: α-D-glucuronate-1-phosphate uridylyltransferase (UDP glucuronate pyrophosphorylase).

This pathway provides an alternative route to the sugar nucleotide oxidation pathway for UDP-glucuronate formation from glucose-6-phosphate, and in studies with D-glucose labeled specifically with both ^{3}H and ^{14}C a tritium isotope effect was found at C_5, indicating that a major portion of the uronic acid and pentose residues of cell walls may be derived by way of a flux of carbon from glucose-6-phosphate along this pathway.

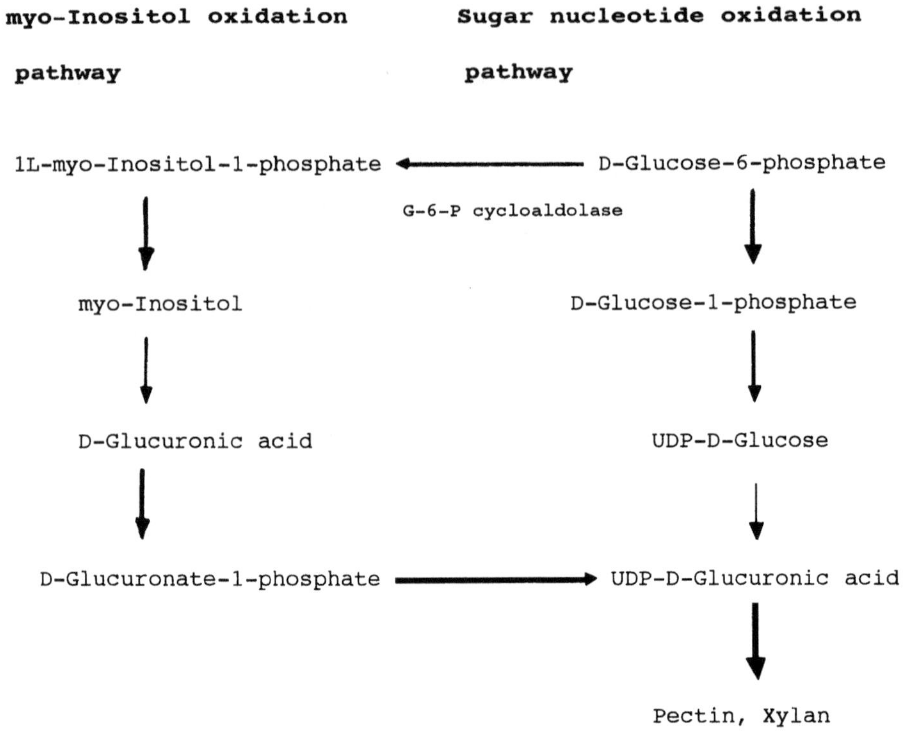

Fig. 71. *myo*-Inositol oxidation pathway leading to biosynthesis of pectin

In rapidly growing tissues, the activity of UDP-D-glucuronate pyrophosphorylase is much higher than that of UDP-glucose dehydrogenase, and in some tissues the dehydrogenase cannot be detected. In these tissues, the *myo*-inositol oxidation pathway may serve to provide uronic acid and pentose residues for cell wall synthesis from glucose. The precise flux down either the *myo*-inositol oxidation or the sugar nucleotide oxidation pathway will depend on such factors as the availability of *myo*-inositol from storage components, the availability of UDP-glucose from sucrose breakdown catalyzed by sucrose synthase, and the stage of development of the tissue in terms of cell wall synthesis (Smith 1993).

4.2.3 Biosynthesis of Cellulose

Cellulose is a linear polymer composed of β-1,4-glycosidic linkage of D-glucose, where the glucose residue is rotated approximately 180°. As a result, cellobiose is the structural repeating unit of the glucose chains in cellulose. X-ray diffraction patterns indicate the presence of hydrogen bonds between

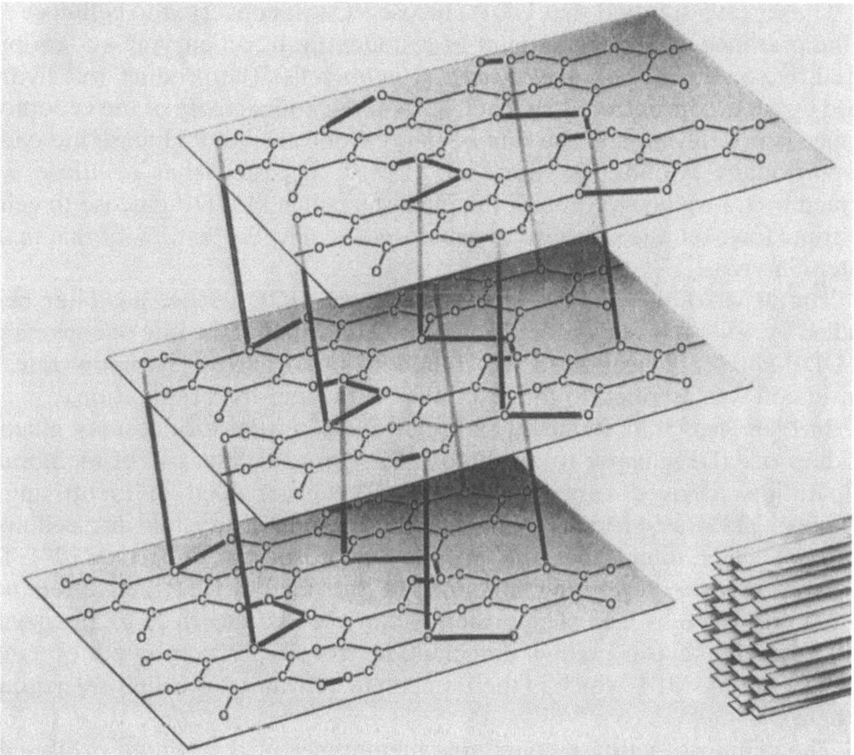

Fig. 72. A model structure of cellulose microfibrils. *Solid lines* Covalent bonds; *thick lines* hydrogen bond. (Albersheim 1975)

the ring oxygen of one glucose residue and the C_3 hydroxyl of an adjacent residue in the same chain and also between the C_2 and C_3 hydroxyl groups (intramolecular bonding) to afford a long stable molecule.

Bonds between cellulose chains (intermolecular bonding) occur between hydroxyl groups at C_6 and C_3 in adjacent chains (intermolecular bonding) to afford a microfibril composed of 30–80 cellulose molecules (Albersheim 1975; Fig. 72).

4.2.3.1 Microorganisms

In 1957 Greathouse (Greathouse 1959) found that cellulose-^{14}C was produced from D-glucose-1-^{14}C and ATP by mediation of a cell-free enzyme system isolated from *Acetobacter xylinum*. Ninety-six per cent of the label was found to be in the original position of one of the glucose residues of the cellulose. He suggested that the energy for this possible direct polymerization of the glucose or glucose phosphate to cellulose by the cell-free enzyme system is probably obtained from the added ATP.

Glaser (1958) found that UDP-glucose-^{14}C was converted to cellulose-^{14}C in the presence of a small amount of cellodextrin by an enzyme system prepared by sonification of *Acetobacter xylinum* cells. The product was hydrolyzed by superconcentrated HCl at 0 °C, and the radioactivity of the cellobiose formed was determined. The rate of the reaction for UDP-glucose increased proportionally within 0.5–7 µmol/ml. It was suggested that cellulose was formed by transglucosylation of the glucose residue of UDP-glucose to cellodextrin. However, the synthesis of cellulose was only less that 1% of that in the system in vivo.

The in vitro biosynthesis of cellulose from UDP-glucose has later been studied by several investigators. However, until recently the rate of conversion of UDP-glucose to cellulose was found to be very low. This low rate of conversion was ascribed to the instability of the enzyme preparation.

In 1986 Ross et al. found an extremely efficient transformation of glucose residues of UDP-glucose to a cellulosic (1–4)-β-glucan by use of membrane preparations derived from *A. xylinum*. They succeeded in identifying a complex GTP-Ca^{2+}-protein factor-mediated regulatory system for cellulose synthase. They found that the protein factor specifically activated GTP. In the presence of limiting amounts of the factor, GTP activation was further enhanced by Ca^{2+}. The protein factor was characterized as an enzyme that formed a cyclic guanyl dinucleotide from GTP composed of GMP residues (c-di-GMP), which is the immediate activator for cellulose synthase (Fig. 73).

The cellulose synthase-containing membranes of *A. xylinum* exhibited a phosphodiesterase-like activity that rapidly degraded the nucleotide activator into 5′-GMP. This activity, however, was strongly inhibited by Ca^{2+}. Based on these results they suggested that intercellular levels of the nucleotide activator, in conjunction with calcium ions, may regulate the rate of cellulose synthesis in vivo as shown in Fig. 74 (Ross et al. 1987). c-di-GMP is connected allosterically to the inactive cellulose synthase and converts the inactive enzyme to the active form, producing cellulose from UDP-glucose.

The cellulose synthase of *A. xylinum* could be solubilized and purified 350-fold by digitonin and trypsin treatments. The enzyme requires Mg^{2+} and c-di-GMP as cofactors, and the yield of cellulose synthesis at optimal conditions (30 °C and pH 7.5) attained 50% of that in vivo. K_m for UDPG was 0.125 mM, and UTP, UDP and UMP competitively inhibited the reaction.

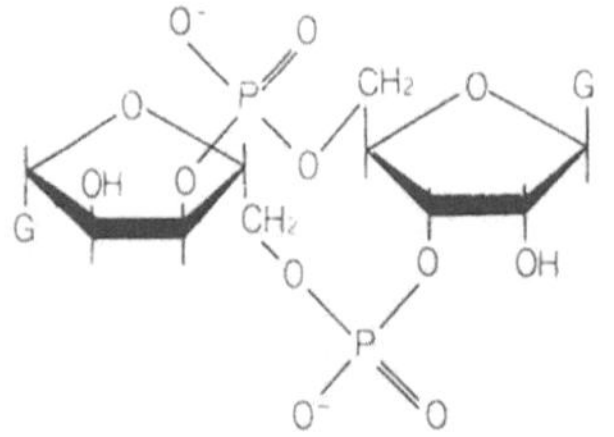

Fig. 73. Chemical structure of cyclic diguanylic acid (c-di-GMP)

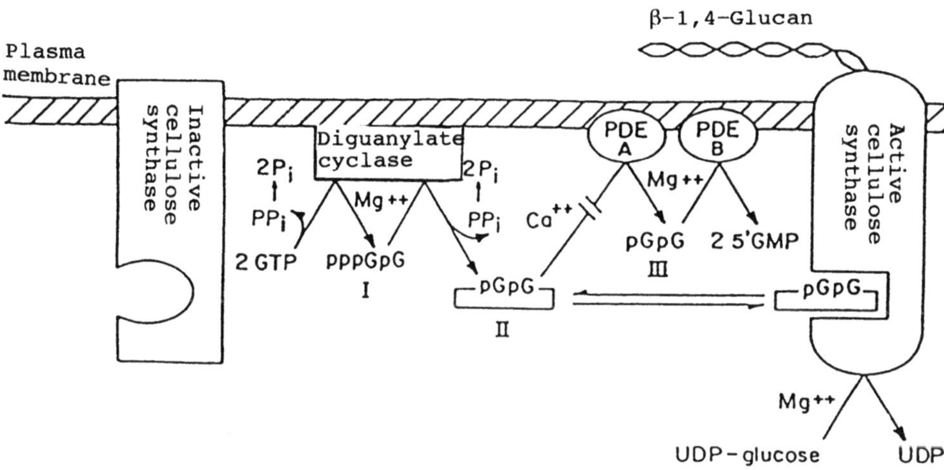

Fig. 74. Cellulose synthesis and its regulation by *Acetobacter xylinum*. P_i Orthophosphate; *pppGpG* diguanosine tetraphosphate; *GMP* guanosine monophosphate; PP_i pyrophosphate; ⌐pGpG¬: cyclic di-GMP; *PDE A* Ca^{2+} sensitive phosphodiesterase; *GTP* guanosine triphosphate; *pGpG* linear di-GMP; *PDE B* phosphodiesterase; *UDP* uridine diphosphate. (Ross et al. 1987)

Benziman and his coworkers (Wong et al. 1990) found that the structural gene (*bcsB*) of cellulose synthase of *A. xylinum* 1306-3 is composed of 2409 bp by which 90 kDa of protein is synthesized. The protein was subjected to processing with protease to afford 67- and 54-kDa proteins. *bcsB* was incorporated into *Escherichia coli*, and a large amount of cellulose synthase was formed. The antibody for the synthase was prepared, using purified cellulose synthase, and submitted to membrane proteins from cellulose synthesizing fungi and plants. The investigation showed that the antibody reacted with both the 90- and the 54-kDa proteins.

In addition, a *bcs* operon (9217 bp in length) comprised of the genes *bcsA, bcsB, bcsC* and *bcsD*, which encode proteins of molecular masses 84.4, 85.3, 141.0 and 17.3 kDa, respectively, was isolated by complementary experiments with noncellulose-synthesizing mutants (Volman et al. 1995; Fig. 75).

Additional complementation analyses indicated that all four *bcs* genes are required for maximal cellulose synthesis in vivo. In vitro cellulose synthase activity was found to be dependent both on *bcsA* and *bcsB*, and their respective gene products have been characterized as the substrate (UDPG)-binding and activator (c-di-GMP)-binding subunits of the synthase complex. The *bcsC* and *bcsD* genes provide functions essential for cellulose synthesis in vivo, and may be involved with microfibril crystallization or extrusion of microfibrils from TCs.

These four genes appear to be translationally coupled and transcribed as polycistronic mRNA with an initiation site 97 bp upstream of *bcsA*, the first gene of the operon.

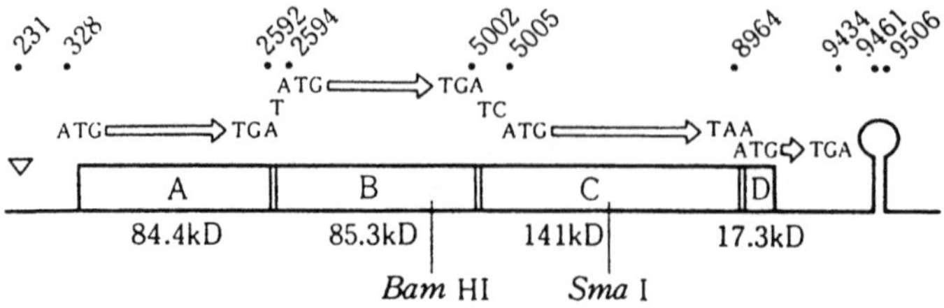

Fig. 75. Structure of the operon of cellulose synthase (*bcs*) from *A. xylinum*. The operon is 9217 bp in length, and is comprised of the genes *bcs*A, *bcs*B, *bcs*C, and *bcs*D, which encode proteins of molecular masses 84.4, 85.3, 141.0 and 17.3 kDa, respectively. The gene products by *bcs*A and *bcs*B have been characterized as the UDPG-binding and activator (c-di-GMP)-binding subunits of the cellulose synthase complex. *bcs*C and *bcs*D genes provide functions essential for microfibril crystallization or extrusion. *ATG* Initiation codon for translation; *TGA* and *TAA* termination codons for translation. (Volman et al. 1995)

Brown and his coworkers (Lin et al. 1990) demonstrated, on the other hand, that the 83-kDa polypeptide of purified cellulose synthase from *A. xylinum* ATCC 53582 is exclusively labeled with [β-^{32}P]5-azidouridine 5′-diphosphoglucose (UDP-glucose). Photoinsertion of [β-^{32}P]5-azido-UDP-glucose was stimulated by the cellulose synthase activator, bis-(3′–5′) cyclic diguanylic acid. Addition of increasing amounts of UDP-glucose prevented photolabeling of the 83-kDa polypeptide (Lin et al. 1990). These studies demonstrated that the 83-kDa polypeptide is the catalytic subunit of the cellulose synthase in *A. xylinum* strain ATCC 53582.

The gene for the catalytic subunit of the cellulose synthase was cloned (Saxena et al. 1990) using an oligonucleotide probe designed from the N-terminal amino acid sequence of the catalytic subunit of the cellulose synthase purified from trypsin-treated membranes of the *A. xylinum*. The gene is composed of an open reading frame of 2169 base pairs coding for a polypeptide of 80 kDa. Fifteen amino acids in the N-terminal region of the amino acid sequence match the N-terminal amino acid sequence obtained for the 83-kDa polypeptide, confirming that the cloned DNA sequence codes for the catalytic subunit of cellulose synthase that transfers glucose from UDP-glucose to the growing glucan chain.

In addition, an open reading frame coding for a polypeptide of 86 kDa was identified by DNA sequencing of the region downstream of the cellulose synthase catalytic subunit gene of the *A. xylinum*. The deduced amino acid sequence of this polypeptide matches from position 27 to 40 with the N-terminal amino acid sequence determined for a 93-kDa polypeptide that is tightly associated with the cellulose synthase catalytic subunit.

Sequence analysis of the cellulose-synthesizing (*acs*) operon showed the presence of three genes (*acsAB*, *acsC* and *acsD*), in which the first gene (*acsAB*) codes for a polypeptide with a molecular mass of 168 kDa, and which was identified as the cellulose synthase. A single base change in the DNA

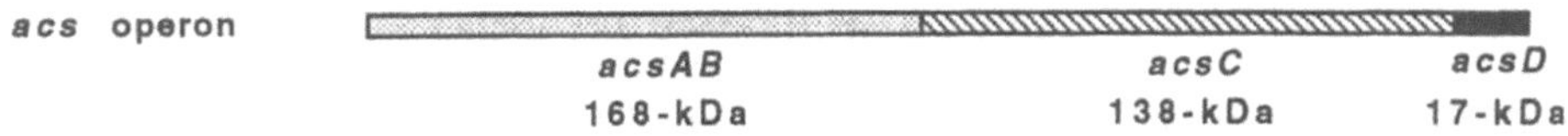

Fig. 76. *acs* operon of cellulose synthase from *A. xylimnum* ATCC 53582 showing organization of the *acs*AB, *acs*C, and *acs*D genes and the predicted sizes of polypeptides encoded by these genes. (Saxena et al. 1994)

sequence of this gene resulted in a frameshift and synthesis of a larger protein (Fig. 76).

Mutant analysis showed that while the *acsAB* and *acsC* genes were essential for cellulose production in vivo, the *acsD* mutant produced reduced amounts of two cellulose allomorphs (cellulose I and cellulose II), suggesting that the *acsD* gene is involved in cellulose crystallization. Further studies showed that the 83- and 93-kDa polypeptide genes in the strain ATCC 53582 were similar to the first two genes (*bcsA* and *bcsB*) of the *bcs* operon that was first isolated by Benziman's group from *A. xylinum* 1306-3, and in which the second gene (*bcsB*) was identified as the cellulose synthase catalytic subunit gene.

4.2.3.2 Plants

In 1964, Barber et al. reported that β-1,4-glucan is synthesized from GDP-glucose by the mediation of a membrane enzyme prepared from mung bean seedlings. Later, Carpita and Delmer (1981) investigated in detail the synthetic pathway of cellulose using cotton fiber cells, and concluded that UDP-glucose and not GDP-glucose, was the true sugar donor for cellulose synthesis in both plants and microorganisms.

An early investigation using a membrane preparation from cotton and UDP-glucose showed that glucan products synthesized in vitro were composed predominantly of β-1,3-glucan with less than 5% of β-1,4-glucan identified by methylation analysis (Heiniger and Delmer 1977). Delmer (1987) improved the isolation methods of plasma membranes from cotton fibers and found that mixtures of cellobiose, Mg^{2+}, and Ca^{2+} may regulate the glucan synthase activity in vitro. When concentrated (mM level) UDP-glucose was used with membrane enzymes, β-1,3-glucan synthesis was activated, while low concentrations of UDP-glucose, (μM level) and Mg^{2+} (10 mM) activated β-1,4-glucan synthesis. Various trials using detached fibers, etc. showed the formation of less than 1% of β-1,4-glucan products after treatment with the AN reagent (acetic–nitric acid; Updegraff 1969, Bacic and Delmer 1981).

Many methods such as acid hydrolysis, acetolysis, specific enzymic digestion, and linkage analysis by methylation and periodate oxidation have been used for characterization of enzymatic glucan products. As a criterion for β-1,4-glucan, insolubility of the product in concentrated alkali (e.g. 24% KOH) has been used, but β-1,3-glucan synthesized in vitro also remained as the alkali-insoluble fraction (Hayashi et al. 1987).

Callaghan and Benziman (1984) reported that cellulose was synthesized from UDP-glucose by the mediation of membrane fractions from mung bean seedlings. However, analysis of hydrolyzates of the methylated product showed that the product was a β-1,3-linked polymer (Hayashi 1994).

Brown and his coworkers (Lin et al. 1985, Bureau and Brown 1987) used AN reagent for the selection of the crystalline cellulose that remains after treatment with this reagent. X-ray and electron diffraction analyses were used to identify crystalline cellulose. Purified cellulase conjugated to colloidal gold was also used as a probe to identify cellulose by electron microscopy. Brown et al. (Li and Brown, 1993, Li et al. 1993) have demonstrated that an ANIP (AN-reagent-insoluble product) is synthesized in vitro by cell-free membrane fractions from cotton fiber cells. They tested various effectors for the cellulose synthesis, and demonstrated the optimal reaction conditions for the synthesis of β-1,3-glucan and β-1,4-glucan by the combination of the effectors, such as cellobiose, cyclic nucleotides, Mg^{2+}, Ca^{2+} and digitonin.

The ethanol-insoluble fraction from this reaction mixture contained a β-1,4-glucan and a β-1,3-glucan in an approximate ratio of 25:69. About 16% of the β-1,4-glucan was resistant to the AN reagent. The x-ray diffraction pattern of the AN-reagent-treated product resembled that of cellulose II. Methylation analysis, enzymic hydrolysis, and estimation of DP showed that the ANIP was a β-1,4-linked polymer with an average DP of 141. In addition, autoradiography proved that the product was really synthesized in vitro (Okuda et al. 1993).

Their further studies showed that β-1,3-glucan synthesis is enhanced up to 31-fold by cellobiose with a K_a of 1.16mM, whereas cellulose synthesis is increased 12-fold by a combination of cellobiose (K_a = 3.26mM) and cyclic-3′:5′-GMP (K_a = 100μM). Mg^{2+} (K_a = 0.89mM) is essential for cellulose synthase. A low concentration of Ca^{2+} (K_a = 90μM) was also required for cellulose synthesis. The common components in the reaction mixture required by both enzymes are cellobiose, Ca^{2+}, and digitonin (Table 14).

The optimal pH for cellulose synthase (7.6–8.0) is slightly higher than that for β-1,3-glucan synthase (7.2–7.6). The K_m for UDP-glucose for cotton cellulose synthase is 0.40mM, while the K_m for UDP-glucose for β-1,3-glucan synthase is 0.43mM (Li and Brown 1993).

Delmer and coworkers (Amor et al. 1991) identified polypeptides of 84 and 48kDa from a membrane protein of cotton fiber cells by photolabeling with ^{32}P-c-di-GMP as activator of the cellulose synthase of *A. xylinum*. Both polypeptides bind c-di-GMP with a high affinity and specificity and show antigenic relatedness to the bacterial synthase. The authors suggested that the 48kDa polypeptide was derived by proteolytic cleavage of the 84-kDa polypeptide.

Delmer and Amor (1995) have recently found that more than 50% of total sucrose synthase of cotton is tightly associated with the cell membrane together with cellulose synthase. In the presence of Ca^{2+} and cellobiose, the synthesis of β-1,4-glucan from sucrose by the cell membrane is favored over the synthe-

Table 14. Effect of different combinations of effectors on in vitro β-glucan synthesis by cotton fiber glucan synthases. (Li and Brown 1993)

Effectors	ANIP	EIP
Mg^{2+}, Ca^{2+}, CB, c-3′:5′-GMP digitonin (complete condition for favoring the synthesis of β-1,4-glucan)	2.4	nmol min^{-1} mg^{-1} [a] 24.1
$-Mg^{2+}$	0.1	– [b]
$-Ca^{2+}$	0.2	–
–CB	0.4	–
–c-3′:5′-GMP	0.5	26.7
–Digitonin	0.2	–
$-Mg^{2+}$, -c-3′:5′-GMP	–	28.4
$-Ca^{2+}$, -c-3′:5′-GMP	–	2.0
–CB, -c-3′:5′-GMP	0.2	0.9
$-Mg^{2+}$, $-Ca^{2+}$, -c-3′:5′-GMP	n.d. [c]	Trace
No effectors	n.d.	Trace

ANIP, AN-reagent-insoluble product; EIP, ethanol-insoluble product.
[a] nmol min^{-1} mg^{-1}, nmol of [^{14}C] glucose incorporated min^{-1} mg^{-1} of protein.
[b] Not tested.
[c] n.d., Not detectable.

sis of callose in the absence of Ca^{2+}. The fact that either callose or cellulose can be synthesized from sucrose suggested that both glucan synthases might be in the same complex. In addition, sequencing of several tryptic fragments from the UDP-binding polypeptide (84 kDa) showed it to be a form of sucrose synthase, but not cellulose synthase (see Sect. 4.2.2.2).

The coupled reaction from sucrose has been demonstrated only with freshly detached cotton fibers permeabilized with digitonin, but not in membrane preparations, suggesting that preservation of some higher level of organization and regulatory factors is critical for obtaining good activity.

Brown and his coworkers (Li et al. 1993) identified a 37-kDa polypeptide as the best candidate of the UDP-glucose-binding polypeptides for the catalytic subunit of cotton cellulose synthase by using differential product entrapment and photolabeling under specifying conditions. This polypeptide was enriched by entrapment under conditions favoring β-1,4-glucan synthesis. It was Mg^{2+} dependent and sensitive to unlabeled UDP-glucose.

A recent review paper by Brown et al. (1996) reported that product entrapment and photoaffinity labeling have failed to definitely identify cellulose synthase from plants. Polypeptides in the range of 52–57 kDa, previously isolated by the methods from higher plants, were not catalytic subunits for callose or cellulose synthase but annexins, an ATPase subunit, a membrane-integrated protein and a membrane-associated sucrose synthase. Thus, the field of cellulose biosynthesis in higher plants is still open with regard to the identification of the catalytic subunits for cellulose synthase, and the genes that control assembly.

Saxena and Brown (1996) have recently found expressed sequence tags (ESTs) from *Arabidopsis thaliana Brassica campestris*, and *Oryza sativa* that exhibit a limited sequence similarity to the *A. xylinum* cellulose synthase. These ESTs contain some of the conserved residues identified in the processive β-glycosyltransferases. All the conserved residues in the derived polypeptide, which appears to be part of a putative cellulose synthase, have been identified. Sequence comparison of proteins with known function and several unidentified proteins have the "D,D,D35Q(R,Q)XRW" motif which is considered a strong predictor of β-glycosyltransferases, such as cellulose and chitin synthases. Using site-directed mutagenesis and their replacement by another amino acid led to a loss of cellulose synthase activity in *A. xylinum*, indicating their essential role as catalytic residues.

In contrast, five cDNA clones isolated by John and Crow (1992) from cotton fiber cells were preferentially expressed in the fiber. One of the cDNA clones, *pCKE6*, corresponded to an abundant mRNA in the fiber. Immunoprecipitation of in vitro translation products and Western blot analysis of fiber proteins showed two polypeptides in the range of 30–32 kDa as the products of E6 mRNA. Concentrations of *E6* mRNA and protein were highest during the late stages of primary cell wall and the early stages of secondary cell wall synthesis. *E6* or a homologous gene(s) is conserved in several members of the Malvaceae as well as in one other fiber-producing plant, kapok, but is not found in several other plants examined nor in *Acetobacter xylinum*. A genomic clone corresponding to *pCKE6* was isolated, and the promoter element of the *E6* gene was shown to direct the expression of a carrot extensin mRNA in a tissue-specific and developmentally regulated fashion in transgenic cotton plants.

Meier et al. (1981) suggested earlier that β-1,3-glucan of cotton fiber cell walls is an intermediate of cellulose, and that β-1,4-glucan could be synthesized by the transfer reaction of glucose in hydrolysis of β-1,3-glucan with β-1,3-glucanase. However, such a reaction could not be confirmed.

The biochemistry and molecular biology of cellulose biosynthesis in plants have suggested an intimate relationship between cellulose synthase and callose synthase. The evidence, however, is inconclusive, and separation and characterization of these enzymes have not yet been accomplished. Further characterization of the cellulose synthase gene, and the elucidation of its expression mechanism would clarify the mechanism of cellulose biosynthesis.

4.2.4 Cellulose Microfibrils

4.2.4.1 Structure of Microfibrils

Cellulose, a β-1,4-glucan, is widely distributed in the biological world, and produced not only by plants but also by fungi, bacteria, myxomycetes, and tunicates. In cell walls, many cellulose molecules assemble to give bundles of microfibrils. Cross sections of microfibrils are known to be squares in algae such as valonia. The length of microfibrils is not clear because of difficulties in observing both ends of microfibrils under the electron microscope.

Gardner and Blackwell (1974) showed by X-ray analysis of native cellulose that the intramolecular hydrogen bond between O_3–O_5' of cellobiose residues is 0.275 nm and the angle of glucosidic linkage (C_1–O–C_4) is 114.8°. They proposed that the cellulose chain at the center of the unit cell and those of the four corners are oriented in the same direction (parallel). Two intrahydrogen bonds, O_3–O_5' (0.275 nm) and O_6–O_2, and one interhydrogen bond, O_3–O_6, are present in the molecular chain, and the angle of glucosidic linkage is 114.8° (Fig. 77a,b,c).

In the same year, Sarko and Muggli (1974) proposed that all cellulose chains within one microfibril were parallel, and suggested the triclinic system for the unit cell of native cellulose with the following parameters: a = 0.941 nm, b = 0.815 nm, c = 1.034 nm (chain axis), $\alpha = 90°$, $\beta = 57.5°$, $\gamma = 96.2°$.

Recent studies have shown that the parameters of the unit cell vary significantly by species. IR absorption analysis of cellulose has indicated that crystalline structures of bacterial and algal celluloses are different from those of cotton, ramie and tunicates, and that native cellulose I is divided into two groups, cellulose IA and IB (Gröbe 1989). Okano et al. (1989), and Okano and Koyanagi (1986) found that the variation of the unit cell by X-ray analysis is related to the evolution and phylogeny of plants (Table 15). They found that the γ-value of the unit cell decreases in the cellulose of the more advanced species.

Recent solid-state NMR analyses of celluloses showed that native cellulose I, classified IA and IB by IR analysis, is a mixture of Iα and Iβ, and the ratio of the two types is different among species (Atalla and Vander Hart 1984, Horii et al. 1987a,b). Bacterial and algal celluloses are mostly the α type, but cotton and ramie cellulose contains larger amounts of the β type.

Sugiyama and his coworkers (Sugiyama et al. 1991, Sugiyama 1994) have proposed a triclinic system as a unit cell with one cellulose chain for Iα and a monoclinic system with two cellulose chains for Iβ (Fig. 78). The occurrence of two crystalline phases (Iα and Iβ) with different ratios in different species explains minute differences in spacing of lattice planes and γ-values in plant species so far examined. Recently, Wada et al. (1995) have reported that native celluloses are of two types, one older, bacterial type rich in Iα and one later, cotton–ramie type, where Iβ dominates. Celluloses in advanced, vascular plants all belong to the latter category.

The major constituent of plant cell walls is cellulose, which constitutes the skeleton substance of the microfibril, but in some plants mannan and xylan form microfibrils (Preston 1974). Microfibrils of some algae, such as Codiaceae and Dasycladaceae are composed of mannan, while those of Bryopsidaceae, Caulerpaceae, Udotaceae, and Dichotomosiphonaceae (Siphonales) are composed of xylan.

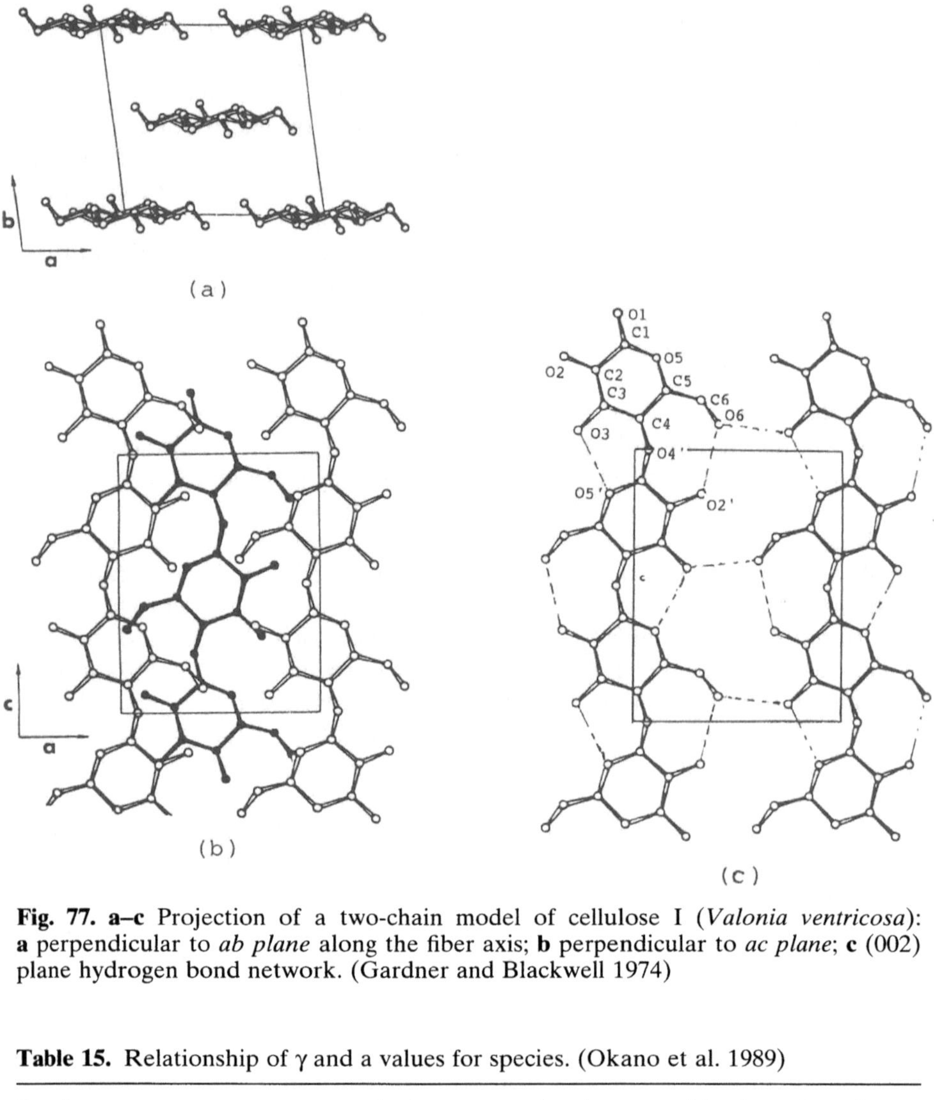

Fig. 77. a–c Projection of a two-chain model of cellulose I (*Valonia ventricosa*): **a** perpendicular to *ab plane* along the fiber axis; **b** perpendicular to *ac plane*; **c** (002) plane hydrogen bond network. (Gardner and Blackwell 1974)

Table 15. Relationship of γ and a values for species. (Okano et al. 1989)

Species	γ (deg)	a (nm)	Number of specimens
Bacterial cellulose	98.15	0.791	5
Valonia	97.04	0.794	6
Nostoc	96.97	0.796	1
Chaetomorpha sp.	96.45	0.796	1
Conifers	97.70	0.798	16
Hardwoods	96.36	0.802	61
Cotton	96.53	0.787	2
Ramie-1	96.40	0.785	4
Ramie-2	95.40	0.778	5
Trachycarpus fortunei	95.48	0.801	1
Bamboo	94.70	0.807	1

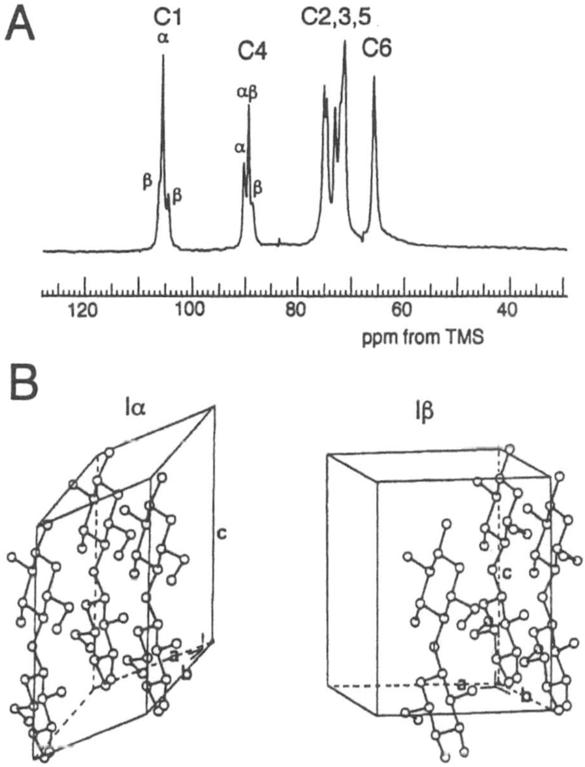

Fig. 78. A,B Unit cell model corresponding to 1α and 1β crystalline phases of cellulose I of *Chaetomorpha* sp. obtained by X-ray diffractometry. **A** CP-MAS spectrum of the cellulose from *Chaetomorpha* sp. **B** Unit cell models corresponding to 1α and 1β crystalline phases. (Sugiyama 1994)

4.2.4.2 Formation of Microfibrils

It has been suggested that UDP-glucose, a precursor of cellulose, is transported from the cytoplasm to the plasma membrane, where cellulose is synthesized by enzyme complexes (TCs) located on the membrane (Brown and Montezinos 1976).

TCs move on the cell membrane, which is composed of double layers of lipids, and the synthesized cellulose microfibrils are deposited in their tracks.

The unicellular alga *Pleurochrysis scherffelii* has a cell wall consisting of characteristic plates composed of cellulose microfibrils oriented in both radial and concentric directions. Brown et al. (1970) found that in this organism the cellulosic plates are synthesized inside the Golgi cisternae, which are transported to the cell wall, where the plates are released. This is one of the few examples known where the synthesis of cellulose takes place in the Golgi apparatus.

TCs can only be observed by the freeze–fracture method. The occurrence of TCs on the cell membrane has been observed in red, brown and green algae, Charophytes, Bryophytes, Pteridophytes and higher plants. They can be classified into two kinds, rosettes and linear types, based on their arrangements and shape (Fig. 79).

There are several types of rosette TCs, namely, individually distributed globules (see below) and rosettes, rectilinearly distributed rosettes, and hexagonally distributed rosettes. Since the globules appear on the E-fracture face, but rosettes are found on the complementary P-fracture face, the rosettes and globules are supposed to constitute TCs as one entity. The rosette-type TCs occur in higher plants, mosses, ferns, and freshwater green algae, Charophytes etc. Single rosettes are observed in land plants and Charophytes, while hexagonally distributed rosettes are found in green algae (Itoh 1989). Linear TCs are distributed in many saltwater algae such as *Chaetomorpha* and *Cladophora* (Fig. 80). These results suggest that the rosette-type TCs, which produce an Iβ-type cellulose, are distributed in evolutionarily more developed plants.

It has been shown recently that the relative amounts of Iα and Iβ crystalline phases in native celluloses are correlated with the types of terminal complexes (TCs): advanced plants which bear rosette-type TCs produce the

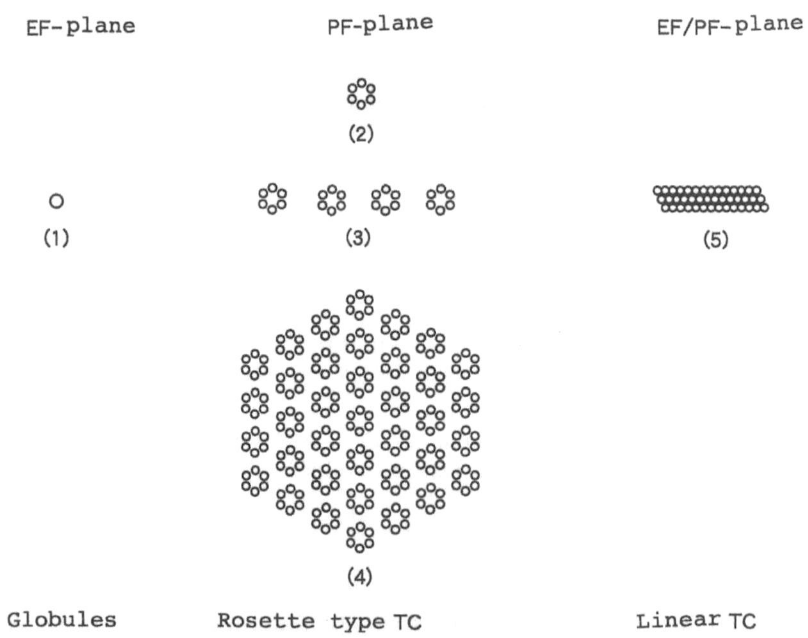

Fig. 79. Type of terminal complexes (TCs) in plants. *1* Globule; *2*, *3*, *4* rosette TCs; *5* linear TCs. In rosette TCs single (*2*), rectilinear (*3*), and hexagonal (*4*) distribution patterns which depend on species occur. (Itoh 1994a)

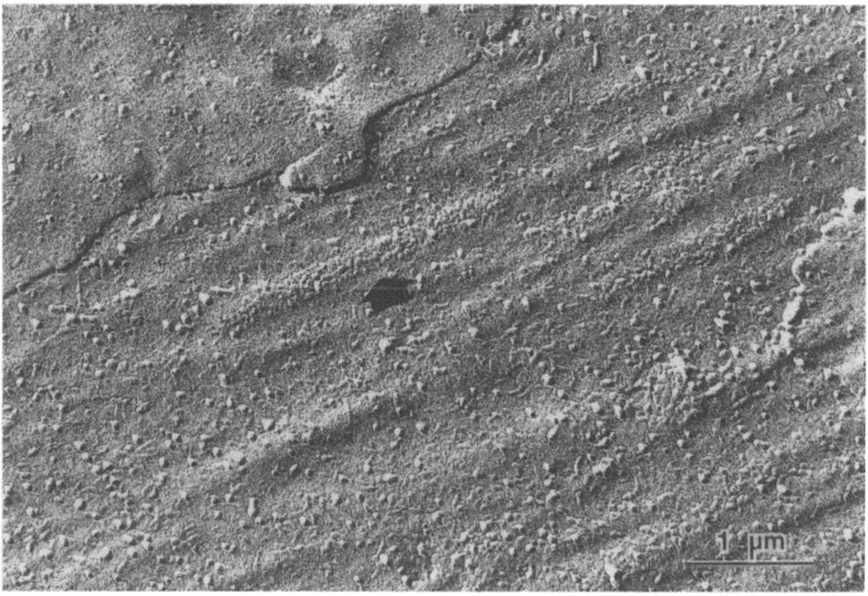

Fig. 80. Linear TCs of *Boodlea composita* on EF face. *Arrow* shows a terminal complexes (TCs). (Courtesy of Dr. T. Itoh)

cotton–ramie type of cellulose, whereas lower-order plants which bear the linear-type TCs produce algal–bacterial-type cellulose. A phylogenetic boundary between the two types exists in the chlorophyta (Wada et al. 1995).

It has been suggested that microtubules are connected with TCs in cell membranes and play a role in guiding TCs in regular deposition of microfibrils. However, recent investigations have shown that TCs are not directly connected with microtubules but distributed between two microtubules. Therefore, TCs held between microtubules cannot move outside, but move in the directions of the microtubules (Giddings et al. 1980; Fig. 81). The connection of the microtubules to the cell membrane is essential, and such a connection was recently confirmed in *Oocystis*. However, the regulation mechanism of the orientation of the microtubules is still unclear.

It was observed that when certain substances are added at the stage of cellulose crystallization, the crystallization and the formation of normal microfibrils are inhibited. For example, *Acetobacter xylinum* cultured in a medium containing fluorescent brightening agents such as FBA-28 or Congo red did not produce crystalline microfibrils but produced a glucan band composed of separated glucan chains (Fig. 82). This suggests that in the formation of bacterial microfibrils, crystallization follows polymerization of cellulose chains. The fluorescent brightening agents are hydrogen-bonded to the newly synthesized glucan chains, leaving no sites along the chains where they can form hydrogen bonds with one another and crystallize.

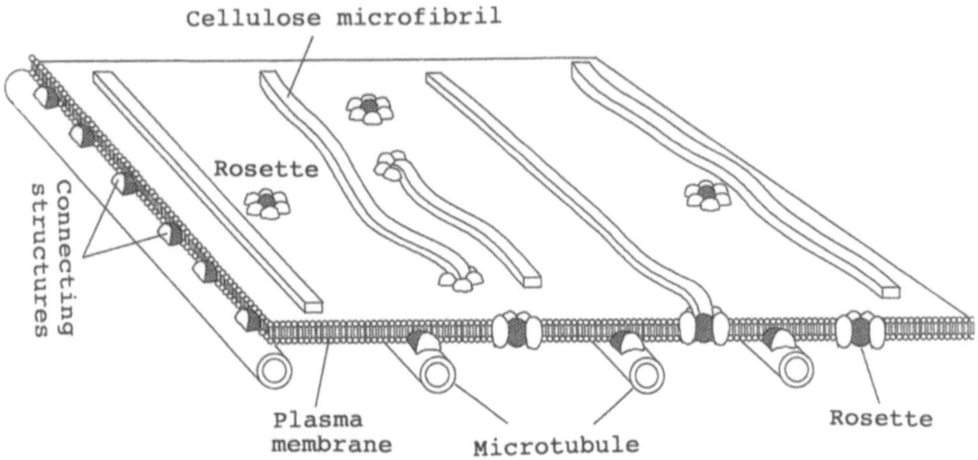

Fig. 81. Control of moving direction of cellulose-synthase complexes on cell membrane by microtubules. (Itoh 1994)

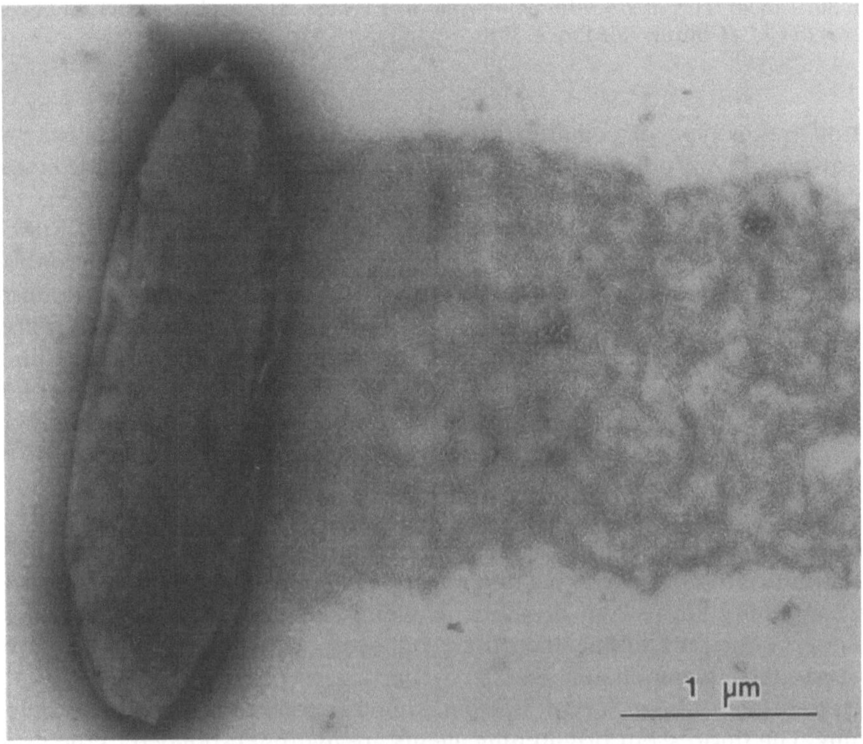

Fig. 82. A glucan band composed of separate glucan chains formed by *A. xylinum* grown in the medium containg Congo red (0.01 mg/ml). (Courtesy of Dr. T. Itoh)

4.2.4.3 Cell Growth and Microfibril Orientation

Growth of plant cells occurs in a certain direction. The length of mature wood cells, cotton and ramie fibers are 100 times, 5000 times, and 25 000 times longer than their diameters, respectively. Since turgor pressure acts equally in all directions within cells, cell growth in a particular direction must be forced by the binding power of cellulose microfibrils (Brett and Waldron 1990). Without reinforcement by laterally winding microfibrils, cylindrical cells might expand spherically.

Cell walls with laterally winding microfibrils occur in cotton fibers, internodal cells of Charophytes, and parenchyma. Epidermal cells, parenchyma cells near the epidermis, collenchyma cells, and filamentous algal walls contain microfibrils oriented in different directions. Such cells are naturally limited in growth, and function to strengthen thicked cell walls. The mechanism of selective growth of polylamellated walls built up of microfibrils with different directions is not known. In the growing apex of root hairs, microfibrils are deposited randomly only at the apex. In other parts of the root, the outside of the cell walls contains randomly oriented microfibrils while in the inside of the walls the microfibrils are oriented parallel with the longitudinal cell axis. Cell growth without lateral reinforcement by microfibrils is limited only to the terminal growth cells.

In round cells growing in the radial direction, initially formed cell walls are composed of randomly oriented microfibrils, after which lamellae of microfibrils oriented in several directions are deposited. Tissues of *valonia* and related species are composed of round unicellular cells. These cells are covered with multilamellar walls that are stronger than walls composed of microfibrils oriented in a single direction.

4.2.5 Biosynthesis of β-1,3-Glucan

β-1,3-Glucan (Fig. 83) is synthesized by the mediation of enzyme prepared from cell membranes. However, β-1,3-glucan is not a cell wall constituent of normal wood. Only compression wood contains small amounts of β-1,3-glucan (laricinan). Another β-1,3-glucan called callose occurs in the secondary phloem and is formed in response to wounding of cell walls and cell membranes or to plasmolysis and chemical and physical stress. Cell walls regenerated from

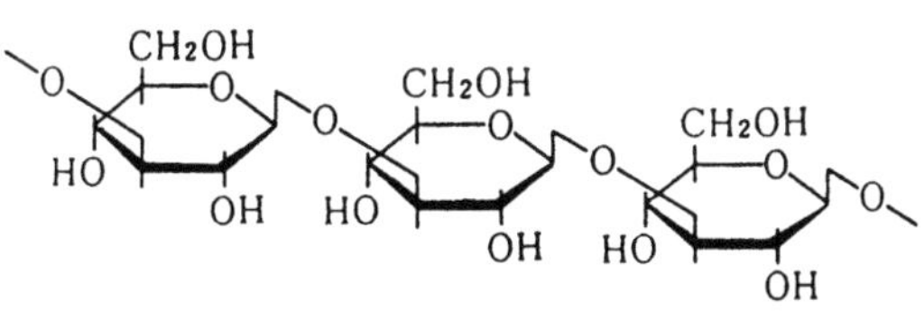

Fig. 83. Chemical structure of β-1,3-glucan

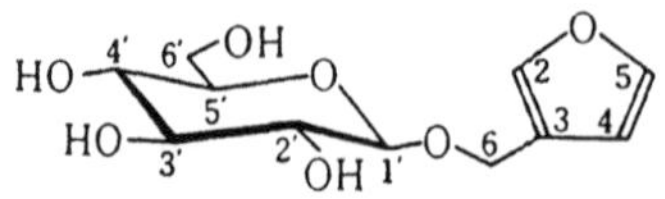

Fig. 84. Chemical structure of β-furfuryl-β-glucoside

protoplasts contain larger amounts of β-1,3-glucan than cellulose in the earlier stages of the regeneration.

β-1,3-Glucan synthase is activated by addition of Ca^{2+} and cellobiose. Except for cellobiose, the enzyme has its highest affinity for laminaribiose (Glcβ1-3Glc), suggesting that laminaribiose functions as an acceptor in the synthesis of β-1,3-glucan. In addition, β-furfuryl-β-glucoside (Fig. 84) was isolated from higher plants as an endogenous activator (Ohana et al. 1991).

Electronmicroscopy of partially purified β-1,3-glucan synthase has shown that the enzyme is a doughnut-like complex of 20-30 nm in diameter, which is similar to cellulose synthase complexes on cell membranes. It has therefore been suggested that the conformation of cellulose synthase changes to give β-1,3-glucan synthase in vitro (Hayashi et al. 1987).

Delmer et al. (1991) identified a 52-kDa polypeptide from cotton fibers as the catalytic subunit of callose synthase, by a combination of glycerol gradient centrifugation and direct photolabeling with [^{32}P]UDP-glucose.

Brown and his coworkers (Li et al. 1993) also identified 32-, 52-, and 66-kDa polypeptides from cotton fibers by using differential entrapment and photolabeling under favorable conditions for β-1,3-glucan synthesis. These were photolabeled under conditions where Ca^{2+} was present and Mg^{2+} absent. The 52-kDa polypeptide was identified as the most likely candidate for the catalytic subunit of β-1,3-glucan synthase. Because it is the most abundant protein in the entrapment fraction, it is photolabeled to the greatest degree among these three bands (32, 52, 66 kDa), under the above-mentioned conditions. The photolabeling of the 52-kDa polypeptide was depressed by Mg^{2+}. The synthesis of β-1,3-glucan is not inhibited by Mg^{2+}. Frost et al. (1990), on the other hand, reported that a 57-kDa polypeptide of red beet is the most likely candidate for the catalytic subunit of β-1,3-glucan synthase, based on its enrichment during product entrapment, its pH optimum for labeling, and its requirement for effectors.

Dhugga and Ray (1991a,b) also showed that a 55-kDa polypeptide was associated with β-1,3-glucan synthase activity from pea. They used a combination of glycerol gradient centrifugation, product entrapment, and immunological techniques. A 31-kDa polypeptide was further identified as the catalytic subunit of β-1,3-glucan synthase from *Glycine max* (Fink et al. 1990) and from *Lolium multiflorum* (Meikle et al. 1991).

It seems that a 57-kDa polypeptide from red beet and a 55-kDa polypeptide from pea are probably analogous to the 52-kDa polypeptide from cotton fibers. These three polypeptides are probably the catalytic subunit in callose biosynthesis (Li et al. 1993). However, recent investigations (Brown et al. 1996, Delmer and Amor 1995) have suggested that the polypeptides in the range of 52–57 kDa, isolated on the basis of binding activities of UDP-glucose, are not involved in cellulose or callose synthesis. These proteins were possibly isolated together with the cellulose or callose synthase activity during membrane solubilization, but were never originally associated with the cellulose-synthesizing complex in situ (see Sect. 4.2.3.2).

4.2.6 Biosynthesis of Xyloglucan

Xyloglucan is a major hemicellulose in primary cell walls of higher plants. The basic structure of the polymer is characterized by a β-D-(1→4)-glucan backbone variously substituted (up to 75%) with xylosyl, galactosyl, and fucosyl residues (Fig. 85).

Xyloglucan is synthesized from UDP-glucose, UDP-xylose UDP-galactose and GDP-fucose by the mediation of multienzyme complexes. Xyloglucan synthases have been studied with Golgi membrane preparations from pea seedlings (Ray 1980, Camirand and Maclachlan 1986, Gordon and Maclachlan 1989) and suspension cultured soybeans (Hayashi and Matsuda 1981a,b). Enzymes from both systems utilize UDP-glucose for synthesis of the glucan backbone and UDP-xylose for addition of the xylose side chains. These studies also suggested a precise biosynthetic mechanism, in which glucose and xylose are added to existing xyloglucan chains concomitantly. The presence of UDP-xylose is essential for significant elongation of the glucan backbone.

The xyloglucan synthase complex catalyzes alternative transfer of glucose and xylose to form xyloglucan polymers. In hydrolysis of the biosynthesized product with cellulase, oligosaccharides composed of nine monosaccharides, (glucose/xylose/fucose/galactose, 4:3:1:1), seven monosaccharides, (glucose/xylose, 4:3), and five monosaccharides (glucose/xylose, 3:2) were obtained, respectively.

It has been suggested that glucan synthase I (GS I) may make glucan chains similar to those in cellulose, while xyloglucan glucosyltransferase (XGT, EC 2.4.1.168) may produce the glucan backbone for xyloglucan (Ray 1980, Dhugga et al. 1991). GS I and XGT activities may originate from the

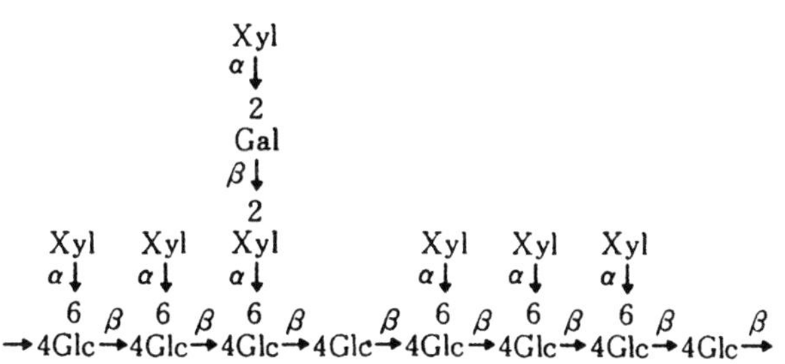

Fig. 85. Basic chemical structure of xyloglucan

same enzyme complex, although they exhibit differences in K_m and Mg^{2+} or Mn^{2+} stimulation.

Xyloglucan xylosyltransferase (XXT, EC 2.4.1.169) uses UDP-xylose as a substrate to add xylosyl side chain residues to the C_6 position of the glucan backbone (Hayashi and Matsuda 1981a, Hayashi et al. 1984), producing 1,6-linked xylosyl and 4,6-linked glucosyl residues. The transfer of xylose to xyloglucan is increased by the addition of UDP-glucose (Ka = 24 µM), which indicates that xylose uptake is coupled with glucose uptake, but not with the transfer of xylose residues previously synthesized β-D-1,4-glucan residues (Hayashi 1989; Fig. 86). In addition, the transfer of glucose is activated by the addition of UDP-xylose (Ka = 10 µM), although high concentrations of UDP-xylose inhibit the glucose transfer (Hayashi and Matsuda 1981b).

Xyloglucan forms a network with cellulose in growing cell walls. It has been suggested that cells regulate their growth by controlling hydrolysis of xyloglucan which binds with high specificity and affinity to cellulose microfibrils. In vitro reconstitution studies on xyloglucan–cellulose binding indicates that pea xyloglucan can bind to cellulose even in the presence of $(1 \rightarrow 2)$-, $(1 \rightarrow 3)$-, $(1 \rightarrow 3, 1 \rightarrow 4)$- and $(1 \rightarrow 6)$-β-D-glucans, arabinogalactan and pectins (Hayashi et al. 1987).

The specificity of the interaction has been attributed to the chemical similarity between the β-1,4-glucan backbone of xyloglucan and the β-1,4-glucan chains in cellulose. Hayashi (1989) suggested that xyloglucan excreted from cells is immediately connected to cellulose and the cellulose microfibrils are linked together by xyloglucan to form large complexes of cell wall polysaccharides (Fig. 87).

Recently, a macromolecular complex composed of xyloglucan and cellulose was isolated from elongating regions of etiolated pea (*Pisum sativum* L. var Alaska) seedlings, and binding of a xyloglucan-specific antibody was inves-

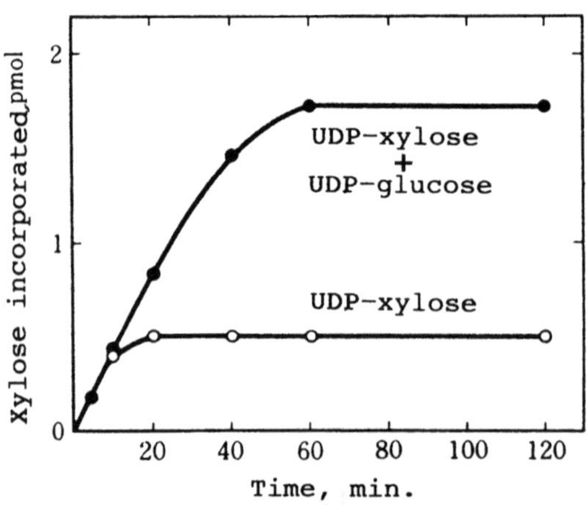

Fig. 86. Time course of xyloglucan synthesis. Incorporation of xylose in xyloglucan is increased by addition of UDP-glucose (*line with solid dots*). (Hayashi and Matsuda 1981b)

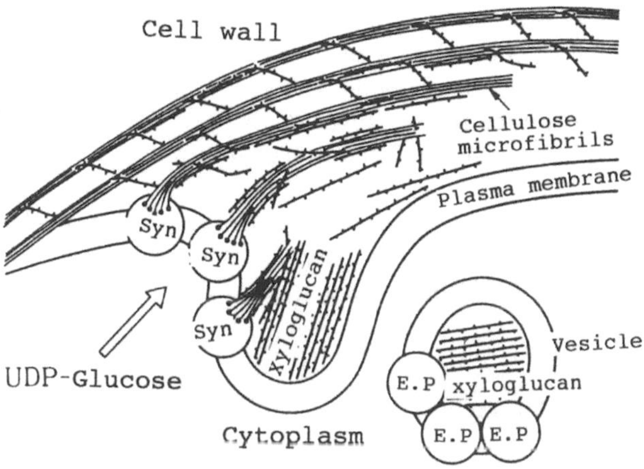

Fig. 87. A model showing the formation of bridges between xyloglucan and cellulose in primary cell walls. *Syn* Cellulose synthase; *E.P.* pre-xyloglucan synthase. (Hayashi 1989)

tigated after treatment of the complex with endo-1,4-β-glucanase or of 24% KOH (Baba et al. 1994). The results showed that the antibody was bound to the complex, but that the extent of binding was reduced after treatment of the complex with endo-1,4-β-glucanase and was hardly detected after treatment with 24% KOH. The molecular weight of the xyloglucan that remained (5%) in the glucanase-treated complex was less than 9200. Immunogold labeling indicated that most of the antigen appeared to be confined to the microfibril. The results suggested that a part of each xyloglucan molecule is strongly associated with cellulose microfibrils while the rest is free of the microfibril in the native complex. It seems that xyloglucan is partially interwoven among the amorphous parts of cellulose microfibrils rather than bound to the surface of the microfibril.

There are several mechanisms by which the xyloglucan–cellulose network could loosen to allow for tugor-driven cell growth. Hydrogen bonds could be broken and the glucan backbone of xyloglucan could be enzymatically cleaved. The recent finding that a xyloglucan endoglycosylase activity exists in the walls of a variety of suspension-cultured cells including sycamore maple, and the cell walls of growing dicot and graminaceous monocot plant tissues (Hayashi 1994).

The xyloglucan endotransglycosylase (XET) recently discovered cuts a xyloglucan chain and reanneals the cut end to another xyloglucan molecule (Fry et al. 1992). XET first cuts one xyloglucan chain and probably forms a transient polysaccharide–enzyme complex. The enzyme-linked portion of the polysaccharides is then transferred to the end of an acceptor substrate. The acceptor can be either another xyloglucan polysaccharide molecule or a xyloglucan oligomer as small as the pentasaccharide. Xyloglucan endotransglycosylation involves cleavage of the xyloglucan backbone and subsequent attachment of the reducing end to an available xyloglucan polymer. This transient loosening of xyloglucan–cellulose network would permit some growth in incremental steps, and the regulated transglycosylation may result in controlled cell expansion.

Genes for the seed enzyme, which catalyzes both hydrolysis and transglycosylation (De Silva et al. 1993), and the stem enzyme which is specific for transglycosylation (Nishitani et al. 1993) have been sequenced.

4.2.7 Biosynthesis of Xylan (Arabinoxylan)

Xylan is a major secondary wall component and the most abundant noncellulosic polysaccharide in the angiosperms. It is a linear polymer composed of (1 → 4)-linked β-D-xylose residues. Side chains of arabinose and 4-*O*-methyl-α-D-glucuronic acid are linked to the xylan backbone (Fig. 88).

Xylan (arabinoxylan) formation by xylose transfer from UDP-xylose is mediated by granular enzyme preparations from sycamore maple and maize. The xylan formed is a β-D-1,4-linked polymer and contains arabinose residues linked by α-L-1,3-linkages. When particulate enzyme preparations from immature corn cobs were incubated with UDP-[^{14}C]xylose, the radioactive product resembled native arabinoxylan and contained radioactivity in both xylose and arabinose (Bailey and Hassid 1966), indicating that the particular enzyme contained the UDP-xylose: UDP-arabinose 4-epimerase.

The incorporation of 4-*O*-methylglucuronic acid into xylan has also been demonstrated in extracts of corn cobs (Kauss and Hassid 1967). In these studies, glucuronic acid from UDP-glucuronic acid was incorporated into the polymer. The glucuronic acid residues became methylated by the transfer of methyl groups from *S*-adenosyl-L-methionine to the 4-position of the glucuronic acid.

Dalessandro and Northcote (1981) found that the activity of the xylan synthesizing enzyme was high in differentiating xylem tissues of sycamore maple. In addition, they found that the enzyme activity increased during differentiation of the vascular cambium of sycamore maple and poplar. The activity of the xylan synthesizing enzymes increased during secondary wall

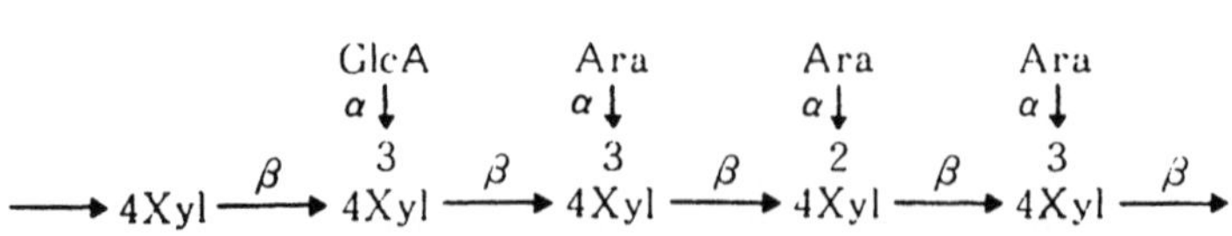

Fig. 88. Basic chemical structure of arabinoxylan

formation, and the transfer activity of the glucuronic acid residues from UDP-glucuronate was also found in the plant (Waldron and Brett 1983).

4.2.8 Biosynthesis of Glucomannan

A galactoglucomannan with a variable content of galactose residues is the major hemicellulose (20%) in the softwoods. In the hardwoods, a glucomannan is a minor constituent (5%). Both hemicelluloses contain 1,4-linked β-D-glucose and β-D-mannose residues. The α-D-galactose residues in the softwood hemicellulose are attached as single side chains to C_6 in either glucose or mannose. There are also acetyl groups present (Fig. 89).

The biosynthesis of glucomannan from GDP-glucose and GDP-mannose was studied by using granular enzyme preparations from mung bean (*Phaseolus aureus*; Elbein 1969). The radioactive polymer formed from GDP-[^{14}C]mannose was compared to that produced from GDP-[^{14}C]glucose. Partial acid hydrolysis of these two products, or enzymatic hydrolysis with a β-mannanase, liberated a similar series of radioactive oligosaccharides from both biosynthetic products. These labeled oligosaccharides contained both D-glucose and D-mannose, indicating that both glucose and mannose were incorporated into the same polymer. Both mannose and glucose became labeled when incubation mixtures contained only GDP-[^{14}C]mannose or GDP-[^{14}C]glucose, suggesting that the particulate enzyme contained a 2-epimerase capable of interconverting GDP-glucose and GDP-mannose. The labeled oligosaccharides produced by hydrolysis were characterized by methylation analysis and periodate oxidation. One of the major disaccharides, 4-*O*-β-D-mannopyranosyl-β-D-glucopyranoside as well as 4-*O*-β-D-glucopyranosyl-β-D-mannopyranose were identified.

These experiments indicate that the synthesized polymer was a linear (1 → 4) glycan composed of β-D-mannose and β-D-glucose (Hori and Elbein 1985). The incorporation of glucose was activated by GDP-mannose, and the molecular weight of the glucomannan increased. The glucosyl transferase involved in the glucomannan synthesis was activated by GDP-mannose in lower substrate concentrations, although the affinity for the substrate decreased. When the concentration of GDP-glucose was higher than that of GDP-mannose the uptake of glucose was inhibited. The activity of mannosyl transferase, which catalyzes the transfer of mannose residues from GDP-mannose, was inhibited by GDP-glucose. Thus, the biosynthesis of glucomannan was regulated by the concentration and ratio of both substrates.

$\xrightarrow{\beta}$ 4Man $\xrightarrow{\beta}$ 4Man $\xrightarrow{\beta}$ 4Glc $\xrightarrow{\beta}$ 4Man $\xrightarrow{\beta}$ 4Man $\xrightarrow{\beta}$

Fig. 89. Basic chemical structure of glucomannan

4.2.9 Biosynthesis of Galactan

There are two kinds of galactans composed of β-1,4-linked or β-1,3-linked residues of galactose in plants. Enzyme preparations from mung bean catalyzed the transfer of galactose residues from UDP-galactose and afforded a galactan with a molecular weight higher than 4600 (Panayotato and Villemez 1973). The galactan formed was found to contain mainly the β-1,4-linkage but also contained a small amount of the β-1,3-linkage. The enzyme activity of the mung bean increased up to 6 days after germination, and then decreased.

4.2.10 Biosynthesis of Arabinan

Fry and Northcote (1983) found that an enzyme prepared from a membrane fraction of cultured cells of spinach catalyzed the formation of arabinan from UDP-arabinose. They also found that configuration of arabinose was a six-membered ring (pyranose ring) in UDP-arabinose, but that it changed to five-membered ring (furanose ring) at the stage of sugar transfer to arabinan.

4.2.11 Biosynthesis of Pectin (Polygalacturonic Acid)

The pectic substances refer to a group of closely associated polysaccharides isolated from the primary cell walls and intercellular regions of higher plants. The main constituent of pectins is polygalacturonate composed of α-1,4-linkages of galacturonic acid. Various sugars, such as L-rhamnose, L-arabinose, L-fucose, D-galactose and D-xylose, are linked to the polygalacturonate in pectins.

Biosynthesis of pectins occurs by the transfer of monosaccharides from their sugar nucleotide donors to an acceptor polysaccharide. A particulate enzyme preparation from mung bean seedlings catalyzed the incorporation of ^{14}C-galacturonic acid from UDP-^{14}C-galacturonic acid into a polygalacturonic acid that appeared to be labeled throughout the chain (Villemez et al. 1965).

The formation of methyl esters of galacturonic acid was examined with cell-free extracts of mung bean seedlings and *S*-adenosyl-L-methionine (SAM), which is the direct methyl donor in this reaction. The result showed that methyl groups were attached by ester linkages to the product. In addition, UDP-galacturonic acid chemically methylated at position 6 of the sugar was not incorporated (Kauss and Hassid 1967).

It was concluded that methylation of carboxyl groups of galacturonic acid residues of pectins occurs after completion of polymer formation by mediation of *O*-methyltransferase using *S*-adenosyl methionine as methyl donor. The synthesis of the polygalacturonic acid backbone and its subsequent methylation depended on distinct intracellular compartmentation, and exogenously added polygalacturonic acid could not be methylated by the particulate methyl transferase.

The synthetic activity of polygalacturonate in growing cells of sycamore maple decreased during secondary cell wall formation (Bolwell et al. 1985). Apiose was transferred from UDP-apiose into an apiogalacturonan by a particulate enzyme preparation from duckweed (Kindel 1973).

4.2.12 Biosynthesis of Sugar Chains of Glycoproteins

Glycoproteins in the outer layer of plant cells are composed of extensin, proteins linked with arabinogalactan and the mannose series. Mannose chains of mannoproteins are synthesized via lipid intermediates as in animals.

Karr (1972) found using cultured cells from sycamore maple that arabinose residues of UDP-arabinose are transferred to hydroxyproline residues of extensin, and that the arabinosyl transferase involved is activated by Mg^{2+} and Mn^{2+}. The galactosyltransferase involved in arabinogalactan-protein synthesis was found in a membrane enzyme preparation from cultured cells of *Lolium multiflorum*. The enzyme catalyzed transfer of the galactose residue of UDP-galactose to the C_6 hydroxyl group of arabinose residues. The molecular weight of the product was 130kDa (Mascara and Fincher 1982).

4.2.13 Cell Organelles Involved in Polysaccharide Synthesis

Pathways involved in polysaccharides biosynthesis are summarized in Fig. 90 (Higuchi 1992).

What organelles, and what mechanisms are involved in the synthesis of polysaccharides and their precursors? To elucidate these problems both radioautography of incorporation processes of ^{3}H- and ^{14}C-labeled glucose to cell walls and activity analyses of granules isolated by centrifugal fractionation of cell homogenetes have been conducted. These investigations indicated that Golgi vesicles are involved in the hemicellulose biosynthesis, whereas granules on plasma membranes are involved in the cellulose and β-1,3-glucan syntheses. By using electron microscopy of plasma membranes of the saltwater algae, *Chaetomorpha* and *Cladophora*, Preston (1964) first proposed that cellulose is formed by granules on the plasma membranes.

Recent electron microscope observations by Brown and Montezinos (1976), and Ito (1989) on plasma membranes of *A. xylinum*, green algae, *Oocystis*, and higher plants by the freeze–fracture method showed that cellu-

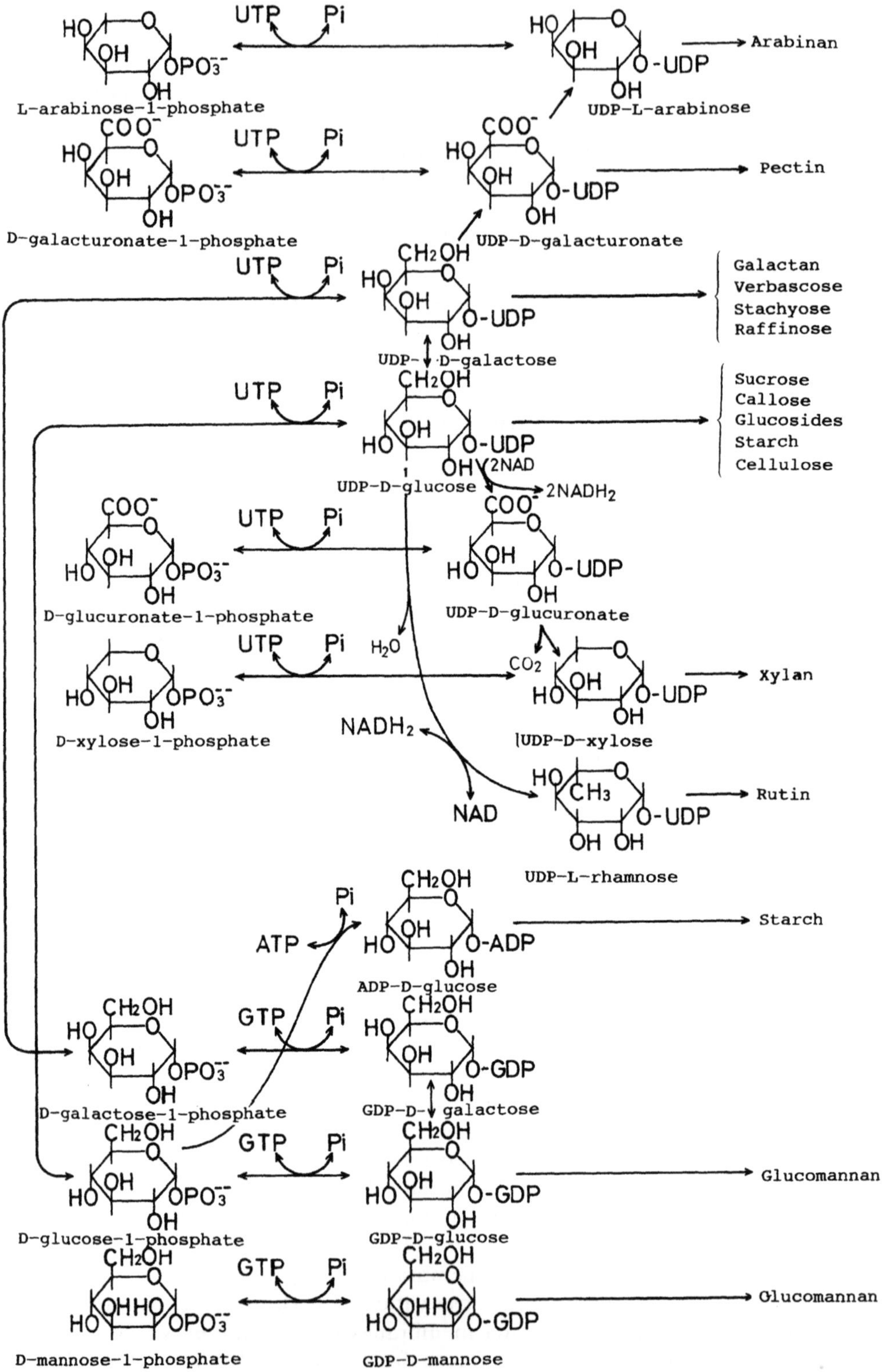

Fig. 90. Synthetic pathways of polysaccharides via nucleotide diphosphate sugars

lose microfibrils are formed by a rosette assembly of cellulose-synthesizing enzyme complexes (TCs).

Haigler and Brown (1986) observed rosettes in the PF faces of Golgi cisternae and Golgi vesicles in mesophyll cells of *Zinnia elegans* in suspension cultures previously induced to differentiate into tracheary elements. The rosettes were observed in the plasma membrane and Golgi vesicles throughout the enlargement of the secondary wall thickening, suggesting that the Golgi apparatus is involved not only in the synthesis and export of matrix components but also in the export of rosettes to the plasma membrane. It seems that the protein components of TCs are synthesized in ribosomes and transported to the plasma membrane via Golgi vesicles.

Northcote and Pickett-Heaps (1966) steeped wheat root tips in a radioactive glucose solution for short periods and then transferred them to a nonradioactive glucose solution. They prepared sections of the treated root tips and subjected them to autoradiography. Radioactivity was detected in cisternae of Golgi vesicles of root tip cells treated for only several minutes. When the root tips were transferred to nonradioactive glucose solutions, radioactivity was detected in the Golgi vesicles at the beginning, but the activity was finally detected in cell walls (Fig. 91). The result indicated that the Golgi apparatus synthesized cell wall hemicelluloses which were transported to the cell wall via Golgi vesicles. Analysis of the polysaccharides formed showed that they consisted of pectin and hemicelluloses. Northcote further found that the membrane fraction containing the Golgi apparatus from root tips of pea seedlings catalyzed the synthesis of glucans containing β-1,3 and β-1,4 linkages from UDP-glucose. They suggested the formation of sugar–lipid complexes as intermediates and that the membrane enzymes were derived from vesicles in the cytoplasm. Nucleoside diphosphate sugars such as UDP-glucose are synthesized in Golgi vesicles, and transported to plasma membranes. Cellulose is synthesized by TCs on the plasma membranes, while pectin and hemicelluloses are synthesized in the Golgi apparatus. The illustration on polysaccharide biosynthesis by Northcote in 1968 is modified as shown in Fig. 92.

4.3 Phenylpropane Derivatives

Phenylpropane derivatives are compounds composed of a C_6–C_3 carbon skeleton comprised of an aromatic ring with a propane side chain. There are several excellent reviews of the biosynthesis of phenylpropanoids (Stumpf and Conn 1981, Conn 1986, Hahlbrock and Scheel 1989, Bentley 1990, Lewis and Yamamoto 1990, Stafford and Ibrahim 1992).

Phenylpropanoids are considered to be essential for plant life. For example, dehydrodiconiferyl alcohol glucoside is present in dividing plant cells and acts as a cytokinin, a certain flavonoid affects the polar transportation of auxin, and flavonoid pigments protect growing meristems against ultraviolet rays.

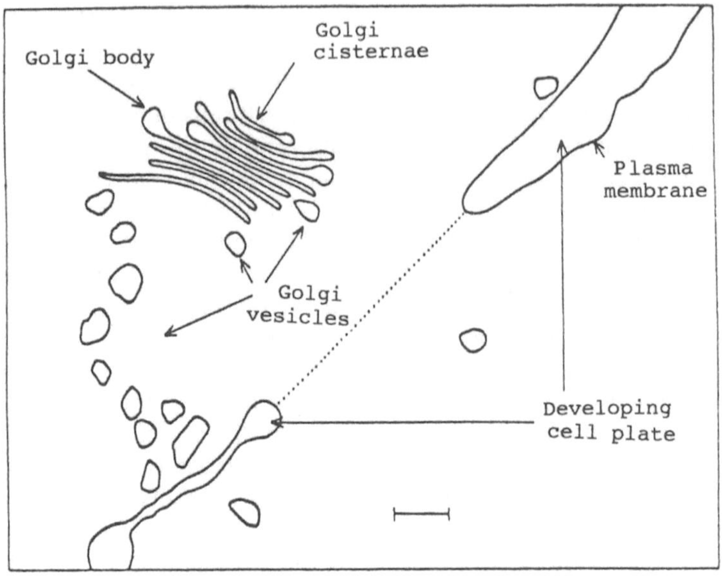

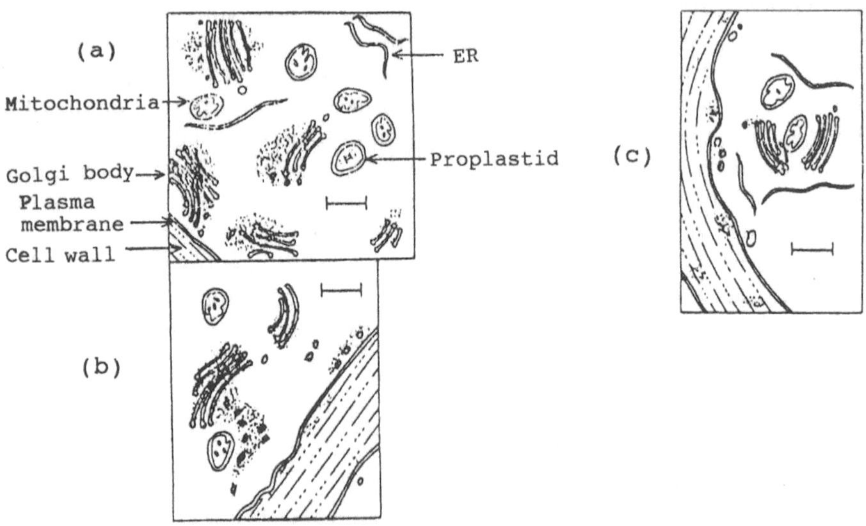

Fig. 91. a–c Incorporation of ^{14}C-glucose into cell wall polysaccharides of wheat root tips. **a** Root tips steeped in ^{14}C-glucose solution for 10 min. ^{14}C (*black spots*) is incorporated into Golgi apparatus. **b** Root tips are steeped in ^{14}C-glucose solution for 10 min and then transferred into a nonradioactive glucose solution for 10 min. ^{14}C incorporation is observed in Golgi vesicles. **c** Root tips are steeped in ^{14}C-glucose solution for 10 min and then transferred into nonradioactive glucose for 30 min. Vesicles are not radioactive, and activity is observed in the cell wall. (Northcote and Pickett-Heaps 1966)

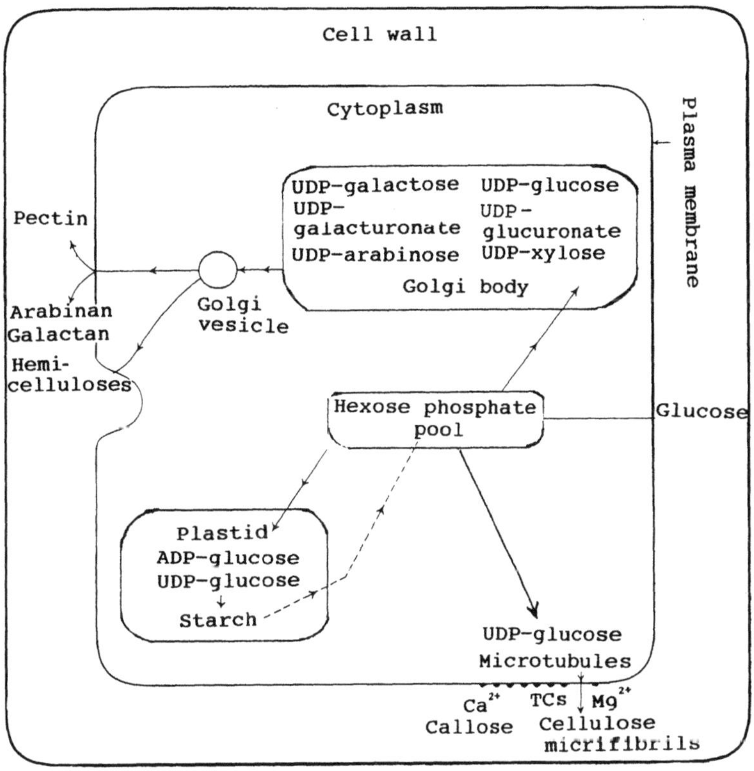

Fig. 92. Sites of synthesis of sugar nucleotides and polysaccharides in plant cells

Isoflavonoids and furanocoumarine derivatives exhibit antibiotic and phytoalexin activities and protect plants from diseases. In addition, acetosyringone has been found to act directly on genes required for infection (the *vir* domain) of *Agrobacterium*.

Lignin is the second most abundant biomass next to cellulose on earth. Lignin gives rigidity to the cell walls and enable terrestrial plants to develop upright forms resistant to various stresses, and assist in the smooth transportation of solutes by decreasing the permeability of cell walls in the conductive xylem tissues. It also imparts resistance to attacks by microorganisms. Fig. 93 shows a scheme of biosynthetic pathway of phenylpropanoids.

The phenylpropanoid pathway can be classified into four routes: (1) the shikimate pathway commonly involved in the biosynthesis of many aromatic compounds (Haslam 1974, Conn 1986, Kaufman 1987, Bentley 1990); (2) biosynthesis of phenylalanine and tyrosine (Conn 1986, Bentley 1990); (3) the general phenylpropanoid pathway to afford 4-coumaroyl-CoA (Hahlbrock and Scheel 1989, Lewis and Yamomoto 1990, Stafford and Ibrahim 1992); and (4) pathways for lignin and lignans etc. associated with the general phenylpro-

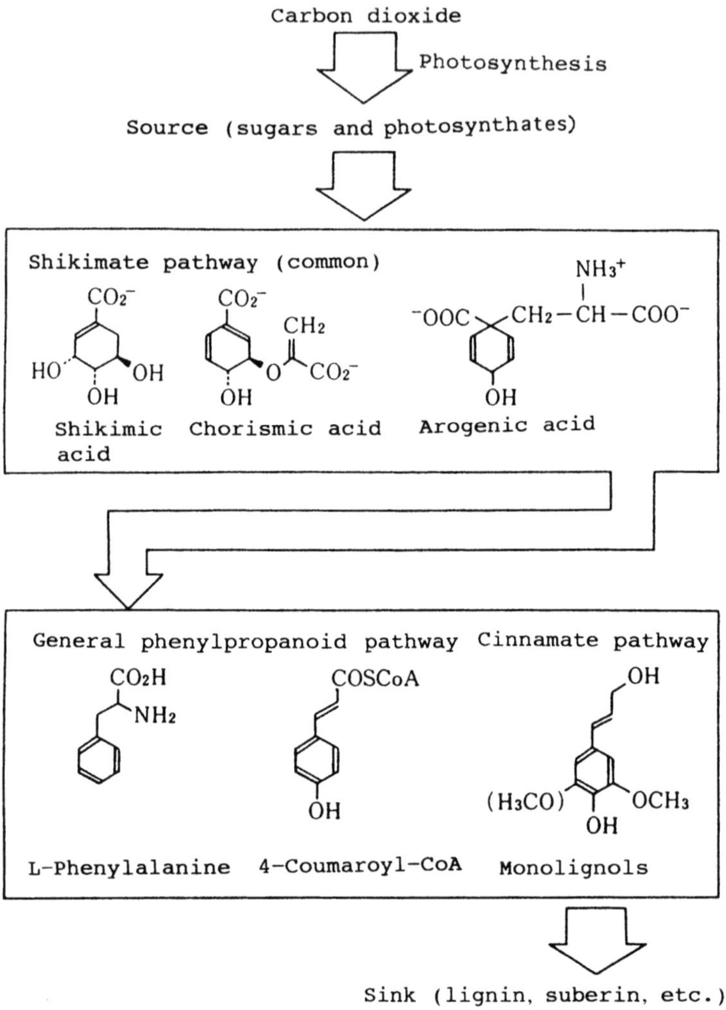

Fig. 93. Outline of the biosynthetic pathway of phenylpropanoids

panoid pathway (Freudenberg 1968, Higuchi 1985, Umezawa 1994; see Sect. 4.3.5).

4.3.1 The Shikimate Pathway Commonly Involved in the Biosynthesis of Aromatic Compounds

Eijkman (1885) identified a rare organic acid comprised of a cyclohexene ring from seeds of a shrubby tree, *Illicium religiosum* Sieb. et Zucc. (Japanese

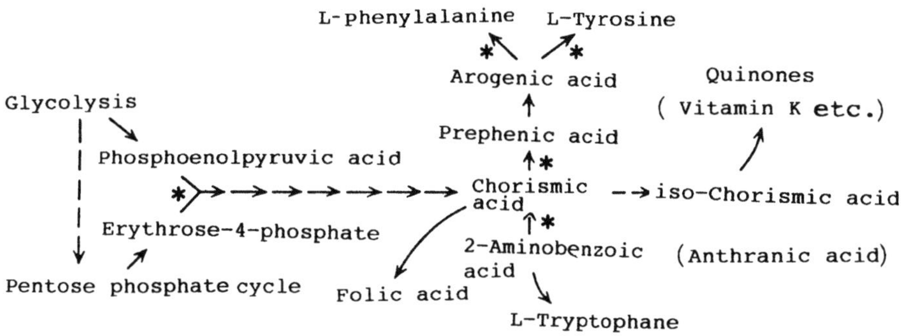

Fig. 94. Biosynthesis of aromatic amino acids and aromatic compounds. → One step in the shikimate pathway; * steps affected by metabolic regulation; *arrows* denote the flow of compounds. Respective steps are shown in Figs. 95 and 96, respectively

name, shikimi) during his stay in Japan. The acid was named shikimic acid. The name of the shikimate pathway is ascribed to shikimic acid, whose physiological properties were not known at that time.

Davis (1955) established the shikimate pathway in microorganisms using various mutants of *Escherichia coli* with deficient synthetic abilities for aromatic amino acids. The shikimate pathway and major metabolites involving the pathway are shown in Fig. 94. The pathway is involved in biosynthesis of many important aromatic compounds including vitamins and aromatic amino acids which are essential nutrients for microorganisms and plants, but not in animals. Animal have no ability to synthesize aromatic compounds from sugars.

Mutant strains of *Escherichia coli* that were deficient in some of their genetic abilities were obtained by UV irradiation of the wild strain (Davis 1955). By a penicillin screening method, several strains that required aromatic nutrients for growth were selected and used for elucidation of the shikimate pathway. A growth inhibitor that intercepts the shikimate pathway in plants was developed recently and used as a herbicide (Steinrücken and Amrhein 1980, Amrhein 1991).

It should be noted that the flow of metabolites in the shikimate pathway in Fig. 94 diverges from chorismic acid. Seven synthetic steps (→) before chorismic acid are always involved in the biosynthesis of aromatic compounds, but the later steps from chorismic acid are branched, leading to different aromatic amino acids. In biosynthesis of phenylpropanoid derivatives, chorismic acid is converted to L-phenylalanine and L-tyrosine via prephenic acid.

4.3.1.1 The Enzyme System in Aromatic Biosynthesis

The shikimate pathway involved in biosynthesis of aromatic compounds is shown in Fig. 95 and Table 16. The first step in Fig. 95 represents the initiation

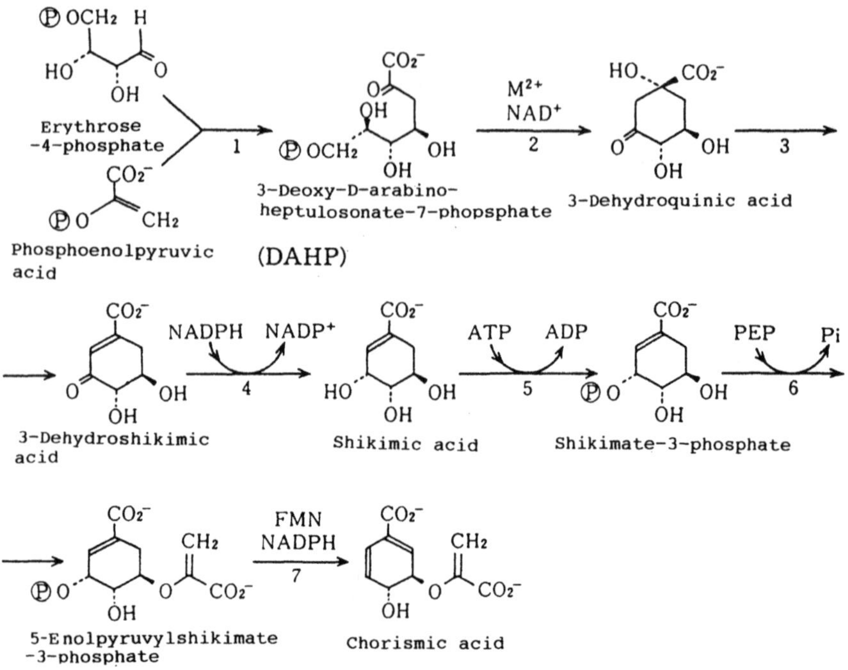

Fig. 95. Enzyme system involved in biosynthesis of aromatic compounds. Compounds and the *number* of reaction steps in figure correspond to the enzymes in Table 16. *PEP* Phosphoenol pyruvate

of the shikimate pathway, and the carbons for the aromatic ring are prepared in this step. It has been found that in a wild strain of *Escherichia coli* three isozymes of DAHP synthase, which are metaloenzymes (Fe^{2+} and Zn^{2+}) and mainly involved in biosynthesis of aromatic amino acids as a protein component, are regulated by feedback inhibition with different aromatic amino acids. Feedback inhibition of the three individual isoenzymes is absolutely specific for the cognate amino acid, e.g., the phenylalanine-sensitive isoenzyme is not affected by tryptophane and tyrosine, or a combination of the two. It was found that the three *E. coli* genes encoding DAHP synthases are repressed by the *tyr*- and *trp*-repressors correspond to the relevant aromatic amino acids (Garner and Herrmann 1985, Herrmann 1995).

Purified plant DAHP synthases from carrot roots and potato tubers, are oligomeric metalloenzymes (Mn^{2+}) like the bacterial enzymes, and are involved in biosynthesis of phenylpropanoids and aromatic amino acids as protein components. These DAHP synthases were not feedback inhibited but were activated by tryptophan and to lesser degree also by tyrosine in a hysteretic fashion (Pinto et al. 1986). An isoenzyme of DAHP from mung bean was inhibited by both arogenic acid and prephenic acid but not by aromatic amino acids (Rubin and Jensen 1985).

Rabbit antibodies raised against the pure potato DAHP synthase were used to screen a cDNA library from potato cells in suspension culture (Dyer et al. 1989). The resulting cDNA (*shkA*) has been used as a probe to clone a second potato cDNA (*shkB*; Zhao and Herrmann 1992), and also to clone

Table 16. The enzyme system involved in aromatic biosynthesis and their abbreviations

1. Preparation of six carbons for aromatic ring (C_7-saturated keto acid)
   2-Dehydro-3-deoxyphosphoheptonate aldolase (EC 4.1.2.15)
   Synonym: 3-deoxy-D-arabinoheptulosonate 7-phosphate synthase (DAHP synthase, DHAP synthase)
   Product: 2-dehydro-3-deoxy-D-arabinoheptulosonic acid-7-phosphate (DAHP or DHAP)
   Source: *Escherichia coli*, yeast, plants

2. Preparation of cyclohexane ring (NAD^+), dephosphorylation
   3-Dehydroquinate synthase (EC 4.6.1.3)
   Product: 3-dehydroquinic acid
   Source: *Escherichia coli* (aroB), yeast (arom), plants

3. Introduction of a double bond into the saturated cyclic carbon chain, dehydration
   3-Dehydroquinate dehydratase (EC 4.2.1.10)
   Synonym: 3-dehydroquinase
   Product: 3-dehydroshikimic acid (DHS)
   Source: *Escherichia coli* (aroD), yeast (arom), plants (EC 4.2.1.10/EC 1.1.1.25)

4. Reduction of the cyclic keto group to a cyclic alcohol group (NADPH) shikimate dehydrogenase (EC 1.1.1.25)
   Product: shikimic acid
   Source: *Escherichia coli* (aroE), yeast (arom), plants (EC 4.2.1.10/EC 1.1.1.25)

5. Addition of a substituent group to the alcohol group (ATP), Phosphorylation, shikimate kinase (EC 2.7.1.71)
   Product: shikimic acid-3-phosphate
   Source: *Escherichia coli* (aroL/SK 2), yeast (arom), plants (EC 2.7.1.71)

6. Addition of side-chain carbons (phosphoenolpyruvate) for phenylpropanoids, dephosphorylation
   Shikimic acid-3-phosphate 1-carboxyvinyltransferase (EC 2.5.1.19)
   Synonym: 5-enolpyruvylshikimate-3-phosphate synthase
   Product: 5-enolpyruvylshikimate-3-phosphate (EPSP)
   Source: *Escherichia coli* (aroA), yeast (arom), plants (EC 2.5.1.19)

7. Introduction of a double bond into the unsaturated cyclic carbon chain (FMN, NADPH), dephosphorylation (β-elimination) Chorismate synthase (EC 4.6.1.4)
   Product: chorismic acid
   Source: *Escherichia coli* (aroC)

The number corresponds to the enzymatic reactions in Fig. 95.
The enzyme name and EC number: IU-BMB (1992) *Escherichia coli* (aroB), etc. indicates gene locus and multifunctional enzymes.

cDNAs from tobacco (Wang et al. 1991), *Arabidopsis thaliana* (Keith et al. 1991) and tomato (Görlach et al. 1993).

Predicted protein sequences from two distinct genes of DAHP synthase, *DHS1* and *DHS2* from *Arabidopsis thaliana*, and a third DAHP synthase gene from potato were highly homologous, but were not similar to the sequence of the microbial DAHP synthase proteins, *DHS1* RNA levels increased in *Arabidopsis* leaves subjected either to physical wounding or to infiltration with pathogenic *Pseudomonas syringae* strains. But *DHS2* RNA levels were not increased by these treatments, suggesting that the DHS1 and DHS2 proteins fulfill different physiological functions. Other enzymes in the *Arabidopsis* aromatic pathway were also encoded by duplicated genes that may allow independent regulation of aromatic amino acid biosynthesis by substances involved in distinct physiological pathways such as protein synthesis and secondary metabolism (Keith et al. 1991).

The two gene families (*shkA, shkB*) encoding DHAP synthase in potato also respond differently to wounding and pathogen attack. Only the shk A isoenzyme responds to wounding. The levels of *shkA* mRNA are much higher in stems and roots where the *shkA* gene is probably involved in secondary metabolism, while in leaves, flowers, or cotyledons the level of *shkB* mRNA is much higher, and the *shkB* gene is probably involved in protein production (Görlach et al. 1993). It seems that a highly regulatable function is involved in the synthesis of *shkA* products required for secondary metabolites, in contrast to a basic housekeeping function for the *shkB* products to satisfy amino acid biosynthesis for protein production (Herrmann 1995).

Activity of DAHP synthase was present in the cytoplasm and plastids, but the cDNAs of DAHP synthases hitherto isolated have chloroplast transit peptides (Dyer et al. 1990, Keith et al. 1991), suggesting that the precursors of the enzyme are synthesized in the cytoplasm by the genetic information from the nucleus and are then transferred to plastids and mature. In parsley, DAHP synthase in chloroplasts was induced by a fungal elicitor (Last et al. 1991).

In bacteria such as *Escherichia coli*, individual enzymes are involved in the reactions from steps two to six. In yeast and other fungi, multifunctional enzymes, which are named arom complexes, exist in the cytoplasm. All the reactions between steps two and six are catalyzed by the one multifunctional enzyme.

The genes of yeast (Waldron and Brett 1983) and *Aspergillus* (Hawkins 1987) do not have introns, and the respective domains are homologous with the corresponding monofunctional enzyme in *Escherichia coli*. The arom complex of *Neurospora crassa* is a homodimer composed of a 165-kDa subunit with five functional domains, which are similar to those of yeast and *Aspergillus*. It is suggested that the gene encoding the arom complex was probably formed by fusing of five genes into a chimera complex in the process of evolution.

The arom complexes found in yeast and other fungi have not been reported in plants, but the involvement of two functional enzymes that catalyze the reactions of steps three and four was demonstrated. Such enzymes exist in pea, mung bean seedlings and mosses, and have been shown to consist of a single polypeptide with two catalytic activities (Bentley 1990).

In steps two, five and six, single enzymes are involved as in bacteria. The purified dehydroquinate synthases (DHQ) in step two from mung bean and pea are dimer enzymes with a molecular mass of 66 to 67 kDa, while the molecular mass of the purified shikimate kinase in step five from spinach is 31 kDa (Schmidt et al. 1990).

A PCR fragment of the kinase cDNA was obtained from tomato by using a primer designed on the homogeneous amino acid sequences of *Erwinia, Escherichia coli*, and yeast enzymes. The cDNA clone obtained by using the PCR fragment as a probe had chloroplast transit sequences. The homogeneity of the amino acid sequences was about 30% of those of bacteria and yeast kinase. The enzyme precursor formed by transcription and translation of the cDNA in vitro showed the activity of shikimate kinase (Schmid et al. 1992).

By the reaction in step six, three carbons of the phenylpropanoid side chain are added. It is well known that the EPSP synthase involved in this reaction is affected by a herbicide, *N*-phosphonomethylglycine (Steinrücken and Amrhein 1980). A poplar tree resistant to this herbicide was produced by the introduction of the nonaffective gene derived from *Escherichia coli* (Fillatti et al. 1987).

The EPSP gene of *Petunia hybrida* consists of eight exons, and the homogeneity of the amino acid sequences is about 54% of that of *Escherichia coli* (Gasser et al. 1989). The molecular mass is 50 kDa which is similar to that of *Escherichia coli* (Keith et al. 1991). It has been suggested that the enzyme precursor is synthesized in the cytoplasm and transported into the stroma of plastids where it matures (Della-Cioppa et al. 1986). Northern analysis showed that the gene is highly expressed in the purple flower petals of the wild strain of the *Petunia*.

The enzyme involved in step seven, chorismate synthase, has been investigated with *Escherichia coli*, *Bacillus subtilis* and *Neurospora crassa*. The enzyme requires NADPH and FMN. By β-elimination of the phosphate group of EPSP, a double bond is introduced. The enzymes of *Escherichia coli* and *Bacillus subtilis* are monofunctional enzymes, while the *Neurospora* enzyme has been suggested to be a two-functional enzyme with diaphorase activity.

The cDNA clone of chorismate synthase from *Cordailis sempervirens* (a plant) has been analyzed (Schaller et al. 1991). The amino acid sequence of the enzyme in this plant is 48% homologous to that of *Escherichia coli*. The enzyme precursor (M_r 48100) is probably synthesized in the cytoplasm and transported into the chloroplasts where it matures.

4.3.2 Biosynthesis of Phenylalanine and Tyrosine

4.3.2.1 Phenylalanine and Tyrosine

After the formation of prephenic acid via chorismic acid, transamination and aromatization occurs, resulting in formation of L-phenylalanine and L-tyrosine (Fig. 96, Table 17).

The enzymes involved in this pathway are classified into three groups, namely aminotransferases, dehydratases, and dehydrogenases.

In *Escherichia coli, Bacillus subtilis, Neurospora crassa* and yeast, the aromatic ring is formed in steps nine and ten, and the resulting phenylpyruvate and *p*-hydroxyphenylpyruvate are aminated to L-phenylalanine and L-tyrosine, respectively, in steps 12 and 13.

Aminotransferases catalyzing step 11 have been found in some blue green algae (Stenmark et al. 1974). Since the time of that paper the formation of L-phenylalanine and L-tyrosine by amination of arogenic acid (step 11) has been observed in *Euglena* and plants (Jung et al. 1986, Siehl and Conn 1988, Bentley

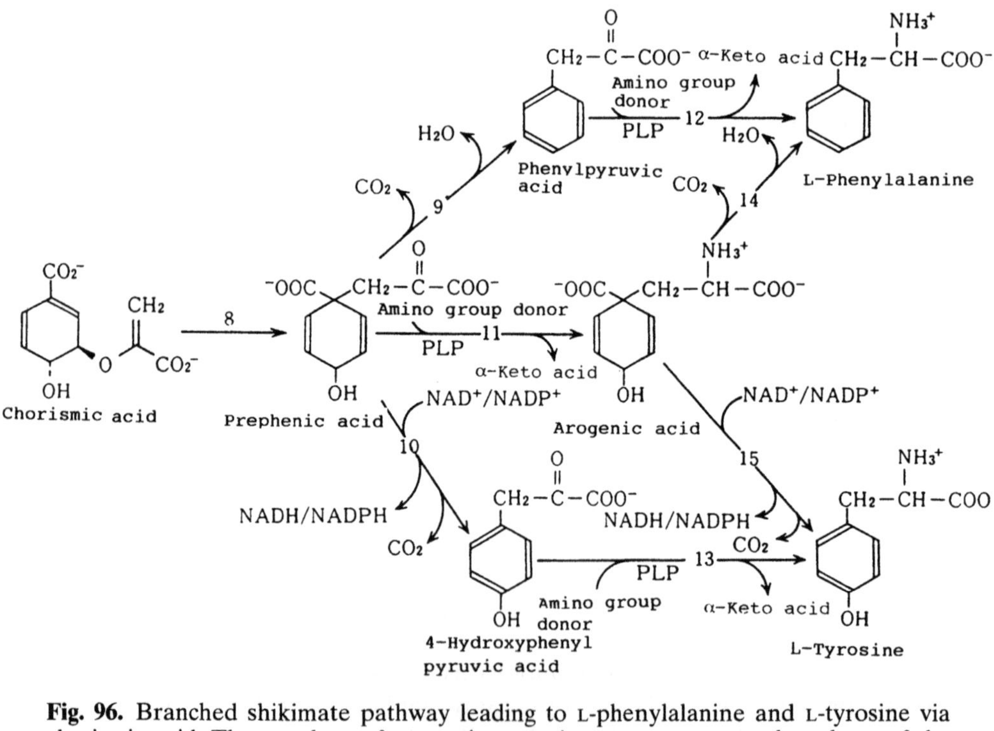

Fig. 96. Branched shikimate pathway leading to L-phenylalanine and L-tyrosine via chorismic acid. The *numbers* of respective reaction steps correspond to those of the enzymes in Table 17

Table 17. The enzymes involved in the biosynthesis of phenylalanine and tyrosine

8.	Chorismate mutase (EC 5.4.99.5) Two-functional enzyme (*Escherichia coli*), inhibited allosterically *Escherichia coli* (pheA) (EC 5.4.99.5/EC 4.2.1.51) *Escherichia coli* (tryA) (EC 5.4.99.5/EC 1.3.1.12)
9.	Prephenate dehydratase (EC 4.2.1.51) Specific for the step 9 reaction, not present in plants *Escherichia coli* (pheA) (EC 5.4.99.5/EC 4.2.1.51)
10.	Prephenate dehydrogenase (EC 1.3.1.12/NAD, EC 1.3.1.13/NADP) Specific for the step 10 reaction, present only in a few plants including mung bean *Escherichia coli* (tryA) (EC 5.4.99.5/EC 1.3.1.12)
11.	Prephenate transaminase (EC 2.6. –) Selectively catalyzes step 11 reaction, a kind of aspartate transaminase
12, 13.	Aromatic amino acid transaminase (EC 2.6.1.57) Main reaction in *Escherichia coli* (typB)
12, 13.	Aspartate transaminase (EC 2.6.1.1) Subordinate reaction in *Escherichia coli* (aspC)
14.	Arogenate dehydratase (EC 4.2. –)
14, 9.	Carboxycyclohexadienyldehydratase (EC 4.2.1.91)
15.	Arogenate dehydrogenase (EC 1.3. –) Specific for step 15 reaction, feedback inhibition by tyrosine
15, 10.	Cyclohexadienyldehydrogenase (EC 1.3.1.43) In EC, the enzyme on step 15 reaction is included

The number corresponds to the enzyme reactions in Fig. 96. The enzyme name and EC number: IU-BMB (1992) *Escherichia coli* (phe A), etc. indicates gene locus.

1990). In many bacteria, such as *Pseudomonas* and *Xanthomonas*, both pathways are involved (Bentley 1990, Saxena et al. 1991). In plants, the dehydrogenase activity in step ten is found in some legume plants including mung bean, albeit only rarely (Gamborg 1966). However, the dehydratase that catalyzes step nine has not been detected in plants.

The affinity of aminotransferases to aromatic keto acids (steps 12 and 13) is low. The aminotransferases involved in step 11 use aspartic and glutamic acids as amino donors, and 2-oxoglutaric acid as well as prephenic acid as amino acceptors. The enzyme exists as an isoform of aspartic aminotransferase. By heat treatment of the fraction of aspartic aminotransferase, the aminotransferase activity for aromatic keto acids is almost lost (Bentley 1990). The purified enzyme from *Anchusa officinalis* is a tetramer comprised of subunits with a mass of 44 and 57 kDa and exhibits a weak activity for aromatic keto acids (De-Eknamkul and Ellis 1988).

Arogenate dehydratase from sorghum, which mediates the step 14 reaction is inhibited (K_i = 24 μM) by the product, L-phenylalanine, but activated (K_a = 2.5 μM) by L-tyrosine (Siehl and Conn 1988). The arogenate dehydrogenase of sorghum involved in step 15 is inhibited (K_i = 61 μM) by the product, L-tyrosine, but not affected by prephenate or L-phenylalanine (Connelly and Conn 1986). The enzyme from *Nicotiana silvestris* is also inhibited by L-tyrosine (Gaines et al. 1982). The arogenate pathway via step 11 could therefore be a major pathway in plants. The gene encoding the enzymes involved in this branched pathway has not been reported.

The mutase involved in step eight mediates the intramolecular rearrangement of the enolpyruvyl side chain. It has been found that enzymes from alfalfa, pea, mung bean, sorghum, tobacco, and *Quercus* are feedback inhibited by aromatic amino acids. For example, an isoenzyme isolated from mung bean seedlings, CM-I, is a 50-kDa protein with pI 4.6. The isoform enzyme is inhibited allosterically by L-tyrosine (K_i = 31 μM) and L-phenylalanine (K_i = 37 μM). Another isoform CM-II is a 36-kDa protein with pI 5.6 and is not inhibited allosterically. The former is present in the plastids while the latter ocuurs in the cytoplasm. A similar result has been reported with *Nicotiana silvestris* (Goers and Jensen 1984). In potato the mutase is induced by wounding (Kuroki and Conn 1988).

In *Escherichia coli*, steps eight and nine as well as steps eight and ten are catalyzed by two functional enzymes, respectively, but in plants no such enzymes have been reported.

4.3.2.2 Diversity of the Shikimate Pathway

As stated above the enzyme system in the shikimate pathway is diverse. This diversity could be caused by (1) enzyme organization; (2) localization of enzymes; (3) regulation mechanism of enzymes; (4) occurrence of isozymes; and (5)biosynthetic pathways via the metabolic grid.

4.3.2.2.1 Enzyme Organization

It is interesting to know whether a series of enzymes in a certain biosynthetic pathway is located at random in cells or in a certain order as multienzyme complexes. Multienzyme complexes could be further classified into multifunctional polypeptide chains, oligomer enzymes, and complexes with components of cell organelles. For example, in yeast and other fungi one polypeptide chain constitutes a multifunctional enzyme that catalyzes five step-reactions continuously, whereas in *Escherichia coli* and plants such an extremely multifunctional enzyme has not been found. One enzyme catalyzes the reactions in steps three and four in plants.

4.3.2.2.2 Localization of Enzymes

All enzymes involved in the shikimate pathway are located in chloroplasts or plastids in plants. In addition, genes of four enzymes in the shikimate pathway have so far been found to have transit peptides for chloroplasts. This fact indicates that chloroplasts are one of the sites where aromatic amino acids are synthesized. However, enzyme activities have also been frequently detected in the cytoplasm. It has not been elucidated whether or not another shikimate pathway is involved in the cytoplasm.

4.3.2.2.3 Regulation Mechanism of Enzymes

In Fig. 94 the reaction steps regulated by feedback inhibition etc. of plant enzymes are marked by an asterisk. It has been found that some isozymes in plants are affected by a feedback inhibition, but others are not affected, and the mode of gene expression by external stimuli is different.

When tryptophan is excessively formed in cells the mutase in step eight is activated, and the pathway is switched to biosynthesis of phenylalanine and tyrosine (Fig. 96). When the level of phenylalanine and tyrosine increases, the reactions in steps 14 and 15 are inhibited, leading to an enhanced level of arogenic acid, and a termination of the reaction in step 11.

4.3.2.2.4 Occurrence of Isoenzymes

Isoenzymes are present in the cytoplasm and plastids in plants, and one to four copies of genes encoding the enzymes in each step have been found in the genomes. The physiological function of these isoenzymes is not known, but specialization of the isoenzymes for the synthesis of aromatic amino acids for proteins, or secondary metabolites has been suggested (Garner and Herrmann 1985).

4.3.2.2.5 Biosynthetic Pathway Via Metabolic Grid

Figure 97 shows a metabolic pathway via a grid. The letters indicate metabolic intermediates, and solid lines show the flow of metabolites. There are several metabolic pathways from A to Z in the model. It has been found that intermediary pathways in the secondary metabolism sometimes make a lattice depending on species or family. These lattice metabolic pathways are named the metabolic grid. The preferential use of a certain pathway in the metabolic grid is named channeling of the metabolic grid.

Plants and microorganisms have specific channeling for the synthesis of phenylalanine and tyrosine. It is interesting to note, in relation to plant evolu-

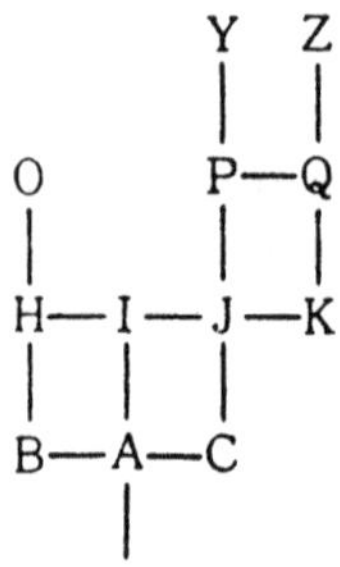

Fig. 97. A scheme of a metabolic pathway via a grid

tion, that blue-green algae have a plant-type channeling and that plant plastids, which may be derived from the blue green algae, have the enzyme system involved in the shikimate pathway.

4.3.3 The General Phenylpropanoid Pathway

The pathway derived from L-phenylalanine to phenylpropanoids is a biosynthetic pathway specific to vascular plants, except that it is also found in a few basidiomycetes.

From 1950 to 1960 Brown, Neish and Higuchi established the pathway of lignin biosynthesis from L-phenylalanine (the cinnamate pathway) mainly by feeding experiments involving ^{14}C-labeled precursors (Brown and Neish 1955a,b, 1956, 1959, Brown 1961, Higuchi 1962, 1990, Higuchi and Brown 1963a,b, Higuchi and Barnoud 1964). Since then lignin biosynthetic studies have resulted in the isolation and characterization of the respective enzymes involved in each reaction step in the pathway (Higuchi 1985, Hahlbrock and Scheel 1989). In addition, the expression of the genes encoding several enzymes in the pathway and their regulations have been investigated since 1990.

The biosynthetic pathway from L-phenylalanine to coniferyl alcohol (monolignol) is shown in Fig. 98 and Table 18.

The name of the general phenylpropanoid pathway was first proposed by Hahlbrock and Grisebach (1975). The pathway involves three enzymes, phenylalanine ammonia-lyase, cinnamate 4-hydroxylase, and 4-coumarate:CoA ligase (steps 16 to 18) in the cinnamate pathway.

The reasons that these three enzymes are distinguished from other enzymes involved in the cinnamate pathway are that (1) the reactions are a common pathway in the biosynthesis of various phenylpropanoids; (2) these three enzymes are induced simultaneously by UV irradiation and fungal elicitors; (3) the microsome fraction contains all activities to convert phenylalanine to 4-coumaric acid (sometimes up to caffeic acid); and (4) these enzymes seem to be regulated by a common gene regulation mechanism.

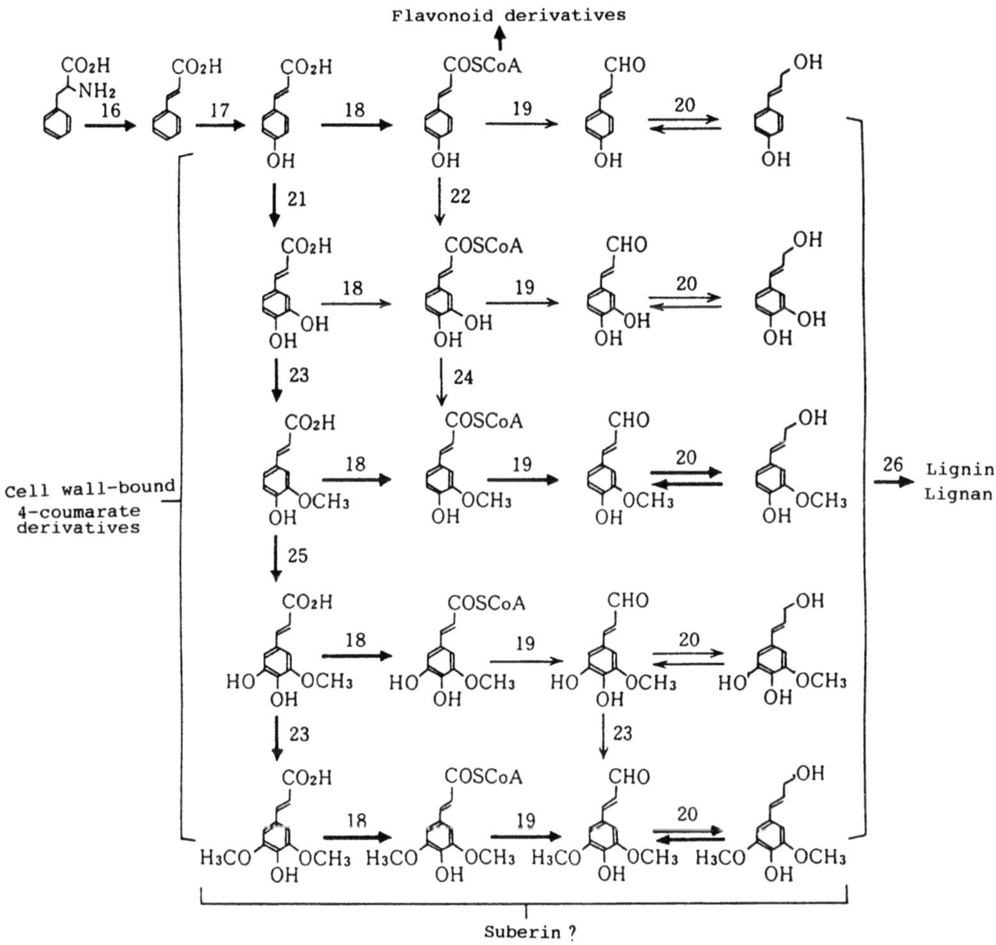

Fig. 98. Biosynthetic pathway from phenylalanine to monolignols. *Thick arrows* denote the generally accepted pathway to lignin biosynthesis. *Thin arrows* denote reaction steps of the enzymes for which there is some evidence. The *numbers* of respective reaction steps correspond to those of the enzymes in Table 18

The name general phenylpropanoid pathway also suggests that the same enzymes (including isoenzymes) are used for biosynthesis of various phenylpropanoids such as lignin and flavonoids.

4.3.3.1 Phenylalanine Ammonia-Lyase

Phenylalanine ammonia-lyase (PAL) and tyrosine ammonia-lyase (TAL) which cause formation of phenylpropanoids from aromatic amino acids were first reported in 1961 by Kokoul and Conn, and Neish, respectively. TAL

Table 18. Enzymes involved in monolignol biosynthesis

Number in Fig. 98	Enzyme name	Cord number
16	Phenylalanine ammonia-lyase	EC 4.3.1.5
17	*Trans*-cinnamate 4-monooxygenase	EC 1.14.13.11
18	4-Coumarate: CoA ligase	EC 6.2.1.12
19	Cinnamoyl-CoA reductase	EC 1.2.1.44
20	Cinnamyl alcohol dehydrogenase	EC 1.1.1.195
20	Coniferyl alcohol dehydrogenase	EC 1.1.1.194
21	Monophenol monooxygenase[a]	EC 1.14.18
23	Caffeate *O*-methyltransferase	EC 2.1.1.42
24	Caffeoyl-CoA *O*-methyltransferase	EC 2.1.1.104
25	Ferulate 5-hydroxylase	EC 1.14. –
26	Peroxidase	EC 1.11.1.7
26	Laccase	EC 1.10.3.2

[a] The former 4-coumarate 3-hydroxylase (EC 1.14.17.2) has been cancelled.

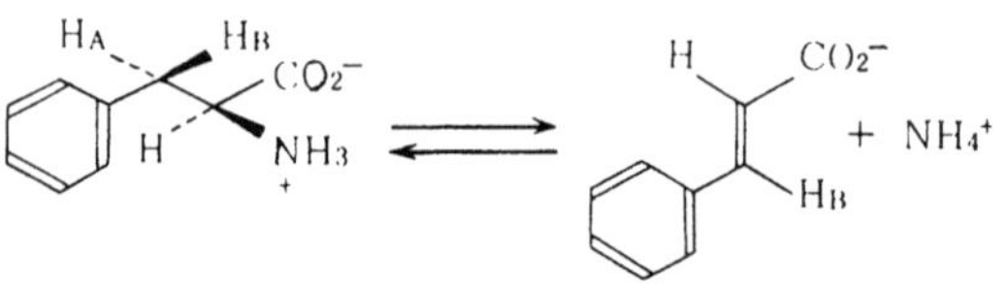

Fig. 99. Formation of *trans*-cinnamic acid from L-phenylalanine by mediation of phenylalanine ammonia-lyase

occurs mainly in grasses, and was initially thought to be a different enzyme from PAL. However, the purified enzymes from maize and yeast have been shown to be single enzymes that have common catalytic sites for L-phenylalanine and L-tyrosine (Hanson and Havir 1981).

PAL is distributed in clubmosses, ferns, and seed plants, but does not occur in mosses and horsetails (Camm and Towers 1973). PAL is exceptionally distributed in some families of the basidiomycetes, in some bacteria (*Streptomyces verticillatus*), in yeast (*Rhodotorula glutinis*), and in algae (*Dunaliella*). PAL is not found in a wild strain of *Escherichia coli*, but the transgenic strain incorporated with a plant PAL gene exhibits PAL activity (Tanaka et al. 1989).

The reaction mechanism of PAL has been investigated in detail by Hanson and Havir (1981). The pro-3S proton and amino group of L-phenylalanine are *trans*-eliminated to afford *trans*-cinnamic acid (Fig. 99). According to the mechanism proposed, L-phenylalanine is first connected to the methylene group of a dehydroalanine residue at the active center of PAL. *Trans*-cinnamic acid is then released, and finally the amino–enzyme complex is hydrolyzed to give ammonia. When the complex reacts with cinnamic acid, L-phenylalanine is formed. As expected from the mechanism, PAL is inhibited by carbonyl reagents such as sodium borohydride, semicarbazide, and by potassium cyanide.

PAL is present as a tetramer with proteins of molecular mass 240 to 330 kDa. In many cases, isozymes have been found. The enzymes are encoded by a small gene family. As in other enzymes in the cinnamate pathway, the base sequence of the PAL gene does not suggest any modification of the enzyme by the ER membrane, or the occurrence of the enzyme in cell organelles.

However, when labeled L-phenylalanine and cinnamic acid are added together to a cell-free system incorporation of L-phenylalanine into 4-coumaric acid is five times greater than that from cinnamic acid (Czichi and Kindl 1977). This result suggests that PAL is not distributed in the cytoplasm at random but localized near the plasma membrane that connected with cinnamate 4-hydroxylase (step 17 in Fig. 98).

Effective turnover of substrates by the enzymes located outside or inside of the membrane in association with membrane-connected enzymes is also named channeling (different from channeling in a metabolic grid). Data to support such channeling in the secondary metabolism have been frequently reported.

Recently, gene families encoding plant PAL have been characterized in *Phaseolus vulgaris* (Cramer et al. 1989), alfalfa (Dixon et al. 1992), parsley (Lois et al. 1989), sweet potato (Tanaka et al. 1989), rice (Minami et al. 1989), tobacco and aspen (Yamakado et al. 1991, Osakabe et al. 1995).

PAL of cell cultures of *Phaseolus vulgaris* is encoded by a small family consisting of three genes. Two genes (*PAL2* and *PAL3*) were cloned and sequenced (Cramer et al. 1989). The degree of the gene expression depends on the organs (vascular bundles) and external stimuli such as wounding and infection. The results suggested that the *PAL2* gene is involved in lignin biosynthesis (Liang et al. 1989a,b, Mavandad et al. 1990). However, the same gene is expressed in epidermal cells and petals, suggesting that the gene is also involved in biosynthesis of flavonoids and anthocyanin pigments.

PAL from alfalfa cell suspension cultures has been resolved into three distinct forms by chromatofocussing or hydrophobic chromatography. These forms are composed of identical subunits of M_r 79000, and the active alfalfa PAL enzyme is a tetramer of M_r 31100. By using a polyclonal antiserum prepared with purified alfalfa PAL, the λ ZAP cDNA expression library was screened. A full-length alfalfa PAL cDNA consisting of a 2175-bp open reading frame, 96-bp 5′-untranslated leader and 128-bp 3′ noncoding region was isolated (Gowri et al. 1991). The amino acid sequence deduced is 86.5% similar to that of the *PAL2* gene from *Phaseolus vulgaris*.

Alfalfa PAL transcripts were rapidly and strikingly induced in cell cultures treated with fungal elicitors, reaching maximum levels approximately 2 h post-elicitation and then declining to near basal level within 24 h. Northern blot analysis of alfalfa seedlings at various stages of development revealed maximum levels of PAL transcripts in roots, stems, and petioles, consistent with the requirement for PAL activity in lignifying tissues.

PAL of parsley is encoded by at least four genes (Lois et al. 1989). cDNA sequences of these genes and the deduced amino acid sequences are highly conserved in many plants.

Aspen (*Populus kitakamiensis*) PAL genes have highly conserved sequences, and are very similar to those of other plants (Yamakado et al. 1991). A genomic clone, designated *pal g1* and corresponding to cDNA1, was isolated from the genomic library of *P. kitakamiensis* and characterized (Osakabe et al. 1995).

pal g1 consists of two exons separated by one intron and encodes a polypeptide of 714 amino acid residues. Another genomic clone, *pal g2a*, which has one intron and encodes a polypeptide of 682 amino acid residues, was also isolated. *pal g1* mRNA was expressed in a tissue-specific manner and was detected at the highest levels in young tissues near the shoot bud. By contrast, *pal g2a* mRNA was mainly accumulated in mature stems and appeared not to be expressed in young tissues.

These results suggested that both *pal g1* and *pal g2a* (including an unidentified *pal g2a*-like gene) or the *pal g2a*-class alone may play a significant role in the lignification of the stems of *P. kitakamiensis*. However, in a conifer (*Pinus taeda L.*) a single gene encoding one enzyme has been reported (Whetten and Sederoff 1991).

It has been reported that in transgenic tobacco plants with a *PAL* gene from a different species there is a decrease of the lignin content in tracheary elements, and the formation of abnormal plants (Elkind et al. 1990).

In phenylpropanoid metabolism, the regulation mechanism of PAL gene expression has been much investigated. Cinnamic acid as a product of this gene inhibits the synthetic rate of this enzyme at the transcription level (Mavandad et al. 1990) and promotes the decomposition of the enzyme. The promoter domain of the *PAL2* gene of *Phaseolus vulgaris* has been investigated in detail.

Analysis of a transgenic tobacco plant with the reporter genes isolated from the promoter gene of *P. vulgaris* showed that the domain contains a *cis* element (about 16 bp) which enhances the expression of PAL in the xylem, but inhibits it in the phloem (Leyva et al. 1992). It is interesting that the domain of the expression regulation of the *PAL* gene has several functions related to tissue-specific gene expression.

The role of PAL in pathway regulation was investigated with near-isogenic transgenic tobacco plants exhibiting a range of PAL levels. In leaf tissues, PAL level is the dominant factor regulating accumulation of the major product chlorogenic acid and the overall flux into the pathway. In stems, PAL at wild-type levels contributes, together with downstream steps, to the regulation of lignin deposition and becomes the dominant, rate-determining step at levels three- to fourfold below wild type. These data suggest that PAL is a key step in the regulation of the overall flux into the phenylpropanoid pathway, and that downstream steps control partitioning into different branch pathways (Bate et al. 1994).

Logemann et al. (1995) isolated a complete gene family encoding PAL from parsley (*Petroselinum crispum*). The *PAL* gene family comprises two closely related members, *PAL1* and *PAL2*, whose TATA-proximal promoter and coding regions are almost identical, and two additional members, *PAL3* and *PAL4* with less similarity. The respective mRNA levels in parsley leaves, in cell cultures treated with UV light or fungal elicitor, and in wounded leaves and roots suggested various degrees of differential responsiveness of *PAL4* relative to the other *PAL* gene family members. The only significant sequence similarities shared by all four *PAL* gene promoters are a TATA-proximal set of three putative *cis*-elements (boxes P, A, and L). The elements appear to be necessary but not sufficient for elicitor- or light-mediated *PAL* gene activation.

4.3.3.2 Cinnamate 4-Hydroxylase (C4H)

The aromatic ring of cinnamic acid derived from phenylalanine is hydroxylated (step 17, Fig. 98). *Trans*-cinnamate 4-monooxygenase catalyzes hydroxylation of the C_4 of cinnamic acid in the presence of O_2 and NADPH (Russell 1971, Potts et al. 1974). The product is 4-coumaric acid. Upon hydroxylation a proton at C_4 is transferred to C_3. This transfer is called NIH shift and suggested to occur via an oxenoid intermediate (Diesperger and Sandermann 1978; Fig. 100).

Cinnamate 4-hydroxylase is widely distributed in the microsome fraction of higher plants. It is inactivated by carbon monoxide and gives a maximum absorption at 450 nm. The activity is recovered by releasing carbon monoxide upon illumination of the inactivated enzyme (Russell 1971). As illustrated in Fig. 101 the hydroxylase is a multienzyme complex belonging to cytochrome P450 monooxygenase. The enzyme is located from the surface to the inner site of the ER membrane, and consists of cytochrome P450 as the terminal oxidase (hemoprotein) and NADPH–cytochrome P450 reductase (flavoprotein: Fpt; Benverniste et al. 1991).

It has been shown that only this enzyme is bound to the ER membrane in the enzymes of the cinnamate pathway. The activities of other enzymes are

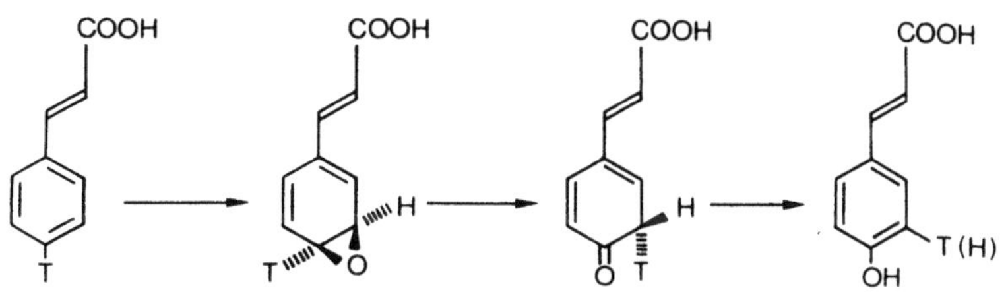

Fig. 100. The hydroxylation of cinnamic acid via an intermediary arene oxide and the NIH shift

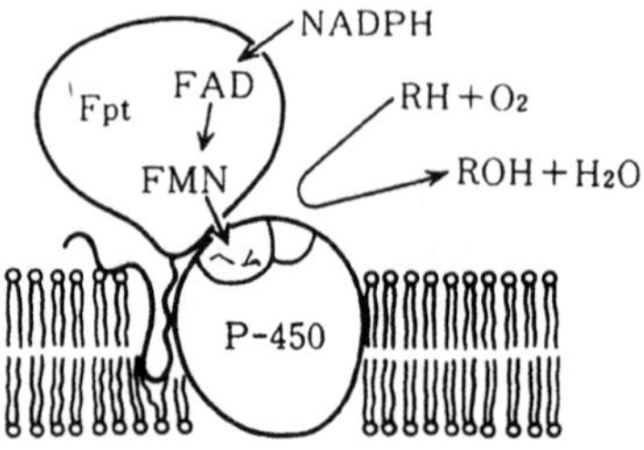

Fig. 101. Multienzyme complex, a model of cytochrome P-450-dependent monooxyganase. *Fpt* Flavoprotein; P-450 hemoprotein

detected in cytoplasm in agreement with the base sequences of their genes. However it is reasonable to suspect that these enzymes are located near the ER membrane to keep intermediates properly in place for effective biosynthesis.

The membrane-connected P450 is unstable to solubilizers and identification of its multiple forms is difficult. Purification of P450 to homogeneity could not be achieved for a long time. However, the solubilization procedure has recently been improved. By taking advantage of the measurement of a minute spectral change in the formation of a P450–substrate complex, P450 of *Helianthus tuberosus* was purified to homogeneity (Gabriac et al. 1991).

The activities of cinnamate 4-hydroxylase with respect to combination with lauric acid, β-pinene, nerol etc. in the microsomal fraction can be separated from that those with respect to cinnamic acid. The antibody for the purified enzyme protein inhibits cinnamate 4-hydroxylases from several plants (Gabriac et al. 1991). Base sequences of the cDNAs from *H. tuberosus*, mung bean, and alfalfa have been identified using the antibodies and the PCR method (GenBank Accession, Meijer et al. 1993).

Two cDNA clones were isolated with the aid of partial amino acid sequences determined for cinnamate 4-hydroxylase (*P450C4H*) purified from mung bean seedlings (Mizutani et al. 1993). The nucleotide sequences of the two clones were nearly identical and contained an open reading frame predicted to encode a polypeptide consisting of 505 amino acid residues. The partial sequences determined from the purified *P450C4H* closely corresponded to the primary structures deduced from the cDNA sequences. By comparison to known cytochrome p450 *p450C4H* was found to belong to a novel P450 gene family.

Teutsch et al. (1993) purified and characterized cinnamate 4-hydroxylase from wound- and Mn^{2+}-induced tissues from tubers of Jerusalem artichokes (*Helianthus tuberosus* L.). A 1130-bp insert was isolated by screening a λgt11 cDNA expression library from wound-induced Jersalem artichokes with highly selective polyclonal antibodies raised against the purified protein. Typical P450 domains identified in this incomplete sequence were used as a probe for the isolation of a 1.7-kb clone in a λgt10 library.

A full-length open reading frame of 1515 bp, encoding a P450 protein of 505 residues (M_r = 57927), was identified and sequenced. The *N*-terminus, essentially composed of hydrophobic residues, matches perfectly the microsequenced N-terminus of the purified protein. Synthesis of the corresponding mRNA which was induced in wounded plant tissues was correlated with the cinnamate 4-hydroxylase activity. This P450 protein was similar (28% amino acid identity) to avocado P450 protein, and was classified as a member of a new P450 family.

4.3.3.3 4-Coumarate:Coenzyme A Ligase (4CL)

In the conversion of 4-coumarate to flavonoids or lignin, the carboxyl group of the acid should be activated (step 18, Fig. 98). Mansell et al. (1972) proved that the coenzyme A (CoA) thioester of 4-coumaric acid is the activated intermediate. 4-Coumarate:CoA ligase, which catalyzes the reaction, requires ATP, CoA, and Mg^{2+}. Cinnamate derivatives are converted to the corresponding CoA ester via AMP–cinnamate derivatives (Fig. 102).

The first step in the reduction of ferulic acid is the activation of the carboxyl group via formation of CoA thioester. Ferulic acid is converted to feruloyl adenylate in the presence of ATP, and feruloyl adenylate subsequently reacts with CoA to afford feruloyl-CoA and AMP (Fig. 102).

Generally, *trans* forms of cinnamates are preferable as substrates. 4-Coumarate and CoA as substrates, and AMP as product inhibit the enzyme reaction depending on their concentrations (Gross 1977).

These reactions are catalyzed by 4-coumarate:CoA ligase (4CL), which is distributed in various higher plants, especially in young lignifying stems. The enzymes isolated from *Forsythia* and *Brassica* have similar substrate specificities. Hydroxycinnamic acids, such as 4-coumaric acid and ferulic acid are effective substrates. Methylated or glycosylated ferulate is not effective, and sinapic acid is also ineffective (Gross 1977).

However, Hahlbrock and Grisebach (1979) found that the ligase activity of soybean cell cultures could be separated into two isoenzymes. Isoenzyme 1 has relatively low K_m and high V/K_m values only for the three typical lignin precursors, 4-coumaric, ferulic and sinapic acids. Isoenzyme 2, on the other hand, has relatively high affinities for 4-coumaric and caffeic acids and does not activate sinapic acid. They suggested that isoenzyme 1 is involved in lignin biosynthesis, while isoenzyme 2 takes part in flavonoid biosynthesis.

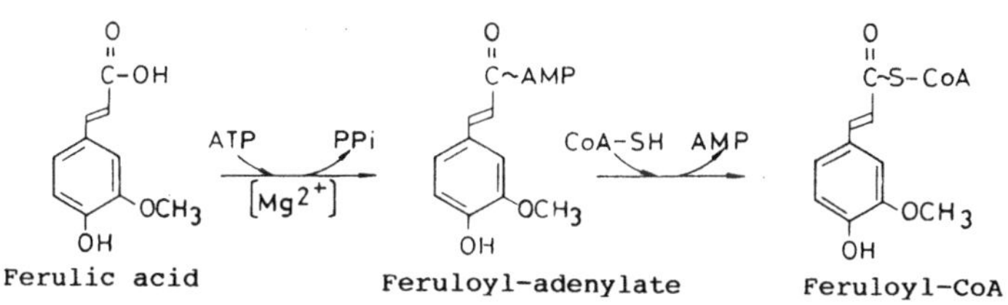

Fig. 102. Formation of feruloyl-CoA from ferulic acid by mediation of CoA ligase

Table 19. Substrate specificities of 4-hydroxycinnamate: CoA ligases from gymnosperms and angiosperms. (Kutsuki et al. 1982a)

Plant species	Activity (p Kat/mg)				
	PA	FA	SA	CA	5-HFA
Gymnospermae					
Chamaecyparis pisifera	–*	73	0	–*	–*
Juniperus chinensis	–*	51	0	–*	–*
Thuja orientalis	206	213	0	144	100
Metasequoia glyptostroboides	337	289	0	179	91
Angiospermae					
Erythrina crista-galli	128	70	30	55	15
Robinia pseudoacacia	301	143	57	112	49
Sophora japonica	–*	44	0	–*	–*
Poplus euramericana	155	137	0	95	51
Paulownia tomentosa	190	141	0	110	48
Liliodendron tulipifera	–*	63	0	–*	–*
Acer buergeriana	–*	260	0	–*	–*
Prunus yedoensis	–*	43	0	–*	–*
Angiospermae (Monocotyledoneae)					
Phyllostachys bambusoides	27	34	13	–*	–*

PA, 4-Hydroxycinnamic acid; FA, ferulic acid; SA, sinapic acid; CA, caffeic acid; 5-HFA, 5-hydroxyferulic acid.
* Not tested.

Kutsuki et al. (1982a) showed that most of angiosperm and gymnosperm CoA ligases are active with ferulic acid but not with sinapic acid, whereas the enzymes of bamboo shoots, *Robinia pseudoacacia*, and *Erythrina crista-galli* are quite active with sinapic acid (Table 19).

Grand et al. (1983) observed three isoenzymes in young shoots of poplars. The isoenzyme that activates sinapic acid is present in xylem, parenchyma cells, and sclerenchyma cells rich in syringyl lignin, while two other isoenzymes, which do not activate sinapic acid, occur in xylem and parenchyma cells rich in guaiacyl lignin. These three isoenzymes exhibited different substrate specificities and tissue distribution, correlating with the monomeric composition of lignins isolated from xylem and sclerenchyma. The results suggested that the hydroxycinnamate:CoA ligase isoenzymes could play an important role in the control of the monomeric composition of lignins in poplars.

Acyl-CoA ligase from the liver microsome of cattle has a low substrate specificity and converts 4-coumarate derivatives to their CoA esters (Knobloch and Hahlbrock 1975). A certain strain of *Pseudomonas putida* effectively produces the CoA ligase when soil microorganisms are cultured in liquid medium containing ferulic acid as the sole carbon source (Gross 1977).

A wild strain of *Escherichia coli* does not have CoA ligase activity, but transgenic strains with CoA ligase exhibits the activity (Douglas et al. 1987). The CoA ligase gene was found to be significantly homologous with the genes encoding proteins to form AMP–substrate–enzyme complexes such as peptide synthase, ester synthase, and luciferase from insects (Becker-André et al. 1991).

Base sequences of two isoenzymes from parsley (Douglas et al. 1987) and potato (Becker-André et al. 1991) were highly homologous, suggesting that the isoenzymes are the same in function and could have branched late in evolution. The genes were expressed by a fungal elicitor or by UV irradiation. Expression of the genes required the base sequence in the coding DNA as well as a promoter sequence composed of 174 bp (Douglas et al. 1991).

It was found that the parsley cinnamoyl-CoA ligase gene (*ps4CL-1*) is expressed strongly in vascular bundles of leaf and flower, and especially in differentiating tracheary elements in xylem. The intact copy of *ps4CL-1* in transgenic tobacco plants correctly responded to endogenous tobacco developmental signals, and demonstrated a complex temporal and spatial pattern of expression during floral organ differentiation (Reinold et al. 1993).

Analysis of the promoter domain of *ps4CL-1* showed (1) the presence of three independent domains (B1, C, and D) containing *cis* elements regulating vascular bundle-specific expression between –244 and –78, which cooperate with domain E2 located near the TATA box; (2) a negative *cis* element inhibiting the gene expression in phloem present at domain C, allowing xylem-specific gene expression; (3) a sequence conserved in *PAL* (*pal* box 1) contained between –137 and –126 in domain C, and (4) a *trans*-acting factor to connect with 42 bp between –120 and –78 that plays an important role for tissue-specific expression (Hauffe et al. 1993). These results are in agreement with the promoter analysis of *fbPAL2* from *Phaseolus vulgaris*. Both *ps4CL-1* and *fbPAL2* are expressed only in the xylem of vascular bundles. This shows that double-sided control, activation of the expression of specific genes in xylem cells but inhibition of the same genes in other cells, occurs to accomplish differentiation of a limited number of cells within the tissue of vascular bundles.

Lois et al. (1989) found that a series of enzymes involved in flavonoid biosynthesis in parsley cell cultures were induced simultaneously by UV irradiation or a fungal elicitor. They found a common *cis* element in the base sequences of upstream regions of the genes of PAL, 4-coumarate:CoA ligase (4CL), and chalcone synthase, suggesting that the element is involved in the simultaneous gene expression of these enzymes.

The promotor of parsley *4CL-1* was expressed in several organs of the transgenic tobacco plants as in the case of PAL (Hauffe et al. 1991).

Recently 4-coumarate:CoA ligase (64 kDa) was purified from differentiating xylem of loblolly pine (*Pinus taeda*) and characterized (Voo et al. 1995). The enzyme used 4-coumaric acid, caffeic acid, ferulic acid, and cinnamic acid as substrates but not sinapic acid. In contrast to angiosperm CoA ligases, no isoform of the enzyme was found even in differentiating xylem that was forming compression wood rich in *p*-hydroxyphenyl lignin, suggesting that the increased synthesis of *p*-hydroxyphenyl lignin in compression wood could not be regulated by differential expression of 4-coumarate:CoA ligase (4CL) isoenzymes with different substrate specificities.

cDNA clones for the ligase (4CL) were isolated from a xylem expression library. The cDNA sequences matched pine xylem 4CL protein sequences and showed 60 to 66% DNA sequence identity with 4CL sequences from herbaceous angiosperms.

4.3.4 Lignin

Lignins are generally classified according to the chemical structure of their monomer units into three major groups, namely softwood lignin, hardwood lignin, and grass lignin. Softwood lignin is an aromatic polymer in which the monomeric guaiacylpropane units are the major component (>90%) and are connected by both ether and carbon–carbon linkages (Fig. 103). Several substructures are involved in lignin macromolecules, of which (1) the guaiacylglycerol-β-aryl ether substructure is the most abundant interphenylpropane linkage (40–60%); followed by the substructures (2) phenylcoumaran (10%); (3) diarylpropane (<5%); (4) pinoresinol (<5%); (5) biphenyl (5–10%); and (6) diphenyl ether (5%), among others (Fig. 104).

Exceptionally, compression wood lignins (*Picea excelsa*, *Pinus radiata*, *Pseudotsuga menziesii* and others) contain remarkably low content of the methoxyl group in comparison with normal conifer wood lignin, and consist of guaiacylpropane and a significant amount of the *p*-hydroxyphenylpropane unit (about 30%; Sarkanen and Ludwig 1971).

Hardwood lignin is composed of guaiacyl and syringylpropane units connected by linkages similar to those found in conifer lignin; the ratio of the syringyl unit to the guaiacyl unit is different among species.

Grass lignin is composed of guaiacyl-, syringyl-, and *p*-hydroxyphenylpropane units also connected by similar linkages to those found in conifer lignin. *p*-Coumaric acid (5–10% of lignin) is mostly esterified at the γ-position of the propyl side chains but is also partly etherified in the lignin (Nakamura and Higuchi 1976, 1978a,b, Iiyama et al. 1990, Lam et al. 1991, 1992, Ralph et al. 1994a,b).

Recent studies have shown that *p*-coumaric and ferulic acids in grasses are connected by ester or ether linkages to hemicelluloses and lignin to form a bridge structure between the two components (Jung and Deetz 1993).

The origin of lignins in plant tissues is closely related to plant evolution. Lignin occurs in cell walls of true vascular plants, ferns, and clubmosses but not in cell walls of mosses, algae, and microorganisms (Kawamura and Higuchi 1964). The land plants, which are believed to have been derived from the

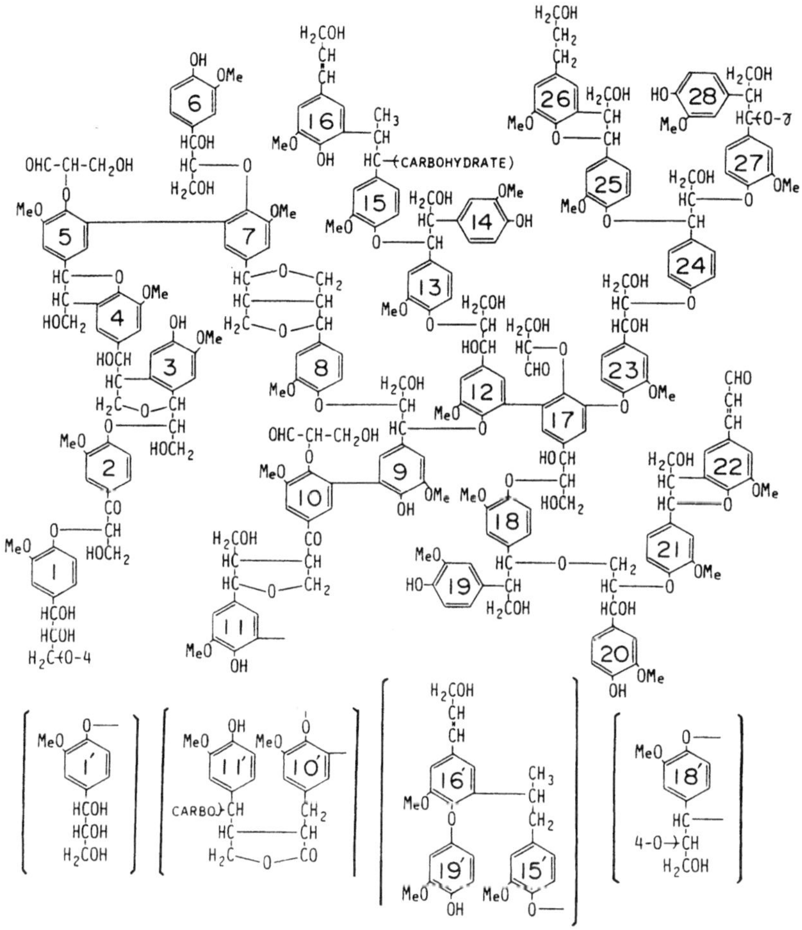

Fig. 103. A structural model of a softwood lignin. (Sakakibara 1983)

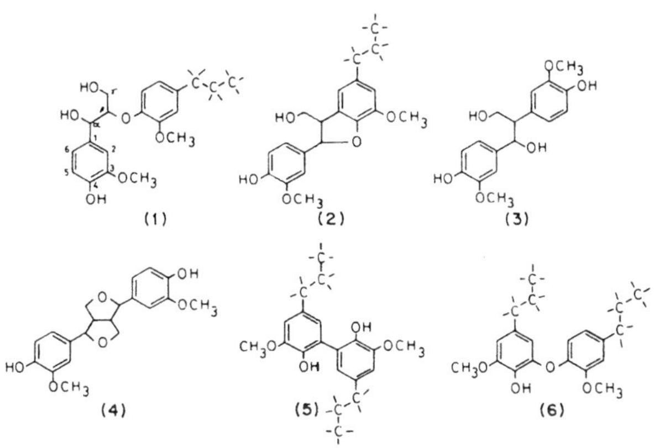

Fig. 104. Dimeric structures containing the principal linkages between monomeric phenylpropane units in lignin macromolecules. **1** Guaiacylglycerol-β-aryl ether; **2** phenylcoumaran; **3** diarylpropane; **4** pinoresinol; **5** biphenyl; **6** diphenyl ether substructures

Psilophytales during the Silurian to Devonian eras, are continually subjected to the mechanical stresses of gravity, wind, and rain. They have therefore acquired strong supporting organs, such as vascular bundles composed of phloem and xylem encrusted by lignin. Lignin generally occurs as an encrusting component with the matrix hemicelluloses between the cellulose microfibrils in primary and secondary walls, and in the middle lamellae. Its function is to connect cells to one another and to reinforce the cell walls of xylem tissues.

Recent studies on selective cleavage of carbohydrate–lignin bonds in lignin-carbohydrate complexes from a conifer (*Pinus densiflora*) have shown that the C_6 hydroxyl group and some of the C_2 and C_3 hydroxyl groups of hexose residues as well as the C_2 and C_3 hydroxyl groups of xylose residues of hemicelluloses are linked by ether linkages to lignin (Watanabe et al. 1986, 1987). In hardwoods (*Fagus crenata* and *Albizia falcata*) the C_2 and C_3 hydroxyl groups of xylose residues and the carboxyl group of glucuronic acid residues of glucuronoxylan are linked to the lignins by ether and ester linkages, respectively (Watanabe et al. 1992, 1993).

4.3.4.1 Formation of Monolignols

Tracer experiments with isotope-labeled lignin precursors and the associated enzyme experiments have elucidated the synthetic pathway for *p*-hydroxycinnamyl alcohols such as coniferyl (8), sinapyl (9), and *p*-coumaryl alcohol (7), which are direct precursors of lignins (monolignols; Fig. 105).

4.3.4.1.1 Hydroxylation of 4-Coumarate Derivatives

Phenoloxidases activate molecular oxygen and hydroxylate 4-coumaric acid to caffeic acid (step 21, Fig. 98). However, phenoloxidases normally catalyze two different reactions, hydroxylation of monophenols to *o*-diphenols (monophenolase), and the oxidation of *o*-diphenols to orthoquinones (oxidase), and generally exhibit very low substrate specificity. Therefore, the role of phe-

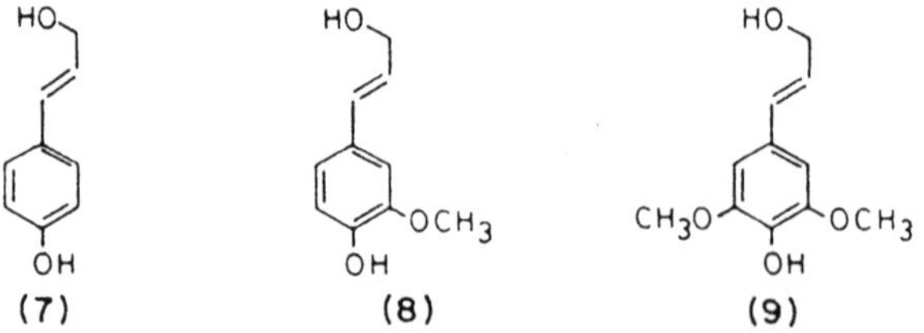

(7) (8) (9)

Fig. 105. Chemical structure of monolignols. **7** *p*-Coumaryl alcohol; **8** coniferyl alcohol; **9** sinapyl alcohol

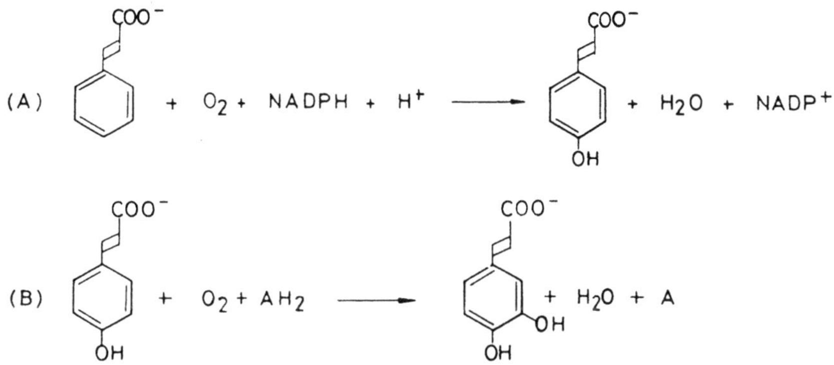

Fig. 106. A,B The reaction sequence catalyzed by *p*-coumarate 3-hydroxylase (phenolase). NAD(P)H, ascorbate, reduced pteridines, or photosynthetic electron transport chains can serve as the external reductant AH_2

noloxidases in the hydroxylation of 4-coumaric acid has been a moot question for long time (Gross 1985).

Vaughan et al. (1975) found that a purified phenoloxidase from spinach beet (*Beta vulgaris*) was a monomeric protein with a molecular mass of 40 kDa, and catalyzed hydroxylation of 4-coumaric acid to caffeic acid. The enzyme is bound to cell particles, especially to chloroplast lamellae, and solubilized by detergents. No changes in the hydroxylase to oxidase ratio were observed during the purification of the enzyme (Vaughan and Butt 1970). The oxidized enzyme remaining after the release of caffeic acid was subsequently reduced by catalytic amounts of the product (caffeic acid). The resulting caffeoyl *o*-quinone was recycled by external reductants such as ascorbic acid, NADPH and reduced puteridines, thus making continued hydroxylation possible (Fig. 106).

On the other hand, investigations with mung bean (*Vigna mungo*) seedlings suggested the presence of a specific *p*-coumarate 3-hydroxylase which is not inhibited by tentoxin, a fungal toxin inhibiting phenol oxidase. Only *p*-coumaric acid served as substrate, and caffeic acid was identified as the product spectrometrically (Kojima and Takeuchi 1989). The enzyme is present in a membrane fraction sedimenting in a sucrose gradient between the mitochondria and the endoplasmic reticulum. However, it is not known whether this specific hydroxylase is distributed in higher plants and is generally involved in phenylpropanoid biosynthesis.

4.3.4.1.2 Methylation of Caffeate Derivatives

O-Methyltransferase (OMT) is required for the formation of guaiacyl and syringyl units of lignins (step 23, Fig. 98). The C_3-hydroxyl group of caffeic acid is methylated to afford ferulic acid by mediation of OMT using *S*-adenosyl

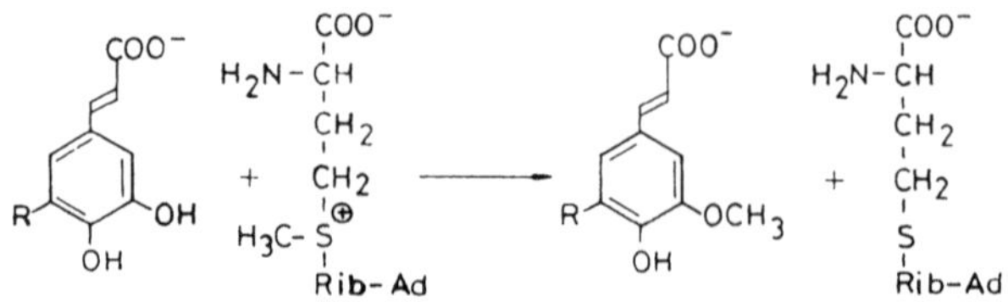

Fig. 107. Enzymatic methylation of caffeic and 5-hydroxyferulic acids *Rib* Ribose; *Ad* adenine; R=H: caffeic acid; R=OCH_3: 5-hydroxyferulic acid

Table 20. Substrate specificity of various plant OMTs. (Shimada et al. 1973a,b)

Substrate	Relative methylation, %		
	Pine	Poplar	Bamboo
Caffeic acid	100	100	100
5-Hydroxyferulic acid	10	320	124
3,4,5-Trihydroxycinnamic acid	25	60	50
Chlorogenic acid	10	46	3
Iso-ferulic acid	0	0	5
m-Coumaric acid	0	0	0
p-Coumaric acid	0	0	0
3,4-Dihydroxyphenyl acetic acid	54	0	0
3,4-Dihydroxymandelic acid	0	0	0
Protocatechuic aldehyde	68	46	45
5-Hydroxyvanillin	0	190	59
Protocatechuic acid	20	0	28
Gallic acid	0	0	0
Pyrocatechol	3	30	0
Pinosylvin	10	–*	–*
D-Catechin	0	0	0
Catechylglycerol-β-guaiacyl ether	0	–*	–*

* Not tested.

methionine as methyl donor. In this reaction, *S*-adenosyl methionine is converted to *S*-adenosyl homocysteine which strongly inhibits OMT activity (Fig. 107).

Bivalent metal ions such as Mg^{2+} are essential for animal catechol-*O*-methyl transferase, but in plants only some plants require the metal ions (Poulton 1981). OMT, which catalyzes the formation of ferulic acid from caffeic acid, was first isolated from apple cambium, and pampas grass (Finkle and Nelson 1963, Finkle and Masri 1964). The properties of caffeate-*O*-methyltransferase, which gives ferulic acid as a lignin precursor, have been well characterized by Higuchi and his coworkers (Higuchi 1990). Flavonoid-specific OMTs, and OMTs which catalyze the C_4 hydroxyl group of the aromatic ring have been reported in plants and basidiomycetes (Poulton 1981, Ohta et al. 1991).

Shimada et al. (1973a,b) investigated the substrate specificity of OMTs from seedlings and young shoots of gymnosperms and angiosperms. The results are summarized in Table 20.

Both caffeic and 5-hydroxyferulic acids are good substrates and 3,4,5-trihydroxycinnamic acid, 5-hydroxyvanillin, protocathuic aldehyde and chlorogenic acid are also fairly good substrates for angiosperm OMTs (Shimada et al. 1970).

The substrate specificity of the gymnosperm OMTs, on the other hand, is completely different, and caffeic acid is the most favorable substrate, followed by protocatechuic aldehyde and 3,4-dihydroxyphenyl acetic acid, whereas 5-hydroxyferulic acid is a poor substrate. Catechylglycerol-β-guaiacyl ether as a dimeric lignin model compound is not methylated by pine (*Pinus thunbergii*) OMT, suggesting that the methylation of lignin takes place at the monomeric stage prior to polymerization of coniferyl and sinapyl alcohols.

The relationship between the ratio of sinapic acid to ferulic acid (SA/FA) formed by OMTs, the ratio of syringaldehyde to vanillin (S/V) on alkaline nitrobenzene oxidation of the lignins, and use of the Mäule reaction of syringyl groups showed that plants with a higher SA/FA ratio, such as angiosperms, give a higher S/V ratio and a positive Mäule reaction, whereas plants with lower SA/FA ratio, such as conifers and ferns, give a lower S/V ratio and a negative Mäule reaction (Table 21; Kuroda and Higuchi 1982).

The results suggested an intimate correlation between the evolution of lignin and the two types of OMTs. Some gymnosperms, such as *Ephedra*, *Gnetum*, and *Podocarpus*, which give a positive Mäule reaction, might have OMTs with higher SA/FA ratios similar to angiosperm enzymes (Shimada et al. 1973a, Higuchi et al. 1977).

The SA/FA ratio of bamboo and poplar wood OMTs remains constant during purification by chromatography on DEAE-cellulose, Sephadex G 200, and hydroxyapatite, and on polyacrylamide gel electrophoresis, and the two methylating activities for FA and SA formation are due to a single enzyme protein (Shimada et al. 1973b, Kuroda et al. 1981). The enzymic methylation of caffeic acid to ferulic acid is competitively inhibited by 5-hydroxyferulic acid, suggesting a preferential formation of syringyl lignin in angiosperms.

The purified OMT from Japanese black pine seedlings (Kuroda et al. 1975), on the other hand, mainly catalyzes the formation of FA. K_m values for FA and SA are 51 and 280 μM, respectively, and SA formation is competitively inhibited by caffeic acid, indicating a preferential formation of guaiacyl lignin in conifers.

In contrast, purified OMT from xylem of *Thuja occidentalis* by chormatography on DEAE-cellulose, hydroxyapatite, and Sephadex G-200 has an SA/FA ratio of 0.4, and the two methylating activities for FA and SA formation are not separable. It seems that such a high SA/FA ratio in *Thuja orientalis* (conifer) OMT is related to the synthesis of dihydroxythujaplicatin methyl ether (a lignan composed of syringyl units) specifically found in *Thuja* species

Table 21. Relationship between the SA/FA ratio, the S/V ratio, and Mäule reaction of lignins. (Kuroda and Higuchi 1982)

Plant species	SA/FA (OMT)	S/V (lignin)	Mäule reaction
Pteridophyta			
Psillotum nudum	0.2	0	–
Angiopteris lygodifolia	0.3	–*	–
Gymnospermae			
Ginkgo biloba	0.1	0	–
Taxus cuspidata	0.1	–*	–
Pinus densiflora	0.1	0	–
P. thunbergii	0.1	0	–
P. taeda	0.3	0	–
P. strobus	0.4	0	–
Thuja orientalis	0.8	0	–
T. standishii	0.03	0	–
Podocarpus macrophylla	0.6	0	–
Angiospermae (Dicotyledoneae)			
Magnolia grandiflora	3.0	2.2	+
Liriodendron tulipifera	2.5	2.7	+
Trocodendron aralioides	1.6	2.6	+
Cercidiphyllum japonicum	3.2	2.9	+
Euptelea polyandra	3.6	–*	+
Nuphar japonicum	2.3	–*	+
Ranunculus acris	3.6	–*	+
Poplus euramericana	3.2	2.1	+
Betula nigra	3.1	2.5	+
Quercus myrsinaefolia	2.5	–*	+
Ulmus americana	3.2	2.6	+
Viscum album	2.2	1.7	+
Erythrina japonica	2.3	0.1	–
Tilia japonica	2.3	–*	+
Paulownia tomentosa	2.9	1.4	+
Pueraria thunbergiana	2.5	–*	+
Angiospermae (Monocotyledoneae)			
Aloe araborescens	1.1	–*	+
Tradescantia virginiana	1.6	–*	+
Oryza sativa	0.9	1.0	+
Triticum aestivum	1.0	1.0	+
Zizania latifolia	1.0	–*	+
Phyllostachys pubescens	1.3	1.1	+
Spaganium stoloniferum	1.5	–*	+
Scirpus triqueter	2.2	–*	+

SA, FA, Activity for sinapic and ferulic acids, respectively; S, syringaldehyde; V, vanillin.
*Not tested.

(Kutsuki et al. 1981). The *Thuja* enzyme efficiently methylates 5-hydroxyferulic acid to sinapic acid, but dihydroxythujaplicatin is not methylated to its methyl ether, suggesting that the latter is formed not by direct methylation of dihydrothujaplicatin but by coupling of coniferyl alcohol with sinapic acid, which cannot be reduced to sinapyl alcohol in conifers (Nakamura et al. 1974).

It is remarkable in relation to plant phylogeny that the angiosperm OMT is a difunctional enzyme that catalyzes the formation of both SA and FA from caffeic and 5-hydroxyferulic acids, and is favorable to the formation of syringyl lignin, while gymnosperm and fern OMTs are essentially monofunctional and preferentially catalyze the formation of FA leading to a guaiacyl lignin. In addition, 5-hydroxyconiferyl aldehyde and caffeoyl-CoA were methylated by the angiosperm OMTs (steps 23 and 24, Fig. 98; Kutsuki 1981, Kuroda 1983).

Recently, OMT cDNAs have been isolated from alfalfa (Gowri et al. 1991), maize (Collazo et al. 1992), and tobacco (Jaeck et al. 1992). Lignin-specific OMT cDNAs of aspen and poplar were also isolated and characterized from the developing xylem tissue and leaves (Bugos et al. 1989, 1991, Gowri et al. 1991, Dumas et al. 1992). Transgenic strains of *Escherichia coli* with the OMT expressed the OMT activity. Genomic hybridization showed the presence of one or two *OMT* genes per haploid poplar genome (Van Doorsselaere et al. 1993a).

A full-length OMT cDNA clone isolated from a poplar leaf cDNA library was sequenced and the amino acid sequence was established. The cDNA insert is 1368 nucleotides in length and contains one open reading frame of 1090 nucleotides encoding a protein of 364 amino acids (calculated M_r 39720). The amino acid sequence of the poplar enzyme is 99% homologous to that of the aspen which was sequenced by Gowri et al. (1991). The poplar *OMT* sequence contains one ATP-binding site (consensus sequence GXGXXG) which may play a role in *S*-adenosylmethionine binding (Dumas et al. 1992).

Xylem specific promoter and transcriptional regulatory sequences for aspen *bi-OMT* have been investigated. Based on the nucleotide sequence of aspen *bi-OMT* cDNA, specific primers, GCTCTAGAGCATGGGTTCAA-CAGGTGAA and GTTGGAAGCTTAAGGCCAATAGG, that are adjacent to a full-length *bi-OMT* cDNA, were used to amplify a 2.7kb DNA fragment from aspen genomic DNA through PCR. Sequence analysis of this 2.7-kb fragment confirmed an exact match of the exon sequences with the cDNA sequence.

The aspen *bi-OMT* gene contains four exons, and three introns between the translation start and stop sites. Within this 1.2kb promoter region, TATA boxes and the transcription start site were identified. Based on electrophoretic mobility shift assay on DNA–nuclear protein binding, three *cis*-acting regulatory sequences were identified in this 1.2-kb promoter region (Chiang et al. 1994).

Jaeck et al. (1992) found that the mRNAs encoding OMTs, which are involved in the biosynthesis of lignin precursors, are highly induced in tobacco leaves during the hypersensitive reaction to tobacco mosaic virus (TMV). OMT messengers were fractionated on a sucrose gradient and translated in vitro. Protein A-Sepharose columns adsorbed with specific antisera raised against purified OMTs were used to select translation products, and the translatable activity of OMT mRNA was measured at different stages of infection. Polypeptides with a molecular mass of 39 kDa (OMT I), which display a strict specificity for phenylpropanoid-type substrates, and minor amounts of polypeptides with molecular masses of 42 kDa (OMT II) and 43 kDa (OMT III), which have a broader substrate specificity, were detected by electrophoretic analysis of the translation products.

Oligonucleotides derived from peptide sequences of purified OMT I were subjected to PCR; total RNA was used as a template to allow the isolation of an OMT I clone. RNA blots, hybridized with OMT I probe, revealed a unique messenger of 1.7 kb. The kinetics of accumulation of OMT I mRNAs during the hypersensitive reaction to TMV parallels the kinetics of translation, suggesting that an increase in mRNA controls the increase in the rate of enzyme synthesis.

In healthy plants, RNA blot hybridizaion showed that the steady-state level of OMT I mRNA is very high in vascular tissues mainly involved in lignification compared to the level in leaves.

The OMT activity for caffeic acid of young poplar trees (*Populus trichocarpa* × *P. deltoides*) was purified both from leaves and xylem by adenosine agarose affinity chromatography and Mono Q chromatography, successively. The molecular mass of the purified enzyme was 38 kDa. The specific activity of the purified 38 kDa enzyme was 0, 30, and 15 nkat/mg for catechol, caffeic acid, and 5-hydroxyferulic acid, respectively. Surprisingly the purified 38 kDa poplar OMT showed only a weak cross-reaction with antibodies of tobacco OMT I, which is supposed to be specific for lignification. The enzyme gave a strong cross-reaction with antibodies of tobacco OMT II which is believed to be involved in the formation of a lignin-like barrier upon infection by TMV.

The results indicate that both the 38 kDa poplar OMT and the tobacco OMT II must have similar epitopes recognized by the anti-(OMT II) antibodies. Northern blots of total RNA isolated from different tissues indicated that the OMT mRNA levels in xylem of trees is approximately 50-fold higher than that in the leaves of the same plant. Both the high OMT activity and the mRNA levels in xylem support the role of the cloned OMT in lignification.

On the other hand, an OMT which catalyzes the methylation of caffeoyl-CoA to feruloyl-CoA (CCoAOMT, step 24, Fig. 98) was purified (Pakusch et al. 1989), and its gene was cloned (Schmitt et al. 1991). The activity of the OMT was not affected by UV irradiation or a fungal elicitor. However, it was shown that at least one of the OMT genes of alfalfa cell cultures is involved in the increase of the transcription product by elicitor treatment (Gowri et al. 1991).

The activity and gene expression of CCoAOMT in parsley cell cultures were induced and then declined in the same pattern as in the enzymes involved in the general phenylpropanoid pathway. CCoAOMT has been detected in several angiosperms, but it is not clear whether CCoAOMT, is involved in lignin biosynthesis (Schmitt et al. 1991).

In addition, cDNAs encoding *S*-adenosylmethionine (adomet) synthase have been isolated from poplar (Van Doorsselaere et al. 1993b), *Arabidopsis thaliana* (Pelman et al. 1989) and parsley (Kawalleck et al. 1992). In *Arabidopsis*, tissue-specific adomet synthase gene expression occurred mainly in vascular tissues, suggesting that the synthase is involved in lignin biosynthesis (Pelman et al. 1989).

In parsley and alfalfa cells, adomet synthase gene expression was induced in response to a fungal elicitor and yeast cell walls, respectively. (Gowri et al. 1991, Kawalleck et al. 1992).

The cDNA clone isolated from poplar leaf (*sam.PdxPt.2*) contains one open reading frame of 1185 nucleotides encoding a protein of 395 amino acids (calculated M_r, 43271). The synthase has 90, 91 and 92% similarity at the protein level with the adomet synthase from rice, *Arabidopsis* and parsley, respectively (Van Doorsselaere et al. 1993b).

4.3.4.1.3 Hydroxylation of Ferulic Acid

For the formation of sinapic acid from ferulic acid, C_5 of the latter acid must first be hydroxylated (step 25, Fig. 98). The enzyme catalyzing this reaction requires molecular oxygen and NADPH. The enzyme in young poplar twigs is located in the microsomal fraction and has been shown to be a cytochrome P450-dependent monooxygenase (Grand 1984). The enzyme activity is higher in sclerenchyma cells of phloem than in xylem. The enzyme is different from cinnamate 4-hydroxylase, but it is unstable and has not been well characterized. 5-Hydroxyferulic acid, the product of this enzyme, is rather unstable but has been identified as an ester in grasses (Ohashi et al. 1987).

Recently a mutant of *Arabidopsis*, defective at a locus designated *FAH1* (*fah1*) was isolated (Chapple et al. 1992). In the *fah1* mutant, the conversion of ferulic acid to 5-hydroxyferulic acid in the general phenylpropanoid pathway appears to be blocked. As a result, the lignin of the mutant lacks the sinapic acid-derived components and the syringyl lignin, typical of the wild-type.

The rosette of wild-type *Arabidopsis* appears pale blue-green under long wave UV light due to the fluorescence of sinapoyl malate in the leaf epidermis. In contrast, the *fah1* mutant which does not contain sinapoyl malate in the leaf epidermis appears dark red under the same conditions. Using the red fluorescent phenotype of the *fah1* mutant, a T-DNA tagged allele was identified in the collection of *Arabidopsis* lines. It was found that the T-DNA insert is very tightly linked to the *fah1* mutation in the tagged line. Using plasmid rescue to retrieve the genomic sequences that flank the T-DNA insertion in the tagged line, a right border-containing plasmid was obtained that contains DNA sequences that hybridize to wild type *Arabdopsis* DNA in Southern analysis. Sequence analysis of the DNA flanking the right border in the rescued plasmid indicated that this region of genomic DNA encodes a protein with high homology to other cytochrome P450-dependent monooxygenases. These data indicate that the *fah1* mutant is defective in the structural gene for F5H (Chapple 1994).

4.3.4.1.4 Reduction of Cinnamate Derivatives

It was found by a tracer experiment that ferulic acid is reduced to coniferyl alcohol via coniferyl aldehyde (Higuchi and Brown 1963b). In addition, feeding experiments with ferulic and sinapic acids to the shoots of poplar, cherry, Japanese red pine, and ginkgo showed that gymnosperms reduce only ferulic acid to coniferyl aldehyde and alcohol, while angiosperms reduce both ferulic and sinapic acids to the corresponding aldehydes and alcohols (Nakamura et al. 1974). The subsequent enzyme studies by Mansell et al. (1972) proved that *p*-hydroxycinnamic acids are converted to the corresponding *p*-hydroxycinnamyl alcohols via *p*-hydroxycinnamaldehydes.

Ferulic and sinapic acids are reduced to the corresponding cinnamyl alcohols by successive mediation of three enzymes, 4-coumarate:CoA ligase, cinnamoyl-CoA reductase and cinnamyl alcohol dehydrogenase, which were first isolated from *Salix* and *Forsythia* (Mansell et al. 1972), and from cell suspension cultures of *Glycine max* (Ebel and Grisebach 1973). The same enzyme system was found in *Brassica* (Rhodes and Wooltorton 1974, 1975) and many other plants (Gross 1977).

4-Hydroxycinnamoyl-CoAs are reduced to the corresponding aldehydes (step 19, Fig. 98) by mediation of 4-hydroxycinnamoyl-CoA reductase (Ebel and Grisebach 1973, Stöckigt et al. 1973, Rhodes and Wooltorton 1975; Fig. 108).

The enzyme requires NADPH as hydrogen donor, and the best substrate is feruloyl-CoA followed by 4-coumaroyl-, sinapoyl-, and 5-hydroxyferuloyl-CoAs. The enzyme does not activate other aromatic or aliphatic esters. It has

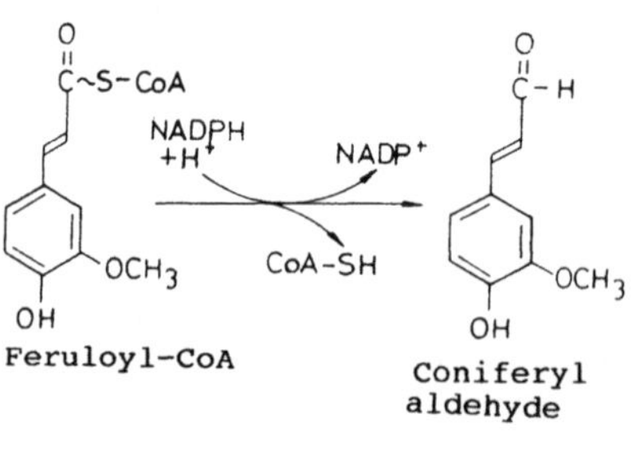

Fig. 108. Enzymatic reduction of feruloyl-CoA to coniferyl aldehyde

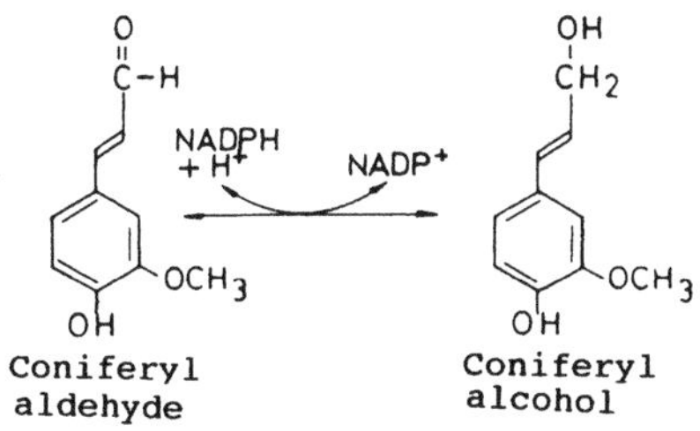

Fig. 109. Enzymatic reduction of coniferyl aldehyde to coniferyl alcohol. The reduction corresponds to step 20 in Fig. 95 and Table 18

been isolated from soybean cell cultures (Wengenmayer et al. 1976, Lüderitz and Grisebach 1981), the cambial zone of spruce (Luderitz and Grisebach 1981), and poplar stems (Sarni et al. 1984).

Cinnamoyl-CoA reductase (CCR) was purified to apparent homogeneity from poplar stems (*Populus euramericana*) and characterized. The reductase was composed of only one subunit of molecular mass 36 kDa, a result which is in agreement with most of the data obtained for the same enzyme isolated from spruce and soybean cells. The reductase exhibited a decreasing affinity towards feruloyl-CoA, sinapoyl-CoA and *p*-coumaroyl-CoA. In spite of differences in the monomeric composition of lignins from xylem and sclerenchyma, the reductase from these two lignified tissues exhibited the same substrate specificity, suggesting that the reductase does not play an important role in the qualitative control of lignins in poplar tissues (Sarni et al. 1984).

CCR activity was feedback inhibited by NADP and CoA. Cloning of the cDNA encoding eucalypt CCR was recently achieved by Lacombe et al. (1994) using oligonucleotides derived from the peptide sequences obtained after digestion of the enzyme by an endoproteinase. A full-length cDNA clone (1193 bp) encoding a polypeptide of 335 amino acids was isolated by oligoscreening of a eucalypt xylem cDNA library. The active recombinant CCR was produced in *E. coli*. A CCR genomic clone has been isolated and the promoter region is being characterized (Van Doorsselaere et al. 1994).

The last step in the formation of 4-hydroxycinnamyl alcohols is the reduction of 4-hydroxycinnamyl aldehydes to the corresponding alcohols mediated by cinnamyl alcohol dehydrogenase (CAD, step 20, Fig. 98; Fig. 109).

The enzyme isolated from *Forsythia* and *Glycine max* (Mansell et al. 1974) showed rather broad substrate specificities for 4-coumaraldehyde, coniferyl aldehyde, sinapyl aldehyde, and their methyl derivatives and required NADPH as hydrogen donor.

Kutsuki et al. (1982b) showed that angiosperm CADs reduce both coniferyl and sinapyl aldehydes to their corresponding alcohols almost equally, but the gymnosperm CADs were remarkably specific for the reduction of coniferyl aldehyde (Table 22).

The purified CAD of Japanese black pine showed that the K_m to NADPH and coniferyl aldehyde are 6.8 and 9.1 μM, and the V_{max} to sinapyl aldehyde is only 2.2% of that for coniferyl aldehyde.

Table 22. Specific activities of cinnamyl alcohol dehydrogenase from gymnosperms and angiosperms. (Kutsuki et al. 1982b)

Plant species	Activity (p Kat/mg)		
	Calc	Salc	Salc/Calc
Gymnospermae			
Pinus thunbergii	330	37	0.11
P. densiflora	–*	–*	0.16
Larix leptolepis	–*	–*	0.22
Ginkgo biloba	300	15	0.05
Chamaecyparis obtusa	–*	–*	0.07
Thuja orientaris	1500	500	0.33
Cryptomeria japonica	680	290	0.43
Metasequoia glyptostroboides	3300	1800	0.55
Angiospermae			
Poplus euramericana	29	43	1.50
Liriodendron tulipifera	460	380	0.83
Robinia pseudoacacia	1600	1900	1.20
Erythrina crista-galli	200	130	0.65
Prunus yedoensis	860	1320	1.53
P. persica	1300	840	0.65
Zelkova serrata	–*	–*	1.20
Angiospermae (Monocotyledoneae)			
Phyllostacys bambusoides	–*	–*	1.69

Calc, Coniferyl alcohol; Salc, sinapyl alcohol.
*Not calculated.

These results suggest that CAD is one of the regulating enzymes like OMT that control the formation of guaiacyl and syringyl lignins. The enzyme is specific for cinnamyl aldehyde derivatives, is distributed in several organs of vascular plants, and is especially active in the cambial zone. CADs were purified and characterized from several higher plants, and the molecular mass was reported to be 69 to 82 kDa (Wyrambik and Grisebach 1975, Lüderitz and Grisebach 1981, Kutsuki et al. 1982b, Sarni et al. 1984).

It was previously reported that the enzyme consists of two identical subunits with molecular masses of approximately 40 kDa. However, recent studies have shown that the enzymes purified from tobacco (Halpin et al. 1992), loblolly pine (O'Malley et al. 1992), *Eucalyptus gunnii* (Goffner et al. 1992) and *Aralia cordata* have two nonidentical subunits of slightly different sizes (Hibino et al. 1993a,b).

Grima-Pettenati et al. (1994) recently isolated and characterized two isoforms of CAD (CAD1 and CAD2) from bean pods, which differ in terms of subunits, M_r, specific activity, substrate affinity and antigenicity. CAD1 is abundant in bean pods, organs with very limited lignification, and has a low affinity for CAD substrates. This enzyme is distinct from benzyl alcohol dehydrogenase, which has a broad substrate specificity.

The enzymatic analyses suggested that the CAD1 isoform may be involved in the synthesis of monolignol derivatives other than lignin, while CAD2 exhibits biochemical properties (subunit M_r, cross-reactivity with *Eucalyptus* CAD2 antibodies), suggesting that it belongs to the very well-conserved CAD2 family that includes all the CAD clones containing the Zn long-chain alcohol dehydrogenase signature. Genes of CADs have been isolated and characterized from tobacco, alfalfa, *Aralia cordata*, *Pinus taeda*, *Eucalyptus botroydes*, *Eucalyptus gunnii* and other plants.

Knight et al. (1992) isolated a pure CAD from tobacco stems and found two closely related polypeptides of 44 and 42.5 kDa. Based on the peptide sequences, mixed oligonucleotides were used to screen a tobacco stem cDNA library, and CAD cDNA clones encoding the two polypeptides were identified. DNA sequence comparisons indicated very high sequence identity between these two clones. The close similarity between the two *CAD* genes suggested that they do not represent different isoforms but are the same from each of the two parental lines of *Nicotiana tabacum*.

In both CAD cDNA clones, the Zn-containing alcohol dehydrogenase signature was found at amino acids 68 to 82. A putative NADP-binding site was also identified at amino acids 188 to 193. No N-terminal targeting sequence was identified, but a putative glyoxysomal targeting sequence "SKL" was found at the C terminus, suggesting that CAD is present as a soluble enzyme and/or located in the glyoxysome.

A full-length cDNA clone encoding a subunit of CAD was isolated from a perennial dicot, *Aralia cordata* (Hibino et al. 1993a, 1993b). The amino acid sequences of peptides derived from the purified CAD protein were highly homologous to regions of the amino acid sequence deduced from the nucleotide sequence of the cDNA. A fusion protein expressed from λgt 11 that carried part of the cDNA reacted with an antibody raised against purified CAD protein. In addition, amino-terminal sequencing of the protein indicated that the mature protein is shorter by two amino acids at the amino terminus than the protein predicted from the nucleotide sequence. A comparison of amino acid sequences indicated that *A. cordata* CAD is a member of the family of alcohol dehydrogenases, and contains zinc-binding residues. The level of expression of the gene was higher in stems than in leaves and roots, consistent with the degree of lignification of these tissues.

The cDNA of the gene controlling CAD in *Eucalyptus botryoides* was recently isolated using *A. cordata* cDNA from a genomic library derived from shoot primodia of *Eucalyptus botryoides* (Hibino et al. 1994). The gene controlling CAD contains four introns and five exons that encord a polypeptide of 355 amino acids. The molecular weight is 38765. The region upstream of the initiation codon contains a putative TATA box sequence, three CAT box sequences (CAAT), and two direct repeat sequences.

Goffner et al. (1992) also isolated and characterized two CAD proteins from *Eucalyptus gunnii* (CAD1 and CAD2). The CAD1 with a molecular mass of 38kDa is probably a monomer and has a low substrate affinity for coniferyl and *p*-coumaryl alcohols and their corresponding aldehydes. No activity with sinapyl aldehyde and alcohol was detected. The more abundant CAD2, whose molecular mass is 83kDa, is composed of two subunits of slightly different molecular masses (42–43kDa). The CAD2 has a high substrate affinity for all the substrates. The two isoforms are immunologically distinct, suggesting their involvement in a specific pathway of monolignol utilization. A full-length cDNA clone (1392bp) was later isolated, using a tobacco CAD cDNA as a probe. The cDNA clone encodes a protein of 356 amino acids, which corresponds to the subunit molecular mass of the *E. gunnii* CAD (CAD2; Feuillet et al. 1993, Grima-Pettenati et al. 1993). Sequence analysis revealed that CAD2 is very well conserved among species, and that it was 97% homologous with the amino acid sequence derived from the *E. botryoides* CAD cDNA.

RNA blot analysis revealed a differential expression pattern of CAD2 in various eucalyptus tissues and organs. The messenger RNA is abundant in developing xylem where lignin synthesis occurs, but it is also present in significant amounts in young stems and leaves, suggesting that CAD2 is also involved in the synthesis of lignans and their glucosides, or suberin.

A quantitative cytochemical method has been devised to determine the localization and level of activity of CAD in plant tissues. The enzyme activity was detected not only in the developing xylem elements and phloem fibers, but also in the epidermal and subepidermal layers of both roots and shoots. This localization pattern suggested that CADs exist at several distinct sites and that the role of this enzyme could be different according to the cell types in which it is located (Baudracco et al. 1993).

A *Cad* gene was isolated from *Pinus taeda* and characterized (MacKay et al. 1995). The identity of the full-length *Cad* cDNA sequence was confirmed by alignment with peptide sequences obtained from a purified active enzyme and by extensive similarity to *Cad* sequences from other species. Southern blot analysis of genomic DNA using the *Cad* cDNA as a hybridization probe gave simple patterns, confirming that the *Pinus taeda Cad* is a single-copy gene as in *4-CL* and *PAL* genes in the same species. The presence of a single gene suggests that different regulatory mechanisms for a single *Cad* gene, rather than differential regulation of several genes as in angiosperms, can account for its expression in response to different factors.

4.3.4.2 Dehydrogenative Polymerization of Monolignols to Lignins

Freudenberg et al. (1951) found that coniferyl alcohol is dehydrogenatively polymerized to a high polymer material (dehydrogenation polymer, DHP), whose properties are closely related to those of conifer lignin, both chemically and spectrometrically (Freudenberg 1965). They isolated several oligolignols including all major substructures of lignin as intermediates in the dehydrogenation reaction, and found that these oligolignols are further polymerized enzymatically to DHP. These results and chemical and spectrometric characterization of the DHP indicated that conifer lignin could be formed by dehydrogenative polymerization of coniferyl alcohol.

Further investigations by Freudenberg (1968) showed that the radicals of monolignols formed enzymatically (Fig. 110) couple in a random fashion to yield quinone methides, which are converted to various dilignols by the addition of water or by intramolecular nucleophilic attack by primary alcohol or quinone groups on the benzyl carbons (Fig. 111).

The dilignols are further dehydrogenated by the enzyme to their radicals, which are finally converted to lignin and lignin–carbohydrate complexes (LCCs) via radical couplings followed by nucleophilic attack on the benzyl carbons of the oligomeric quinone methides by water, by aliphatic and phenolic hydroxyl groups of lignols, and by hydroxyl- and carboxyl groups of sugar residues of cell wall polysaccharides (Fig. 112).

Based on the extensive studies on dehydrogenative polymerization of respective monolignols by peroxidase/H_2O_2 (Freudenberg 1965) and analyses of natural lignins (Adler 1977, Sakakibara 1983), it was established that gymnosperm lignin is mainly formed by dehydrogenative polymerization of coniferyl alcohol, while angiosperm lignin is formed by dehydrogenative

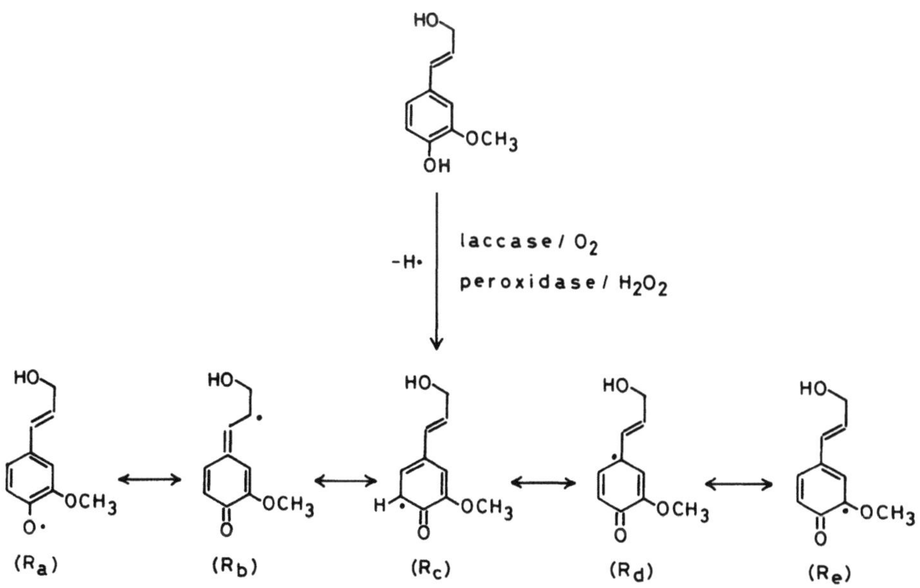

Fig. 110. Mesomeric forms of the phenoxy radical of coniferyl alcohol formed by the mediation of laccase/O_2 or peroxidase/H_2O_2

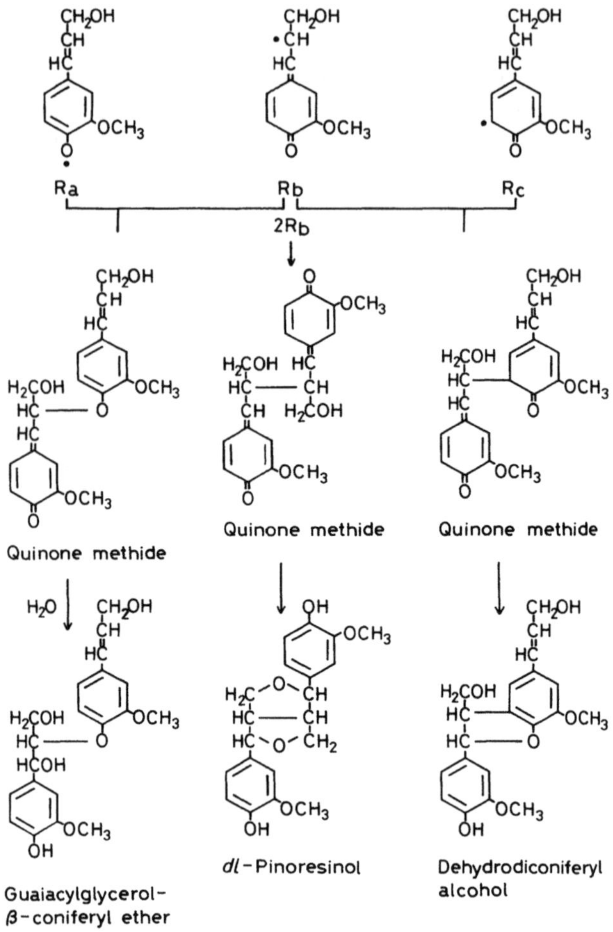

Fig. 111. Formation of dilignols via the coupling of monolignol radicals and subsequent nucleophilic attack by hydroxyl and quinone groups on the benzyl carbons of the quinone methides

polymerization of a mixture of coniferyl and sinapyl alcohols, and grass lignin by the dehydrogenative polymerization of a mixture of coniferyl, sinapyl, and *p*-coumaryl alcohols.

Freudenberg et al. (1951) first found that coniferyl alcohol is dehydrogenatively polymerized to a pale brown precipitate by a crude enzyme preparation from a mushroom (*Psalliota campestris*) under aeration, and that the precipitate (DHP) is closely related to spruce milled wood lignin (MWL).

In the study of the enzymes involved in dehydrogenative polymerization of coniferyl alcohol, Higuchi (1958b) found that the mushroom enzyme prepared by Freudenberg's method was a mixture of laccase (*p*-diphenol oxidase) and tyrosinase (*o*-diphenol oxidase), and that only the former enzyme/O_2

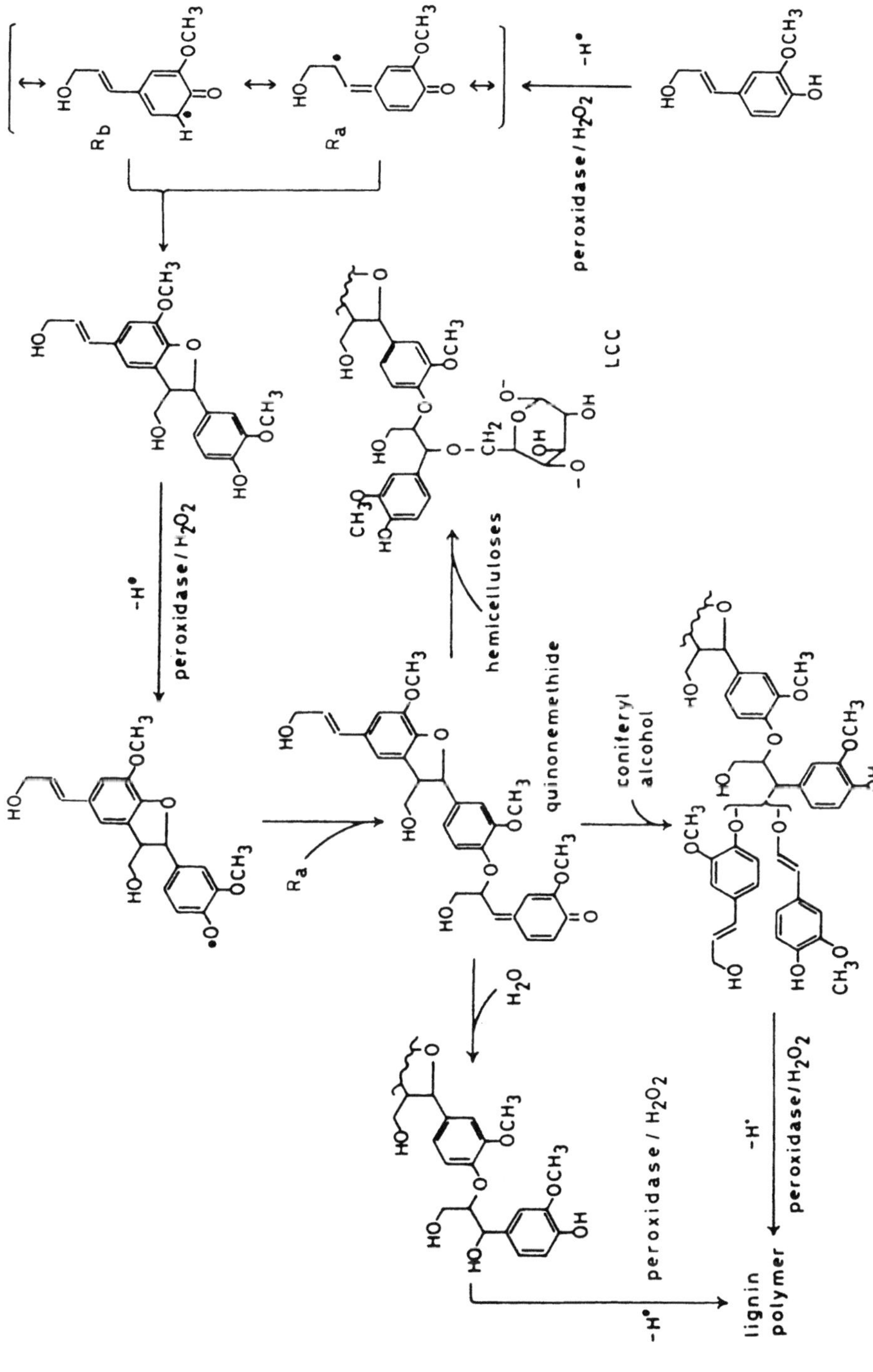

Fig. 112. Formation of lignin polymers and lignin-carbohydrate complexes (*LCC*) via oligolignol quinone methide

mediates the formation of the DHP. Higuchi and Ito (1958) further found that plant peroxidase/H_2O_2 which is widely distributed in higher plants mediates the formation of a DHP of coniferyl alcohol, whose chemical properties resemble those of the DHP obtained with mushroom laccase/O_2.

Higuchi (1959) therefore proposed that peroxidase is involved in lignification of plant tissues. Nakamura (1967) found that a purified plant laccase from the latex of Japanese lacquer trees (*Rhus vernicifera*) is incapable of catalyzing the dehydrogenative polymerization of coniferyl alcohol, but pure bamboo shoot peroxidase dehydrogenates coniferyl alcohol to yield a DHP. These results indicate that peroxidase but not laccase is actually involved in the dehydrogenative polymerization of monolignols in higher plants. This finding was supported later by Harkin and Obst (1973) who found that syringaldazine, a chromogenic substrate for fungal laccase, does not stain tree stem tissue sections unless H_2O_2 is added.

If peroxidase is really involved in lignification, the enzyme should be localized in tracheary elements, especially in the secondary walls, which become heavily lignified. A positive correlation of the localization of peroxidase with lignified walls has been found by several investigators.

Helper et al. (1972) found that peroxidase activity in wound vessel elements of *Coleus* was located in the reticulate secondary wall and in the primary wall where secondary thickening was in progress. They also noted that the enzyme activity was located in the plasmalemma of differentiating vessel elements during secondary thickening, and in the Golgi bodies and their associated vesicles.

Mäder et al. (1977, 1980) characterized four groups of peroxidase isozymes in tobacco: two anionic groups that were located in the cell walls, and two cationic groups that were located in the vacuoles. They showed that the anionic isozymes catalyze the oxidation of hydroxycinnamyl alcohols more effectively than the cationic isozymes.

Sato et al. (1993, 1995) isolated and characterized five cell-bound peroxidase isozymes from differentiating *Zinnia* cells. They found that one of the isozymes, P5, is specifically located in the cell walls of differentiating cultures and is involved in lignin formation from coniferyl alcohol. These results strongly support the involvement of peroxidase in dehydrogenative polymerization of monolignols to lignin in higher plants.

Peroxidases that oxidize various phenols are distributed in plants higher than green algae, while brown and red algae peroxidases oxidize only simple phenols but not guaiacyl compounds (Siegel 1969). cDNAs and genomic clones for peroxidase have been isolated from many different plants and characterized (Simon 1993).

Dehydrogenative polymerization of coniferyl alcohol, localization of the enzyme in tissues, and response against wounding have been investigated in detail with tobacco peroxidase. The gene for acidic peroxidase bound to cell walls was cloned from a cDNA library of tobacco leaves. It was found that the gene was expressed specifically in the stem (Lagrimini and Rothstein 1987, Lagrimini et al. 1987).

Similar genes have been isolated and characterized from *Arabidopsis* and poplar (*Poplus kitakamiensis*). In the domain of their structure genes, four exons and three introns were identified. Comparison of the deduced amino acid sequences of the two cDNAs, designated *pA2* and *pA3*, with those encoded by other known genes for peroxidases revealed several already identified regions of homology (the regions of the two His residues, the distal and proximal His residues that are bound to protoheme). Southern blot analysis indicated that *pA2* and *pA3* correspond to two or three genes in each class of isozyme. The transcripts of the *pA2* and *pA3* groups revealed that both groups were expressed in the lignifying tissue of *P. kitakamiensis* (Osakabe et al. 1994).

In contrast, Driouich et al. (1992) recently showed by immunochemical methods that a laccase is localized in the cell wall of sycamore maple (*Acer pseudoplatanus*) and that the enzyme is most abundant in xylem tissue. In addition, Sterjiades et al. (1992, 1993) demonstrated that laccase could polymerize the monolignols to form lignin-like polymers in vitro. A laccase purified from lignifying xylem of loblolly pine (*Pinus taeda* L.) was also shown to polymerize monolignols (Bao et al. 1993). Furthermore, enzymes capable of catalyzing these reactions in the absence of added H_2O_2 have been detected in other gymnosperms (Savidge and Udagama-Randeniya 1992), and in *Forsythia suspensa* (Davin et al. 1992).

Liu et al. (1994) stained stem tissues of *Zinnia elegans* with chromogenic substrates of laccase, and analyzed the zymogram of the phenoloxidase activities of the enzymes. They found that a single enzyme is predominantly responsible for the laccase-like phenoloxidase activity in the stems of *Z. elegans*. Some of this enzyme is released from cell wall residues by washing with a high ionic strength buffer, although substantial amounts of the enzyme could be recovered only after treatment of the residue with cell wall-degrading enzymes. They suggested that this phenoloxidase shares significant characteristics with the coniferyl alcohol oxidase isolated from developing secondary xylem in pines.

Dean and Eriksson (1994), O'Malley et al. (1993) and Liu et al. (1994) concluded that plant laccase, which mainly catalyzes dimerization of monolignols, is involved in an earlier stage (formation of oligolignols), and that the oligolignols are subsequently polymerized to lignin by peroxidase. The base sequence of cDNA of a plant laccase has been elucidating by Eriksson (per. comm. 1995). It is now considered that both peroxidase and laccase are involved in dehydrogenative polymerization of monolignols.

The origin of H_2O_2 required for the peroxidase reaction was investigated using horseradish cell walls (Elstner and Heupel 1976) and cell walls isolated from *Forsythia* xylem by Gross et al. (1977). They found that H_2O_2 is formed in a complex reaction that involves a dismutation reaction of superoxide radicals ($O_2^{\cdot -}$), generated by the reduction of O_2 with $NAD^{\cdot}$. The $NAD^{\cdot}$ is postulated to be formed via oxidation of NADH, which is provided by the oxidation of malate with NAD: malate oxidoreductase bound to the cell walls, by the phenoxy radicals generated by an oxidase reaction of some phenols

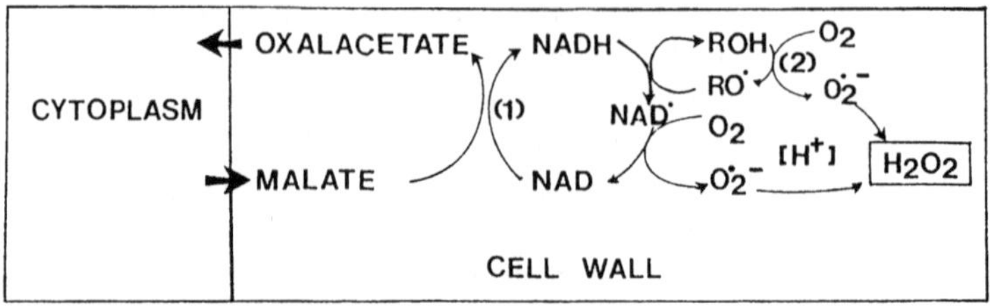

$$2\ \text{MALATE} + 2\ O_2 \longrightarrow 2\ \text{OXALACETATE} + 2\ H_2O_2$$

Fig. 113. A hypothetical scheme for the formation of hydrogen peroxide in the plant cell walls. *1* NAD: malate oxidoreductase; *2* peroxidase-Mn^{2+} complex

with a peroxidase-Mn^{2+} complex (Fig. 113). The formation of H_2O_2 is stimulated by various monophenols and especially by coniferyl alcohol, which indicates that the monophenols could be directly involved in the regulation of H_2O_2 required for the dehydrogenation.

Gross et al. (1977) found that coniferyl alcohol is dehydrogenated, without addition of H_2O_2, in the presence of added malate by lignifying *Forsythia* cell walls, that contain a bound malate dehydrogenase and a peroxidase. It seems that the scheme adequately explains both the dehydrogenative polymerization of lignols and the formation of H_2O_2 at the lignification site, without requiring transport of toxic H_2O_2 from the cytoplasm to the cell wall. It is considered that oxaloacetate formed from malate is transported into the cytoplasm and converted to malate again by the malate–oxaloacetate shuttle and then repeatedly supplied to the cell wall.

Czaninski et al. (1993) demonstrated the occurrence of hydrogen peroxide in lignifying cell walls cytochemically, and found that peroxidase is actually involved in lignin biosynthesis.

4.3.4.3 Structural Differences in Dehydrogenation Polymers

Freudenberg (1968) found that the yields of dehydrodiconiferyl alcohol and *dl*-pinoresinol were considerably higher than the yield of guaiacylglycerol-β-coniferyl ether when a coniferyl alcohol solution is added at once to the peroxidase/H_2O_2 solution (Zulaufverfahren, or high flux), whereas the yield of guaiacylglycerol-β-coniferyl ether increased when the substrate was added dropwise over long periods of time to the enzyme solution (Zutropverfahren, or low flux).

The difference between the polymerization processes in the two methods was discussed in detail by Sarkanen (Sarkanen and Ludwig 1971), who suggested that in the former method the coupling of monolignol radicals, whose concentration is higher, would be predominant, and that the dilignols formed were dehydrogenatively polymerized to tetramers and finally to a large lignin molecule (bulk polymer) via mainly 4–O–5 and 5–5 linkages (Fig. 114). In the latter method, the coupling of monolignol radicals occurs initially to give oligolignols, after which monolignol radicals are successively coupled to the oligolignol radicals initially formed, mainly via β–O–4 linkages, to produce an endwise polymer. This suggestion was made because the concentration of monolignol radicals is lower in the second method (Fig. 115).

Tanahashi and Higuchi (1981) found that DHPs with a higher molecular weight, which are less soluble in organic lignin solvents, are formed when a cellulosic dialysis tube containing horseradish peroxidase is added to a flask containing H_2O_2 and coniferyl alcohol solutions. The molecular weight is about 6000. The results suggest that the concentration of polymer radicals in the tube is high; smaller molecules of oligolignols diffuse through the cellulose membrane into the flask. Couplings between polymer radicals and monomer radicals in the tube would proceed in preference to monolignol–monolignol coupling, resulting in higher molecular weight polymers. In the lignification of

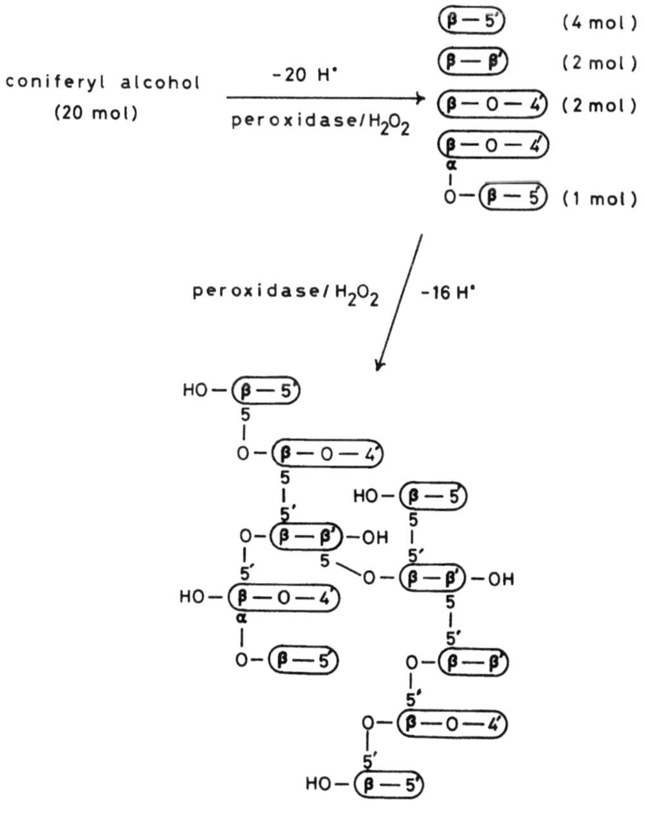

Fig. 114. Formation of a bulk polymer of coniferyl alcohol by the "Zulauf" (high flux) method

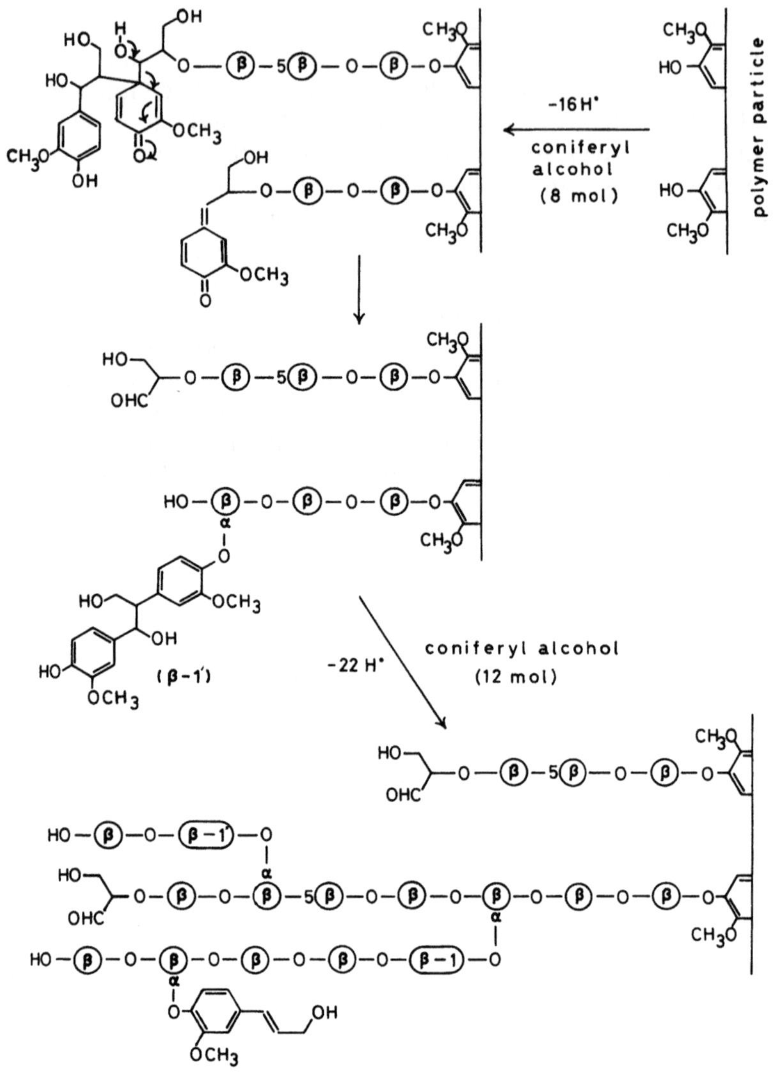

Fig. 115. Formation of an end-wise polymer of coniferyl alcohol by the "Zutropf" (low flux) method

plant cell walls, polymerization of lignins may proceed in a similar way, involving diffusion of lignols across plasma membranes and cell walls.

Tanahashi and Higuchi (1990) also found that large amounts of syringylglycerol-β-sinapyl ether are produced in the dehydrogenation of sinapyl alcohol by $FeCl_3$ when the dioxane content or acidity of the reaction medium is increased. A β–O–4 rich DHP was also formed from sinapyl alcohol by dehydrogenative polymerization in a hemicellulose matrix gel. These results could be explained by the suppression of the E-effect (electron release) and involvement of the I-effect (electron induction) of methoxyl groups in a low polarity solvent (hydrophobicity), and by a low pH of the reaction medium (Tanahashi et al. 1976).

The hydrophobicities of hemicelluloses were recently estimated by a fluorescent probe method (Shigematsu et al. 1994). The maximum wave lengths in emissions (λ_{max}) of 8-anilino-1-naphthalenesulfonic acid (ANS) in hemicellulose solutions were measured, and the blue shift of λ_{max} was estimated as a function of the hemicellulose concentration. The results showed that the hydrophobicity of softwood hemicelluloses was higher than that of hardwood hemicelluloses and pectin. The hydrophobic region in softwood hemicelluloses was notably more hydrophobic than 1,4-dioxane. These results suggest that endwise-type lignin, which is mainly connected by β–O–4 linkages, could be preferentially formed in the hydrophobic regions of hemicelluloses under slightly acidic conditions during lignification of the cell walls.

It has been demonstrated that a basic peroxidase bound to cell walls of poplar callus cultures catalyzed the oxidation of sinapyl alcohol faster than coniferyl alcohol (Tsutsumi and Sakai 1994, Tsutsumi et al. 1994). The enzyme is significantly hydrophobic, and this property seems to be related to its substrate specificity for sinapyl alcohol.

Bruno et al. (1989), and Sipilä and Bruno (1991a,b,c) studied the effect of pH on the formation of noncyclic benzyl aryl and alkyl ethers, benzyl esters, and benzyl glucuronate from a β–O–4-type quinone methide in water–dioxane and organic solutions. The results were compared with structural analyses of milled wood lignins (MWL). The investigators concluded that the lignin biosynthesis might take place in more acidic media than is commonly believed.

4.3.4.4 Differences in Biosynthesis of Lignins Between Tissues and Plants

Musha and Goring (1975) showed by UV spectral analyses of birch lignin in situ that chemical structure of birch lignin is different in wood fibers and vessels. A syringyl lignin occurs in the secondary walls of the fibers, whereas a guaiacyl lignin is present in vessel walls. Monties (1985) postulated a macromolecular and cytochemical inhomogeneity of poplar lignin with respect to guaiacyl and syringyl units, and differences in lignin substructures involving phenolic acids in wheat and poplar lignins. Chemical differences between protolignin of wheat and synthetic lignin (DHP) have also been pointed out based on solid-state ^{13}C NMR analysis of protolignin derived from administered ^{13}C-ferulic acid (Lewis et al. 1987).

Higuchi and Barnoud (1964), and Wolter et al. (1974) found that tissue cultures of hardwoods that are mostly composed of parenchyma cells with small amounts of tracheary elements are rich in guaiacyl lignin. Terashima and Fukushima (1988) showed by radioautography of differentiating pine xylem administered with ^{3}H-labeled monolignol glucosides that *p*-hydroxyphenyl lignin is formed in the compound middle lamellae and cell corners at an early stage of cell wall differentiation. Guaiacyl lignin is deposited as a major component in the compound middle lamellae at an early stage and in the secondary walls at a later stage. The content of condensed guaiacyl lignin is higher in the middle lamellae than in the secondary wall lignin. Syringyl lignin is a minor component of conifer lignin and is formed in the inner layer of the secondary walls at a late stage. It was suggested that lignins in xylem tissue are chemically heterogeneous, that the lignification is controlled by the individual cell, and that the mode of lignin biosynthesis changes with the age of the cell.

Tracer experiments showed that ^{14}C-ferulic acid administered to gymnosperms is mainly converted to guaiacyl lignin, but when ferulate is administered to angiosperms it is converted to guaiacyl-syringyl lignin (Higuchi 1990).

Higuchi (1983, 1985, 1990) investigated the mechanisms for the formation of guaiacyl lignin in gymnosperms and guaiacyl-syringyl lignin in angiosperms and demonstrated that several factors are involved in differentiation of guaiacyl and guaiacyl-syringyl lignins as illustrated in Fig. 116 and given below:

1. OMT of gymnosperms primarily catalyzes the ferulate formation from caffeate, while angiosperm OMTs catalyze not only the ferulate formation but also sinapate formation from 5-hydroxyferulate.
2. Ferulate 5-hydroxylase, a key enzyme in the differentiation of lignin biosynthesis from guaiacyl to syringyl lignins, is distributed only in angiosperms.

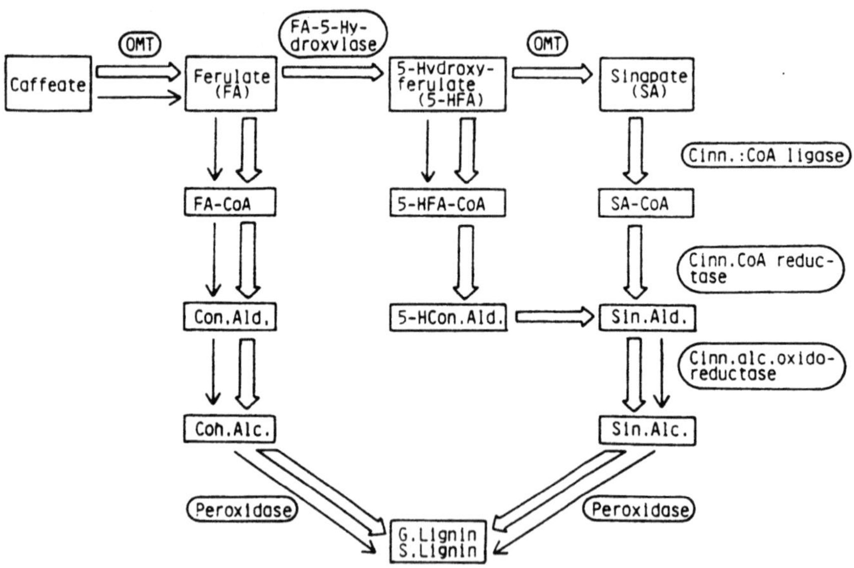

Fig. 116. Regulatory enzymes involved in the biosynthesis of guaiacyl lignin in conifers, and guaiacyl-syringyl lignin in angiosperms. →, Pathway in gymnosperms; ⇒, pathway in angiosperms

3. *p*-Hydroxycinnamate: CoA ligase of gymnosperms is active for only ferulate and *p*-coumarate, whereas some of the angiosperm enzymes are active for *p*-coumarate, ferulate and sinapate.
4. Sinapyl alcohol may synthesized by reduction of the sinapaldehyde formed via successive reactions of 5-hydroxyferulate, 5-hydroxyferuloyl-CoA, 5-hydroxyconiferyl aldehyde, as an alternative pathway in some angiosperms.
5. *p*-Hydroxycinnamyl alcohol dehydrogenase, which mediates the last step of monolignol formation, varies in substrate specificity; gymnosperm enzymes primarily catalyze the formation of coniferyl alcohol from coniferyl aldehyde, whereas angiosperm enzymes catalyze not only the formation of coniferyl alcohol but also the formation of sinapyl alcohol from sinapaldehyde.

It seems that the formation of sinapate: CoA ligase is intimately related to xylem differentiation of angiosperm tissues. This is because poplar and cherry shoots convert ferulate to coniferyl alcohol, and sinapate to sinapyl alcohol, respectively, but the callus of angiosperms reduces ferulate to coniferyl alcohol but not sinapate to sinapyl alcohol (Nakamura et al. 1974). It was also found that the activities of PAL, OMT, *p*-hydroxycinnamate: CoA ligase, and cinnamyl alcohol dehydrogenase are several times higher in compression wood, which contains a larger amount of lignin, than in wood on the opposite side of the tree from the compression wood (Kutsuki and Higuchi 1981).

Grand et al. (1983) found three different forms of hydroxycinnamate: CoA ligase, which exhibit different substrate specificities in poplar stems, and proposed that the isozymes play an important role in the control of the monomeric composition of poplar lignins.

It is obvious that the enzymes involved in the synthesis of monolignol intermediates after ferulate are fundamentally different in gymnosperms and angiosperms. Gymnosperms are programed to synthesize guaiacyl lignin via reactions mediated by the enzymes which preferentially activate guaiacyl intermediates such as ferulate, feruloyl-CoA, and coniferyl aldehyde. In contrast, angiosperms synthesize a guaiacyl-syringyl lignin by mediation of enzymes that adequately activate both guaiacyl and syringyl intermediates, such as sinapate, sinapoyl-CoA, and sinapaldehyde.

Grasses which synthesize a guaiacyl-syringyl-*p*-hydroxyphenyl lignin, contain enzymes whose substrate specificities are almost identical to those of common angiosperms, catalyzing the formation of both guaiacyl and syringyl intermediates. The formation of *p*-hydroxyphenyl lignin and esterified *p*-coumarate characteristically present in grass lignin seems to derive from a high concentration of *p*-coumaric acid additionally supplied from L-tyrosine by tyrosine ammonia-lyase activity of the difunctional grass PAL (Higuchi 1966; Fig. 117).

In both angiosperms and gymnosperms, the activities of the various enzymes in lignin biosynthesis differ according to the stage of xylem differentia-

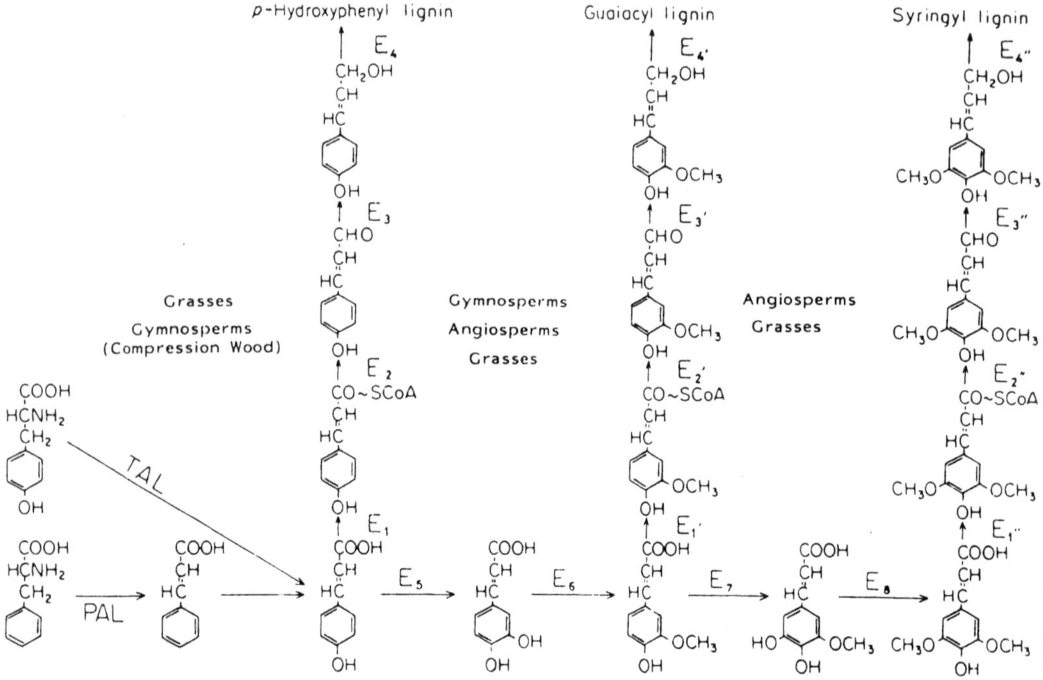

Fig. 117. Biosynthetic pathways of *p*-hydroxyphenyl, guaiacyl, and syringyl lignins. Enzymes: (E_{1}), ($E_{1'}$), and ($E_{1''}$), hydroxycinnamate:CoA ligase; (E_{2}), ($E_{2'}$), and ($E_{2''}$), hydroxycinnamoyl-CoA reductase; (E_{3}), ($E_{3'}$), and ($E_{3''}$), cinnamyl alcohol dehydrogenase; (E_{4}), ($E_{4'}$), and ($E_{4''}$), peroxidase; (E_{5}), *p*-coumarate 3-hydroxylase; (E_{6}) and (E_{8}) hydroxycinnamate *O*-methyltransferase; (E_{7}), ferulate 5-hydroxylase; *PAL* phenylalanine ammonia-lyase; *TAL* tyrosine ammonia-lyase

tion. Certain enzymes such as hydroxycinnamate: CoA ligase form isoforms with different substrate specificity in different tissues and mediate the synthesis of lignins with different monomeric compositions.

2-Aminooxy-3-phenylpropionic acid has been developed as an inhibitor for PAL. Treatment of plants with this inhibitor caused a decrease in lignin content as well as a growth inhibition of the plants (Amrhein and Gödeke 1977). This is probably due to the inhibition of all the gene family for PAL involved in biosynthesis not only of monolignols but also of other essential components for growth.

Inhibition of CAD with *N*-(*O*-hydroxyphenyl)-sulfinamoyltertiobutyl acetate or *N*-(*O*-aminophenyl)-sulfinamoyltertiobutyl acetate has also been reported. The lignin content decreased to 55% of that in the normal plant (Grand et al. 1985).

Excellent review articles on biochemistry and molecular biology of lignin biosynthesis and its regulation (Boudet et al. 1995, Whetten and

Sederoff 1995) have recently been published. Readers should refer to these articles.

4.3.5 Lignans

T. Umezawa

4.3.5.1 Introduction

Lignans constitute an abundant class of phenylpropanoids, and their biosynthetic mechanisms are receiving widespread interest. Many aspects of these mechanisms such as the stereoorganic chemical, the biochemical, and the biotechnological have been studied.

First, the optical rotation of a particular lignan can vary with the plant source (Yamaguchi et al. 1990). No satisfactory explanation, however, has been proposed to account for the stereochemical difference in biosynthesis of the lignans from various plants. The difference in stereochemical mechanisms for lignan and lignin biosyntheses is also of interest. Biosynthesis of optically active lignans may involve enantioselective processes, which sharply contrasts with the nonenantioselective processes in the formation of optically inactive lignins.

Second, a variety of lignans and related compounds have antitumor, antimitotic, antiviral, and insecticidal activities (MacRae and Towers 1984, Ayres and Loike 1990, Gotanda et al. 1993, Sackett 1993). Antitumor *Podophyllum* lignans have been of particular interest, and a number of papers dealing with the biotechnological production of the lignans have been published (Sackett 1993). Detailed knowledge of lignan biosynthesis would be useful for improving this production.

Third, understanding the mechanisms for gene expression in lignan synthesis may lead to an elucidation of the biochemical roles of lignans in higher plants; biosynthetic studies of heartwood lignans may contribute to elucidating the molecular mechanisms of heartwood formation.

Although the biosynthesis of lignans has long attracted much interest among plant chemists, even a single enzymatic step involved in the initial dimerization of phenylpropanoid monomers to give a lignan, or a subsequent functional group modification from one lignan to another had not been reported until 1990. Following the first report of an enzymatic reaction to produce an optically pure lignan from an achiral phenylpropanoid monomer (Umezawa et al. 1990a), much has been learned about the enzymes synthesizing lignans (Chu et al. 1993, Paré et al. 1994), although molecular biological studies have not yet been reported in this field. This chapter summarizes briefly recent developments in this field.

4.3.5.2 Structural Features of Lignans

4.3.5.2.1 Definition and Enantiomeric Composition of Lignans

The term "lignan" was introduced by Haworth to describe a group of phenylpropanoid dimers, where the phenylpropane units were linked by the central carbon (C_8) of their side chains (Haworth 1936). Lignans (Fig. 118) are classified into several subgroups (boldface numbers refer to Fig. 118): dibenzylbutanes (e.g. secoisolariciresinol **1**), dibenzylbutyrolactones (e.g. matairesinol **2** and arctigenin **3**), furans (e.g. lariciresinol **4**), furofurans (e.g. pinoresinol **5** and syringaresinol **6**), aryltetralins (e.g. podophyllotoxin **7**), arylnaphthalenes (e.g. diphyllin **8**), and dibenzocyclooctadienes (e.g. steganacin **9**). Lingans often occur as glycosides (e.g. arctiin **10**). Details of these types of lignans are described in a number of reviews (Rao 1978, Whiting 1985, 1987, 1990, Gottlieb and Yoshida 1989, Nishibe 1989, Ayres and Loike 1990, Ward 1993, 1995).

The C_8 atoms of lignan molecules are asymmetric as shown in Fig. 118, and, therefore, each lignan molecule is chiral, with a small number of exceptions; mesosecoisolariciresinol **11** (Agrawal and Rastogi 1982) and arylnaphthalenes are achiral.

Most lignans isolated from plants are optically active. Thus, either one enantiomer predominates or only one enantiomer is present in a specific lignan specimen. In contrast to amino acids, the optical rotation of a particular lignan can vary with the plant source. For example, (−)-matairesinol **(−)-2** was isolated from *Forsythia koreana*, whereas the (+)-antipode **(+)-2** was obtained from *Wikstroemia sikokiana*. Chiral HPLC analysis established that the lignans are optically pure (Umezawa et al. 1992, Umezawa and Shimada 1996b). Another example is syringaresinol **6** (Yamaguchi et al. 1990); (+)-and (−)-enantiomers were isolated from various plants. In contrast to matairesinol **2**, the syringaresinols isolated from these plants are found to be optically impure with various enantiomeric compositions. Furthermore, even optically inactive lignans have been isolated: (±)-syringaresinols **(±)-6** ($[\alpha]_D \pm 0°$) from *Xanthoxylum inerme* (Ishii et al. 1972) and (±)-eudesmins (dimethyl ether of pinoresinol) ($[\alpha]_D \pm 0°$) from *Carphephorus odoratissiums* (Wahlberg et al. 1972).

Thus, enantiomeric composition of a specific lignan can vary with the plant source from 100% of (+)-enantiomer to 0% of (+)-enantiomer (i.e. 100% of (−)-antipode).

4.3.5.2.2 Differences Between Lignins and Lignans

Lignins and lignans are closely related in terms of their chemical structures. For instance, pinoresinol **5** is a typical lignan, but its structure, i.e., the β-β (pinoresinol or syringaresinol) substructure, is also found in lignin (Sakak-

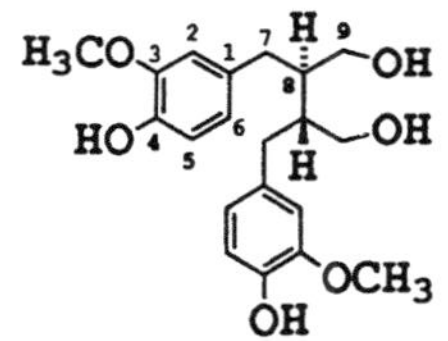

(-)-Secoiso-
lariciresinol (-)-1

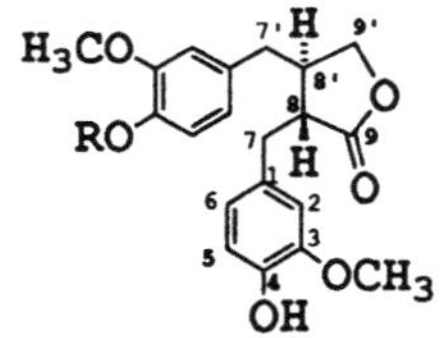

R=H, (-)-Matairesinol (-)-2
R=CH$_3$, (-)-Arctigenin (-)-3

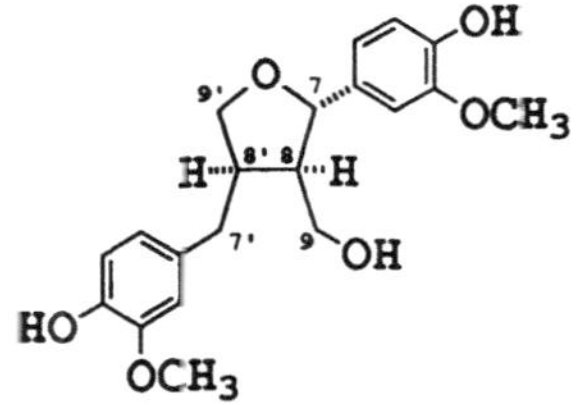

(+)-Lariciresinol
(+)-4

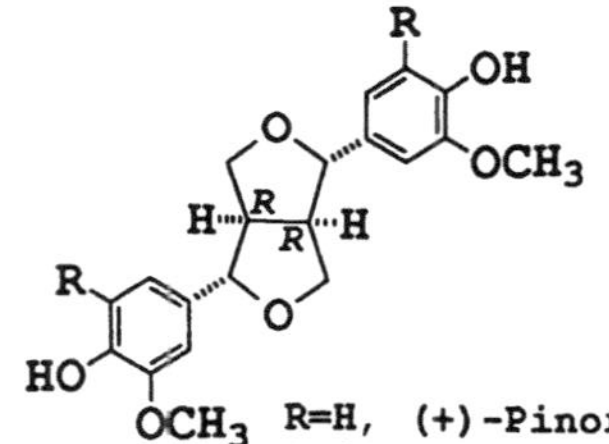

R=H, (+)-Pinoresinol (+)-5
R=OCH$_3$, (+)-Syringaresinol (+)-6

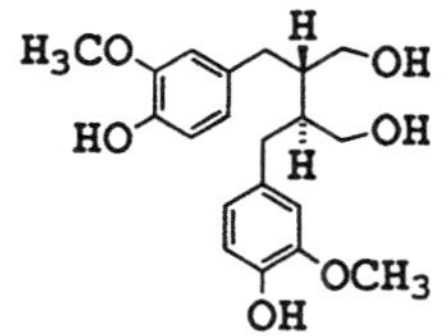

(+)-Secoiso-
lariciresinol (+)-1

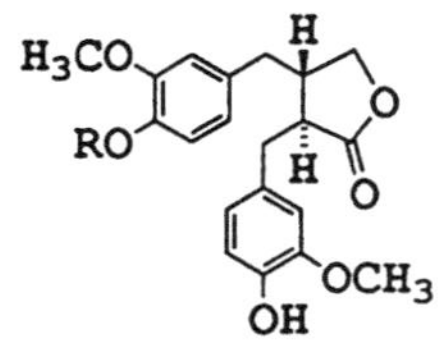

R=H, (+)-Matairesinol (+)-2
R=CH$_3$, (+)-Arctigenin (+)-3

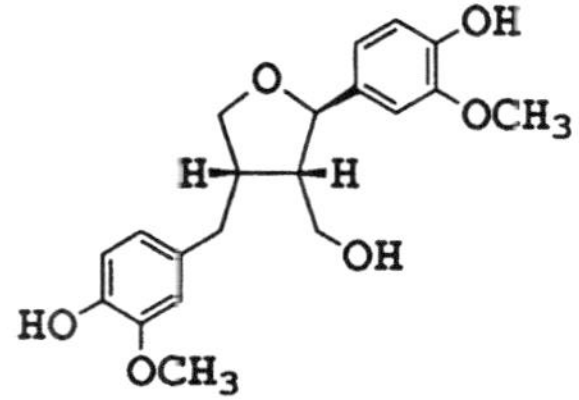

(-)-Lariciresinol
(-)-4

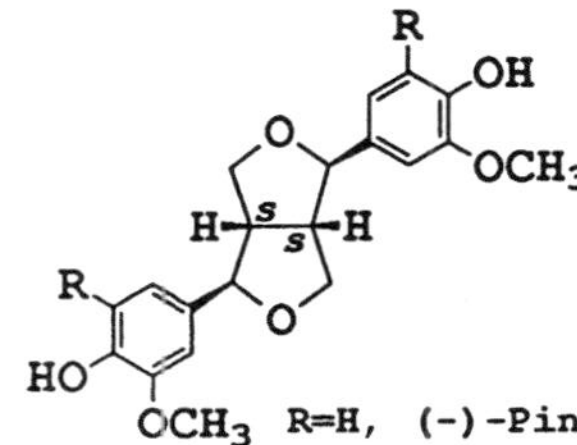

R=H, (-)-Pinoresinol (-)-5
R=OCH$_3$, (-)-Syringaresinol (-)-6

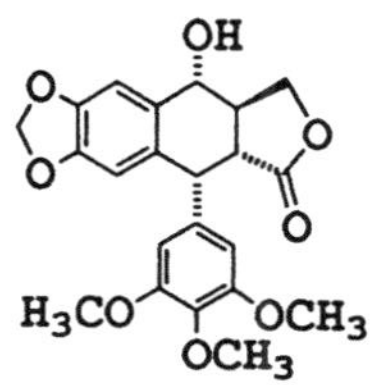

Podophyllotoxin 7

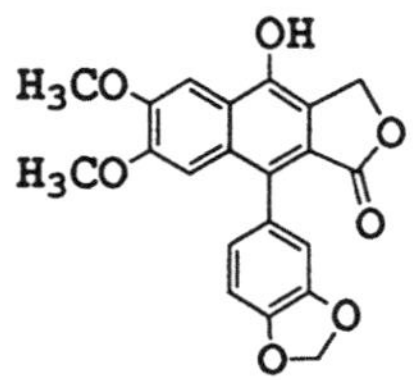

Diphyllin 8

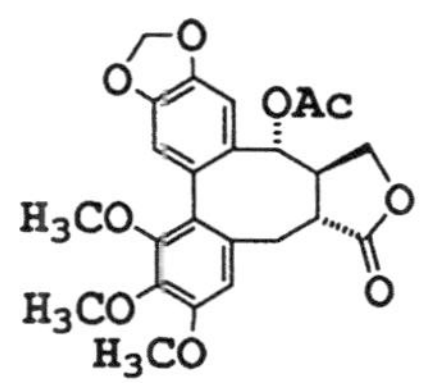

Steganacin 9

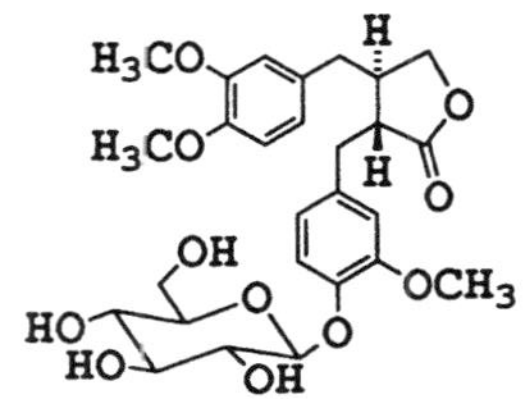

Arctiin 10

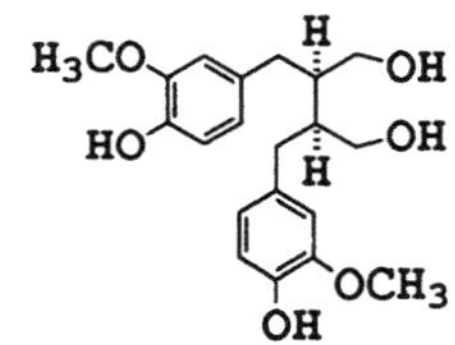

Mesosecoiso-
lariciresinol 11

Fig. 118. Examples of lignans

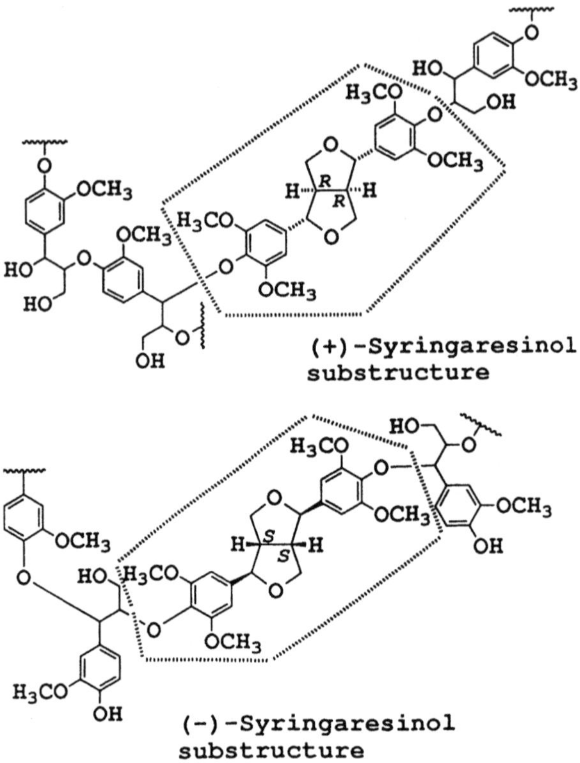

Fig. 119. (+)- and (−)-Syringaresinol substructures in lignin. ~~~~ Continuation of lignan polymer

ibara 1980). Lignins and lignans, however, differ fundamentally in optical activity; the latter are optically active, with rare exceptions, whereas no optical activity has been observed for milled-wood lignin (Björkman and Person 1957) and other lignin specimens (Brauns 1952, Freudenberg 1965). In spite of this optical inactivity, lignin molecules are chiral and not racemic. Lignins can be said to be "racemic-modification-like" with respect to substructures. This may be explained using the β-β substructure as an example (Fig. 119), as follows.

Syringaresinol with the *R* configuration at C8 and C8′ is the (+)-enantiomer **(+)-6**, and with *S* at C8 and C8′ it is the (−)-antipode **(−)-6** (Fig. 118). Similarly, in this chapter a syringaresinol substructure with *R* at C8 and C8′ is referred to as the (+)-syringaresinol substructure, and that with *S* at C8 and C8′ is referred to as the (−)-syringaresinol substructure (Fig. 119). Chemical degradation of lignins gave optically inactive degradation products, e.g., racemic (±)-syringaresinols **(±)-6** after hydrolysis of *Fagus sylvatica* (Nimz and Gaber 1965) and *Fraxinus mandshurica* (Omori and Sakakibara 1971). This indicates that the lignin molecules as a whole contain equal amounts of (+)-

and (–)-syringaresinol substructures. Furthermore, optically inactive di- and trimers derived from other substructures were isolated following hydrolysis (Aoyama and Sakakibara 1978) or hydrogenolysis (Hwang and Sakakibara 1981, Yasuda and Sakakibara 1981) of wood. This strongly suggests that the lignin macromolecules are composed of equal amounts of (+)- and (–)- types of each substructure.

Thus, lignin is not racemic but can be said to be "racemic-modification-like" with respect to substructures. Optical rotation due to each lignin substructure may cancel out, resulting in no observed optical activity.

4.3.5.3 Biosynthetic Pathways of Lignans – In Vivo Experiments

Most of in vivo studies of lignan biosynthesis have employed *Podophyllum* and *Forsythia* lignans as target compounds. One reason is that the *Podophyllum* lignans are of considerable interest as precursors for clinically useful anticancer drugs, and another is that *Forsythia* lignans, which have a simpler substituent pattern than those of *Podophyllum*, were expected to be useful for investigations of the initial coupling of phenylpropanoid monomers.

4.3.5.3.1 Monomeric Precursors of Lignans

Erdtman (1933) first suggested that the lignan structure was formed by coupling of two phenylpropanoid units. Later, this assumption was substantiated by radioisotope and stable isotope tracer experiments. These investigations were reviewed by Ayres and Loike (1990) and by Dewick (1989).

Ayres (1969) initially reported incorporation of [U-^{14}C]phenyl alanine **24** into podophyllotoxin **7** in a whole plant of *Podophyllum emodi* (Fig. 120A). Stökigt and Klischies (1977) showed incorporation of [8-^{14}C, $OCH_2{}^3H$]glucoferulic acid **30**, [8-^{14}C, $OCH_2{}^3H$]glucoconiferyl aldehyde **31**, and [8-^{14}C, $OCH_2{}^3H$]coniferin **32** into arctiin **10** and phillyrin **12** in shoots of *Forsythia suspensa* var. *fortunei* (Fig. 120B). They showed that the precursors are incorporated without degradation of the carbon skeleton of the phenylpropane unit.

In the same year, Fujimoto and Higuchi (1977) reported incorporation of L-[U-^{14}C]phenylalanine **24**, [8-^{14}C]ferulic acid **26**, and [8-^{14}C]sinapyl alcohol **33** into liriodendrin **13** (diglucoside of syringaresinol) in shoots of *Liriodendron tulipifera* (Fig. 120C). Jackson and Dewick (1984a) conducted feeding experiments with various [^{14}C]phenylpropanoid monomers; cinnamic acid **25**, ferulic acid **26**, and 3,4-methylenedioxycinnamic acid **27** were found to be incorporated into podophyllotoxin **7** and demethylpodophyllotoxin **14** in *Podophyllum hexandrum* (syn. *emodi*; Fig. 120A). These investigations unequivocally confirmed that phenylpropanoid monomers are the precursors of lignans.

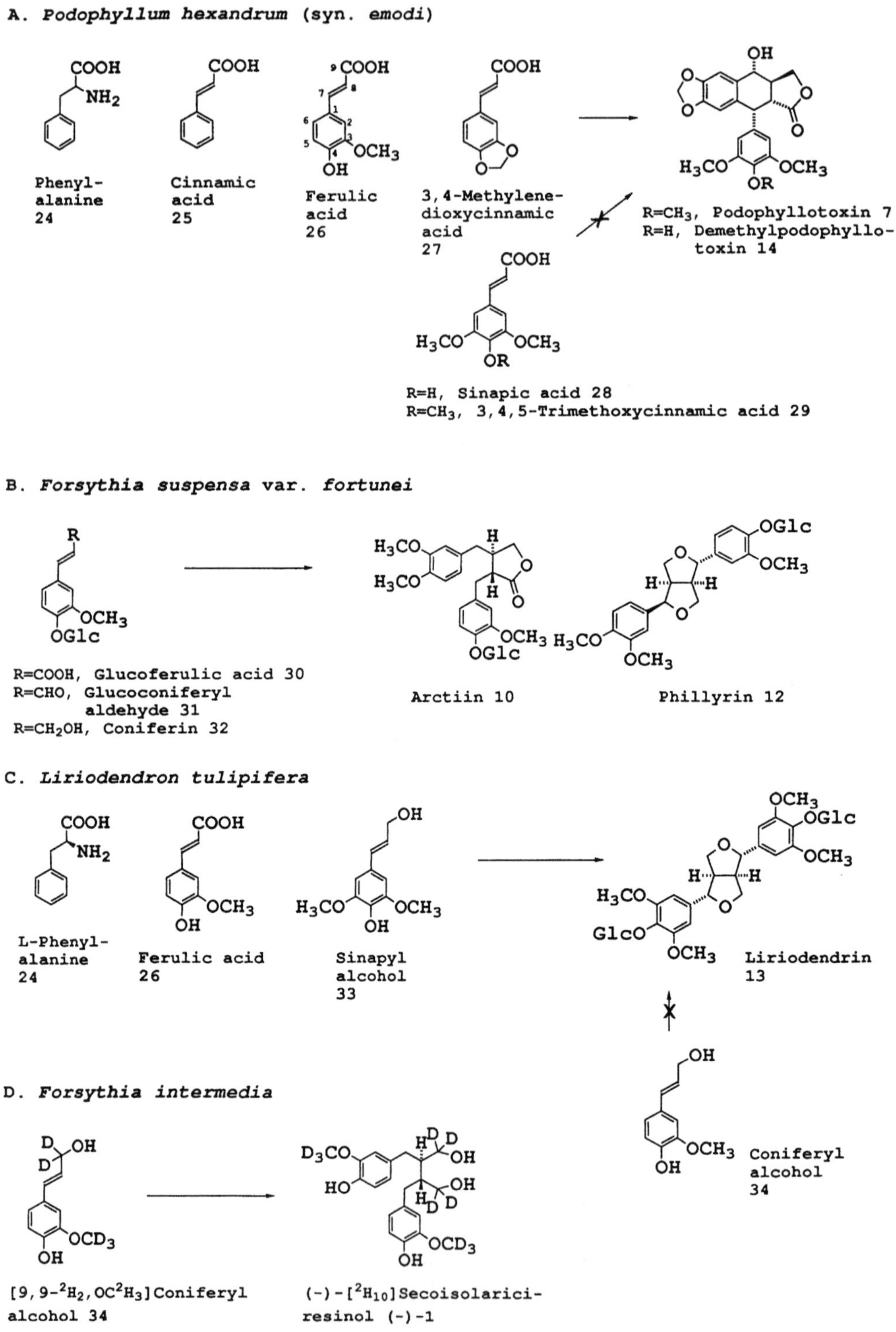

Fig. 120. A–D Feeding experiments with monomeric precursors. *Glc* Glucoside

At the same time, attention was also directed to establishing the chemical identity of the phenylpropanoid monomer(s) undergoing coupling. Stökigt and Klischies (1977) briefly reported, although data were not shown, that when double-labeled [9-^{14}C, 9-^{3}H]coniferin **32** was administered, there was no reoxidation to ferulic acid **26** and subsequent dimerization at this oxidation level, inferring that coupling occurs between two coniferyl alcohol units. Later, this result was confirmed by similar feeding experiments with [8-^{14}C, 9-^{3}H]coniferyl alcohol **34** in *Forsythia* × *intermedia* by Rahman et al. (1990). They showed that the double-labeled coniferyl alcohol **34** was incorporated into arctigenin **3**, phillygenin, and epipinoresinol with a significant increase in the ^{3}H:^{14}C isotopic ratio.

Umezawa et al. (1990b, 1991) found that deuterium atoms at the C9 of the pentadeuterated coniferyl alcohol **34** were retained in optically pure (–)-[^{2}H$_{10}$]secoisolariciresinol **(–)-1**, when [9,9-^{2}H$_2$, OC^{2}H$_3$]coniferyl alcohol **34** was administered to *F.* × *intermedia* (Fig. 120D). These results indicate coupling of two coniferyl alcohol units at least in *Forsythia* plants, and this was substantiated by enzymatic work as described in the following section.

Many lignans have 3,4,5-trimethoxy or 3,5-dimethoxy-4-hydroxy (syringyl) aromatic units. Podophyllotoxin **7** and syringaresinol **6** are typical examples. It was of interest to establish whether these trimethoxy and dimethoxyhydroxy units are built up before or after coupling of the monomers. Jackson and Dewick (1984a) showed that [^{14}C]sinapic (3,5-dimethoxy-4-hydroxycinnamic) acid **28** and [^{14}C]3,4,5-trimethoxycinnamic acid **29** were not significantly incorporated into demethylpodophyllotoxin **14** and podophyllotoxin **7** in *P. hexandrum*, whereas [^{14}C]ferulic acid **26** was incorporated into the lignans, showing that the substitution pattern of the pendent aryl rings is built up after coupling of the two phenylpropane units (Fig. 120A). This was further confirmed by the finding that matairesinol **2** with two 4-hydroxy-3-methoxy aromatic units serves as an precursor of the lignans in the plant (Broomhead et al. 1991).

On the other hand, Fujimoto and Higuchi (1977) proposed coupling of sinapyl alcohol units to afford syringaresinol **6** and liriodendrin **13**; [8-^{14}C]coniferyl alcohol **34** was poorly incorporated into liriodendrin **13** in *L. tulipifera*, whereas significant incorporation of [8-^{14}C]sinapyl alcohol **33** was observed (Fig. 120C).

4.3.5.3.2 Postcoupling Modifications of Lignans

There is a great diversity of aromatic substituents and degrees of oxidation in the chemical structures of lignans (Rao 1978, Whiting 1985, 1987, 1990, Gottlieb and Yoshida 1989, Nishibe 1989, Ayres and Loike 1990, Ward 1993, 1995). Postcoupling modifications of some lignans, especially *Forsythia* and *Podophyllum* lignans, have been well demonstrated.

Biosynthetic sequences of *Forsythia* lignans, coniferyl alcohol **34** → secoisolariciresinol **1** → matairesinol **2** → arctigenin **3**, were established based on feeding experiments using radioisotope- and stable isotope-labeled compounds; Umezawa et al. (1990b, 1991) reported administration of (±)-[Aromatic-^{3}H] and (±)-[Aromatic-^{2}H]secoisolariciresinols **(±)-1** to *F.* × *intermedia* shoots, resulting in a significant conversion into labeled (−)-matairesinol **(−)-2**; no evidence for the formation of the corresponding (+)-enantiomer **(+)-2** was found. Rahman et al. (1990) showed that [^{14}C]matairesinol **2** was an excellent precursor of arctigenin **3** when fed to shoots of *F.* × *intermedia*. Subsequently, these transformations were demonstrated with cell-free extracts as described in the following section (Fig. 123).

Podophyllum aryltetralin lignans can be subdivided into two groups. One contains a 3,4,5-trimethoxy substituted pendent aromatic ring, e.g. podophyllotoxin **7**, while the other has a 3,5-dimethoxy-4-hydroxy substituted one, e.g. demethylpodophyllotoxin **14**. In a series of feeding experiments in *Podophyllum* plants and *Diphylleia cymosa* Dewick and coworkers demonstrated that [^{14}C]matairesinol **2**, [$O^{14}CH_3$]yatein **15**, [$O^{14}CH_3$] and [$OCH_2{}^{3}H$]desoxypodophyllotoxins **16** were incorporated into podophyllotoxin **7**, while [$O^{14}CH_3$]podorhizol **21**, [$O^{14}CH_3$]epipodorhizol **22**, and [$O^{14}CH_3$]anhydropodorhizol **23** were poorly incorporated (Jackson and Dewick 1984b, Kamil and Dewick 1986a,b, Broomhead et al. 1991). Incorporation of [$O^{14}CH_3$] and [$OCH_2{}^{3}H$]desoxypodophyllotoxins **16** into β-peltatin **17** was also reported. Based on these experiments, the authors cited proposed a biosythetic pathway from matairesinol **2** to the trimethoxyaryltetralin lignans in *Podophyllum* plants (Fig. 121).

A similar sequence of reactions was proposed for the corresponding demethyl series based on feeding experiments (Fig. 121; Jackson and Dewick 1984b, Kamil and Dewick 1986a): incorporation of [8-^{3}H]demethyldesoxypodophyllotoxin **19** into demethylpodophyllotoxin **14** in *P. hexandrum*, and of [$O^{14}CH_3$]demethyldesoxypodophyllotoxin **19** into α-peltatin **20** in *P. peltatum*. In addition, the authors concluded that these demethyl compounds did not appear to be methylated to the corresponding 3,4,5-trimethoxyphenyl series, because [8-^{3}H]demethylpodophyllotoxin **14** and [$O^{14}CH_3$] and [8-^{3}H]demethyldesoxypodophyllotoxins **19** were poorly incorporated into podophyllotoxin **7**.

4.3.5.4 Enzymatic Lignan Formation

4.3.5.4.1 Enantiomeric Composition of Enzymatically Formed Lignans

The initial stage of lignin biosynthesis involves formation of pinoresinol **5** and/or syringaresinol **6** as well as other dilignols; these reactions are believed to be catalyzed by nonenantioselective peroxidase and laccase (Freudenberg 1965, Higuchi 1985, Sterjiades et al. 1993). The pinoresinol **5** and syringaresinol **6**

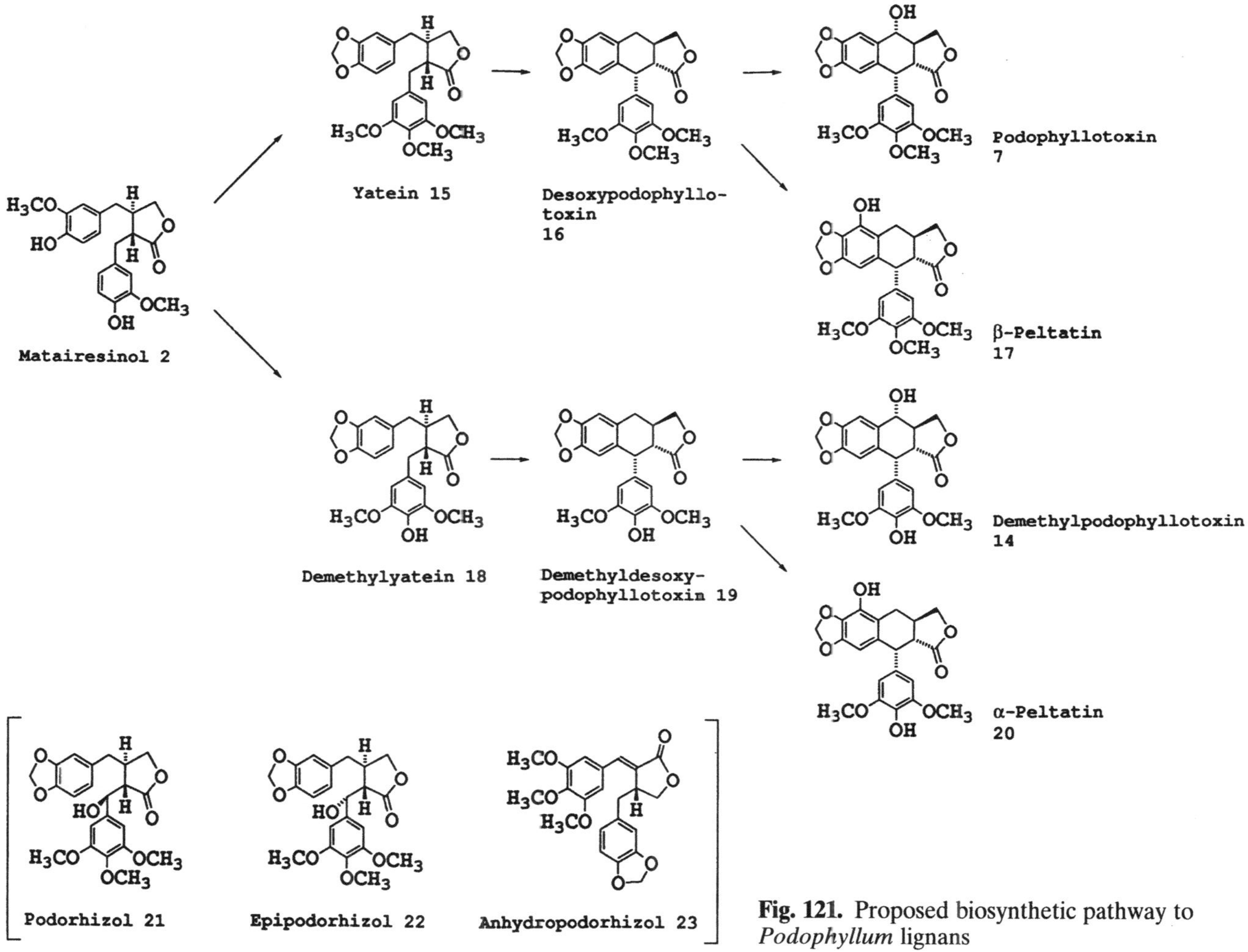

Fig. 121. Proposed biosynthetic pathway to *Podophyllum* lignans

present as intermediates in lignin biosynthesis are racemic, whereas naturally occurring lignans including the two resinols are normally optically active.

Hence, in order to distinguish enzymes for lignan and lignin syntheses, the authenticity of lignan-synthesizing enzymes is being established by optical activity of the product lignans; preferential occurrence of one enantiomer must be demonstrated. This can be done by chiral HPLC and LC-MS techniques (Davin et al. 1991, Umezawa et al. 1992); only a few μg of sample are enough to determine the enantiomeric composition, provided that the chiral HPLC separation is established for the enantiomeric pair of a particular lignan. For example, an enzyme preparation from *Forsythia* × *intermedia* has been found to catalyze enantioselective coupling of coniferyl alcohol **34** to afford (+)-pinoresinol **(+)-5** (>97% enantiomer excess, abbreviated e.e.), as described in the following section (Paré et al. 1994). The enantiomeric composition was established using a chiral HPLC technique.

Thus, optical activity is an important criterion of lignans, although it is still possible that nonenantioselective coupling enzymes might be involved in lignan biosynthesis of plants other than *Forsythia*, as outlined in the following.

First, naturally occurring racemic lignans are probably formed by the action of nonenantioselective enzymes. Second, a racemic lignan could be formed initially in the nonenantioselective coupling, after which preferential conversion of one enantiomer may result in accumulation of the other enantiomer and formation of an optically active product. The second mechanism can account for the cooccurrence in *Xanthoxylum ailanthoides* of (−)-pinoresinol **(−)-5** and (−)-secoisolariciresinol **(−)-1**, which bear opposite configurations to each other at C8 (Ishii et al. 1983); (−)-secoisolariciresinol **(−)-1** can be prepared by catalytic hydrogenation of (+)-pinoresinol **(+)-5**, but not (−)-pinoresinol **(−)-5**, with retention of the C8 configuration (Stevenson 1978). It is likely that (−)-secoisolariciresinol **(−)-1** may be formed from (+)-pinoresinol **(+)-5** in the plant, and (−)-pinoresinol **(−)-5** may accumulate.

The possible involvement of the nonenantioselective coupling enzymes awaits clarification, although it is difficult to distinguish them from lignin synthesizing enzymes.

4.3.5.4.2 Lignan Synthesizing Enzymes

Umezawa et al. (1990a) reported for the first time the enzymatic formation of optically pure lignans with soluble cell-free extracts from *Forsythia* × *intermedia*. In the presence of NAD(P)H and H_2O_2 the enzyme preparation catalyzed the enantioselective formation of an optically pure lignan, (−)-secoisolariciresinol **(−)-1**, from achiral coniferyl alcohol **34** (Fig. 122). This is the naturally occurring enantiomer in this plant. The (+)-antipode was not formed. When (−)- or (±)-secoisolariciresinols **(−)-1** or **(±)-1** were used as substrates, in the presence of NAD(P), only (−)- and not (+)-matairesinol formation occurred (Umezawa et al. 1990b, 1991). The antipode, (+)-secoisolariciresinol

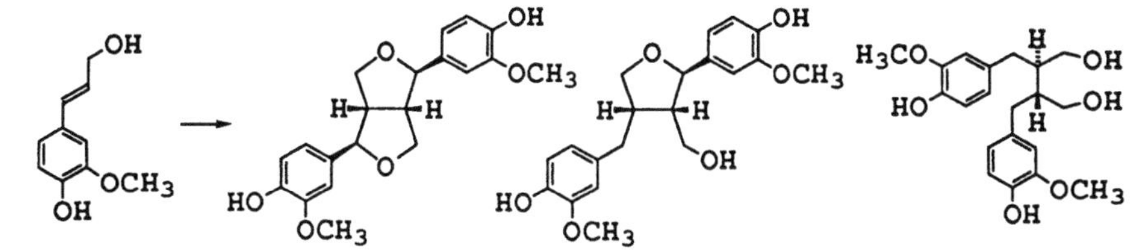

Coniferyl Alcohol **34** | (–)-Pinoresinol (–)-**5** | (–)-Lariciresinol (–)-**4** | (–)-Secoisolariciresinol (–)-**1**

Fig. 122. Lignans formed following incubation of coniferyl alcohol **34** with the soluble cell-free extracts of *Forsythia* plants in the presence of NADPH and H_2O_2

(+)-1, did not serve as a substrate for the formation of either (+)- or (–)-matairesinol (Fig. 123).

Recently, other examples of enzymatic secoisolariciresinol formation from coniferyl alcohol **34** were demonstrated using soluble cell-free extracts from shoots and callus culture of *F. koreana* (Umezawa et al. 1994, Choi et al. 1995, unpubl. results) and petioles of *Arctium lappa* (Umezawa et al. 1996a).

Attention was next directed to elucidating the detailed mechanisms of the secoisolariciresinol formation. In the assay mixture forming (–)-secoisolarici resinol **(–)-1**, (–)-pinoresinol **(–)-5** was found to be accumulated in >96% e.e., even though it is not the naturally occurring antipode present in *Forsythia* spp. (Fig. 122; Katayama et al. 1992). Unnatural (–)-enantiomers of lariciresinol **(–)-4** as well as pinoresinol **(–)-5** occurred when the cell-free extracts from *F. koreana* shoots were incubated with coniferyl alcohol **34** (Fig. 122; Umezawa et al. 1994). Here, it should be noted that the naturally occurring *Forsythia* lignans, (+)-pinoresinol **(+)-5**, (+)-lariciresinol **(+)-4**, (–)-secoisolariciresinol **(–)-1**, (–)-matairesinol **(–)-2**, and (–)-arctigenin **(–)-3**, have the same absolute configuration at C8 (Fig. 118); these lignans can be interconverted chemically with retention of the configuration (Weinges and Spänig 1967, Kato and Munakata 1978, Stevenson 1978, Ayres and Loike 1990).

The accumulation of the unnatural enantiomers was explained by initial formation of both enantiomers followed by preferential conversion of the corresponding natural antipodes. Katayama et al. (1993) reported that the *F.* × *intermedia* enzyme preparation catalyzed the reduction of (+)-pinoresinol **(+)-5** to (–)-secoisolariciresinol **(–)-1** via (+)-lariciresinol **(+)-4**. When (+)- and (–)-pinoresinols **(+)-5** and **(–)-5** were used as substrates, selective reduction of the (+)-enantiomer **(+)-5** occurred to afford (+)-lariciresinol **(+)-4** (i.e., >99:1 in favor of the (+)-antipode; Fig. 123). Reduction of lariciresinol **4** was also selective with reduction of only the (+)-enantiomer **(+)-4** being observed. Similarly, selective reduction of (+)-lariciresinol **(+)-4** into (–)-secoisolariciresinol **(–)-1** was reported with the *F. koreana* enzyme preparation (Umezawa et al. 1994).

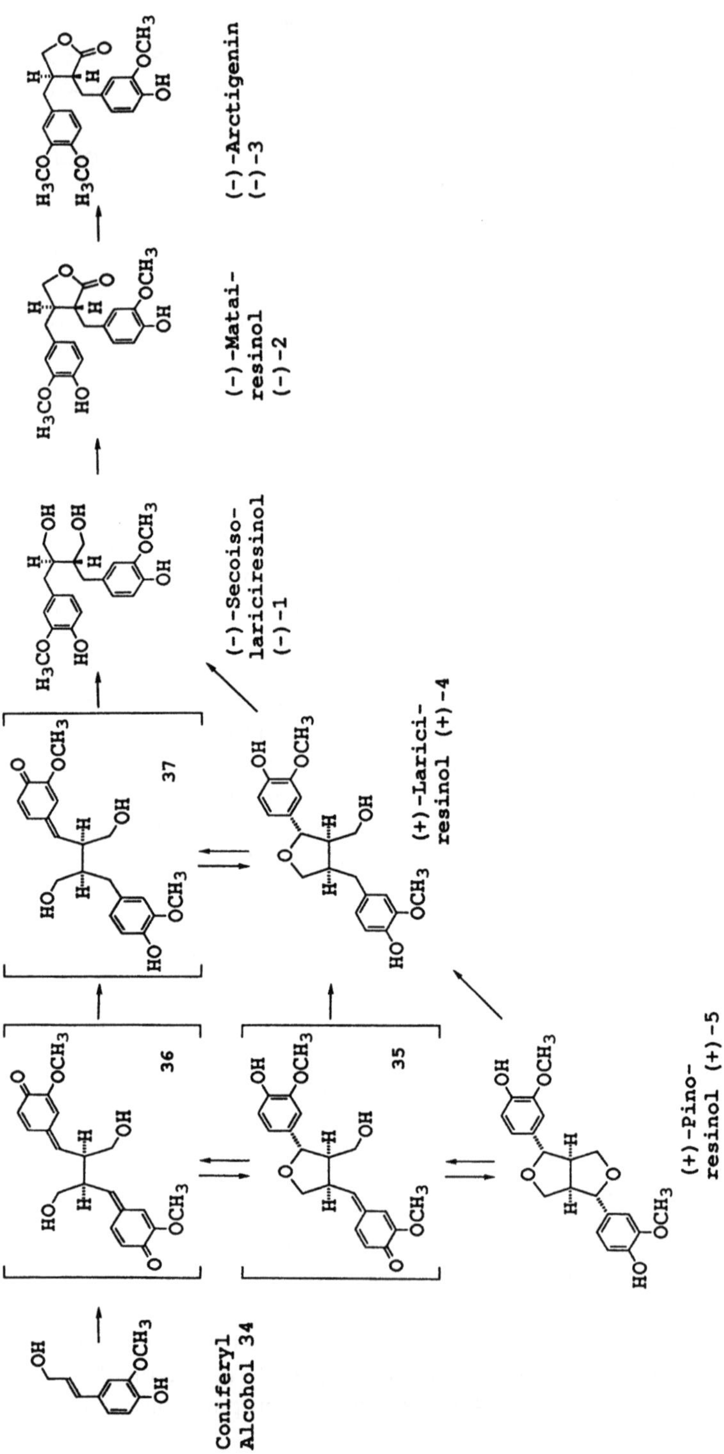

Fig. 123. Enzymatic formation of *Forsythia* lignans

Chu et al. (1993) studied the stereochemistry of the reduction using a partially purified enzyme preparation; *pro-R* hydrogen of NADPH was found to be incorporated into the *pro-R* position at C7 (and/or C7′) in lariciresinol **4** and secoisolariciresinol **1**. Two possible mechanisms were proposed for the pinoresinol reduction (Fig. 123; Katayama et al. 1993, Chu et al. 1993); one is a direct furan ring reduction and the other is via generation of the quinone methides **35** and **36** at the enzyme active site with concomitant reduction.

When the soluble enzyme preparation was incubated with coniferyl alcohol **34** in the presence of only H_2O_2 as cofactor, racemic (±)-pinoresinols **(±)-5** were formed. In contrast, "insoluble residues" from *F. suspensa* were found to catalyze a stereoselective coupling of coniferyl alcohol **34**; Davin et al. (1992) reported the "insoluble residues" obtained after removal of soluble and ionically bound enzymes were able to catalyze enantioselective formation of natural (+)-enantiomer **(+)-5** (ca. 30% e.e.) from coniferyl alcohol **34** in the absence of exogenously provided cofactors. The enantioselectivity was enhanced to ca. 60% e.e. when NAD and malate were added.

Subsequently, Paré et al. (1994) reported >97% e.e. using an enzyme preparation dissolved from *F.* × *intermedia* "cell wall residues". This coupling required O_2 as cofactor. These authors concluded that the enantioselective phenylpropanoid coupling enzyme in lignan formation was established and designated it as (+)-pinoresinol synthase (Fig. 123).

Pinoresinol formation from coniferyl alcohol **34** was also reported using a cell-free preparation of *Larix leptolepis* callus tissues (Nabeta et al. 1991). The unnatural (−)-enantiomer **(−)-5** was found to slightly predominate (6% e.e.), on the other hand, the (+)-antipode **(+)-5** was the major component in the pinoresinols from the callus and sapwood of the plant (14.9% e.e. and 92.3% e.e., respectively).

Enzymatic *O*-methylation of matairesinol **2** was examined with cell-free extracts of *F.* × *intermedia* (Fig. 123; Ozawa et al. 1993). In contrast to the enzymatic steps from coniferyl alcohol **34** to (−)-matairesinol **(−)-2**, the conversion of matairesinol **2** into arctigenin **3** was less selective; both (+)- and (−)-enantiomers of matairesinol **2** served as substrates for methylation to afford arctigenin **3**, with the naturally occurring (−)-antipode **(−)-2** slightly preferred. However, the enzyme preparation also catalyzed the synthesis of (+)- and (−)-isoarctigenins (4-*O*-methyl matairesinol), with (−)-matairesinol **(−)-2** again the preferred substrate. Since no evidence for the presence of either (−)- and (+)-isoarctigenins in *F.* × *intermedia* was obtained, the authors proposed that regioselective glucosylation first occurred to afford matairesinoside (4-*O*-glucoside of matairesinol) in the plant; subsequent methylation followed by deglucosylation gave arctigenin **3**.

As a whole, conversion of coniferyl alcohol **34** to the natural enantiomers of the *Forsythia* lignans by the *Forsythia* enzyme preparations was established as shown in Fig. 123. Each step, except for the final methylation, is well controlled in terms of stereochemistry; (+)-pinoresinol **(+)-5** is enantioselectively formed from achiral coniferyl alcohol **34**, and (+)-enantiomers of pinoresinol **(+)-5** and lariciresinol **(+)-4**, and (−)-secoisolariciresinol **(+)-1** are converted preferentially over the antipodes to give rise to almost optically pure (+)-lariciresinol **(+)-4**, and optically pure (−)-secoisolariciresinol **(−)-1** and (−)-matairesinol **(−)-2**, respectively.

These enzymatic studies strongly suggest that the biosynthetic pathway from coniferyl alcohol **34** to matairesinol **2** via pinoresinol **5**, lariciresinol **4**, and secoisolariciresinol **1** (Fig. 123) is occurring in *Forsythia* plants. Regarding the intermediacy of pinoresinol **5** and lariciresinol **4** in the in vivo transformation, however, an alternative but remote possibility could not be ruled out. As shown in Fig. 123, quinone methides **36** and **37** were proposed as possible intermediates of each step in the enzymatic conversion of coniferyl alcohol **34** to secoisolariciresinol **1** (Chu et al. 1993, Katayama 1993, Paré et al. 1994, Umezawa et al. 1994). If the enzymes are located closely enough together in the plants so that the quinone methides **36** and **37** were trapped by the active site of the next reductase before furan ring formation could take place, secoisolariciresinol **1** might be formed directly through the quinone methides **36** and **37** but not via pinoresinol **5** and lariciresinol **4**. Studies with purified enzymes would be required before final conclusions could be drawn regarding the intermediates.

It is of importance to discuss the steps that determine enantiomeric composition of *Forsythia* lignans. Since more than 97% e.e. was recorded for (+)-pinoresinol **(+)-5** obtained following incubation of coniferyl alcohol **34** with (+)-pinoresinol synthase (Paré et al. 1994), the largest contribution is ascribed to the initial coupling. In contrast, a smaller enantiomer excess value (73% e.e.) was found for (−)-pinoresinol **(−)-5** isolated from *Wikstroemia sikokiana*, and (+)-matairesinol **(+)-2** and (+)-wikstromol from the plant are optically pure (Umezawa et al. 1996b). This suggests that the initial pinoresinol formation is less enantioselective in *W. sikokiana*, and that postcoupling modification also takes part in the selective formation of optically pure lignans of the plant. Differences in stereochemical control mechanisms of lignan biosynthesis between plant species will be a subject of future studies in this field.

4.3.6 Flavonoids

C_6–C_3–C_6 compounds which are composed of the C_6–C_3 (phenylpropane) fragment that contains the B ring, and of the C_6 fragment that contains the A ring are named flavonoids. Flavonoids are present as flower pigments and as bark and heartwood components of conifers and hardwoods, and are classified into flavonoids, isoflavonoids and neoflavonoids based on the substitution pattern of the B ring on the C ring. These are further classified into flavones, flavanones, chalcones, aurones, catechins, and leucoanthocyanidins based on modifications of ring C (Fig. 124).

The flavonoid skeleton is widely distributed in the plant kingdom, but the isoflavonoids and the neoflavonoids occur only in a few higher plants, such as

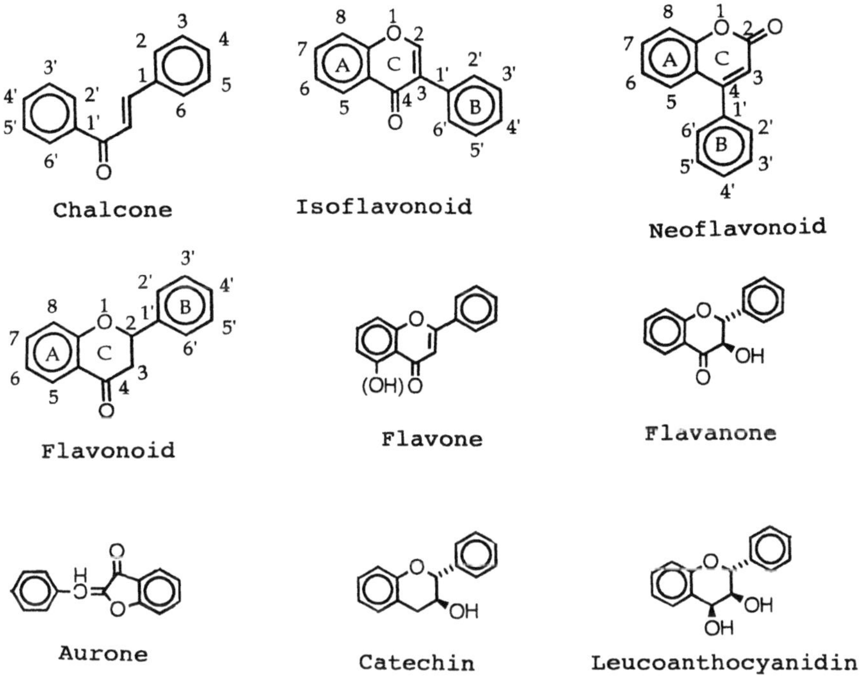

Fig. 124. Basic chemical structures of flavonoid compounds

chickpea (*Cicer arietinum*), soybean (*Glycine max*), alfalfa (*Medicago sativa* L.) etc., (Leguminosae), *Hypericum* (Guttiferae) and *Iris* (Iridaceae). Therefore, the isoflavonoids and neoflavonoids are used as indicator compounds in plant chemotaxonomy (Harborne and Mabry 1982).

In 1953 Birch and Donovan proposed that the A ring of the flavonoid is derived from three molecules of acetic acid, while the B ring and C_3 chain of the C ring originate from cinnamic acid or a related compound. Since then, feeding experiments with ^{14}C-labeled compounds have demonstrated that three molecules of acetic acid are incorporated into the A ring, while shikimic acid, L-phenylalanine, cinnamic acid and *p*-coumaric acid are incorporaterd into the C_3–C_6 (B and C rings) of cyanidin in red cabbage (Grisebach 1957), of rutin (quercetin glucoside) in buckwheat (Geissman and Swain 1957, Shibata and Yamazaki 1957, Underhill et al. 1957, Watkin et al. 1957), of phlorizin in apple (Hutchinson et al. 1959) and of catechin in tea seedlings (Zaprometov 1962).

4.3.6.1 Biosynthetic Pathways

Feeding experiments with radioactive precursors and subsequent enzymatic studies indicated that chalcone is a common precursor of flavonoids and is

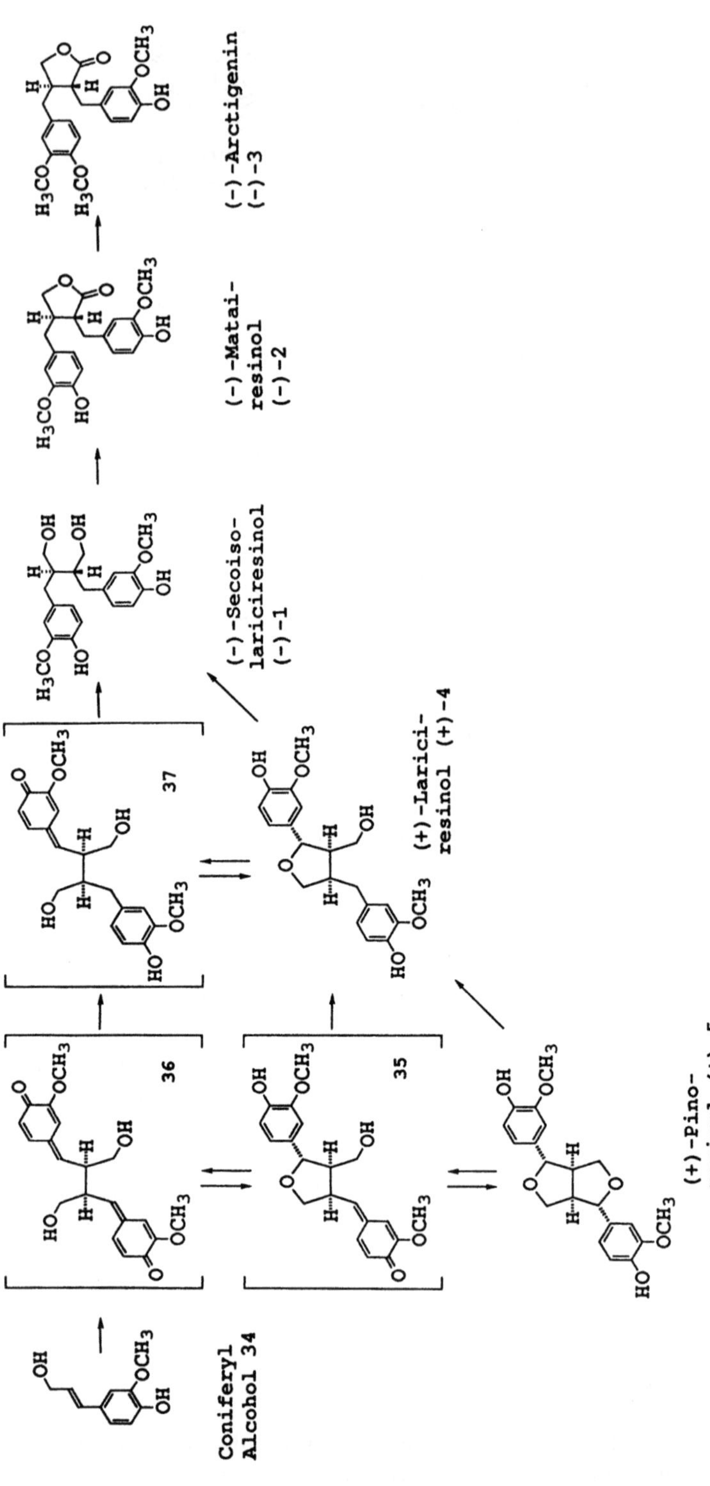

Fig. 123. Enzymatic formation of *Forsythia* lignans

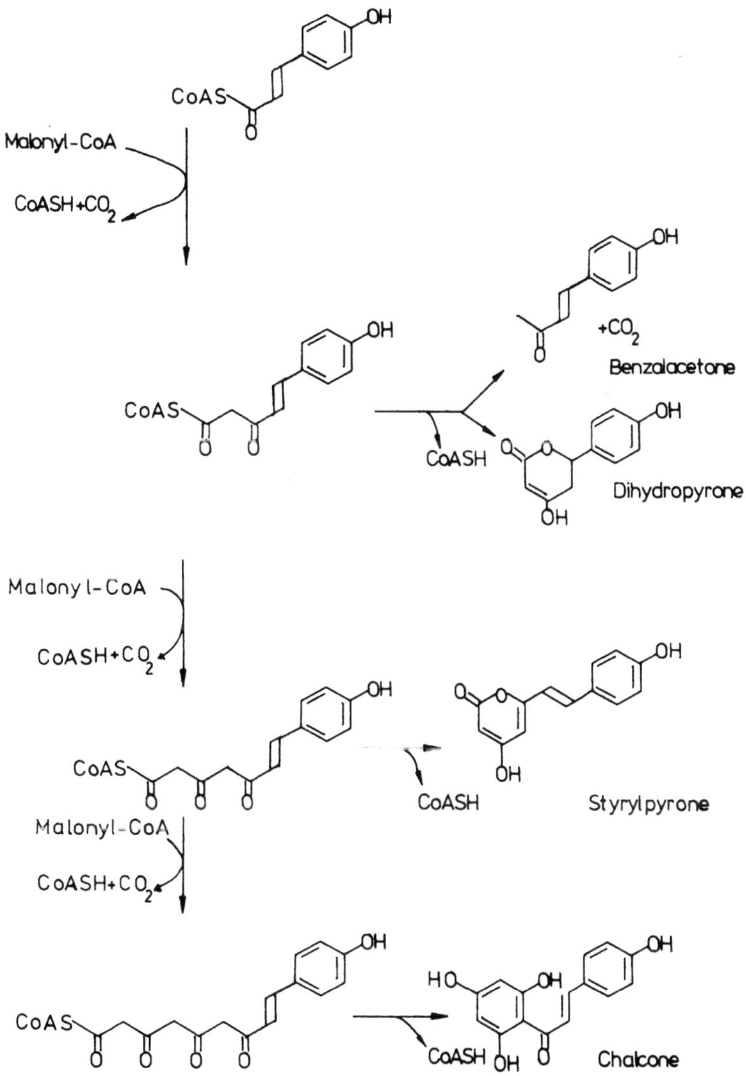

Fig. 126. Mechanism of chalcone synthase from parsley. The released products (benzalacetone or others) are formed from polyketide intermediates with either one or two acetate units, respectively. (Grisebach 1985)

tion of side products that originate from the condensation of one (benzalacetones or dihydropyrones) or two (styrylpyrones) molecules of malonyl-CoA (Kreuzaler and Hahlbrock 1975b, Hrazdina et al. 1976) with enzymes from parsley and *H. gracilis* cell cultures (Fig. 126).

These products in the enzymatic reaction have not been detected as released products in parsley cell cultures, but the occurrence of styrylpyrones in higher plants (Mors et al. 1961) and 3′-hydroxybisnoryangonin in basidiomycetes (*Polyporus hispidus*) has been demonstrated (Perrin and Towers 1973). It seems that such compounds are synthesized by enzymes related to CHS. The molecular mass of the purified CHS from parsley cell cultures was estimated at about 77 kDa. The CHS is composed of two identical subunits with a molecular mass of about 42 kDa and is encoded by CHS-mRNA (about 1.7 kb). The primary structure of the enzyme subunit was elucidated based on the sequence analysis of the cDNA. It is composed of 398 amino acids with a molecular weight of 43682 (Reimold et al. 1983).

In relation to anthocyanidin synthesis in petals, eight genes of CHS were cloned from *Petunia hybrida*. A variety "red star", the petals of which are mottled white and red, was found to contain at least four genes for expression of flower color. The white part of the petal does not contain any active CHS-mRNA. Another variety, V30, involves 8 CHS genes, as well as gene fragments consisting of one of the two CHS exons. These genes are classified into subgroups based on sequence homology and the position of their loci on the chromosomes (Gerats and Martin 1992).

It has been shown that in CHS genes from *Antirrhinum* and parsley, a CACGTG motif lying between –129 bp and –135 bp upstream of the transcription start is involved in light-induction of gene expression. The box (TC-CACGTGGC) located at –169 is highly homologous to the gene of the small subunit of ribulose-1,5-bisphosphate carboxylase involved in photosynthesis.

p-Coumaric acid (4-CA) stimulates expression of the bean (*Phaseolus vulgaris* L.) CHS gene (*chs15*) promoter in electroporated protoplasts of alfalfa (*Medicago sativa* L.). Two distinct *cis*-elements, the H-box (consensus $CCTACC(N)_7CT$) and the G-box(CACGTG), were found to be required for stimulation of the *chs15* promoter by 4-CA. The H-box and G-box in combination establish the complex developmental pattern of the *chs15* expression and are also involved in stress induction. Potential internal pathway regulation through feed-forward stimulation by 4-CA operates by modulation of the signal pathways for developmental and environmental regulation (Loake et al. 1992).

The CHS gene has been cloned from many plants such as *Petunia hybrida*, *Matthiola incana*, *Antirrhinum majus*, *Phaselous vulgaris*, *Cicer arietinum*, *Zea mays*, *Petroselinum sativum*, *Penoculus acer*, *Magnolia liliflora*, *Hordeum vulgare*, *Arabidopsis thaliana* and others. It has been shown that CHS genes isolated from these plants are 70% homologous (Gerats and Martin 1992).

Chalcone isomerase (CHI), which catalyzes the stereospecific isomerization of chalcone to (–)-(2S)-flavanone, has been isolated from several plants (Fig. 125). The substitution pattern of the CHI substrate generally corresponds to that of the flavonoids in the plant from which the enzyme was isolated. For example CHI from parsley is specific for 4,2′,4′,6′-tetrahydroxychalcone, corresponding to the 5,7-dihydroxyflavonoids in this plant (Hahlbrock et al. 1970). The equilibrium constant in isomerization of 2′,4,4′-trihydroxychalcone to 7,4′-dihydroxyflavanone with soybean CHI shifts remarkably to the flavanone formation (Boland and Wong 1975).

Table 23. Genetically defined lines of *Petunia hybrida* and activity of chalcone isomerase in anthers. (Grisebach 1985)

Line pollen genotype	Pollen color	Pollen flavonoids	Specific activity of chalcone isomerase in anthers ($\Delta A\ min^{-1} mg^{-1} protein$)
V10 *PoAn4*[a]	Blue	Anthocyanin	0.572 ± 0.017
V28 *poAn4*	Green	Naringenin chalcone, little anthocyanin	0.026 ± 0.014
V31 *Poan4*	White	Dihydroflavonol and flavonol	0.509 ± 0.070
V33 *poan4*	Yellow	Naringenin chalcone, some flavonol and dihydroflavonol	0.023 ± 0.020

[a] *Po* is the gene for chalcone isomerase, while gene *An4* controls anthocyanin synthesis in pollen (De Vlaming and Kho 1976).

The important role of CHI in the flavonoid biosynthesis in plants was shown by analyses of the flavonoid and CHI activity in flower and pollen of genetic lines of *Petunia hybrida* (Table 23; De Vlaming and Kho 1976, Forkmann and Dangelmayer 1980). Enzymatic investigation on anthers of four lines with different pollen coloration revealed a clear correlation between accumulation of the naringenin chalcone and a lack of activity of chalcone isomerase.

The genetic control of chalcone isomerase in flowers of carnation also showed that yellow and yellow-orange strains are recessive states for CHI synthesis and accumulate tetrahydroxychalcone 2′-glucoside. The carnations, Legion d'honneur (brick-red) and Jeanne Dionis (white), which contain pelargonidin and kaempferol, respectively, showed high CHI activities, whereas yellow and yellow-orange strains which contain tetrahydroxychalcone 2′-glucoside, had no CHI activity (Forkmann and Dangelmayer 1980).

The expression of the *CHI* gene in the anthers of *Petunia hybrida* is regulated by the *Po* gene. In the anthers of *po/po* plants (recessive) the *CHI* gene is not activated, resulting in accumulation of chalcone in the pollen. Two *CHI* genes (*CHI-A* and *B*) were found to be located in different chromosomes of *Petunia*. The *CHI-A* gene does not contain any intron and is induced by UV irradiation in petals and flower stalks. The induction is completely linked to the *Po* gene, which encode *CHI-A* (Gerats and Martin 1992).

Upstream of gene *A* two promoters are present, one of which is involved in pigment formation in petals, flower and stalk. In the seedlings irradiated by UV 1-kb mRNA is formed by expression of the gene. In immature anthers of *po/po* plants no transcripted products are detected. These results suggest that the expression of anther-specific genes is inactivated in the *po* mutant.

The *CHI-B* gene involves three introns. The promoter is one, which is induced by UV irradiation only in immature anthers, to produce a 1-kb mRNA. The *CHI-B* gene is separated from the *Po* gene and independent of it

(Gerats and Martin 1992). The gene for CHI has been cloned from *Petunia hybrida*, *Phaseolus vulgaris*, and *Antirrhinum majus*, among others.

4.3.6.1.1 Modification of Ring C

Grisebach and Kellner (1965) found that dihydrokaempferol isolated from *Chamaecyparis obtusa* previously fed with chalcone labeled at the β-position (the 2 position of naringenin) with ^{3}H and ^{14}C, keeps the same $^{13}H:^{14}C$ ratio as the precursor. This result indicates that dihydrokaempferol is formed by direct hydroxylation of naringenin (flavanone) at the 3 position, and not via flavone or flavonol, which would involve a loss of ^{3}H. The 3-hydroxylation of flavanone was later demonstrated with microsomal preparations of *H. gracilis* cells (Fritsch and Grisebach 1975).

Britsch et al. (1981) also found that a soluble enzyme preparation from irradiated parsley cell cultures catalyzed the conversion of naringenin or eriodictyol to the corresponding dihydroflavonol, flavonol, and flavone. Forkmann et al. (1980) found that a genetic line of white flowered *Matthiola incana* was blocked in the anthocyanidin pathway, after dihydroflavonol formation (recessive *e*), and in 3′-hydroxylation (recessive *b*). They isolated 3-hydroxylase as a soluble enzyme from these flowers and characterized it. The enzyme requires oxoglutarate, ferrous ion, and ascorbate as cofactors as in the 2-oxoglutarate-dependent dioxygenase. The enzyme catalyzes the conversion of eriodictyol (5,7,3′,4′-tetrahydroxyflavanone) to dihydroquercetin (flavanonol). On the other hand, dihydrokaempferol and naringenin were converted to kaempferol and apigenin, respectively with the enzyme preparation in the presence of molecular oxygen, Fe^{2+}, and 2-oxoglutarate. The fact that both flavanone 3-hydroxylation and formation of flavones and flavonols require the typical cofactors for 2-oxoglutarate-dependent dioxygenase suggests that flavone and flavonol syntheses proceed via 2-hydroxyflavanone and 2,3-dihydroxyflavanone, respectively (Britsch et al. 1981; Fig. 127).

These reactions involve three enzymes, flavanone 2- and 3-hydroxylases, and dehydratase for hydroxyflavanones, to give flavonol. 2-Hydroxylases of flavanone and dihydroflavonol were isolated from cell-suspension cultures of parsley and characterized (Grisebach 1985).

The role of dihydroflavonols in anthocyanidin biosynthesis has been elucidated by using genetically defined lines of *Matthiola incana* (Forkmann 1977), *Petunia hybrida* (Kho et al. 1975, 1977, Gerats and Martin 1992), and *Antirrhinum majus* (Harrison and Stickland 1974, Stickland and Harrison 1974a,b).

In flower buds of the white-flowering mutant W19 of *Petunia hybrida*, dihydroquercetin 7-glucoside, dihydroquercetin 4′-glucoside, dihydroquercetin and dihydrokaempferol 7-glucoside are accumulated. When extracts of flower line 19 or dihydroquercetin are supplied to in vitro cultured corollas of the white-flowering mutant W18, cyanidin 3-glucoside is formed as the major

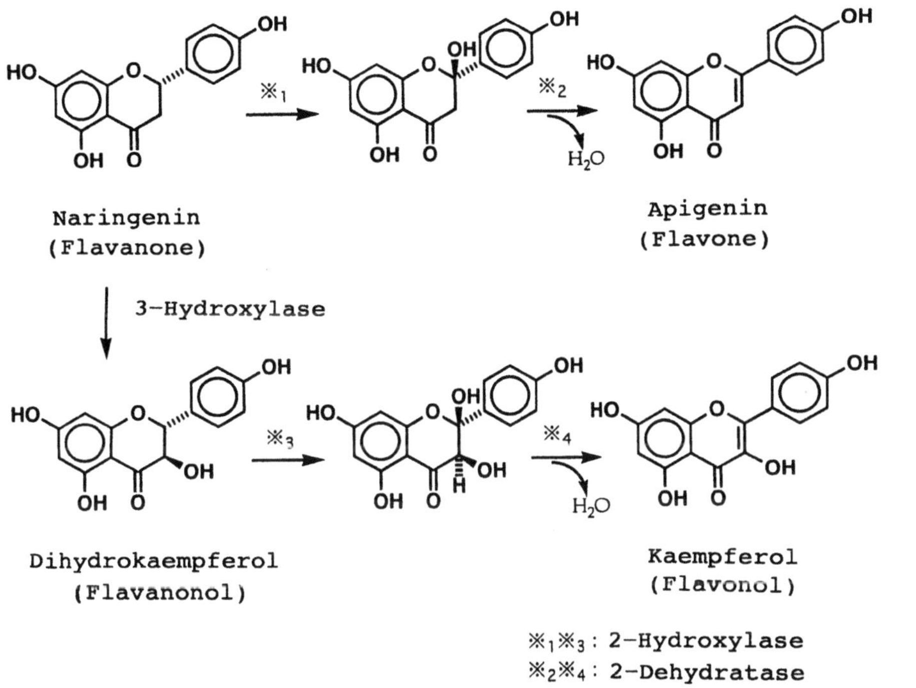

Fig. 127. Conversion of naringenin (flavanone) to apigenin (flavone) and kaempferol (flavonol) with an enzyme preparation from parsley cell cultures in the presence of 2-oxoglutarate, Fe^{2+}, and ascorbate. (Grisebach 1985)

pigment, together with smaller amounts of cyanidin 3-diglucoside and delphinidin 3-glucoside as in flowers of a red-flowering genotype, R3 of *Petunia hybrida* (Forkmann 1977).

It was recently shown that the substrate of flavanone 3-hydroxylase (F3H) in *Petunia* is determined by the allelic status of the genes *Ht1* and *Hf1*. The F3H gene is located at gene locus *An3*. In allele *an3-1*, F3H activity decreases by 90%, and the color of petals and anters of *an3-1* is similar to that of wild plants, but flower stalks are almost colorless. In petals of *an3-1* anthocyanidin synthesis is almost normal, but the formation of flavonol is remarkably reduced. *an3-1* contains a structurally changed gene for F3H, and lacks F3H activity, but the plant synthesizes anthocyanidin from externally added dihydroquercetin (Gerats and Martin 1992). As a result of these investigations the F3H gene locus of mutants was determined in *Antirrhinum majus* (*Inc*), *Gerbera hydrida* (*Eht*), *Zinnia elegans* (*Fht*) and *Verbena* (*A*).

Investigations with a variety of *M. incana* suggested that flav-3-en-3,4-diol as well as flavan-3,4-diol are intermediate to anthocyanidin. In fact, the dihydroflavonol-4-reductase involved in the reduction of this stage has been identified (Heller et al. 1985).

Anthocyanin biosynthesis in the aleurone tissue of maize endosperm has also been investigated by supplementation experiments (McCormick 1978). In this tissue, the complementary genes *CC*, *C2*, *R*, *A*, *Az*, *Bz*, and *Bz2* are required for anthocyanin formation. Recessive alleles at one of the first five loci result in a colorless aleurone. Supplementation of seeds recessive in the *C2* with dihydroquercetin resulted in synthesis of cyanidin; when naringenin was supplied, the pigments formed were cyanidin and pelargonidin.

An enzyme preparation from *Psudotsuga menziesii* has been shown to mediate the conversion of (+)-dihydroquercetin and 3,4-*cis*-leucoanthocyanidin to (+)-catechin, while cell suspension and tissue cultures of *Pseudotsuga menziesii* and *Ginkgo biloba* contain reductases that mediate reduction of (+)-dihydroquercetin and 3,4-*cis*-leucodelphinidin to (+)-gallocatechin (Stafford and Lester 1984, 1985).

Genes controlling the expression of dihydroflavonol-4-reductase (DFR) were identified in *Matthiola* (*e*), *Callistephus* (*f*), and *Dianthus* (*a*), while the structural gene was cloned from *Zea mays* (*A*), *Antirrhinum*, and *Petunia hybrida* (*An6*). Another DFR gene was cloned from *Petunia*, in which three copies of the gene are located in different chromosomes (Martin et al. 1985, Beld et al. 1989). The *DFR-A* gene located at chromosome IV was identified as the *An6* locus. DFR of *Petunia* preferentially converts dihydromyricetin to leucodelphinidin, but dihydrokaempferol is not converted to leucopelargonidin. Therefore, the accumulation of delphinidin and nonformation of pelargonidin are explained by the difference of substrate specificity. In fact, the transformants introduced with the *Zea mays A1* gene synthesized pelargonidin (Gerats and Martin 1992).

In a line of Japanese morning glory bearing variegated flowers, a transposable element of 6.4 kb (*Tpn1*) was found within the genes encoding DFR. The 6.4-kb element carries 28-bp perfect terminal inverted repeats, the outer 13 bp being identical to those of the maize transposable element *Suppressor-mutator/Enhancer*. It is flanked by 3-bp direct repeats within the second intron of the *DFR* gene, 9 bp upstream of the third exon. Cosegregation data of the variegated flower phenotype and the *DFR* gene carrying *Tpn1*, indicated that the mutable phenotype is due to excision of *Tpn1* from the *DFR* gene (Inagaki et al. 1994).

4.3.6.1.2 Hydroxylation and Methylation in Ring B

Feeding experiments with many plants have shown that cinnamic acid and *p*-coumaric acid are preferred substrates of chalcone synthase (CHS). It was suggested therefore, that the hydroxyl and methoxyl groups at the 3′ and 5′ positions of the B ring of flavonoids were introduced secondarily after completion of the C_6–C_3–C_6 skeleton.

In tracer experiments with buckwheat seedlings both 4,2′,4′,6′-tetrahydroxychalcone and 3,4,2′,4′,6′-pentahydroxychalcone were efficiently incorporated into cyanidin (Patschke and Grisebach 1965), suggesting that the 3′-hydroxylation occurs at the chalcone or flavanone stage or later in the pathway.

NADPH-dependent hydroxylation of naringenin, dihydrokaempferol, kaempferol, and apigenin to eriodictyol, dihydroquercetin, quercetin, and luteolin, respectively, was found to be mediated by microsomal fractions from *Haplopappus gracilis* cell cultures (Fritsch and Grisebach 1975), parsley cell cultures (Hagmann et al. 1983) and flowers of defined genotypes of *Matthiola incana* (Forkmann et al. 1980). The enzyme reaction required molecular oxygen and NADPH and was partly inhibited by carbon monoxide, indicating that the enzyme is a cytochrome P-450-dependent monoxygenase (Potts et al. 1974).

On the other hand, chemogenetic analysis showed that in *M. incana*, 3′-hydroxylation occurs only in lines with wild-type alles at the locus *b* which is responsible for 3′-hydroxylation (Forkmann 1980). The correlation between enzyme activity and genotype indicated that this enzyme participated in cyanidin biosynthesis.

Analysis of genotypes in flowers of *Petunia hybrida* (Tabak et al. 1978, Doodeman et al. 1982, Gerats and Martin 1992) and *Antirrhinum majus* (Stickland and Harrison 1974a,b, Harrison and Stickland 1974), and seeds or seedlings of *Meliotus alba* (Specht et al. 1976) also showed the occurrence of specific genes responsible for 3′-hydroxylation of flavanone and dihydroflavonol.

Blue and violet flowers generally contain derivatives of cyanidin or pelargonidin. Differences in hydroxylation patterns of these major classes of anthocyanidins are controlled by the cytochrome P450 enzymes flavonoid 3′-hydroxylase and flavonoid 3′,5′-hydroxylase.

Two genes, *Ht1* and *Ht2* which regulate 3′-hydroxylation of naringenin and dihydrokaempferol were found in *Petunia* (Wiering 1974, Tabak et al. 1978). The expression pattern of these genes is different. *Ht1* is expressed in petals and flower stalks and yields only delphinidin, while *Ht2* is expressed only in petals and yields dihydromyricetin from a part of the precursor, leading to a mixture of cyanidins and delphinidins. Furthermore, two genes, *Hf1* and *Hf2*, have been found to control 3′,5′-hydroxylation. 3′,5′-Hydroxylase mediates the conversion of dihydrokaempferol and dihydroquercetin to dihydromyricetin. Feeding and genetic experiments suggested that naringenin is converted either to dihydroquercetin or to dihydrokaempferol via eriodictyol, and to pentahydroxyflavanone. Dihydrokaempferol and pentahydroxyflavanone are converted to dihydromyricetin (Stotz et al. 1985).

cDNAs of two different flavonoid 3′,5′-hydroxylase genes have been recently cloned from the cDNA library derived from *Petunia* flowers by the PCR method using oligonucleotides designed to the conserved P450 haem-binding domain. The cDNA clones of two different flavonoid 3′,5′-hydroxylase genes were expressed in petunia flowers. Restriction-fragment length polymorphism mapping and complementation of mutant petunia lines showed that the

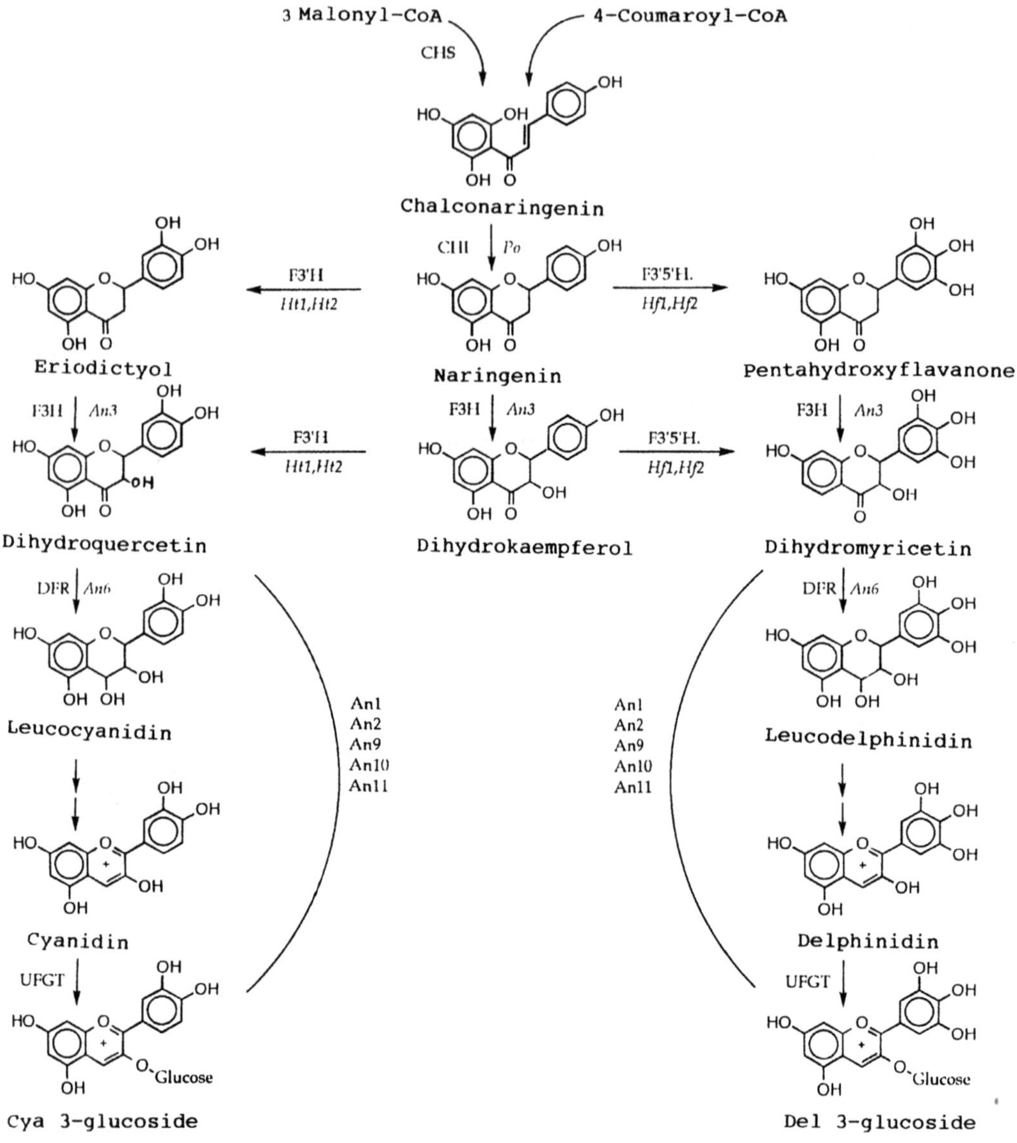

Fig. 128. Anthocyanin biosynthetic pathway in *Petunia hybrida*. Enzyme symbols are placed on the *left* or *above arrows*; gene symbols on the *right* of or *below arrows*. *CHS* Chalcone synthase; *CHI* chalcone isomerase; *F3H* flavanone 3-hydroxylase; *F3′H* flavonoid 3′-hydroxylase; *F3′5′H* flavonoid 3′5′-hydroxylase; *DFR* dihydroflavonol 4-reductase; *UFGT* UDP-glucose:flavonoid 3-*O*-glucosyltransferase; *Po*, *An3*, *Ht1*, *Ht2*, *Hf1*, *Hf2*, and *An6* are structural genes coding for a specific flavonoid enzyme; *An1*, *An2*, *An10*, and *An11* are regulatory genes in the sense that they influence the expression of a number of structural genes. The action of the *An9* gene remains unclear. (Gerats and Martin 1992)

flavonoid 3′,5′-hydroxylase genes correspond to the genetic loci *Hf1* and *Hf2* (Holton et al. 1993). However, in petals of the red campion (*Silene dioica*) hydroxylation in the B ring seems to occur at the cinnamic acid stage. An enzyme that catalyzes hydroxylation of *p*-coumaroyl-CoA to caffeoyl-CoA was found in the petals of genotype *PP* (Kamsteeg et al. 1981). The central part of the anthocyanin biosynthesis pathway is shown in Fig. 128 (Gerats and Martin 1992).

O-Methyltransferase (OMT), specific for the 3′-hydroxyl group of flavonoids, has been isolated from several plants. The enzyme uses *S*-adenosylmethionine as methyl donor (Ebel and Hahlbrock 1982). In soybean cell cultures two OMTs were found. One enzyme is specific for methylation of flavonoid substrates at the 3′-hydroxyl group, whereas the second enzyme is specific for methylation of hydroxycinnamic acids in the 3 and 5 positions.

The methylation of anthocyanins by cell-free extracts from flower buds of *Petunia hybrida* was studied by Jonsson et al. (1982). Cyanidin, delphinidin, and petunidin in the B ring were methylated when the 3-(4-coumaroyl)-rutinoside-5-glucoside derivatives were used as substrates. These results are in agreement with the finding that in *Petunia hybrida* two pairs of duplicate genes (*Mt1*/*Mt2* and *Mf1*/*Mf2*) are involved in methylation at the 3′ and 5′ positions, respectively, and that these genes are only expressed if two other genes (*Rf* and *Gf*) involved in glucosylation, rhamnosylation and acylation of the anthocyanin are present as dominant alleles.

4.3.6.1.3 Glycosylation

Many flavonoids are present as glycosides. UDP-glucose: flavone/flavonol 7-*O*-glucosyltransferase (Sutter et al. 1972) and a UDP-glucose:flavonol 3-*O*-glucosyltransferase (Sutter and Grisebach 1973) were isolated and characterized from parsley. 3-*O*-Glucosyltransferases for cyanidin from red cabbage seedlings (Saleh et al. 1976a), cell cultures of *Haplopappus gracilis* (Saleh et al. 1976b), and petals of *Silene dioica* (Kamsteeg et al. 1978, 1979) have also been reported.

Cyanidin was the best substrate for the transferases from red cabbage and *H. gracilis*. Anthocyanidins such as pelargonidin, peonidin, and malvidin were also glucosylated. The transferases are not specific for anthocyanidins, but also glucosylate flavonols such as kaempferol and quercetin at the 3 position, but not dihydroflavonols such as dihydroquercetin.

In addition, the rhamnosyl moiety of UDP-L-rhamnose was transferred by another transferase from petals of *S. dioica* to the 6-hydroxyl group of the 3-*O*-glucosyl group of anthocyanidin 3-*O*-glucoside. Enzyme activity is controlled by a single gene *N*, and no rhamnosyltransferase activity was found in petals of homozygous recessive *nn* plants. The 3-*O*-glucosides of pelargonidin, cyanidin, and delphinidin serve as acceptors (Kamsteeg et al. 1980b). Another transferase that catalyzes the transfer of the glucosyl moiety of UDP-glucose to the 5-hydroxyl group of cyanidin 3-rhamnosyl glucoside was also detected in these plants (Kamsteeg et al. 1978). This transferase is controlled by a single gene *M*. Pelargonidin 3-rutinoide is also an acceptor.

4.3.6.1.4 Acylation

Two malonyltransferases which catalyze the transfer of the malonyl residue from malonyl-CoA to the primary hydroxyl group of the glucosyl moiety were isolated from irradiated parsley cell cultures (Matern et al. 1981, 1983). One enzyme is most active with flavone and flavonol 7-*O*-glucosides; the other enzyme preferentially malonylated flavonol 3-*O*-glucosides. The malonylated flavonoid glucosides are synthesized in the cytoplasm and are deposited in the vacuoles.

The presence of an enzyme catalyzing the transfer of the 4-coumaroyl or caffeoyl moiety of the corresponding CoA esters to the 4-hydroxyl group of the rhamnosyl moiety of anthocyanidin 3-rutinosides or 3-rutinoside 5-glucosides has been demonstrated in the petals of *Silene dioica*. (Kamsteeg et al. 1980c). General biogenetic relationships among flavonoids derived from the chalcone are illustrated in Fig. 129.

4.3.6.2 Isoflavonoids

Feeding experiments with phenylpropanoid precursors and chalcone specifically labeled with ^{14}C showed that isoflavonoids are formed by 1,2-aryl migration occurring after the stage of the chalcone/flavanone intermediate (Grisebach and Doerr 1960, Hahlbrock and Grisebach 1975, Dewick and Martin 1979a,b).

4.3.6.2.1 Isoflavone Synthase

Hagmann and Grisebach (1984) found that the oxidative aryl migration is mediated by two enzymes. First, a membrane-bound cytochrome P-450 monooxygenase produces a 2-hydroxyisoflavanone with the oxygen atom of the hydroxyl group derived from O_2. The intermediate is then converted by a soluble dehydratase to the isoflavone (Kochs and Grisebach 1968, Hakamatsuka et al. 1990, 1991; Fig. 130).

Feeding experiments with radioactive precursors in alfalfa seedlings showed the following pathway involved in medicarpin biosynthesis (Dixon et al. 1992; Fig. 131). The intermediates have been identified (Dewick and Martin 1979a,b), and the enzymes involved have been characterized (Dixon et al. 1992). The elicitor-induced accumulation of medicarpin in alfalfa cell suspen-

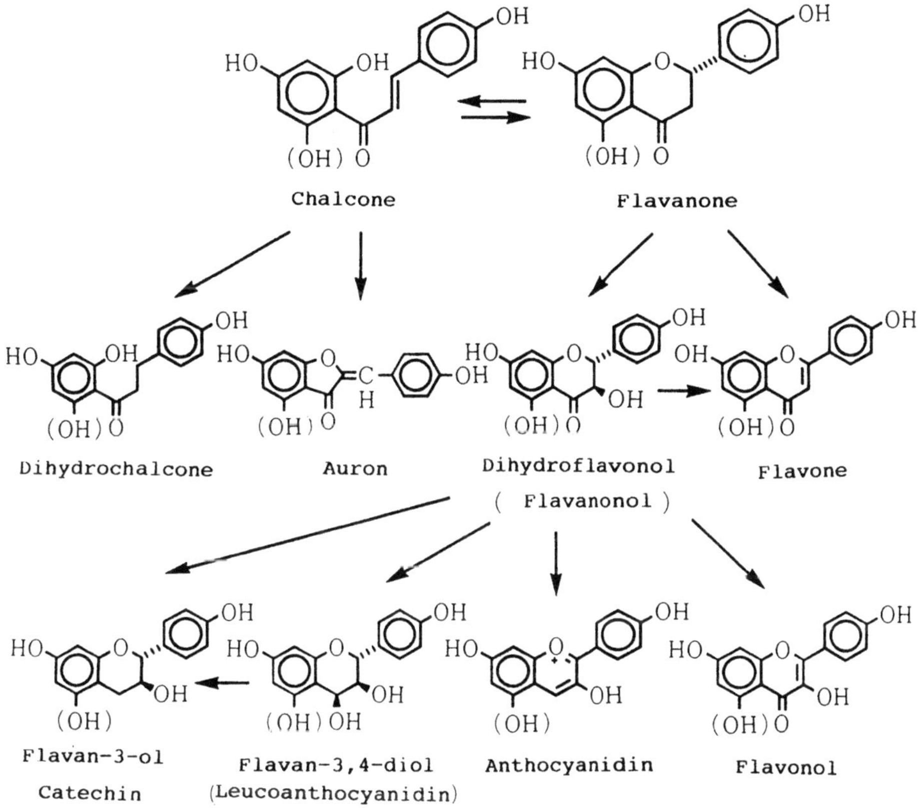

Fig. 129. Biosynthetic pathways from chalcone to various flavonoids

sion cultures is preceded by increases in the activities of all the enzymes involved in the pathway. It was found that isoflavone synthase (IFS), and isoflavone 2′-hydroxylase (IFOH) are cytochrome P-450 enzymes (Kochs and Grisebach 1986). These enzymes are induced in elicitor-treated alfalfa cell-suspension cultures. Several molecular approaches, including polymerase chain reaction (PCR) amplification of cDNA, genomic sequences using series of synthetic oligonucleotide primers based on the sequence of P-450 families of various organisms, and the use of heterologous antiserum raised against P-450 from ripening avocado fruit, have been conducted for characterization of these enzymes (Dixon et al. 1992).

The isoflavones and pterocarpan induced by infection of plant pathogengenic bacteria and fungi in chickpea and soybean (Barz and Roland 1992), and *Pueraria lobata* (Hakamatsuka et al. 1991, 1992) have been extensively investigated. Chickpea plants and cell cultures mainly accumulate formononetin and biochanin A, whereas soybean tissues are rich in daizein and genistein. In both systems these isoflavones occur predominantly in the form of 7-*O*-

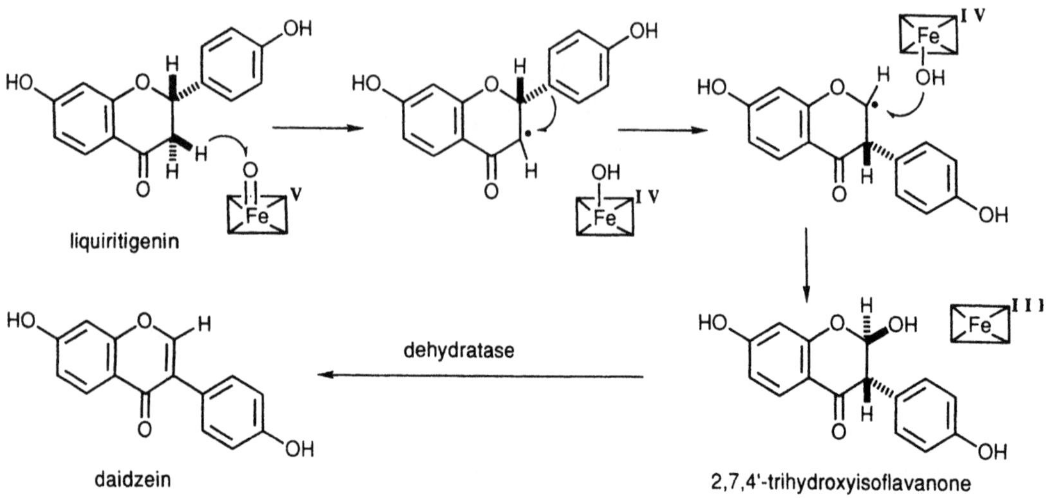

Fig. 130. Mechanism of phenyl migration in isoflavone synthase reaction. (Hakamatsuka et al. 1991)

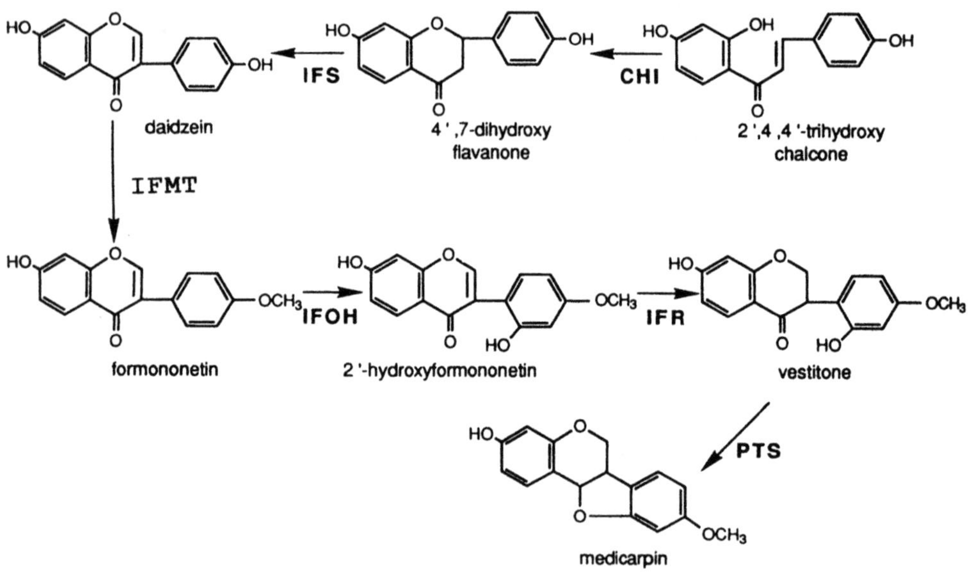

Fig. 131. Biosynthesis of medicarpin in alfalfa. *CHI* Chalcone isomerase; *IFS* isoflavone synthase; *IFMT* isoflavone 4′-*O*-methyltransferase; *IFOH* isoflavone 2′-hydroxylase; *IFR* isoflavone reductase; *PTS* pterocarpan synthase. (Dixon et al. 1992)

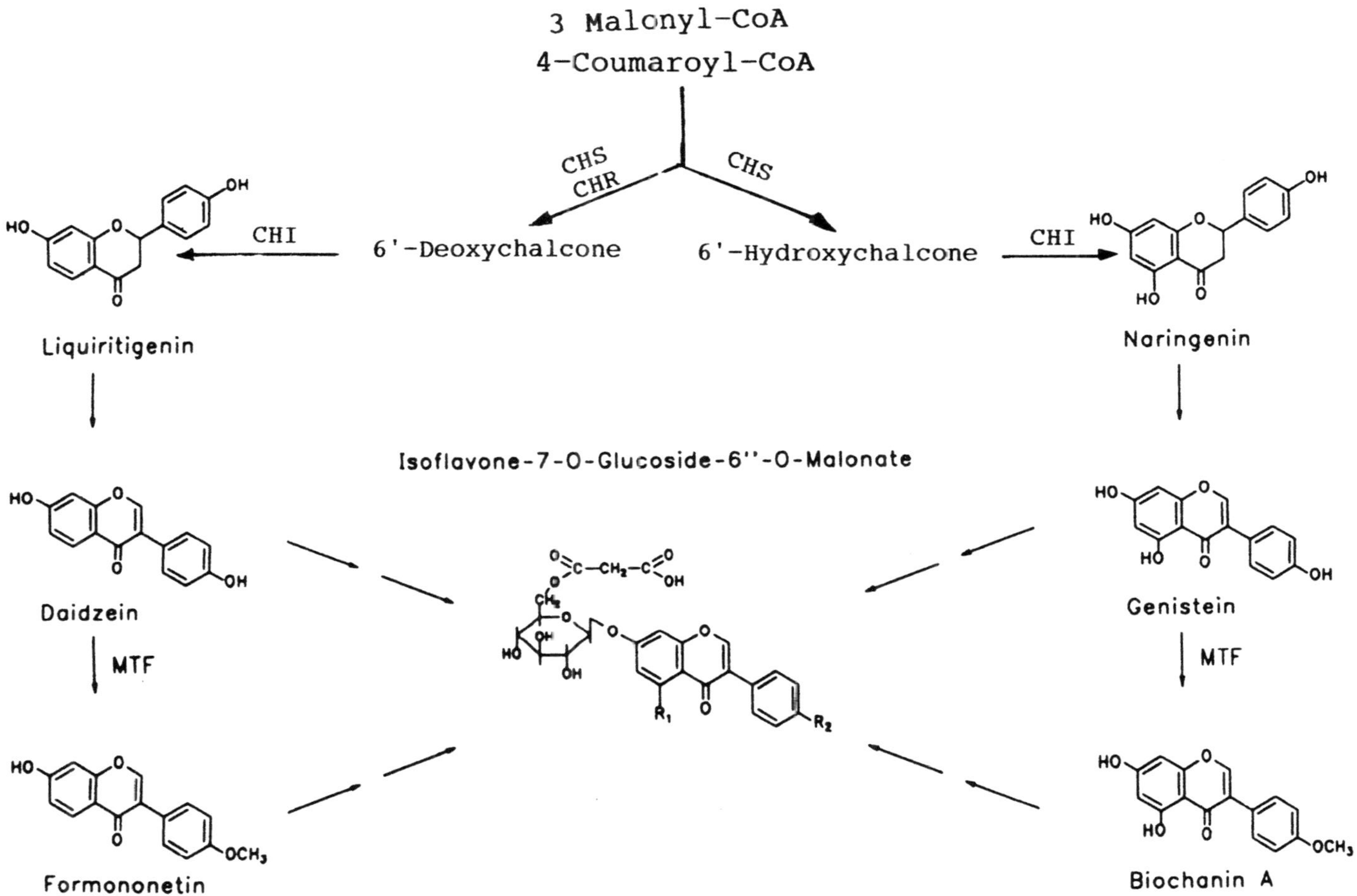

Fig. 132. Biosynthetic pathway to isoflavones and their malonylglucosides in soybean (diadzein and genistein) and chickpea (formononetin and biochanin A). *CHS* Chalcone synthase; *CHR* chalcone reductase; *CHI* chalcone isomerase; *MTF* isoflavone *O*-methyltransferase. (Barz and Roland 1992)

glucoside 6″-*O*-malonate conjugates (Köster et al. 1983, Graham et al. 1990; Fig. 132).

The activity of all enzymes from PAL to isoflavone synthase, including the isoflavone *O*-methyltransferase, increases remarkably in both soybean and chickpea during pterocarpan phytoalexin induction. 6′-Hydroxychalcone is isomerized to naringenin, the intermediate for genistein and biochanin A formation (Barz and Roland 1992). 6′-Deoxychalcone (DOC, 4,2′,4′-trihydroxychalcone) is the intermediate via liquiritigenin for daidzein, formononetin and the pterocarpane phytoalexins. The enzyme involved in the formation of DOC was isolated and identified (Welle and Grisebach 1988).

It was demonstrated that chalcone reductase (CHR) cooperates with chalcone synthase in the formation of DOC, and that DOC is only formed by mediation of a set of CHS and CHR in the presence of NADPH (Barz and Roland 1992).

CHR, the key enzyme in the biosynthesis of 5-deoxyisoflavones and 1-deoxypterocarpans is rapidly induced upon elicitation of chickpea cell cultures with CHS. The molecular mass of CHR is 34 kDa, and the enzyme consists of four isoforms with IEP between 5.55 and 6.10. DNA sequence of the cDNA, the amino acid composition of the enzyme, and the synthesis of the enzyme protein in an *E. coli* expression system has been established (Welle et al. 1991). Hybridization experiments of the soybean CHR cDNA with genomic DNA from other plants revealed homologies in bean and peanut.

4′-*O*-Methylation of isoflavones yielding formononetin and biochanin A is catalyzed by a SAM-dependent methyltransferase from chickpea roots and cell cultures (Wengenmayer et al. 1974).

Pueraria lobata cell cultures elicited with the hydrolyzate of *Pueraria* cell wall by endopolygalacturonase, or yeast extracts produced daidzein and its dimers. The dimers could be formed by oxidative coupling of daidzein by peroxidase induced by the elicitor. CHS and CHR of *Pueraria* cell cultures consist of several isoforms. CHS cDNA was cloned, and the sequence of the full length of the CHS gene was determined. M_r of the deduced amino acid sequence is in agreement with the M_r (43 000) of the purified enzyme. CHR cDNA prepared from the mRNA was introduced into a vector pET (plasmid for expression with T7 phage RNA polymerase) and expressed (Hakamatsuka et al. 1990, 1991).

The biflavonoids (oxidative coupling dimers of flavonoids) such as biflavones, flavanone-flavones, and biflavanones have been identified in Bryales, Psilotales, Selaginellas, primitive gymnosperms, and sporadically in angiosperms. These compounds always carry carbonyl functions at the C-4 position, and are different from the proanthocyanidins which are related to condensed tannin. The registered genes of enzymes involved in lignin and flavonoid biosynthesis are shown in Table 24.

Table 24. Genes involved in lignin and flavonoid biosynthesis

Source	Accession no.	Length (bp)	RNA/DNA
Phenylalanine ammonia-lyase			
Sweet potato	M29232	2335	RNA
Poplar	L11747	2380	RNA
Stylosanthes humilis	L36822	2428	RNA
Avocado	U16130	2044	RNA
Soybean	X52953, S88074	4085	DNA
Rhodosporidium toruloides	M18261	2986	DNA
Tobacco	M84466	6976	DNA
Tomato	M90692	3416	DNA
Tomato	M83314	3939	DNA
Trifolium subterraneum	M91192	3300	DNA
Arabidopsis thaliana	L33677	4921	DNA
Arabidopsis thaliana	L33678	4507	DNA
Cinnamate 4-hydroxylase			
Phaseolus aureus	L07634	1766	RNA
Madagascar periwinkle	Z32563	1713	RNA
Artichoke	Z17369	1620	RNA
Caffeic acid *O*-methyltransferase			
Populus tremuloides	X62096, M72523	1503	RNA
Chrysosplenium americanum	U16793	1101	RNA
Eucalyptus gunnii	X74814	1452	RNA
O-methyltransferase			
Maize	M73235	2512	RNA
Catechol *O*-methyltransferase			
Tobacco	X74452	1418	RNA
Tobacco	X74453	1460	RNA
Tobacco	X71430	1471	RNA
4-Coumarate: CoA ligase			
Soybean	X69954	1041	RNA
Soybean	X69955	1491	RNA
Parsley	X13324	1914	RNA
Parsley	X13325	1927	RNA
Arabidopsis thaliana	Z37656	389	RNA
Arabidopsis thaliana	U18675	1869	RNA
Loblolly pine	U12012	1969	RNA
Loblolly pine	U12013	1972	RNA
Rice	X52623	5225	DNA
Parsley	X05350	1190	DNA
Parsley	X05352	295	DNA
Parsley	X05351	958	DNA
Parsley	X05353	300	DNA
Potato	M62755	4030	DNA
Cinnamyl alcohol dehydrogenase			
Cider tree	X65631	1392	RNA
Alfalfa	Z19573	1321	RNA

Table 24. *Continued*

Source	Accession no.	Length (bp)	RNA/DNA
Tobacco	Z19573	1398	RNA
Tobacco	X62343	1419	RNA
Populus deltoides	Z19568	1305	RNA
Norway spruce	X72675	1278	RNA
Aralia cordata	D13991	1462	RNA
Arabidopsis thaliana	Z34154	373	RNA
Arabidopsis thaliana	Z34557	350	RNA
Arabidopsis thaliana	Z46703	320	RNA
Arabidopsis thaliana	Z46704	334	RNA
Stylosanthes humilis	L36456	1197	RNA
Stylosanthes humilis	L36823	2942	RNA
Pinus taeda	Z37991	1406	RNA
Pinus taeda	Z37992	1355	RNA
Eucalyptus botryoides	D16624	3288	DNA
Arabidopsis thaliana	Z31715	3614	DNA
Peroxidase			
Barley	X62438	740	RNA
Barley	X53384	621	RNA
Rice	X66125	1287	RNA
Rice	D14997	1325	RNA
Pea	X62077	1054	RNA
Wheat	X56011, S55687	1247	RNA
Potato	M21334	1227	RNA
Nicotiana sylvestris	M74103	1179	RNA
Populus kitakamiensis	D30653	1170	RNA
Alfalfa	L36156	1271	RNA
Tomato	X15853	2907	DNA
Arabidopsis thaliana	M58381	3020	DNA
Armoracia rusticana	D90115	3095	DNA
Populus kitakamiensis	D11102	2801	DNA
Chalcone synthase			
Tomato	X55194	1380	RNA
Tomato	X55195	1348	RNA
Common stock	X17577	1378	RNA
Apple tree	X68977	945	RNA
Alfalfa	X68106	1361	RNA
Alfalfa	X68107	932	RNA
Alfalfa	L02901	1386	RNA
Petunia	X14599	1131	RNA
Petunia	X04080	1391	RNA
Pea	X63333	1247	RNA
Pea	X63334	1302	RNA
Pea	X63335	1316	RNA
Kidney bean	X06411	1342	RNA
White mustard	X14315	986	RNA
Pueraria lobata	D10223, D01023	1445	RNA
Daucus carota	D16256	1375	RNA
Camellia sinensis	D26593	1390	RNA
Garden snapdragon	X03710	3574	DNA

Table 24. *Continued*

Source	Accession no.	Length (bp)	RNA/DNA
Soybean	X65636	2208	DNA
Soybean	X54644	2073	DNA
Soybean	X53958	4047	DNA
Soybean	X52097	2323	DNA
Parsley	V01538	1431	DNA
Petunia	X14591	4966	DNA
Petunia	X14592	5973	DNA
Petunia	X14593	2948	DNA
Petunia	X14594	2542	DNA
Petunia	X14595	4680	DNA
Petunia	X14596	2576	DNA
Petunia	X14597	3580	DNA
Arabidopsis thaliana	M20308	3864	DNA
Scotch pine	X60754	1401	DNA
Barley	X58339	5230	DNA
Maize	X60205	3829	DNA
Trifolium subterraneum	M91193	2679	DNA
Soybean	M98871	2688	DNA
Chalcone flavanone isomerase			
Petunia	Y00852	759	RNA
Arabidopsis thaliana	M86358	1999	DNA
Maize	Z22760	2278	DNA
Chalcone isomerase			
Apple	X68978	793	RNA
Kidney bean	X16470, S54703	791	RNA
Alfalfa	M91079	950	RNA
Garden snapdragon	M68326	667	RNA
Garden pea	U03433	986	RNA
Vitis vinifera	X75963	979	RNA
Flavanone 3-hydroxylase			
Barley	X58138	1363	RNA
Vitis vinifera	X75965	1419	RNA
Maize	U04434	1523	RNA
Petunia	X60512	1227	RNA
Petunia	X70786	1502	RNA
Callistephus chinensis	X72593	1309	RNA
Clove pink	X72592	1513	RNA
Garden snapdragon	S60876	376	DNA
Alfalfa	X78994	4300	DNA
Alfalfa	X81812	4089	DNA
Dihydroflavonol 4-reductase			
Arabidopsis thaliana	Z24528	327	RNA
Tomato	Z18277	1651	RNA
Vitis vinifera	X75964	1315	RNA
Petunia	X79723	3879	DNA
Arabidopsis thaliana	M86359	2294	DNA

4.3.7 Stilbenes

Stilbenes were named for α,β-diphenyl ethylene compounds, but the newly discovered bibenzyls and phenanthrenes, which are composed of a C_6–C_2–C_6 skeleton, are generally included in the group of stilbenes. Stilbenes are distributed in the Pinaceae, Moraceae, *Eugenia jambos*, Saxifragaceae etc. as well as in algae and ferns (Harborne 1980, Kindl 1985, Hillis 1987).

Pinosylvin is a stilbene constituent of the heartwood of pine species and is used as a taxonomic marker of the genus *Pinus* (Erdtman 1959). The physiological role of stilbenes as phytoalexins and their role in decay resistance has attracted attention.

4.3.7.1 Biosynthesis

Pinosylvin synthesis in wood (Billek and Ziegler 1962, von Rudloff and Jorgensen 1963) and the synthesis of hydrangeic acid (Billek and Kindl 1962, Ibrahim and Towers 1962) were elucidated by feeding experiments with ^{14}C-labeled compounds. These experiments suggested that stilbenes are synthesized from L-phenylalanine via cinnamic acid. The synthesis involves the condensation of an activated cinnamic acid derivative with acetate units.

Rupprich and Kindl (1978) demonstrated stilbene synthase activity in a cell-free system from the rhizome of *Rheum rhaponticum*. The partially purified enzyme exhibited resveratrol-forming activity. Since then, resveratrol-forming stilbene synthase (type I) has been found in *Picea abies* (Rolf and Kindl 1984), *Pinus griffitii* (Schöppner and Kindl 1984), *Arachis hypogaea* (Rolfs et al. 1981, Schöppner and Kindl 1984), *Vitis* spp. (Fritzemeier and Kindl 1981), and other plants.

The resveratrol-forming enzyme is a dimer with M_r 90000 consisting of subunits with M_r 47000. The *Arachis* and *Rheum rhaponticum* enzymes prefer *p*-coumaroyl-CoA rather than cinnamoyl-CoA (K_m values 2 and 14 μM for *Arachis* and *Rheum* enzymes, respectively), whereas the K_m value toward malonyl-CoA is 10 μM for *Arachis*, but 380 μM for *Rheum*. Another type of stilbene synthase (type II), selectively converting cinnamoyl-CoA to pinosylvin, has been isolated from pine seedlings (*Pinus sylvestris*; Schöppner and Kindl 1979, 1984; Fig. 133).

The stilbene synthase from *Pinus sylvestris* preferentially converts the nonhydroxylated cinnamoyl-CoA to pinosylvin. The selectivity of the enzyme toward nonhydroxylated vs *p*-hydroxylated substrate is 4:1 (Schöppner and Kindl 1979).

The purified stilbene synthase does not exhibit chalcone synthase activity, and anti-stilbene synthase antibodies do not cross-react with chalcone synthase. Purified stilbene synthase incubated with malonyl-CoA as sole substrate

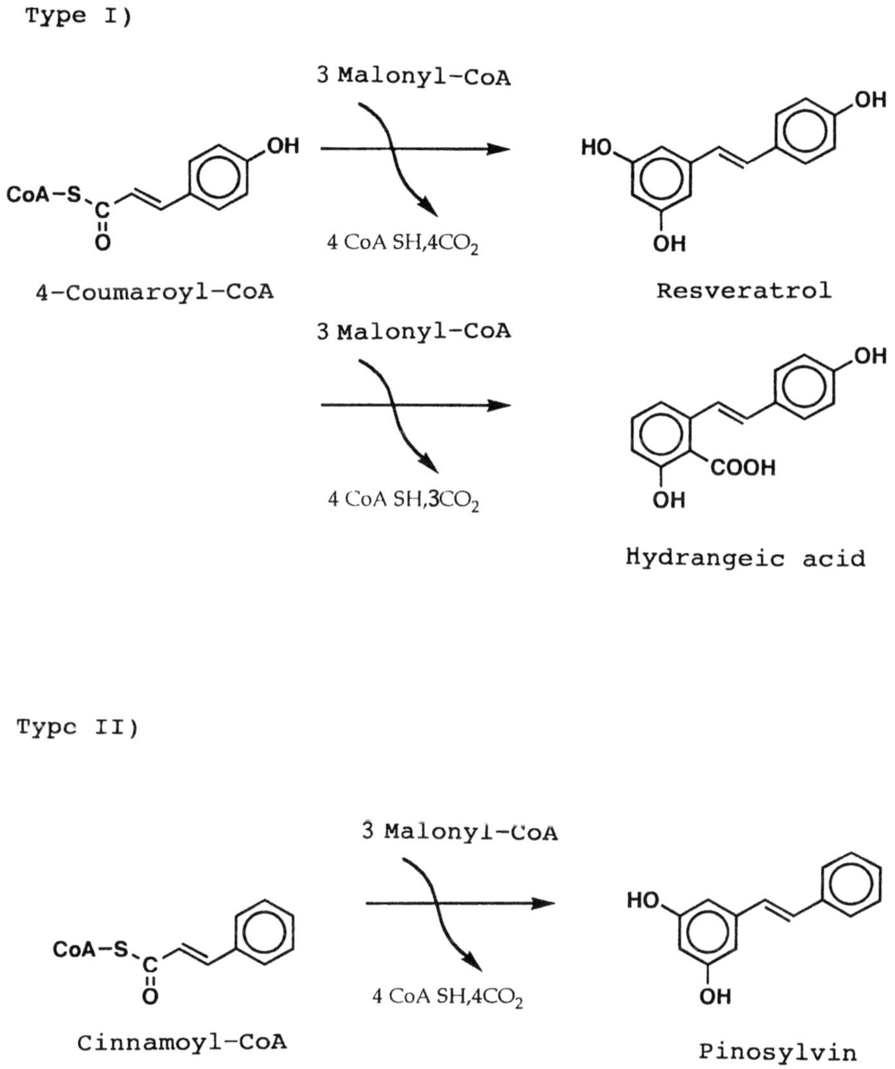

Fig. 133. Biosynthesis of pinosylvin, resveratrol, and hydrangeic acid

shows decarboxylase activity; the decarboxylation of the carboxyl group of malonyl-CoA is reversible. The first condensation step can therefore be formulated as a nucleophilic attack of a carbanion (acetyl-CoA) at the carbonyl carbon atom of the thioester, with the thiolate ion serving as leaving group (Kindl 1985; Fig. 134). In the same manner, two more malonyl-CoA units are introduced, giving rise to a CoA ester of C_6C_9 polyketocarboxylic acid. This compound should be an appropriate substrate for an aldol condensation. The resulting cyclic compound forms enols and thus becomes aromatic.

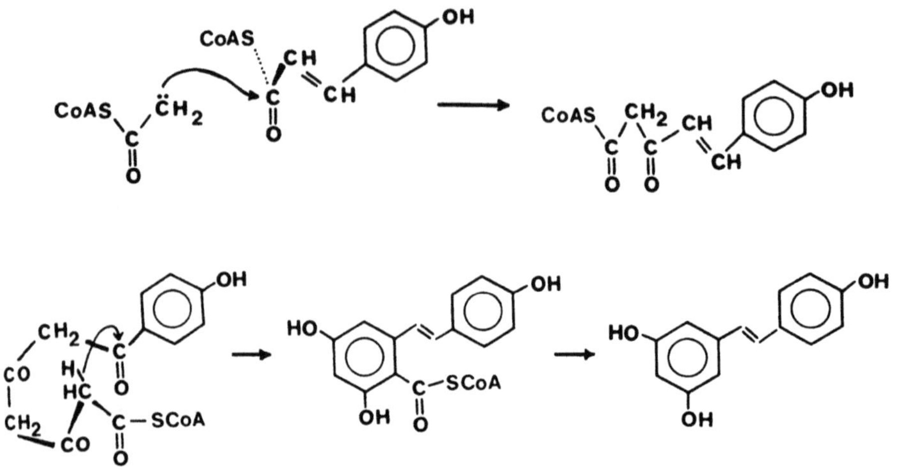

Fig. 134. The first condensation step in the formation of stilbene. Attack of the carbanion (at C_2) on the carbonyl atom (C_7), and the resulting cyclic compound forms enol to result in an aromatic system. (Kindl 1985)

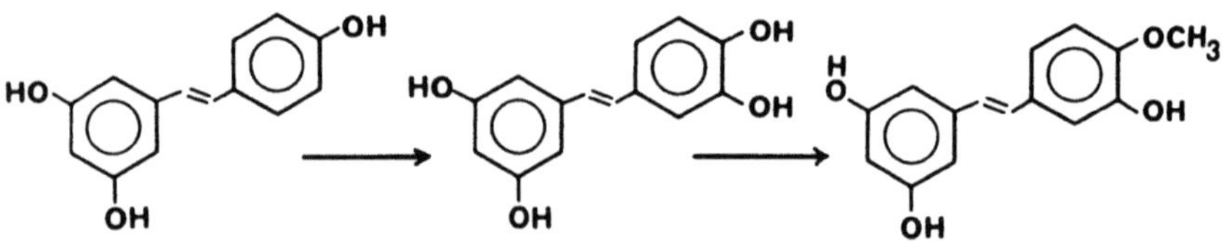

Resveratrol **Piceatannol** **Rhapontigenin**

Fig. 135. Biosynthesis of piceatannol and rhapontigenin

Fritzemeier and Kindl (1983) suggested that dihydrocinnamoyl-CoA and malonyl-CoA are used as substrates for the synthesis of bibenzyl, and that the condensation is catalyzed by dihydrostilbene synthase (DHSS). DHSS is distributed in plants belonging to the Orchidaceae and Dioscoreaceae (Letcher and Nhamo 1972, Letcher et al. 1972). However, recent studies have shown that cinnamoyl-CoA is a preferable substrate rather than phenylpropionyl-CoA for the cloned enzyme from *Pinus sylvestris*, and the participation of DHSS is no longer deemed likely (Fliegmann et al. 1992, Schanz et al. 1992).

Several hydroxylated and methylated derivatives are derived from the stilbene skeleton. Conversion of resveratrol to piceatannol (3′ hydroxylation) in *Rheum rhaponticum* (Polygonaceae) and *Arachis hypogaea* is catalyzed by a hydroxylase. Two *O*-methyltransferases, which specifically catalyze the methylation of the 4′ and 3′ hydroxyl groups of the stilbene, were obtained from roots of *R. rhaponticum* (Rupprich et al. 1980; Fig. 135).

C-alkylation of stilbene with an isoprenyl group to afford 4-isopentenylresveratrol in *A. hypogaea*, and polymerization of the stilbene to dimers and trimers in *Hopea odorata* were observed in feeding experiments. Later, prenyltransferase was isolated from *A. hypogaea* (Aguamah et al. 1981). The presence of a stilbene carboxylic acid synthase (SCS) is suggested by the formation of hydrangeic acid in feeding experiments with *Hydrangea macrophylla*.

4.3.7.2 Induction of Stilbene Formation

The amount of stilbene is larger in young seedings than in mature plants and is increased by several elicitors (Fritzemeier and Kindl 1981). The enhanced formation of stilbene is ascribed to increased activities of phenylalanine ammonia-lyase, cinnamate-4-hydroxylase, cinnamate: CoA ligase and stilbene synthase by the elicitor.

mRNA encoding stilbene synthase increased in the elicitor-treated plant, and the amount of stilbene increased to 100 times that of normal plants (Rupprich et al. 1980). The response is considered to be a defense reaction of plants as a response to fungal attack. On the other hand, a noninduced stilbene formation takes place in roots of *R. rhaponticum* and *Picea excelsa* (Rolfs and Kindl 1984). This is a normal stilbene formation, and the enzyme involved in this stilbene synthesis, such as stilbene synthase, are not activated by external stimuli. It has been observed that different patterns of stilbene formation are involved in various physiological events in plant growth. Stilbene is present in xylem tissue, especially in heartwood and contributes to the durability of wood.

A full-length cDNA encoding stilbene synthase has been isolated from a Scots pine (*Pinus sylvestris*) and sequenced (Schwekendiek et al. 1992). The sequence consists of a 1176-bp open reading frame, which encodes a peptide with 392 amino acid residues. The molecular mass of the peptide was calculated to be 42.6kDa, which closely corresponds to the 43kDa for the resveratrol-forming stilbene synthases of grapevine. The sequence is quite homologous to those of resveratrol-forming stilbene synthase and chalcone synthase, but the cDNA encoding pinosylvin-forming stilbene synthase differs considerably from the cDNA sequence of the known resveratrol-forming stilbene synthase.

The most remarkable property of pinosylvin synthase (PS) is the insertion of three amino acid residues in the N-terminal part. The insertion of a glutamyl residue at position 90 and the changes in charged residues in the region 81–89 reflect the fact that pinosylvin synthase acts on a nonhydroxylated substrate.

The cDNA encoding resveratrol synthase (RS) in *Arachis hypogaea* cell cultures has been characterized (Lanz et al. 1990). Four different genomic sequences (*RS* cDNA, *RS1*, *RS2*, *RS3*) were expressed during growth of a cell cultures from *Arachis hypogaea* after application of various inducers. Transcripts from *RS* cDNA were induced by all of the factors tested; they represented the majority of all identified *RS* RNAs. Expression from *RS*1 and *RS*3 was much lower than from *RS* cDNA and differently induced. It was concluded that *RS* genes in *Arachis hypogaea* represent a gene family, and that some of the members are regulated by different signals.

The induction of stilbene synthase was investigated in cell suspension cultures of grape (*Vitis* cv) by treatment with fungal cell walls. Three independent cDNA clones for stilbene synthases were isolated from a cDNA library of fungal cell wall-induced grape cells and identified by sequence analysis. The cDNA sequence of *pSV21* predicted a protein of 392 amino acids with M_r 42791, similar in size to that observed for immunodetected stilbene synthase. The cDNA sequences of *pSV21* and *pSV25* differed in 76 bp in the coding region. The sequence of grape stilbene synthase cDNAs exhibited significant homology (76%) to the sequence for the peanut stilbene synthase cDNA. The stilbene synthase mRNA was induced within 1 h after addition of fungal cell wall preparations to the cell cultures, rose to a maximum by the sixth hour and then declined slowly over the next 20 h (Melchior and Kindl 1991).

Stilbene synthase genes were isolated from grapevine and transferred into tobacco plants. The regenerated tobacco plants containing these genes were more resistant to infection by *Botrytis cinerea*, showing increased disease resistance in transgenic plants based on an additional foreign phytoalexin.

The increased resistance was attributed to rapid synthesis of large amounts of the phytoalexin in transgenic plants. In transgenic plants containing stilbene synthase genes from grapevine, up to 400 µg resveratrol/g fresh weight were detected (Hain et al. 1993).

A 13 kb DNA fragment was isolated from a grapevine genomic library. After fragmentation with *Eco*RI, subcloning and sequencing, two full-size stilbene synthase genes (*Vst1* and *Vst2*) and the 3′ end of a third stilbene synthase gene (*Vst3*) were found to be located within the 13 kb fragment. *Vst1* and *Vst2*, differing only slightly in the coding region, were distinguished in the intron size and in the structure of the promoter region. The 5′ flanking region of gene *Vst1* contained a TATAA box at nucleotide -48. The substantial structural differences for the promoters of the two genes were paralleled by a striking difference in the expression of the two genes in elicitor-treated cells. The profile of total mRNA obtained after induction of grapevine suspension cultures showed the occurrence of two maxima, an early (3–5 h) and a late (11–16 h) response following the onset of elicitation, suggesting a consecutive expression of at least two types of genes (Wiese et al. 1994).

A comparison of chalcone synthase (CHS) and stilbene synthase (SS) proteins showed that both enzymes are significantly homologous. Recent studies have shown that the two enzyme types posses a single conserved cysteine (Cys-169), which is essential for the catalytic activity and may be required for the condensing reaction. On the other hand, some significant differences exist in the amino acids preceding the cysteine. In the -2 and -3 positions, all CHS

```
CHS P. syl.   VKRVMMY |QQ| G |C| FAGGTVLR
STS P. syl.   VKRVGVF |QH| G |C| FAGGTVLR
STS A. hyp.   VKRYMMY |HQ| G |C| FAGGTVLR
Predicted     ***  ## |##| * |*| ********
structure      ------ |--| - |-| --> --
CHS con       ##R MMY |QQ| G |C| FAGGTVLR
```

Fig. 136. Environment of the active site cysteine (Cys-169) in chalcone synthase (*CHS*) and stilbene synthase (*STS*). *CHS P. syl.* CHS from *P. sylvestris*; *STS P. syl.* STS (dihydropinosylvin forming) from *P. sylvestris*; *STS A. hyp.* STS (resveratrol forming) from *A. hypogaea*. An extended secondary structure (---) with a β-turn (>) is predicted for this part of the proteins. *Boxed* The active site cysteine (Cys-169) and Gln-Gln are conserved in all CHS (*CHS con*). *V* Valine; *K* lysine; *R* arginine; *M* methionine; *Y* tyrosine; *Q* glutamine; *G* glycine; *C* cysteine; *F* phenylalanine; *A* alanine; *T* threonine; *L* leucine; *H* histidine. (Schröder and Schröder 1992)

posses Gln-Gln, while the known SS contain either His-Gln or Gln-His. The resveratrol-forming SS from *Arachis hypogaea* and *Vitis* var. optima, which prefer 4-coumaroyl-CoA, contains His-Gln, but the dihydropinosylvin-forming SS from *Pinus sylvestris* contains the sequence Gln-His (Fig. 136).

Gln-Gln in the CHS, and Gln-His in the SS from *P. sylvestris*, and His-Gln in the SS from *A. hypogaea* were changed by site-directed mutagenesis into all sequence combinations found in the natural enzymes. Assays with the mutant proteins showed that the histidine does not determine the product specificity. With the CHS and SS from *P. sylvestris*, any sequence deviation reduced the activity without marked effects on the substrate preference. However, in the SS from *A. hypogaea* the change from His-Gln to Gln-His suppressed enzyme activity almost completely with all three substrates. Interestingly, the change to Gln-Gln selectively reduced the activity of 4-coumaroyl-CoA, and the enzyme was converted from a resveratrol-forming into a dihydropinosylvin-forming SS (Schröder and Schröder 1992).

4.3.8 Quinones

Quinones occur in bacteria, fungi, lichens, gymnosperms, and angiosperms, but not in mosses (Leistner 1985). It was previously believed that ferns and mosses do not contain quinones, but 3,3′-bi-(2-methyl-1,4-naphthoquinone) was isolated from a fern (*Asplenium laciniatum*; Gupta et al. 1979).

Quinones have a yellow to purple color, confer high durability to wood against microorganisms, termites and shipworms, and are skin-irritating. Various quinones have been isolated from the wood or heartwood of plants belonging to families that are tentatively classified into three groups according

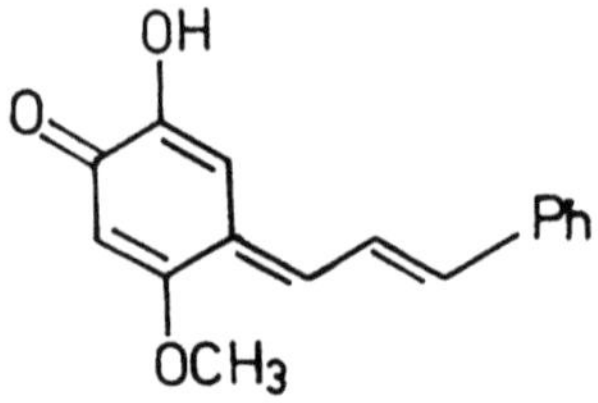

Obtusaquinone

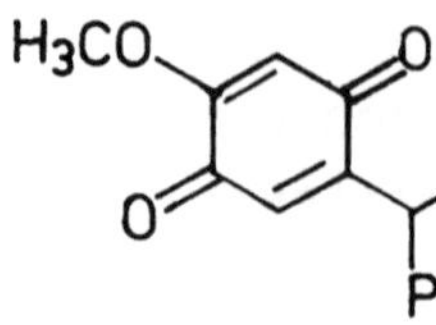

4-Methoxydalbergione

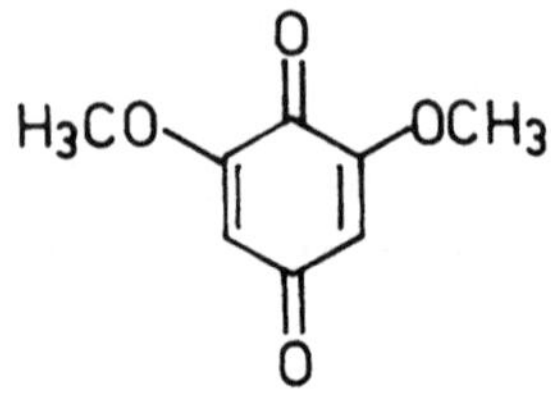

2,6-Dimethoxybenzoquinone

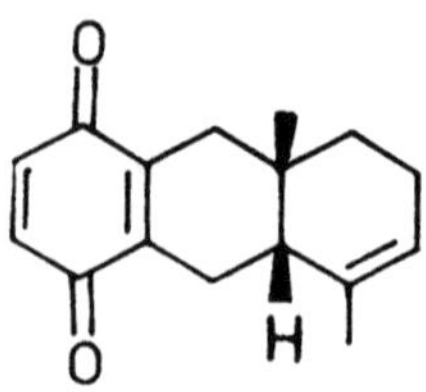

Cordiachrome A

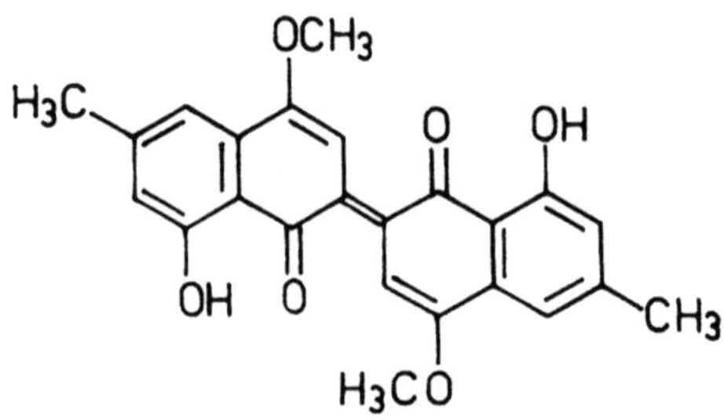

Diosindigo A

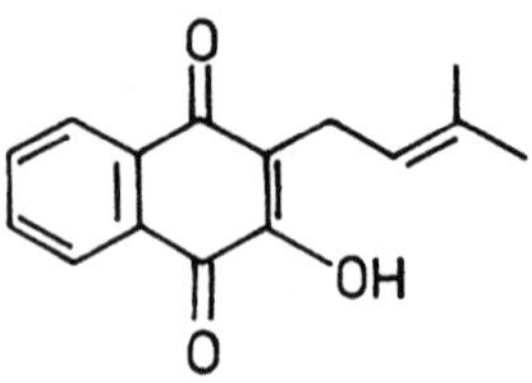

Lapachol

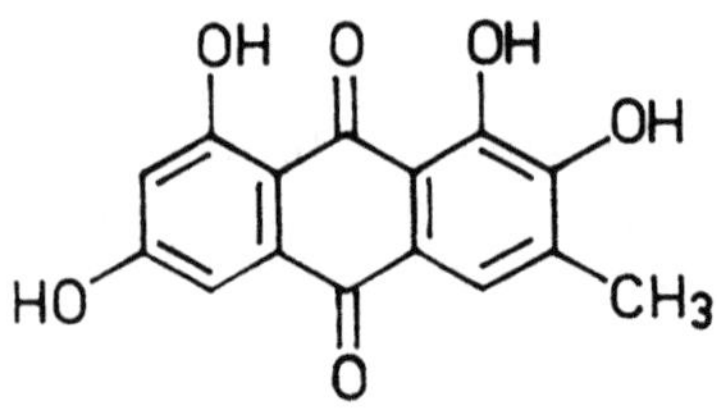

Alternin

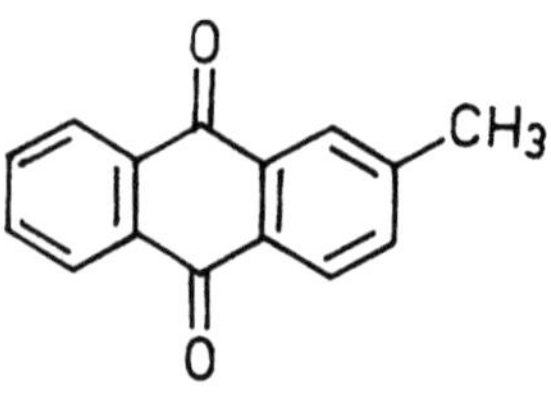

Tectoquinone

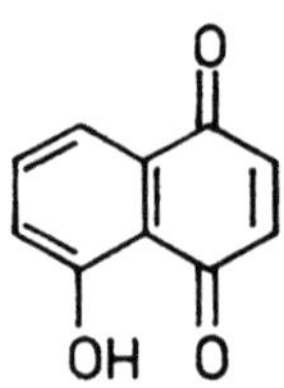

Juglon

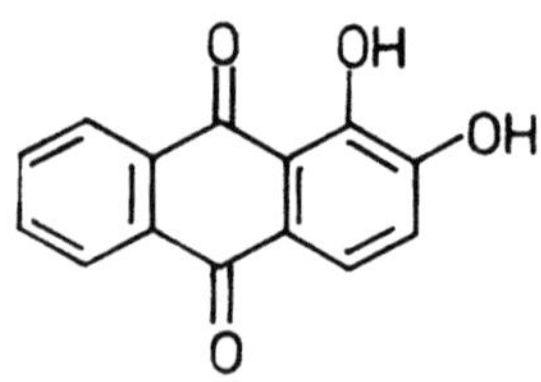

Alizarin

Fig. 137. Major quinones in hardwoods

to the structural types of the quinones present (Leistner 1985; Fig. 137): group (1) Leguminosae, Boraginaceae, Ebenaceae, Rhamnaceae; group (2) Bignoniaceae, Juglandaceae, Proteaceae, Rubiaceae, Verbenaceae, Rutaceae; and group (3) Malvaceae, Sterculiaceae, Ulmaceae.

Dalbergia woods (Leguminosae) contain quinones that show antimicrobial and algicidal activity. The woods are resistant to marine borers and cause allergic reactions. *Dalbergia retusa* wood is highly resistant to the attack by teredos, limnoria, and other marine boring organisms. The heartwood is intensely orange in color, containing mainly two pigments, obtusaquinone, which is highly toxic to fish, and a black pigment, dalbergion (Manners and Jurd 1977), a quinhydrone of racemic 4-methoxydalbergione. Dalbergion is likely to be of a mixed phenylpropanoid/polyketide origin. Other compounds in group 1 including naphthalene and anthracene derivatives are most probably derived via polyketides. The polyketide intermediate is still hypothetical, but it was found that diketonaphtol, a possible intermediate, is converted to islandicin (1,4,8-trihydroxy-3-methyl-9,10-anthraquinone) in cultures of *Penicillum islandicum*. It seems that at least in microorganisms, acetate-derived anthraquinones are synthesized via a bicyclic intermediate (Franck and Stange 1981).

Australian blackwood (*Acacia melanoxylon*, Leguminosae) which is a commercially valuable wood for furniture, contains 2,6-dimethoxybenzoquinone and 2-methyl-6-methoxyfuranobenzoquinone. These quinones cause an eczema allergy and allergic bronchial asthma.

Cordiachrome, and related quinones from *Cordia* species, prenylated naphthoquinones from *Lithospermum, Echium*, etc. (all Boraginaceae), bisnaphthoquinones from *Euclea* and *Diospyros*, whose timbers are of high commercial value (Ebenaceae), have been isolated. Diosindigo A which is responsible for the blue color of the heartwood of some *Diospyros* species has also been isolated and characterized.

Lithospermum erythrorhizon (Boraginaceae) accumulates shikonin (a purple dye, prenylated naphthoquinone) in the root bark. The selected cultured callus of *Lithospermum* synthesized large amounts of shikonin. Feeding experiments with ^{14}C labeled precursors showed that shikonin is synthesized via *m*-geranyl-*p*-hydroxybenzoic acid, which is formed by condensation of *p*-hydroxybenzoic acid derived from L-phenylalanine and two molecules of mevalonic acid (Inoue et al. 1979).

Further experiments (Okamoto et al. 1995) showed that *m*-geranyl-*p*-hydroxybenzoic acid is converted into a key intermediate, geranylhydroquinone, which is subsequently transformed to shikonin via deoxyshikonin. ^{14}C-Labeled deoxyshikonin supplied to cell cultures of *Lithospermum* was incorporated into shikonin derivatives, e.g. acetyl- and β-hydroxyisovalerylshikonin. Crude enzyme preparations catalyzed the conversion of deoxyshikonin into esterified shikonin derivatives, indicating that these derivatives are biosynthesized from deoxyshikonin by hydroxylation at the C_1 position of its side chain by the enzymes (Fig. 138).

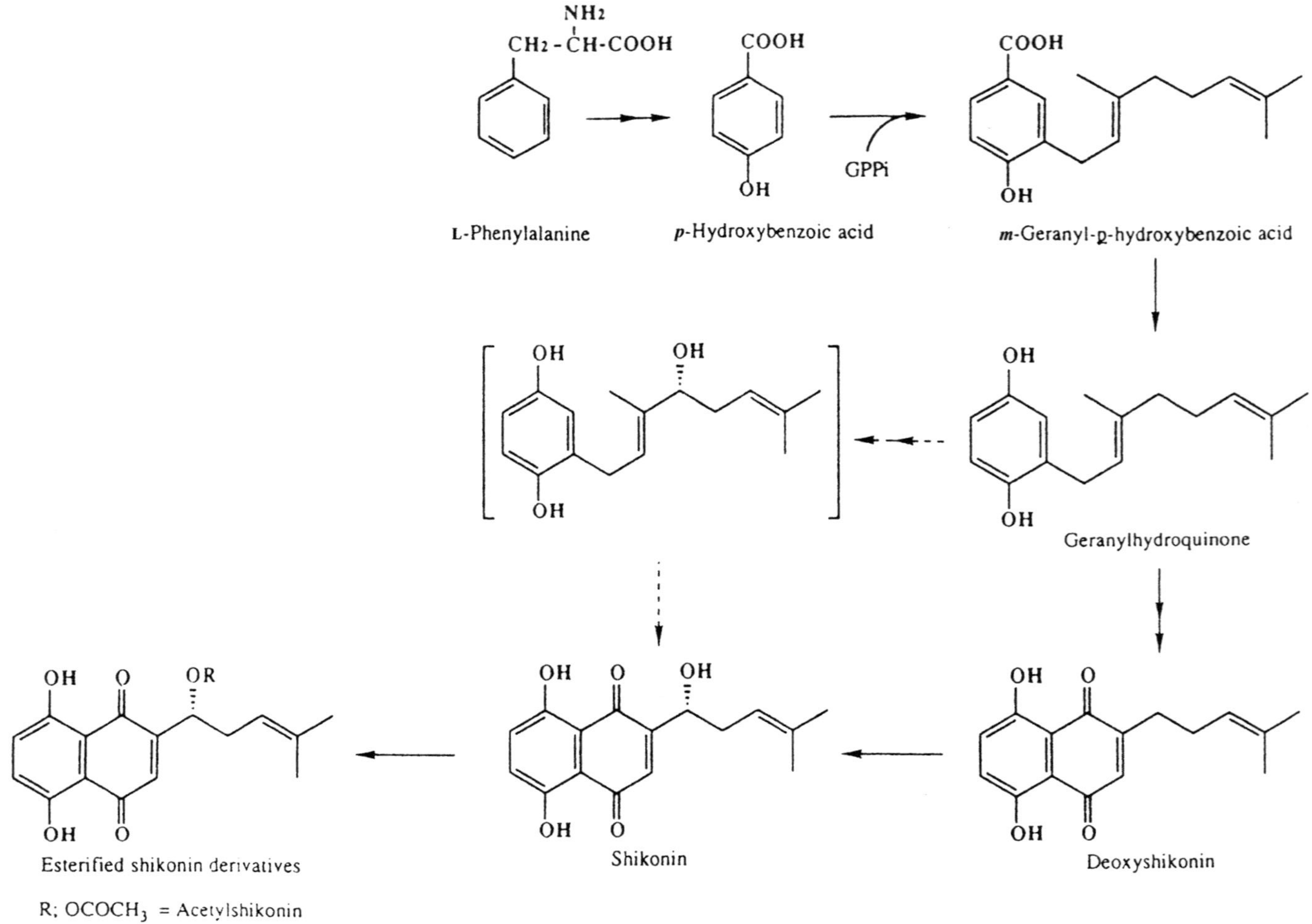

Fig. 138. Biosynthetic pathway of esterified shikonin derivatives via deoxyshikonin in cultured cells of *Lithospermum erythrorhizon*. *GPPi* Geranylpyrophosphate. (Okamoto et al. 1995)

Geranylphosphate: *p*-hydroxybenzoate geranyltransferase was identified from cultured cells of *Lithospermum*. The enzyme is bound to the cell membrane, and irradiation of the culture inhibits the enzyme formation, so that the shikonin synthesis stops completely. The *p*-hydroxybenzoic acid accumulated by the inhibition of prenylation is converted to its glucoside by UDP-glucose glucosyltransferase activated by the irradiation (Heide and Tabata 1987, Heide et al. 1989). Electron microscopic observations of cultured cells of *Lithospermum* showed that shikonin is synthesized in the ER and excreted outside of the cell wall via ER vesicles, accumulating in the outer bark (Tsukada and Tabata 1984).

Plants belonging to the Rhamnaceae contain anthraderivatives like emodin, chrysophanol, physcion and the corresponding dianthrones and heterodianthrones.

Benzoquinones are derived from phenylpropanoids by way of β-oxidation to give benzoic acid, which is converted to the corresponding benzoquinone. In fact, 2,6-dimethoxybenzoquinone is formed from several lignin substructure dimers and lignin by peroxidases and laccase (Higuchi 1985).

As quinones in group 2, lapachol, prenylated naphthoquinone (Bignoniaceae), juglone, a fungitoxic agent, and its derivatives (Juglandaceae), alizarin, a well-known dye, and anththraquinones (Rubiaceae), tectoquinone and 2-methoxyanthraquinone, most effective agents against termites (Verbenaceae), have been isolated and characterized. Tectoquinone and 2-methoxyanthraquinone occur in heartwood of *Tectona grandis*, where their concentration is highest in the center of the heartwood.

o-Succinylbenzoic acid (OSB) derived from chorismic acid and α-oxoglutarate was found to be a precursor of naphthalene derivatives in *Juglans*. Callus cultures of *Juglans* species synthesize juglone, and the glucosides of the corresponding hydroquinones (Müller and Leistner 1978a). β-Hydroxyjuglone and α-oxotetralone are present in the fruit of *Juglans* and have been shown to be derived from OSB (Müller and Leistner 1978b).

It was also shown that anthraquinones are derived from OSB in cell culture of *Marinda citrifolia* L. (Rubiaceae). Feeding experiments with ^{13}C-labeled compounds using cell cultures of *Gallium mollugo* plants showed that OSB is a precursor of anthraquinones, and that during biosynthesis of anthraquinone, 1,4-dihydroxy-2-naphtoic acid (DHNA) is prenylated at C_3 rather than C_2, in agreement with the structure of 2-methoxycarbonyl-3-prenyl-1,4-naphthoquinone isolated from *G. mollugo* plants (Heide and Leistner 1983). It is suggested that anthraquinones are derived from OSB via DHNA. An enzyme preparation from *G. mollugo* cell suspension cultures catalyzes the formation of OSB-CoA ester and that this CoA ester is metabolized to DHNA in the presence of naphthoate synthase, an enzyme involved in bacterial vitamin K biosynthesis (Fig. 139).

Quinones in group 3 are *o*-quinones in *Mansonia* (Sterculiaceae) heartwood, and *o*-naphthoquinones in *Mansonia* and Ulmaceae (Rowe et al. 1972, Thomson 1979). The *o*-quinones in *Mansonia* are violet to yellow dyes, and

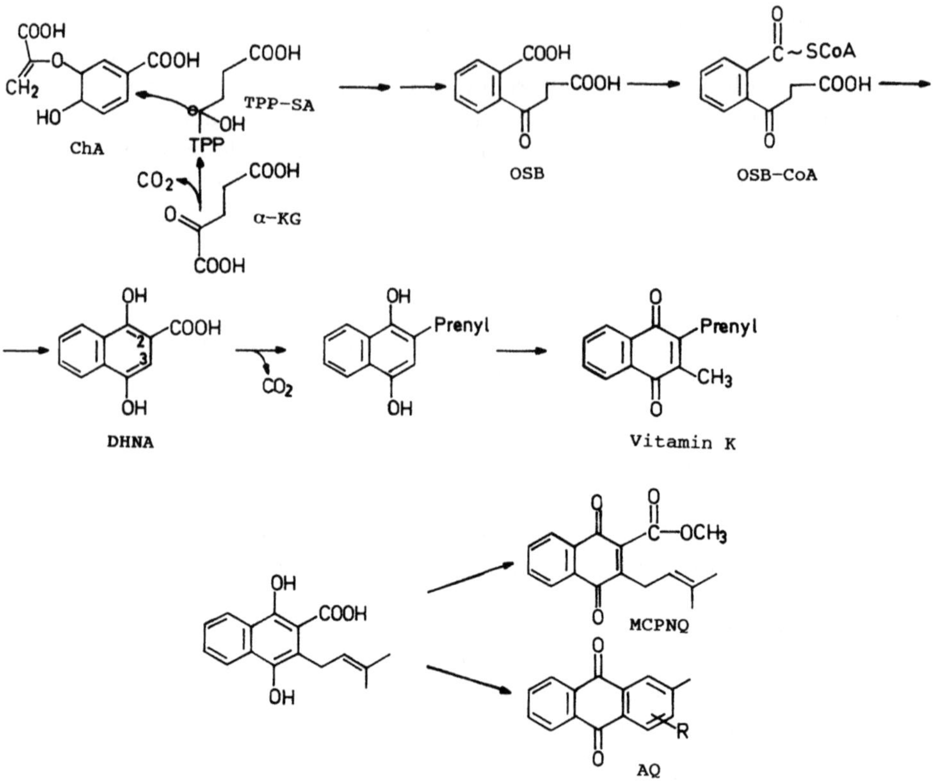

Fig. 139. Biosynthesis of bacterial menaquinones (vitamin K_2) and anthraquinones in Rubiaceae. *ChA* Chorismic acid; *TPP-SA* TPP-succinic semialdehyde; *α-KG* α-ketoglutaric acid; *OSB* *o*-succinylbenzoic acid; *OSB-CoA* *o*-succinylbenzoyl-CoA; *DHNA* 1,4-dihydroxy-2-naphthoic acid; *MCPNQ* 2-methoxycarbonyl-3-prenyl-1,4-naphthoquinone; *AQ* anthraquinones. (Leistner 1985)

their concentration is highest in the outer heartwood and decreases in the center (Sandermann 1966).

Keyakinone A (cadalenic *o*-quinone) and B (dimeric ether of keyakinone A with cadalene) as well as mansonones were isolated from the heartwood of *Zelkova serrata* (Ulmaceae), one of the most important hardwoods for timber and furniture in Japan (Hayashi 1990). The cadalenic quinones (orange color) are formed via oxidation of 7-hydroxycadalenes, and if the 7-hydroxy group is methylated (Tsuki-type, *Zelkova* sp.), the formation of the quinones is inhibited. The methylation rate of the oxycadalenes in *Zelkova* is very important to the color and value of *Zelkova* wood. The biosynthetic pathway of cadalenic quinones in *Zelkova* wood is in Fig. 140.

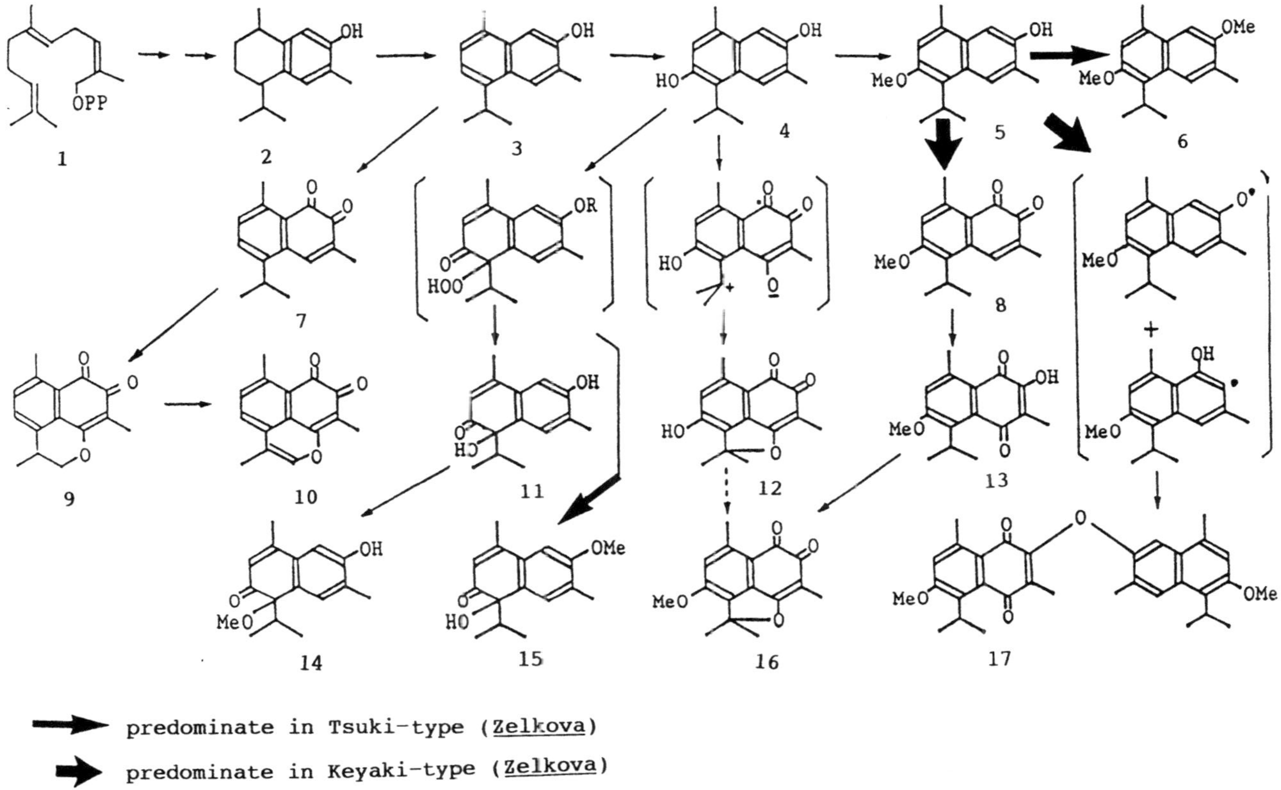

Fig. 140. Possible biogenetic routes of the oxycadalenes and related quinones. **1** Farnesyl pyrophosphate; **2** 7-hydroxycalamenene; **3** 7-hydroxycadalene; **4** 3,7-dihydroxycadalene; **5** 7-hydroxy-3-methoxycadalene; **6** 3,7-dimethoxycadalene; **7** mansonone C; **8** mansonone G methyl ether; **9** mansonone E; **10** mansonone F; **11** zelkoserratone; **12** dehydrooxoperezinone; **13** keyakinone A; **14** 4-*O*-methylzelkoserratone; **15** 7-*O*-methylzelkoserratone; **16** methoxydehydrooxoperezinone; **17** keyakinone B. (Hayashi 1990)

4.3.9 Tannins

Tannins are water soluble polyphenols with molecular weights of 500–3000, and are widely distributed in the plant kingdom. Tannins give a blue color with ferric salts and precipitate alkaloids, gelatin and other proteins (tanning effect). Tannins convert hide into leather. Tannins are classified into hydrolyzable tannins and condensed tannins.

4.3.9.1 Hydrolyzable Tannins

Hydrolyzable tannins are esters of phenolic acids (such as gallic and hexahydroxydiphenic acids) and sugars. By hydrolysis of the glucosidic bond, gallic acid and ellagic acid are formed. Tannins that give gallic acid by hydrolysis are named gallotannins, while tannins that give ellagic acid are named ellagitannins. It was later shown that ellagic acid is an artifact derived from the hexahydroxydiphenic acid moiety of tannins by acid hydrolysis.

A component of insect galls of leaves of *Rhus semialata* is a typical gallotannin. 1,2,3,4,6-Penta-*O*-galloylglucose is obtained from insect galls of oak leaves (Fig. 141), and 3,6-di-*O*-galloyl-1,5-anhydro-D-sorbitol from leaves of *Acer ginnala*.

Ellagitannins are found in the wood and bark of chestnut, oaks and *Eucalyptus*, and in their galls. Large amounts of ellagitannins are present in the heartwood of some *Eucalyptus* and *Quercus* species. Colaridin from dividi-

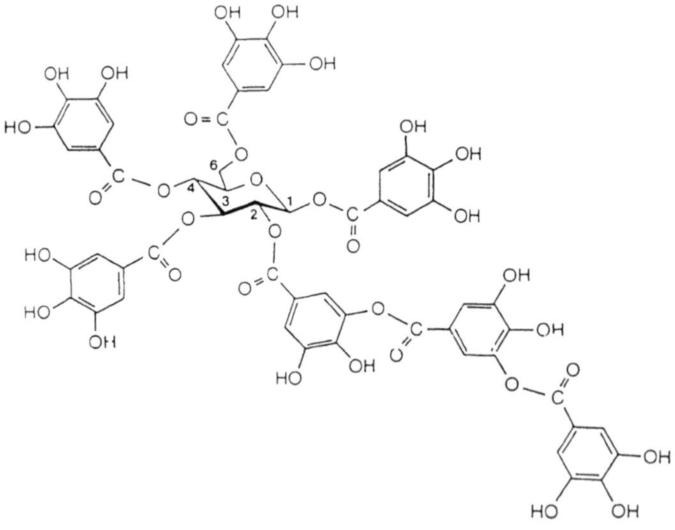

Fig. 141. Chemical structure of 2,3-bis-*O*-digalloyl-1,4,6-tri-*O*-galloylglucose (gallotannin)

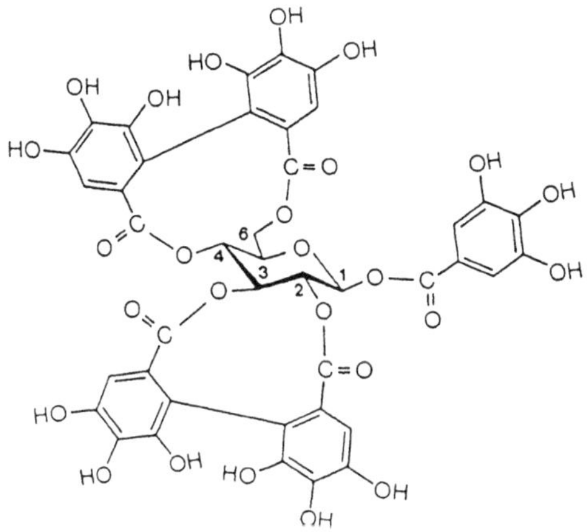

Fig. 142. Chemical structure of 1-*O*-galloyl - 2,3:4,6-di -*O*- hexahydroxydiphenoylglucose (ellagitannin)

vi (*Caesalpinia coriaria*), quebracho (*Shinopsis lorentzii*), and myrobalan (dried fruits of *Terminalia* sp.) are well known as typical ellagitannins. In addition, there are ellagitannins containing quebulic acid and dehydrohexahydroxydiphenic acid (Gross 1992; Fig. 142).

In feeding experiments with *Phycomyces blakesleeanus* Brucker and Drehmann (1958), and Hashem and Brucker 1962 found that glucose-6-^{14}C is incorporated into shikimic acid, protocatechuic acid and gallic acid to the extent of 80, 67, and 12% of the substrate, respectively.

It is remarkable that shikimic acid-^{14}C was incorporated into protocatechuic acid but not into gallic acid. It was concluded that shikimic acid is a precursor of protocatechuic acid, while gallic acid might be formed from aromatic amino acids. Feeding experiments with ^{14}C-labeled L-phenylalanine, L-tyrosine and benzoic acid suggested that gallic acid is formed by β-oxidation of 3,4,5-trihydroxycinnamic acid via L-phenylalanine in *Rhus typhina* (Zenk 1964).

Neish et al. (1964) suggested that β-oxidation occurs at the stage of caffeic acid, and that the protocatechuic acid formed is hydroxylated to give gallic acid. El-Basyouni et al. (1964) found that protocatechuic acid was converted to gallic acid in feeding experiments with *Geranium hortorum*. They confirmed the pathway suggested by Neish where cinnamic acid is converted to caffeic acid which is converted to protocatechuic acid to give gallic acid.

Later, in feeding experiments with *Geranium pyreniacum* (Conn and Swain 1961) and with *Phycomyces blakesleeanus*, *Rhus typhina*, *Acer*, *Geranium* and *Camellia sinensis* (Cornthwait and Haslam 1965, Dewick and Haslam 1969) dehydrogenation of 3-dehydroshikimic acid to give gallic acid was shown to occur in addition to the caffeic acid pathway. Evidently, the formation of gallic acid in higher plants occurs via both the caffeic acid and the 3-dehydroshikimic acid pathways (Fig. 143).

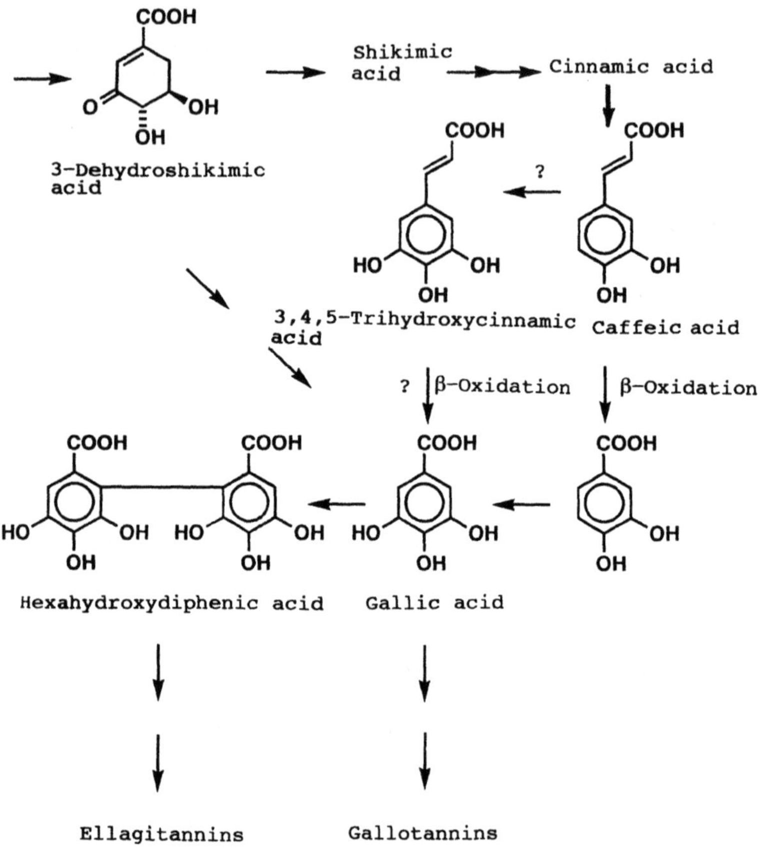

Fig. 143. Biosynthetic pathways of gallic acid

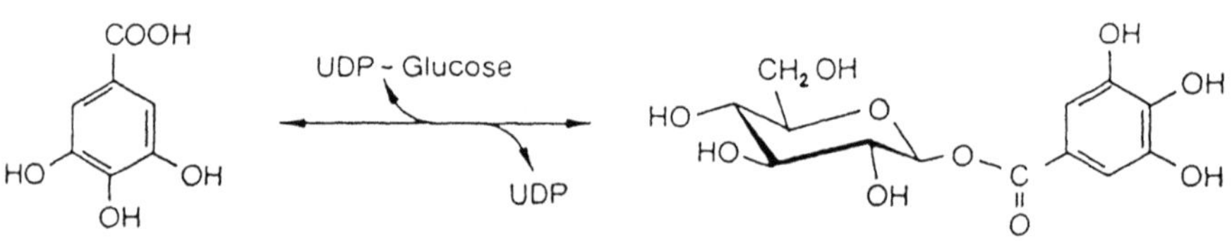

Fig. 144. Synthesis of β-glucogallin by an enzyme preparation from leaves of *Quercus robur*. (Gross 1983a)

Recently Gross (1983a,b) found that an enzyme preparation from the leaves of *Quercus robur* catalyzes the formation of β-D-glucogallin from ^{14}C-gallic acid and UDP-glucose (Fig. 144). Since the isolation of β-D-glucogallin from *Rheum officinale* in 1903 (Haslam 1989), β-glucogallin has been held to be a major intermediate in the formation of the hydrolyzable tannins (Haddock et al. 1982). Gross (1983a) showed that enzyme preparations from young

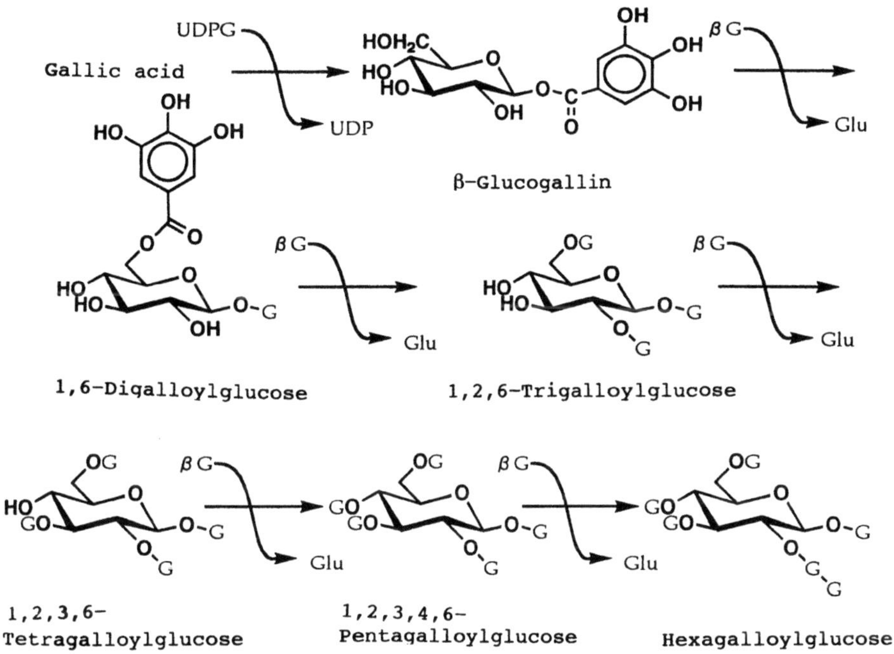

Fig. 145. Synthesis of 1,6-di-, 1,2,6-tri-, 1,2,3,6-tetra-, and 1,2,3,4,6-pentagalloylglucoses, and hydrolyzable tannin by enzyme preparations from young leaves of *Quercus robur* and *Rhus typhina*. *βG* β-Glucogallin. (Gross 1992)

leaves of *Quercus robur* and *Rhus typhina* catalyze the formation of di- and trigalloylglucose from β-glucogallin (Fig. 145). One molecule of β-glucogallin is used as donor of the galloyl group, and another molecule is used as an accepter to give 1,6-digalloylglucose in this enzyme reaction. Furthermore, an enzyme that catalyzes the formation of 1,2,6-trigalloylglucose from 1,6-digalloylglucose as a donor of β-glucogallin has been isolated from the leaves of *R. typhina* (Gross 1983a,b).

1,2,3,6-Tetra and 1,2,3,4,6-pentagalloylglucose are formed via similar reactions from 1,2,6-trigalloylglucose and β-glucogallin, which is the galloyl group donor. In addition, another enzyme, galloyltransferase, which catalyzes the formation of 1,2,6-trigalloylglucose from 1,6-digalloylglucose without the presence of β-glucogallin has been isolated (Denzel and Gross 1991).

4.3.9.1.1 Formation of Hydrolyzable Tannins from Pentagalloylglucose

Hofmann and Gross (1990) found that an enzyme preparation from *R. typhina* catalyzes the formation of *m*-depside linkages of hexa-, hepta-, octa- and nonagalloylglucoses in the presence of β-glucogallin (Fig. 145). However, the formation of ellagitannins from pentagalloylglucose is not known. Gallic acid and its derivatives, such as β-glucogallin, 3,6-digalloylglucose, and pentagalloylglucose give ellagic acid by oxidation with microbial laccase and plant peroxidase, but the formation of hexahydroxydiphenic acid as its glucose ester has not been detected (Gross 1992).

4.3.9.2 Condensed Tannins

The condensed tannins are classified into the phloroglucinol type and the resorcinol type, based on the oxidation pattern of the A rings. The former are those containing flavonoid units with a phloroglucinol pattern A ring: propelargonidins, procyanidins and prodelphinidins. The latter are those containing flavonoid units with a resorcinol pattern A ring: proguibourtinidins, profisetinidins, and prorobinetinidins. It has been shown that a significant proportion of the linkages in procyanidin polymers are (4–8) bonds (Fig. 146a), in addition to the (4–6) bonds (Fig. 146b).

The amount of condensed tannins in wood is much higher than that of hydrolyzable tannins. Tropical and subtropical trees contain especially large amounts of condensed tannins. The condensed tannins give insoluble phlobaphenes on heating with hot hydrochloric acid.

Proanthocyanidins (the phloroglucinol type) are ubiquitously distributed in conifers, occurring in relatively large amounts in the bark, needles, and cones. The proanthocyanidins are present in virtually all woody dicotyledons and some monocotyledon species. In contrast, the resorcinol type is limited to Leguminosae and Anacardiaceae. Wattle bark (*Acacia mearnsii*) and quebracho wood (*Schinopsis* spp.) are the most important sources of the tannins of commerce (Hillis 1985). The genus *Acacia* is the richest source of the resorci-

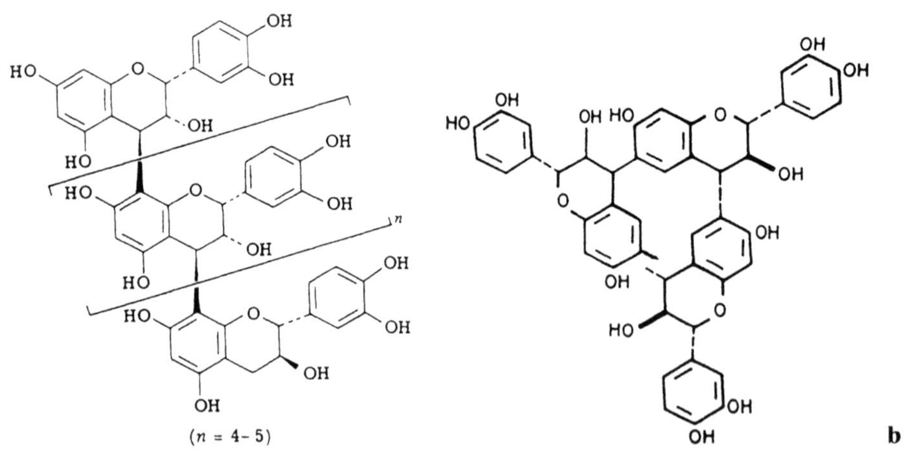

a b

Fig. 146. Structure of condensed tannins derived from epicatechin (**a**, 4–8 linkage), and catechin (**b**, 4–6 linkage). (Porter 1989)

nol type of tannins, and 5-deoxyflavanoids normally occur exclusively in the heartwood of this genus, whereas such flavonoids are often accompanied by procyanidins and prodelphinidins in the bark (Porter 1989).

Ribes species commonly contain prodelphinidins. These are predominantly of the gallocatechin-4 type in the leaves, but of the epigallocatechin-4 type in the fruit. In the outer bark of *Pinus radiata* tannins are predominant over 2,3-*trans* catechin-4 procyanidin polymers, but in the phloem, needles, male cones, and winter buds, a variety of mixed procyanidin and prodelphinidin polymers exist, largely with a 2,3-*cis* stereochemistry. These observation suggest that more than one metabolic pool for the production of various proanthocyanidins can exist in a plant. Condensed tannins are usually concentrated in the epidermal tissue of plants, the leaves, outer bark, and phloem, and in the epidermis of fruit skins and seeds.

Electron microscope studies of tannins in cell suspension cultures of *Picea glauca* showed that the tannin inclusions originated in cytoplasmic vacuoles, possibly derived from the endoplasmic reticulum, and accumulated in the central vacuole. Tannin accumulation in tissue-culture cells of slash pine was found in the smooth endoplasmic reticulum, in vesicles, and in cell vacuoles (Porter 1989).

Proanthocyanidins normally accumulate in plant tissue as polymers (condensed tannins) rather than as oligomers. This implies that the major biosynthetic pathway is directed toward the synthesis of proanthocyanidin units to form those polymers, in contrast to the formation of lesser amounts of monomeric flavan-3-ol.

^{13}C-NMR of oligomeric and polymeric proanthocyanidins showed that 4–8 linkages occur approximately three times more frequently than 4–6 linkages. Studies on an epicatechin homopolymer from the fruits of *Chaenomeles chinensis* suggested that the polymer theoretically consists of four types of unit (Porter 1989; Fig. 147): (1) T (top) units, in which an epicatechin-4 unit is linked through C-4 to C-6 or C-8 of a neighboring unit; (2) M (middle) units, which are doubly linked, both through C-4, and C-6 or C-8 to two neighboring units; (3) J (branching or junction) units, which are triply linked through C-4, C-6 and C-8 to neighboring units; and (4) B (bottom) units, which are singly linked through C-6 or C-8 to neighboring units, and C-4 is unsubstituted.

An average polymerization number of the *Chaenomeles* polymer was 14 per chain. The ^{13}C NMR spectrum showed that both T(4β-6)M and T(4β-8)M are represented as interflavanoid linkages throughout the chain.

In contrast, it was shown that proanthocyanidin in the secondary phloem of *Pinus radiata* mostly contains about equal numbers of epicatechin-4 and epigallocatechin-4 together with minor amounts of catechin-4 units. During bark development, the structure of the extractable tannins undergoes progressive changes across the phloem and outer bark, suggesting structural change caused by the oxidative enzymes of the phloem cells.

Precursors of polymeric polyphenols in the heartwood of *Schinopsis quebracho-colorata* are (+)-catechin and profisetinidin (trihydroxyflav-3,4-diol).

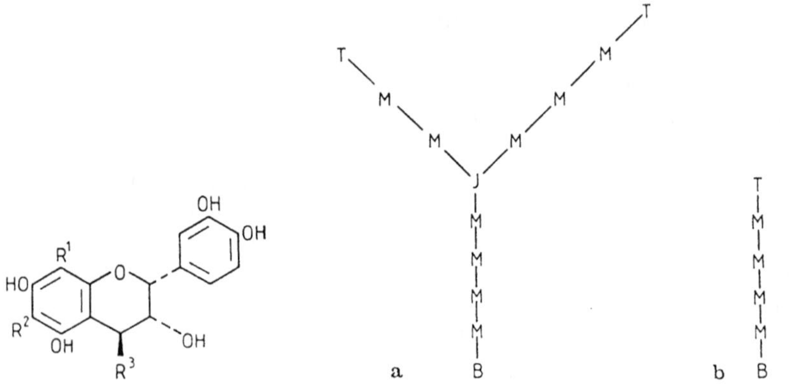

Fig. 147. Structure of an average molecule of the procyanidin polymer from *Chaenomeles chinensis* fruits as deduced from high-field ^{13}C NMR. **a** Branched chain; **b** linear chain. T unit; R^1=R^2=H, R^3=flavonoid. M unit: R^1 or R^2=H, R^1 or R^2=R^3=flavonoid. J unit: R^1=R^2=R^3=flavonoid; B unit: R^1 or R^2=flavonoid, R^1 or R^2=R^3=H. (Porter 1989)

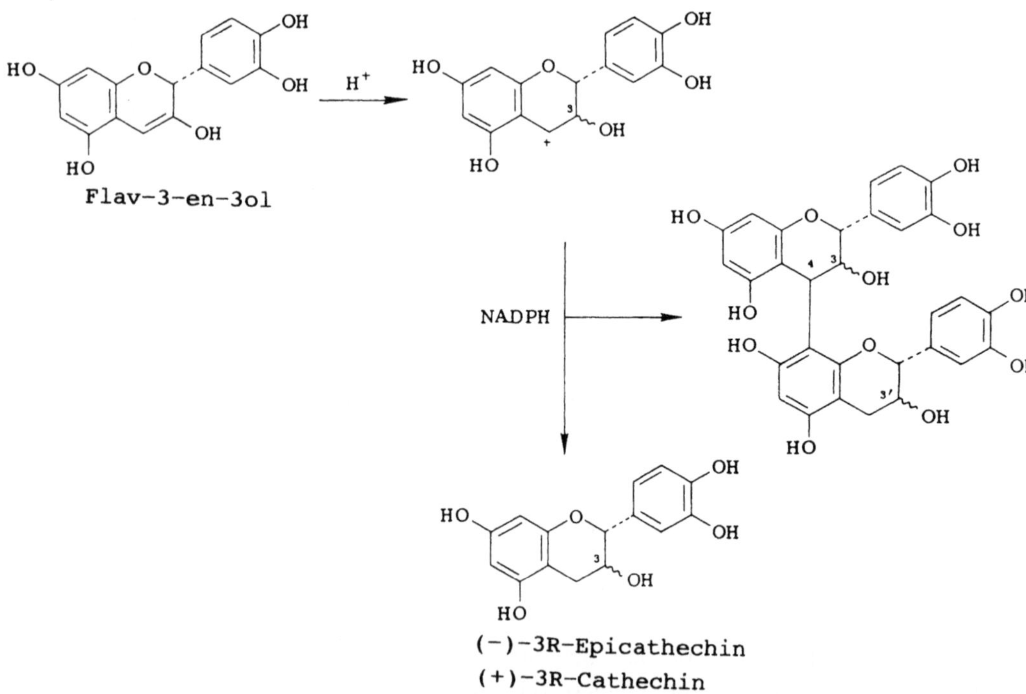

Fig. 148. A possible mechanism of coupling of flavan-3-ols and flavan-3,4-diols into proanthocyanidin polymers. (Hillis 1985)

The concentration of the precursors is high in outer heartwood and decreases toward the inner heartwood. Only polymerized profisetinidin is present in the heartwood, and its molecular weight increases toward the center of the heartwood (Porter 1989).

The coupling mechanism of flavan-3-ols and flavan-3,4-diols to give proanthocyanidindin polymers has not been elucidated, but it is suggested that the polymers are formed by oxidative couplings at the 6 or 8 positions of the A ring of catechins or leucocyanidins via flav-3-en-3-ols (Hillis 1985; Fig. 148).

The biosynthesic pathways of catechins and leucocyanidins were discussed in connection with flavonoid biosynthesis. The A ring is derived from three molecules of malonyl-CoA, while the B ring originates from cinnamoyl-CoA via chalcone.

4.3.10 Suberin

Suberin is a complex polymer composed of a lignin-like phenolic polymer esterified with aliphatic acids ($CH_3(CH_2)mCOOH$, m=18–30), dicarboxyaliphatic acids ($HOOC(CH_2)nCOOH$, n = 14–20), ω-hyrdroxyaliphatic acids ($CH_2OH(CH_2)nCOOH$) and aliphatic alcohols ($CH_3(CH_2)mCH_2OH$; Fig. 149). No method to isolate suberin in its native state has been developed so far.

Table 25 shows the composition of aliphatic acids of suberin released by depolymerization of total bark samples (Holloway 1983). It has been suggested that dicarboxylic acids and ω-hydroxy acids are esterified to phenolic polymers and other ω-hydroxy acids and afford linear polyesters. Suberin contains several hydroxyaliphatic acids and forms large cutin-like polyesters. Long aliphatic acids and aliphatic alcohols contain one terminal functional group, which is esterified. The aliphatic portion of the suberin polymer is attached to a wax layer to form a lamellate structure (Kolattukudy and Espelie 1989).

^{13}C-NMR of cutin from fruits of *Malus pumila* showed that most of the carbons are from methylene (CH_2, 20–30 ppm), while the spectrum of suberin of *Solanum tuberosum* indicated the presence of a large number of CHOH groups derived from sugar, and aromatic carbons (120–160 ppm) with many aliphatic CH_2 groups. In contrast, the suberin from bark of *Pseudotsuga menziesii* contained a large number of aromatic carbons with a lesser amount of sugar carbons (Kolattukudy and Espelie 1985).

It has been suggested that suberin is more condensed than lignin, and it gives a small amount of vanillin and *p*-hydroxybenzaldehyde with a trace of syringaldehyde after alkaline nitrobenzene oxidation. In summary, the following five reactions seem to be involved in the biosynthesis of suberin.

1. ω-Hydroxylation of long aliphatic acids.
2. Oxidation of ω-hydroxy acids to dicarboxylic acids.

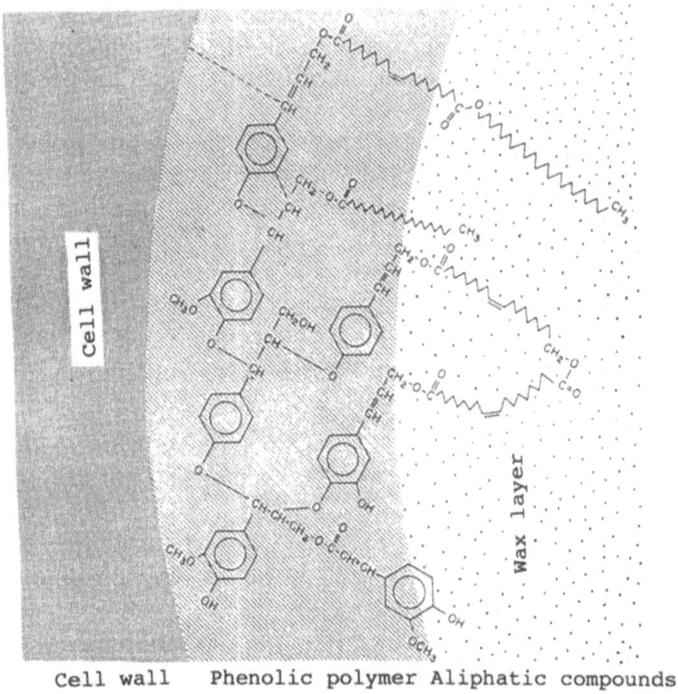

Fig. 149. A model structure of suberin. (Kolattukudy and Espelie 1989)

Table 25. Total ω-hydroxy- and dicarboxylic acid contents of the aliphatic portion of monomers released by depolymerization of total bark samples. (Holloway 1983)

Source	ω-Hydroxy acid (C_{16}–C_{28})	Dicarboxylic acid (C_{16}–C_{26})
Acer pseudoplatanus	45.3	20.5
A. griseum	39.6	25.9
Betula pendula	21.6	8.0
Castanea sativa	13.4	9.9
Cupressus leylandii	51.2	31.6
Fagus sylvatica	18.7	11.9
Fraxinus excelsior	24.3	40.1
Laburnum anagyroides	48.6	21.8
Poplus tremula	54.5	27.8
Quercus robur	21.3	15.1
Q. ilex	23.9	10.5
Q. suber	40.7	32.7
Ribes nigrum	49.7	20.2
Sambucus nigra	50.5	24.6

3. Formation of aliphatic acids with middle hydroxyl groups.
4. Formation of phenolic polymer.
5. Ester formation of phenolic polymer and aliphatic acids.

It has been shown that monooxygenase (P450) from an ER fraction of *Vicia faba* seedlings catalyzes ω-hydroxylation of hexadecanoic acid in the presence of NADPH and oxygen (Soliday and Kolattukudy 1977). In addition, two enzymes, ω-hydroxydehydrogenase and ω-oxo acid dehydrogenase, have been isolated from the epidermis of the leaves of *Vicia faba*, and wounded tissues of potato, respectively. The former enzyme catalyzes the oxidation of ω-hydroxyhexadecanoic acid to ω-oxyhexadecanoic acid, and the latter enzyme catalyzes the oxidation of ω-oxyhexadecanoic acid to C_{16} dicarboxylic acid (Kolattukudy and Espelie 1985; Fig. 150).

Feeding experiment with *Vicia faba* showed that 18-hydroxyoctadecanoic acid is oxidized to 9,10-epoxi-18-hydroxyoctadecanoic acid and 9,10,18-trihydroxyoctadecanoic acid. A hydroxylase (P450) that catalyzes C-10 hydroxylation of 16-hydroxydecanoic acid was isolated from the epidermis and hypocotyl of *Vicia faba* (Kolattukudy and Espelie 1985; Fig. 151).

Phenolic polymers seem to be formed by a biosynthetic pathway similar to that of lignin. L-U-^{14}C-phenylalanine and U-^{14}C-cinnamic acid were incorpo-

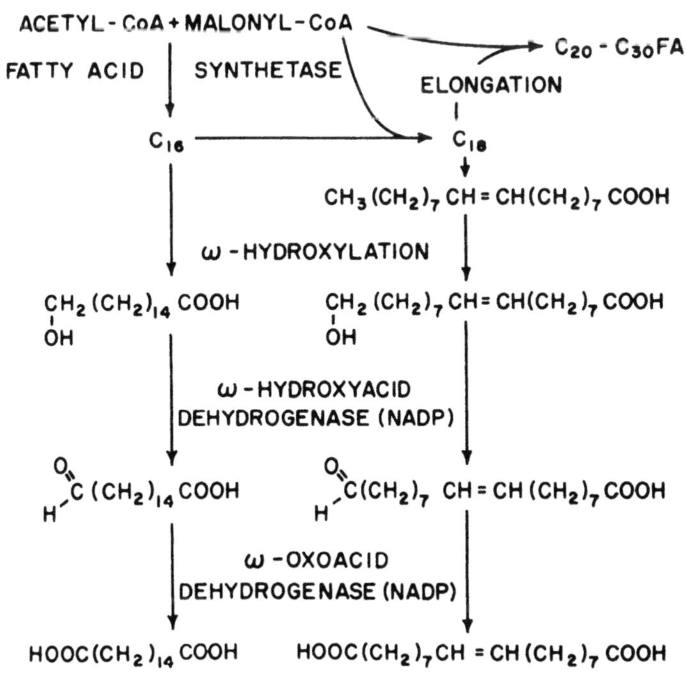

Fig. 150. Biosynthetic pathways for the major aliphatic acids constituting suberin. (Kollatukudy and Espelie 1985)

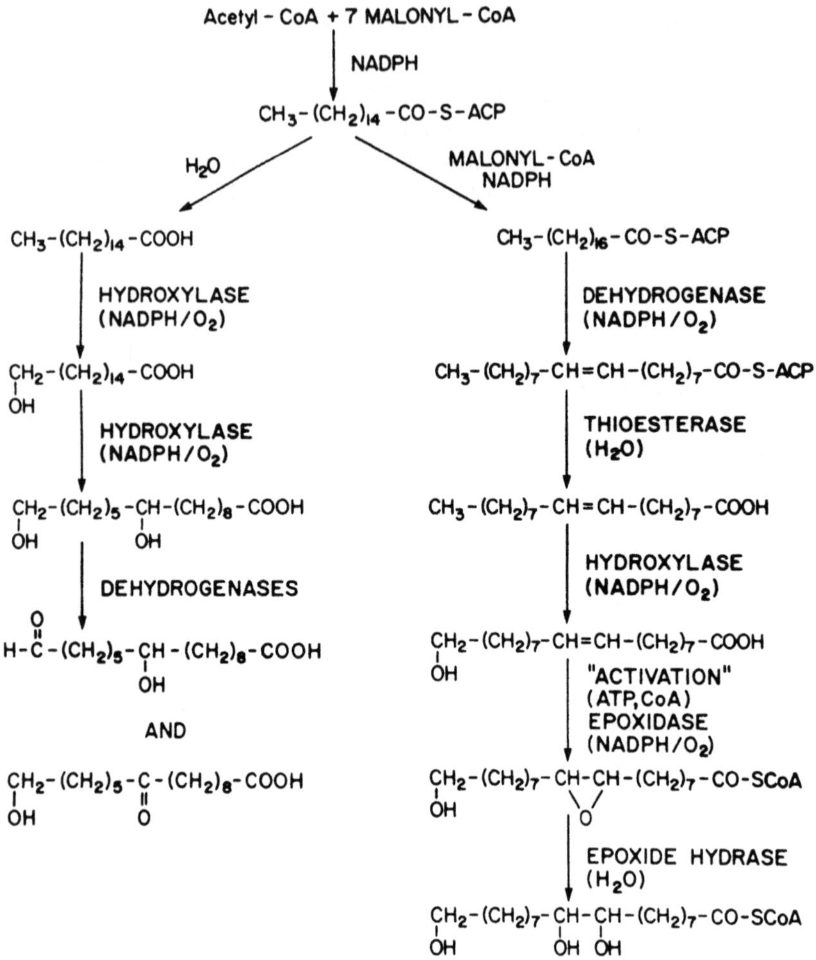

Fig. 151. Biosynthetic pathways for hydroxy fatty acids constituting suberin. (Kollatukudy and Espelie 1985)

rated into the insoluble polymer of suberin of potato, which produced radioactive *p*-hydroxybenzaldehyde and vanillin with a trace of syringaldehyde on nitrobenzene oxidation. The induction of specific anionic peroxidase was also observed in suberin formation (Cottle and Kolattukudy 1982). However, when thioacidolysis was applied to the suberized potato periderm tissue, only guaiacyl and syringyl derivatives were released, and no *p*-hydroxyphenylpropane moieties were detected. The exact nature of the phenolic monomer that constitutes the phenolic suberin is therefore still unkown. It has been suggested that activated ω-hydroxy and dicarboxy acids are involved in the esterification of the phenolic polymer.

4.4 Lipids

Physiologically important lipids are membrane lipids such as phospholipids and galactolipids, which are major components of all biological membranes. Other important lipids are the neutral lipids, such as triacylglycerides (fats), which act as an energy reserve in many organisms. It is known that the thylakoid membranes of chloroplasts are mainly composed of highly polyunsaturated galactolipids.

4.4.1 Membrane Lipids

Formation of membrane lipids begins with phosphatidic acid. In galactolipid synthesis, phosphatidic acid is dephosphorylated to produce diacylglycerols, which accumulate transiently on the chloroplast envelop membrane. By mediation of UDP-galactose: diacylglycerol galactosyltransferase, monogalactosyldiacylglycerol is formed from diacylglycerol and UDP-galactosc. Λ further galactosylation reaction with UDP-galactose by the transferase produces digalactosylglycerol. These two galactolipids are the major structural components of the chloroplast thylakoid membranes.

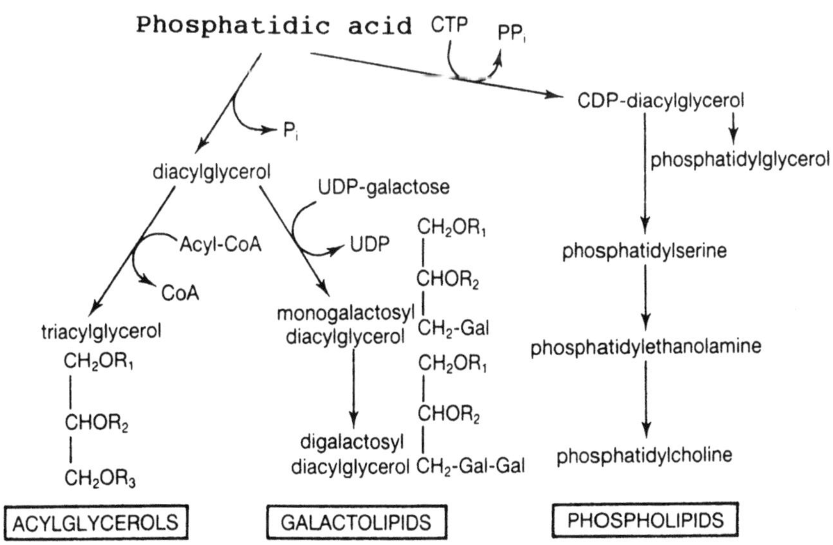

Fig. 152. Biosynthetic pathways for acylglycerols, galactolipids, and phospholipids. The central intermediate is phosphatidic acid. Phospholipids are synthesized directly from phosphatidic acid. Galactolipids and acylglycerols are formed from diacylglycerols, the dephosphorylated form of phosphatidic acid. *CTP* Cytidine triphosphate. (Murphy 1993)

In the biosynthesis of phospholipids in plants, phosphatidic acid is converted to phosphatidylglycerol, or it is dephosphorylated to diacylglycerol. Phosphatidic acid can also sever as a substrate for either phosphatidylcholine, phosphatidylethanolamine or phosphatidylserine synthesis (Murphy 1993; Fig. 152).

4.4.2 Acyltriglycerides (Fats)

Acyltriglycerides (neutral fats) are present in seeds and ray parenchyma cells of wood, such as *Picea excelsa*, *Pinus sylvestris*, *Tilia vulgaris*, *Populus deltoides*, and *Aesculis hippocastanum*. The major fatty acids are C_{18} unsaturated acids, such as oleic acid and linoleic acid, palmitic acid (a C_{16} saturated acid), and small amounts of C_6, C_8, C_{10}, and C_{12}–C_{32} fatty acids with even and uneven carbon numbers.

4.4.2.1 Saturated Fatty Acids

The biosynythesis of fatty acids in plants largely occurs in the plastids. The synthesis requires two enzymes, namely acetyl-CoA carboxylase and fatty acid synthase. The former enzyme consists of a protein complex of between 220 and 240 kDa, and the latter enzyme is a dissociable complex made up of six subunits.

Generally, the carbon number of natural fatty acids is even, and acetic acid is their precursor, as shown in feeding experiments. Decarboxylative condensation of malonyl-CoAs with acetyl-CoA resulting in C_{16}–C_{18} acids is referred to as the malonate pathway (Yamada 1981), and is a major pathway of fatty acid synthesis in animals and plants. This pathway is presented in the following six reactions:

1. Acyl transfer

$$CH_3COSCoA + ACPSH \overset{\text{Acetyltransacylase}}{\leftrightarrows} CH_3COSACP + COASH$$

2. Malonyl transfer

$$HOOCCH_2COSCoA + ACPSH \overset{\text{Malonyltransacylase}}{\leftrightarrows} HOOCCH_2COSACP + COASH$$

3. Condensation

$$CH_3COSACP + HOOCCH_2COSACP \overset{\text{3-Oxoacyl-ACP synthase}}{\leftrightarrows} CH_3COCH_2COSACP + CO_2 + ACP$$

4. 3-Oxoacyl reduction

$$CH_3COCH_2COSACP + NAD(P)H + H^+ \xrightleftharpoons{\text{3-Oxoacyl-ACP reductase}} CH_3CHOHCH_2COSACP + NAD(P)^+$$

5. Dehydration

$$CH_3CHOHCH_2COSACP \xrightleftharpoons{\text{3-Oxoacyl-ACP dehydratase}} CH_3CH{=}CHCOSACP + H_2O$$

6. Enoyl reduction

$$CH_3CH = CHCOSACP + NAD(P)H + H^+ \xrightleftharpoons{\text{Enoyl-ACP reductase}} CH_3CH_2CH_2COSACP + NAD(P)^+ \text{ (Butyryl-ACP)}$$

By repetition of the reactions 1 to 6, two carbons are extended.

In these reactions the acyl group is always connected to ACP (acyl group transporting protein) and the acyl-ACP is involved in the synthesis of fatty acids. First, acetyl-CoA is converted to acetyl-SACP, while malonyl-CoA is converted to malonyl-SACP. Acetyl-SACP is decarboxylatively condensed with malony-SACP to afford acetoacetyl-SACP which is reduced, dehydrated, and reduced successively to afford butyryl-SACP. Repetition of these reactions gives C_{16} or C_{18}-SACP, after which acyl groups are transferred from the SACP to CoA to give acyl-CoA.

Fatty acids with chain length less than 16 carbons are relatively rare in most cells, but are sometimes found as components of seed oils. In this case the fatty acid synthase is probably responsible for an extra activity leading to premature chain termination, resulting in the accumulation of medium- and short- chain fatty acids, such as C_8, C_{10} and C_{12} fatty acids. Malonyl-CoA is synthesized by carboxylation of acetyl-CoA (Fig. 153).

Partial cDNAs of an acetyl-CoA carboxylase (ACCase) of alfalfa were isolated from a cDNA library of suspension-cultured cells that had been elicited for isoflavonoid phytoalexin synthesis (Shorrosh et al. 1994). The full-

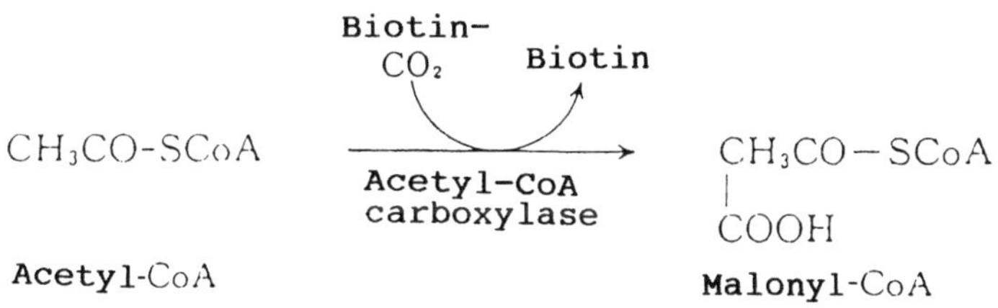

Fig. 153. Enzymatic formation of malonyl-CoA

length sequence was obtained by primer extension and amplification of the cDNA with synthetic primers. The sequence codes for a protein of 2257 amino acids with a calculated M_r of 252039. The biotin carboxylase, the biotin carboxyl carrier protein, and the carboxyltransferase domains show approximately 72, 50, and 65% sequence similarity, respectively, to those of animal, diatom, and yeast ACCase sequences. ACCase enzyme activity and transcripts are induced severalfold upon addition of yeast or fungal elicitors to alfalfa cell cultures.

A set of overlapping clones encoding the entire wheat ACCase was isolated by using the cDNA fragments as a probe. The cDNA sequence contains a 2257-amino acid reading frame encoding a 251-kDa polypeptide. The amino acid sequence of the most highly conserved domain, corresponding to the biotin carboxylases of prokaryotes, is 52–55% identical with ACCase of yeast, rats, and diatoms. Identity with the available C-terminal amino acid sequence of maize ACCase is 66% (Gornicki et al. 1994).

4.4.2.2 Unsaturated Fatty Acids

The synthetic patterns of unsaturated fatty acids are different in animals and plants. In plants, the formation of palmitoleic acid (16:1) from palmitic acid (16:0) and oleic acid (18:1) from stearic acid (18:0) are mediated by desaturases in the presence of NADH and O_2:

$$\text{Stearoyl-ACP} + \text{NADH} + H^+ + O_2 \xrightarrow[\text{Oxidative desaturase}]{} \text{Oleoyl-ACP} + 2H_2O + NAD^+$$

Oleoyl-ACP is regarded as a central metabolite in plant lipid metabolism. All the reactions on the pathway from acetyl-CoA to oleoyl-ACP are present in the plastid.

Stearoyl-ACP desaturase was purified from developing embryos of safflower seed, and the extensive amino acid sequence was determined. This sequence was used to clone a full-length cDNA. The primary structure of the protein, as deduced from the nucleotide sequence of the cDNA, includes a 33-amino acid peptide not found in the purified enzyme.

Expression in *Escherichia coli* of a gene encoding the mature form of stearoyl-ACP desaturase did not result in an altered fatty acid composition. However, an active enzyme and a change in fatty acid composition were detected when assayed in vitro with added spinach ferredoxin, indicating that ferredoxin is a required cofactor for the enzyme, and that *E. coli* ferredoxin functions poorly as an electron donor for the plant enzyme (Thompson et al. 1991).

Full-length cDNA clones of stearoyl-ACP desaturase were isolated from *Ricinis communis* (castor) seed and *Cucumis sativus* (cucumber). The nucleotide sequence of the castor clone *pRCD1* revealed an open reading frame of 1.2 kbp encoding a 396-amino acid protein of 45 kDa. The cucumber clone *pCSD1* encoded a homologous 396-amino acid protein with 88% amino acid identity with the castor clone. Expression of *pRCD1* in *Saccharomyces cerevisiae* resulted in the accumulation of a functional stearoyl-ACP desaturase, demonstrating that the introduction of this single gene product was sufficient to confer soluble desaturase activity to the yeast.

There was no detectable identity between the deduced amino acid sequences of the caster Δ^9-stearoyl-ACP desaturase and either the Δ^9-stearoyl-CoA desaturase from rat or yeast or the Δ^{12}-desaturase from *Synechocystis*, suggesting that these enzymes may have evolved independently. Stearoyl-ACP mRNA was present at substantially higher levels in developing seeds than in leaf and root tissue, suggesting that expression of the Δ^9-desaturase is developmentally regulated (Shanklin and Somerville 1991).

In addition, the formation of oleic acid (18:1) from lauric acid (12:0) is known as an example of the extension reaction of a carbon chain. In this reaction, oleic acid is aerobically formed from saturated fatty acids with carbon numbers of 10:0, 12:0, and 14:0. Oleic acid is aerobically dehydrogenated to linoleic acid (18:2), and linoleic acid is converted to linolenic acid (18:3) by the same reaction mechanism. It has been shown that the conversion of oleoyl-ACP to linoleoyl-ACP and linolenoyl-ACP occurs in the endoplasmic reticulum as follows. There is no desaturation of oleoyl-ACP inside the plastid.

$$\text{Oleoyl-ACP} + \text{NADH} + H^+ + O_2 \rightarrow \text{Linoleoyl-ACP} + 2H_2O + \text{NAD}^+$$

$$\text{Linoleoyl-ACP} + \text{NADH} + H^+ + O_2 \rightarrow \text{Linolenoyl-ACP} + 2H_2O + \text{NAD}^+$$

Polyene formation of fatty acids in higher plants occurs every three carbons from double bonds in the direction toward the methyl group. Elongation of the carbon chain does not occur in this reaction. Therefore, polyene acids with a carbon number higher than C_{20} are not present in the plant kingdom.

Oleoyl-CoA is desaturated inside the plastid to form linoleoyl and linolenoyl fatty acid derivatives.

$$\text{Oleic acid } (18{:}1,\Delta 9) \rightarrow \text{Linoleic acid } (18{:}2,\Delta 9,12) \rightarrow \text{Linolenic acid } (18{:}3, \Delta 9,12,15)$$

4.4.2.3 Acyltriglycerides

Fats are fatty acid esters of glycerol. As glycerol is a tervalent alcohol, mono, di, and triglycerides esterified with 1, 2, and 3 fatty acids are present, respectively. Acyltriglycerides are named fat. Generally, triglyceride is a mixed glyceride composed of three different fatty acids. In animal and plant triglycerides, saturated fatty acids are linked to the primary alcohol groups at both ends of the glycerol, and the secondary alcohol group at the center carries an unsaturated fatty acid, especially in highly unsaturated fatty acids.

Fats composed of saturated fatty acids have higher melting points than unsaturated fatty acids. Animal fats are generally composed of saturated fatty acids, while plant fats contain unsaturated fatty acids.

The synthesis of triglycerides begins with glycerol-3-phosphate. Glycerol-3-phosphate is successively acylated by two separate acyltransferases to produce lysophosphatidic acid and phosphatidic acid. The phosphatidic acid is then dephosphorylated to diacylglycerol. A further acyltransferase reaction produces triacylglycerol (Strickland 1967; Fig. 154). This synthetic pathway was demonstrated in microsomes of the sarcocarp of avocado (Matsumura and Stumpf 1968). In addition, glycerol-3-phosphate is formed by phosphorylation of glycerol, or reduction of dihydroxyacetone phosphate formed in glycolysis (Fig. 155).

4.4.3 Waxes

Waxes are esters of higher monohydric alcohols with higher fatty acids, and are distributed in flowers, stems and on the surface of fruits. Waxes are present

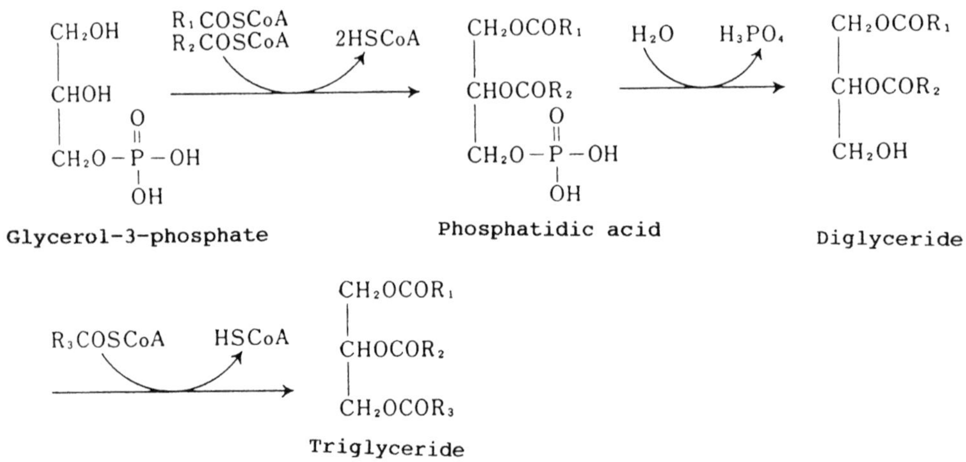

Fig. 154. Biosynthesis of triglycerides

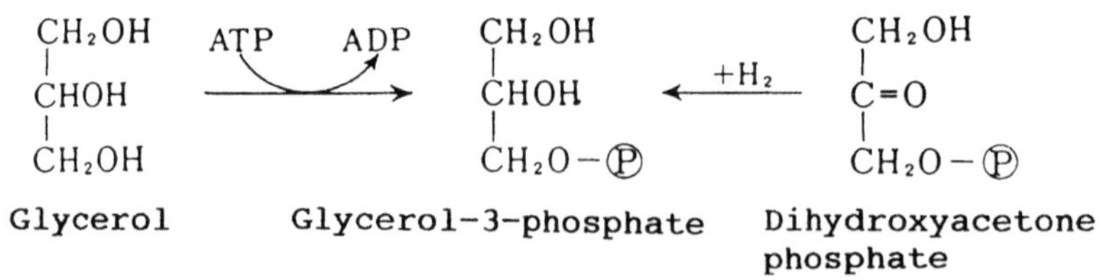

Fig. 155. Biosynthesis of glycerol-3-phosphate

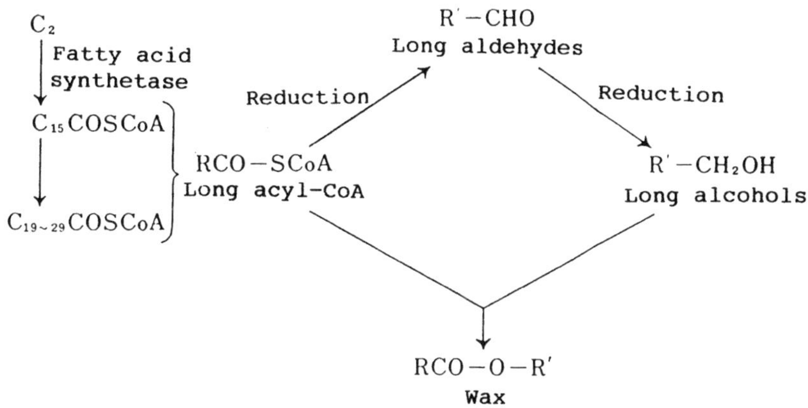

Fig. 156. Biosynthetic pathway of wax

with several higher alcohols, higher fatty acids, and higher hydrocarbons. Japan wax produced from stone fruits of *Rhus succedanea* is not a wax but a fat with a high melting point. Wax is synthesized by combination of long chain acyl-CoAs derived from acetyl-CoA and with long chain alcohols (Kolattukudy and Espelie 1985; Fig. 156). Long-chain acyl-CoAs are reduced to long-chain alcohols via the aldehydes.

4.5 Isoprenoids

Terpenoids such as volatile terpenes, essential oils and resin acids in conifers, and crude rubber in rubber trees have structures built up of isoprene units.

However, isoprene is not incorporated into terpenoids. Tamura (1957), and Wright et al. (1956), discovered mevalonic acid (MVA) as a growth factor of *Lactobacillus acidophilis*, *L. homohiochii* and *L. heterohiochii*, and found that MVA is a precursor of cholesterol. Lynen et al. (1958) found that the active C_5 unit in terpenoid biosynthesis is isopentenylpyrophosphate (IPP) derived from MVA, and proved that IPP is the biological isoprenoid unit leading to polyisoprenoids. Lynen and Henning (1960), and Archer et al. (1961) established the synthetic mechanism of MVA and IPP, and the formation of polyisoprenoids from IPP in sterol biosynthesis in animals and microorganisms (Fig. 157).

Two molecules of acetyl-CoA are condensed to give acetoacetyl-CoA to which one molecule of acetyl-CoA is added, giving 3-hydroxy-3-methylglutaryl-CoA, which is reduced to MVA. In this reduction two molecules of NADPH are required. The MVA formed is converted to mevalonic acid-5-pyrophosphate (MVA-5-PP) by MVA kinase and MVA-5-PP phosphokinase in the presence of ATP. Phosphorylation of the tertiary hydroxyl group of

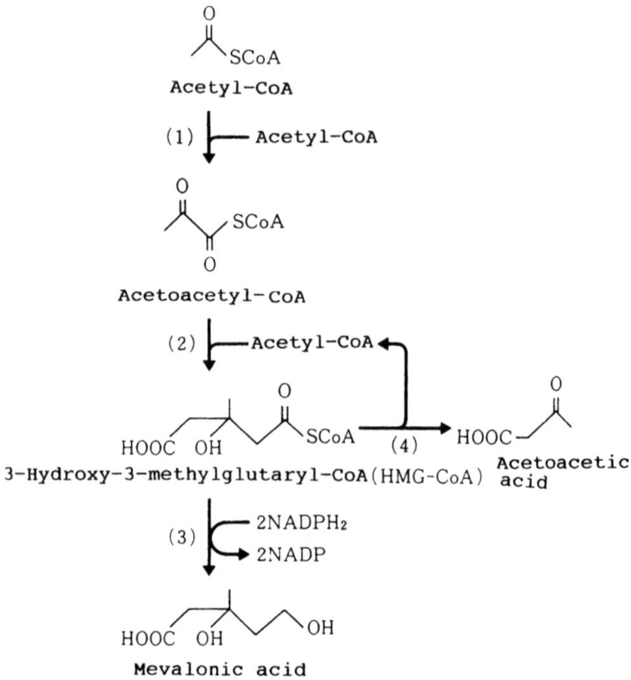

Fig. 157. Biosynthesis of mevalonic acid. *1* Acetoacetyl-CoA thiolase; *2* HMG-CoA synthase; *3* HMG-CoA reductase; *4* HMG-CoA lyase

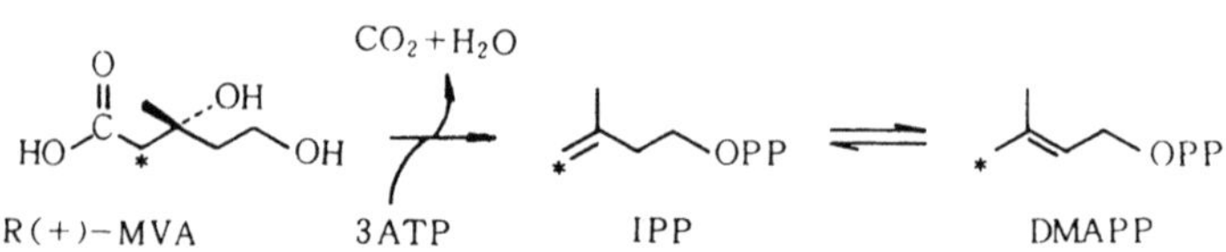

Fig. 158. Formation of isopentenylpyrophosphate (*IPP*) and dimethylallylpyrophosphate (*DMAPP*) from mevalonic acid (*MVA*)

MVA-5-PP follows the release of 1 mol each of phosphoric acid and CO_2 to afford IPP, which is isomerized to γ, γ'-dimethylallylpyrophosphate (DMAPP) by the mediation of isopentenylpyrophosphate isomerase (Fig. 158). IPP is an active alkylation reagent called active isoprene, and gives geranylpyrophosphate (GPP) by stereospecific head-to-tail condensation with DMAPP by prenyltransferase.

It has recently been found that the mevalonic acid pathway plays an important role not only in the biosynthesis of isoprenoids but also in the prenylation of various proteins bound to cell membranes, which are essential for the physiology and activity of cells (Bach et al. 1991, Minamikawa 1994).

4.5.1 Structure of Isoprenoids

Isoprenoids is the generic name of compounds composed of two or more isoprene (C_5H_8) units connected linearly or cyclically. They are also called terpenoids. Isoprenoids are divided into monoterpenes (C_{10}), sesquiterpenes (C_{15}), diterpenes (C_{20}), triterpenes (C_{30}), tetraterpenes (C_{40}), and polyterpenes (C_{5n}) depending on the number of the constituent C_5 units (Fig. 159).

The simplest isoprenoids are monoterpenes, such as linear citral (lemon oil), geraniol (rose oil, geranium oil), cyclic cineole (*Eucalyptus* oil), menthol (mint oil) and camphor (camphor oil), which are all fragrant essential oils. Farnesol (oil from *Chelonopsis moschata*) cadinene (*Chamaecyparis obtusa* (Japanese cypress) oil), zingiberene (oil from *Zingiber officinale*), abscisic acid (a plant hormone) and santonin (a vermifuge) are sesquiterpenes. Diterpenes

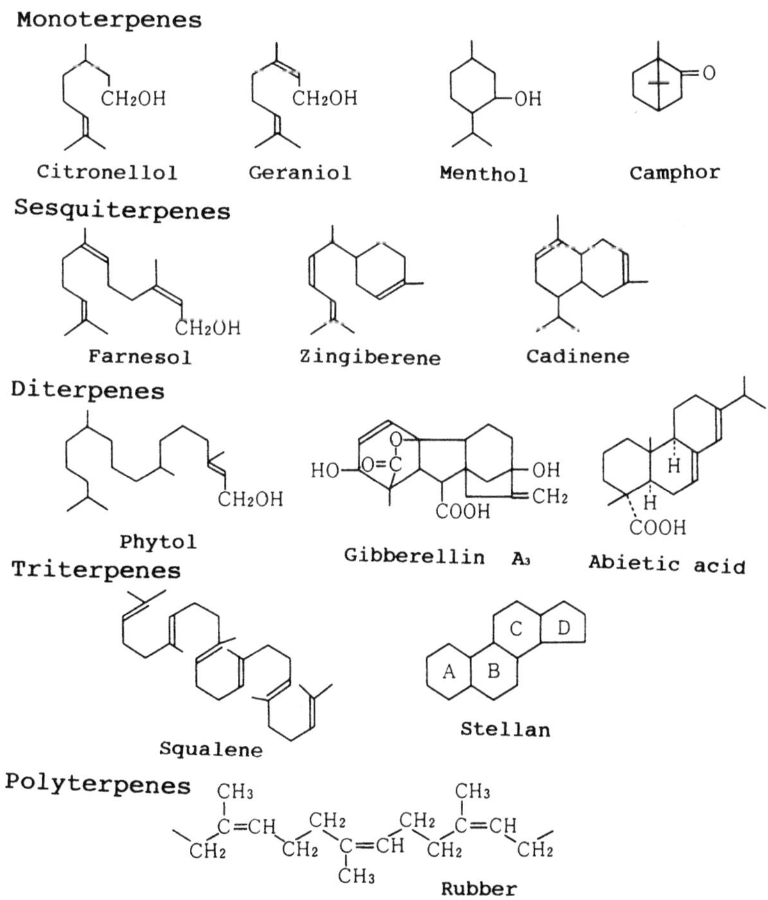

Fig. 159. Chemical structures of major terpenoids

are resins present in the higher boiling fractions of essential oils of conifers. Phytol as a component of chlorophyll, and gibberellins are also diterpenes. Oxidation products of the essential oils, resin acids (diterpene) such as abietic acid and its isomers are also included in sesquiterpenes and diterpenes. Amber is a fossil of conifer resin. A typical triterpene, squalene, is widely distributed in the animal and plant kingdoms, and sterols such as β-sitosterol are synthesized from squalene.

Steroids are compounds containing a four-membered cyclic structure, the sterol nucleus (stellan). Sterols, steroid saponins, cardiac-active glycosides and steroid alkaloids are included in the steroids. Sterols are important components of the cell membrane. Cholesterol in animals, and β-sitosterol and stigmasterol in plants constitute a unit membrane with lipids.

Carotene as a carotenoid pigment and xanthophyll are tetraterpenes, and rubber, guttapercha, and chicle rubber are polyterpenes. Rubber is a linear *cis*-polymer of 500–5000 isoprene units, while guttapercha is a *trans*-polymer and chicle rubber is a mixture of *cis* and *trans*-polymers.

4.5.2 Polymerization of Active Isoprene

Prenyltransferase catalyzes condensation of IPP and prenylpyrophosphate, thereby extending the carbon chains in terpenoid biosynthesis. By this reaction, new allylpyrophosphates containing five more carbons than the substrates are formed. The first condensation of IPP and DMAPP gives geranylpyrophosphate (GPP), which is converted to monoterpenes.

One molecule of isopentenylpyrophosphate is condensed to GPP, giving farnecylpyrophosphate (FPP), a precursor of sesquiterpene. Two molecules of FPP are condensed tail-to-tail by presqualene synthase to give squalene, from which triterpenes such as steroids are derived.

One isopentenylpyrophosphate and FPP are condensed to give geranylgeranylpyrophosphate (GGPP), a precursor of diterpenes. By tail-to-tail condensation of GGPP, phytoene and carotene are formed. One molecule of IPP and GGPP are condensed to give geranylfarnecylpyrophosphate (GFPP), a precursor of sestaterpenes. In these reactions *trans*-linear terpenoids are always formed (Dev Sukh 1989; Fig. 160).

The mechanism of the chain extension and stopping of the reaction at certain chain lengths has gradually been revealed. Several prenyltransferases that use different substrates to form products with different chain lengths have been detected in plants, bacteria, yeast and mammals. These enzymes are divided into the following four groups (Ogura 1993): (1) prenyltransferase I catalyzes the formation of C_{10}, C_{15} and C_{20} terpenopids from C_5 units; (2) prenyltransferase IIA catalyzes the formation of C_{30} and C_{35} from C_{10}, C_{15} and C_{20}; (3) prenyltransferase IIB catalyzes the formation of C_{40} and C_{45} from C_{10}, C_{15} and C_{20}; and (4) prenyltransferase III catalyzes the formation of C_{45}, C_{55} and C_{95}–C_{105} from C_{10}, C_{15} and C_{20}.

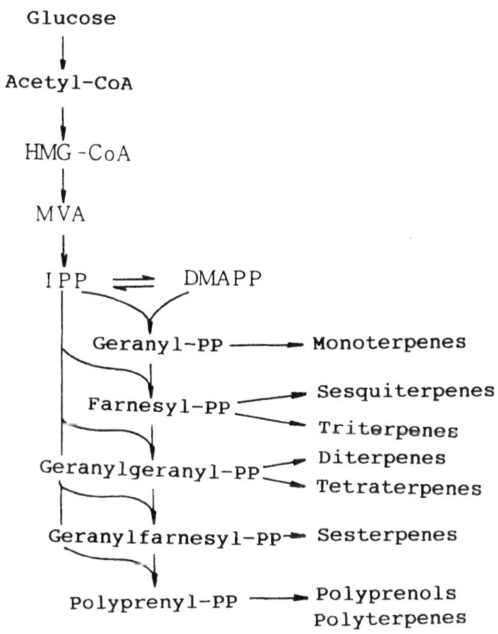

Fig. 160. General pathways for terpene biosynthesis

The longest chain of the product in the extension reaction is determined by the specificity of the prenyltransferase in each group. However, the chain length of the product within the highest chain length in each group depends on the concentration of IPP and Mg^{2+} as substrate.

GPP synthase (from salvia cell culture), FPP synthase and GPP synthase (*Capsicum annuum*) were isolated and characterized. GPP synthase from grape cell culture is located in plastids. It seems that monoterpene synthesis is regulated by the permeability of the plastid membrane for isopentenylpyrophosphate, IPP. In addition, squalene synthase, which catalyzes condensation of two molecules of FPP to afford squalene, was solubilized and partially purified as a 47-kDa polypeptide from a microsome membrane fraction of tobacco cell culture. The primary structure of prenyltransferase has been reported for FPP synthase from humans, rats, yeast and *Escherichia coli*, for GGPP synthase from *Neurospora crassa*, and for hexaprenylpyrophosphate synthase from yeast. These enzymes are highly homologous in several parts of their primary structure and it has been suggested that these parts are involved in the common functions of prenyltransferases (Ogura 1993; Fig. 161).

A site-directed study for yeast farnesyl diphosphate synthase (FPPSase) catalyzing the sequential 1′-4 coupling of IPP with DMAPP and geranyl-PP has recently been reported (Song and Poulter 1994). Ten site-directed mutants were constructed in FPPSase extended by the C-terminal α-tubulin epitope-Glu-Glu-Phe (FPPS::EEF). The six aspartates in domain I (at positions 100, 101, and 104) and domain II (at positions 240, 241, and 244) were changed

		Rigion II (73-103)	Region VI (217-250)	Region VII (285-300)
FPP synthase				
B.s.	N₂H···	LIHDDLPSMDNDDLRRGKPTN···	GLAFQIRDDILD···	ICELVAARDH-COOH
E.c.	N₂H···	LIHDDLPAMDDDDLRRGLPTC···	GLAFQVQDDILD···	LADYIIQRNK-COOH
Yeast	N₂H···	LVADDM··MDKSITRRGQP·C···	GEYFQIQDDYLD···	L·NKVYKRSK-COOH
Rat	N₂H···	LVADDI··MDSSLTRRGQ·TC···	GEFFQIQDDYLD···	LARKIYKRRK-COOH
Human	N₂H···	LVLDDI··MDSSHTRRGQI·C···	GEFFQIQDDYLD···	LANKIYKRRK-COOH
GGPP synthase				
N.c.	N₂H···	LV·DDV··EDNSVLRRGFPVA···	GLIFQIADDYHN···	VAQKNGK·KE-COOH
HexaPP synthase				
Yeast	N₂H···	LLHDDV··IDHSDTRRGRPSG···	GICFQLVDDMLD···	LTNSILTRRK-COOH

Fig. 161. Conserved amino acid sequence of prenyltransferase. *GGPP* Geranylgeranyl pyrophosphate; *HexaPP* hexaprenylpyrophosphate; *B.s. Bacillus stearothermophilus*; *N.c. Neurospora crassa*; *E.s. Escherichia coli*. *L* Leucine; *I* isoleucine; *H* histidine; *D* aspartic acid; *P* proline; *S* serine; *M* methionine; *N* asparagine; *R* arginine; *G* glycine; *K* lysine; *T* threonine; *A* alanine; *F* phenylalanine; *Q* glutamine; *C* cysteine; *E* glutamic acid; *V* valine. (Ogura 1993)

to alanine; three arginine residues were changed, Arg-109 and Arg-110 to glutamine and Arg-350 to alanine; and Lys-254 was converted to alanine.

Mutations of the aspartic residues and the nearby arginine residue in domain I, and the Asp-240 and Asp-241 in domain II drastically lowered the catalytic activity of FPPSase::EEF. Addition of an EEF epitope to the C-terminus of wild-type FPPSase resulted in a 14-fold increase of K_m^{IPP} and a 12-fold decrease of k_{cat}, suggesting that the conserved hydrophilic C-terminus of the enzyme may have a role in substrate binding and catalysis.

4.5.3 Formation of Cyclic Mono- and Sesquiterpenes

Figures 162 and 163a,b show typical mono- and sesquiterpenes in conifers, and various trees, respectively (Croteau and Johnson 1985). It has been shown that GPP is converted to cyclic monoterpenes (the major monocyclic terpenes) by monoterpene cyclase (monoterpene synthase). Studies on the partially purified cyclases for the synthesis of 1,8-cineole, γ-terpinene, α-terpinene, and limonene have revealed these enzymes to possess molecular weights in the 50000–100000 range, a pH optimum between 6 and 7, a requirement for a divalent cation (Mg^{2+} or Mn^{2+}), and a sensitivity to inhibition by thiol-directed reagents and inorganic pyrophosphate (Croteau and Johnson 1985).

Linalyl pyrophosphate (LPP) in most cases, and neryl pyrophosphate (NPP), the *cis* isomer of GPP in some cases are effective precursors of cyclic monoterpenes. The interconversion of NPP, LPP and GPP has not been observed prior to cyclization, and therefore it is suggested that these substrates

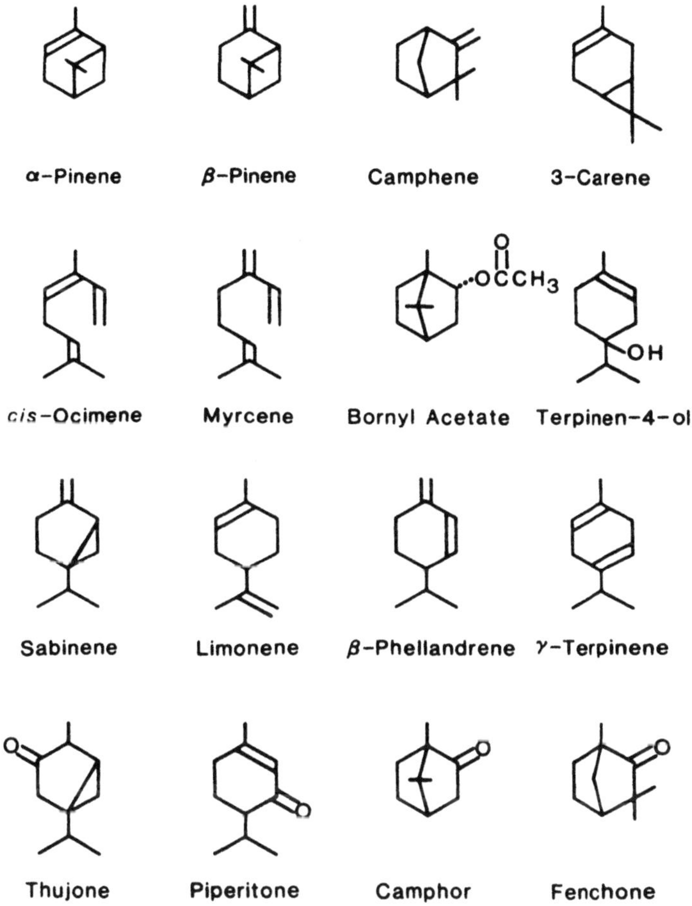

Fig. 162. Major cyclic monoterpenes in conifers. (Croteau and Johnson 1985)

are combined with the enzyme to afford an enzyme–substrate complex (*cis*-cation) which is cyclized stepwise to give cyclic monoterpenes (Croteau and Johnson 1985; Fig. 164).

Limonene synthase, a monoterpene cyclase, was purified from the leaves of *Mentha arvensis*. The enzyme catalyzes the formation of limonene from GPP via enzyme-bound LPP. Cyclic monoterpenes such as menthol, piperitone etc. are formed by a series of reductions with NADPH from limonene. The rabbit antibody of the limonene synthase was very specific, and no cross-reactions were found for limonene synthases of other plants or for other monoterpene cyclases (Minamikawa 1994).

Typical bicyclic monoterpenes, such as pinane, bornane and fenchane types, are distributed in conifers and some other plants. The general properties of these cyclases resemble those of the *p*-menthane series.

The enzymes catalyzing the synthesis of α- and β-pinene have been examined in partially purified preparations from *Citrus limonum* and *Salvia officinalis*. Fractionation of *Salvia* leaf extracts by size exclusion chromatography afforded two regions of pinene cyclase activity (cyclase I of molecular weight 96000 and cyclase II of molecular weight 55000) that catalyzed cyclizations of opposite enantiomeric specificity. Cyclase I catalyzed the conversion of GPP

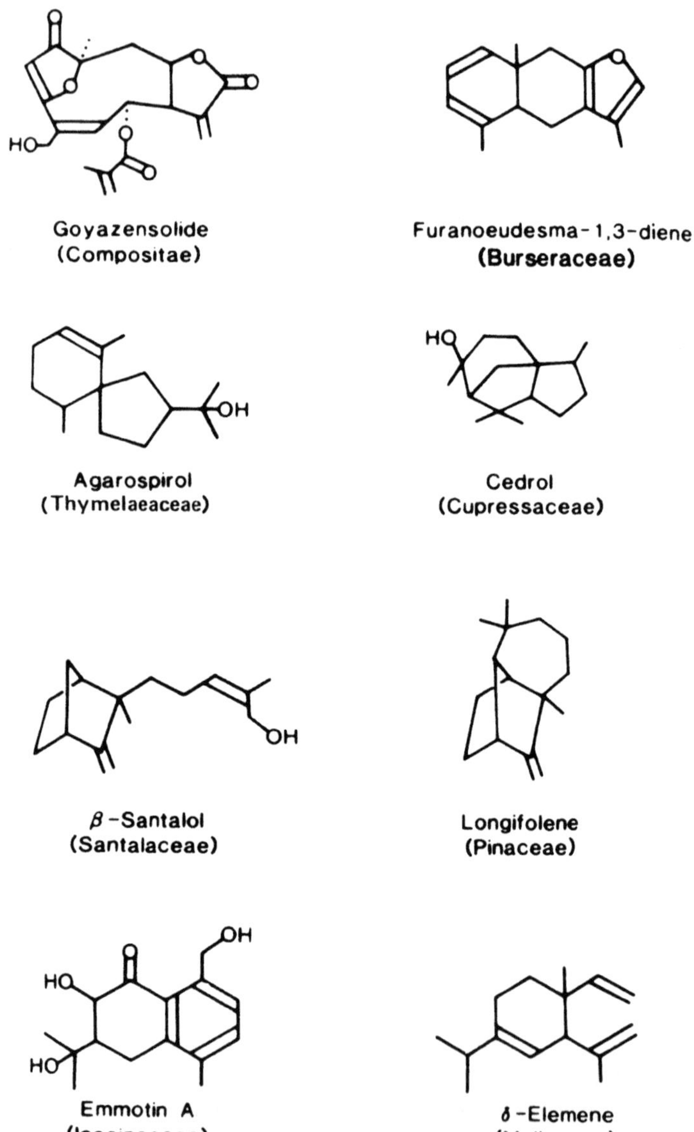

a

Fig. 163a,b. Major sesquiterpenes in wood extractives. (Croteau and Johnson 1985)

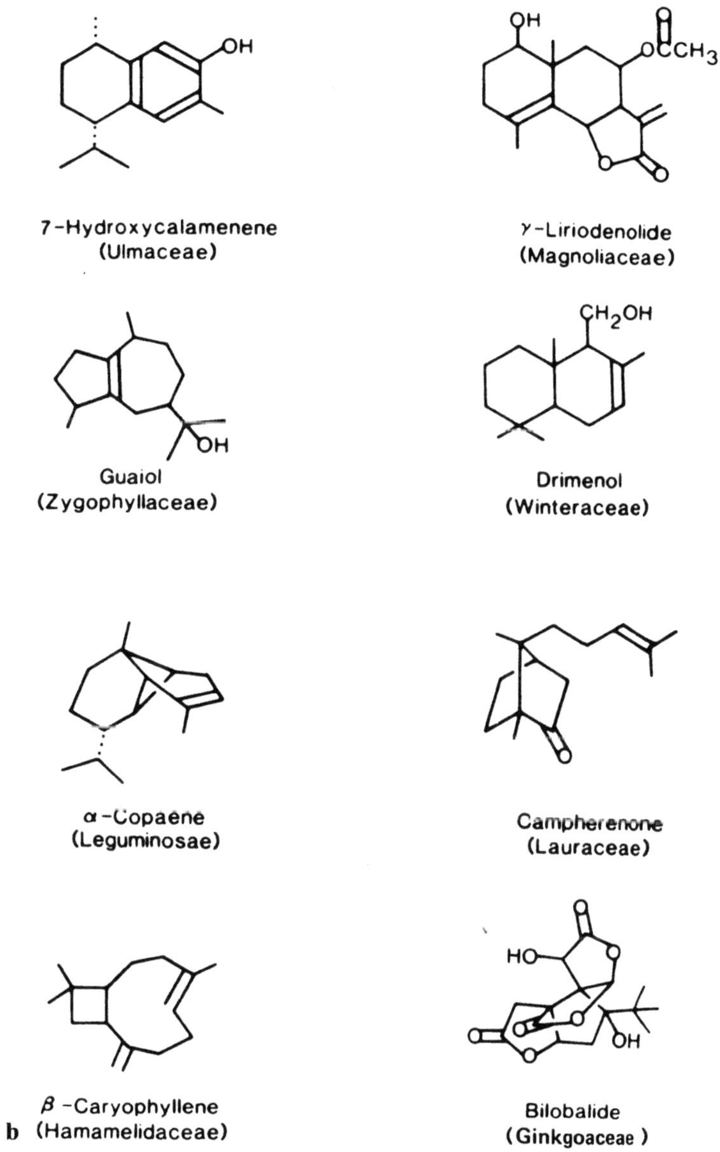

Fig. 163b

to (+)-α-pinene and to smaller quantities of (+)-camphene and (+)-limonene, whereas cyclase II transformed GPP to (−)-α-pinene and (−)-β-pinene, as well as to (−)-camphene, (−)-limonene, and the acyclic olefin myrcene. NPP and LPP served as alternate substrates for both cyclases (Croteau and Johnson 1985).

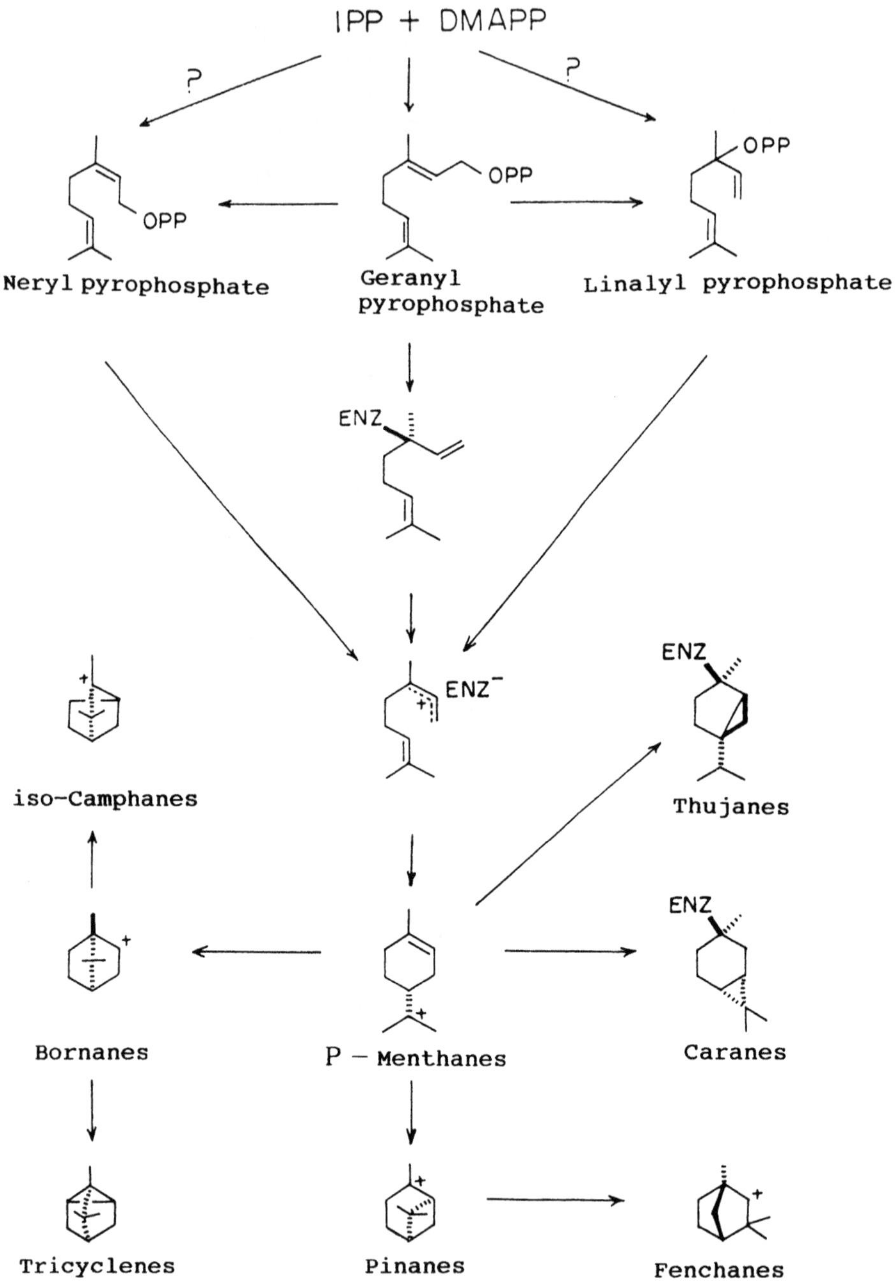

Fig. 164. Proposed biosynthetic pathways for major cyclic monoterpenes. (Croteau and Johnson 1985)

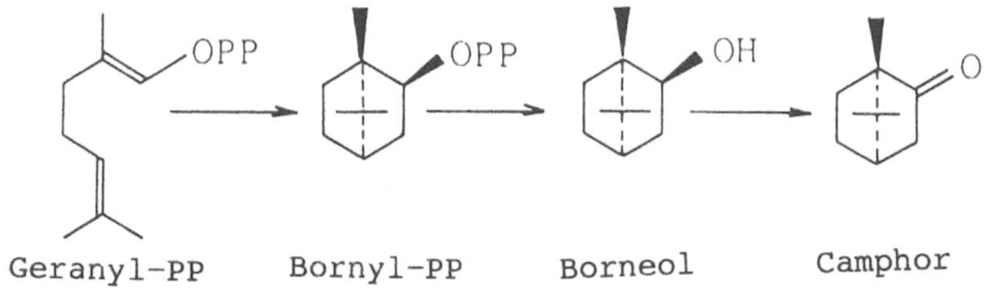

Fig. 165. Biosynthesis of borneol and camphor

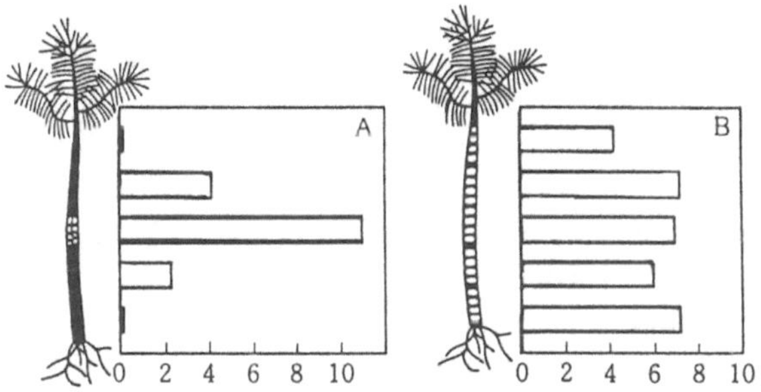

Fig. 166. A Monoterpene cyclase activity induced by wounding of conifer (*Abies grandis*) (central part), or **B** whole stem of *Abies grandis* was wounded artificially; the enzyme activity was measured after 1 week. Increased fold against control is shown. *Horizontal axis* shows the increased ratio of the enzyme activity compared with the activity of control plant. (Lewinsohn et al. 1991)

The biosynthesis of (+)-camphor in *Salvia* was shown to involve the conversion of GPP to (+)-bornyl pyrophosphate, which is subsequently hydrolyzed by a phosphatase to (+)-borneol, followed by the NAD-dependent dehydrogenation of the alcohol to the ketone. In this reaction NPP was not a mandatory intermediate. The pyrophosphate moiety of the substrate was found to retain bornyl pyrophosphate in the bicyclic product (Croteau and Johnson 1985; Fig. 165).

Conifer resin acts as a preventing reagent against attack by bark beetles and pathogenic fungi. The amount of resin formation in secretory tissue of grand fir (*Abies grandis*) is correlated with the activity of monoterpene cyclase. The enzyme is suggested to be a key enzyme regulating the synthesis of cyclic monoterpenes such as α- and β-pinene, etc. When the bark of grand fir is peeling off, rendering the tree liable to attack by insects, cyclase activity in the wounded part increases remarkably (Lewinsohn et al. 1991; Fig. 166). The principal wound-inducible monoterpene cyclase of grand fir stem produces both (−)-α-pinene and (−)-β-pinene in a fixed 2:3 ratio from GPP via a common cationic intermediate.

The cyclase was purified as a protein of 62 kDa. Western blot analysis with a specific antibody showed that the synthesis of the enzyme protein was induced by the wounding, and that the antipinene synthase granuloma antiserum gave strong cross-reactivity with all of the monoterpene cyclases from pine, spruce or other conifer genera, indicating that the pinene synthase is more closely related to other cyclases of fir that synthesize different monoterpene skeletal types than the pinene synthases from related conifer species (Steele et al. 1995).

The enzymes responsible for constitutive turpentine production are very similar in general properties to the inducible form; however, these monoterpene cyclases are distinguishable in several characteristics (pH optimum, metal ion requirement) from their angiosperm counterparts.

It was shown that the increase in monoterpene biosynthesis after wounding of grand fir stems is the result of the apparent enhancement of constitutive activities, principally limonene synthase, and the appearance of distinct, inducible activities, including 3-carene, β-phellandrene and α- and β-pinene synthases (Steele et al. 1995; Fig. 167).

Sesquiterpenes often occur in complex mixtures as minor constituents (ca. 10%) in oleoresins of the Pinaceae. Germacrene D and longifolene are the principal sesquiterpenes in pines, and cedrol and cedrene often occur as major components of wood oils of the Cupressaceae, giving the characteristic pleasant fragrance of the wood.

FPP is an intermediate at the branching point leading to sterols by the mediation of squalene synthase, and to sesquiterpenes by the mediation of sesquiterpenecyclase (Fig. 168).

It is now accepted that all sesquiterpenes are derived from FPP, and extensive biogenetic schemes based on this principle have been put forward. The central theme of these hypotheses is an intramolecular electrophilic attack by the carbon bearing the pyrophosphate on either the central or distal double bond of the farenesyl molecule. Further cyclizations, rearrangements, and redox interconversions lead to the array of some 200 known sesquiterpenoid skeletons. The origin of the sesquiterpenes longiforene (*Pinus*), α-cedrene (*Cupressus*), and α-cubebene (*Ulmus*) has been reviewed (Croteau and Johnson 1985).

FPP is also a precursor of prenyllipids such as ubiquinones and dolichols. When plant tissues are attacked by microorganisms and viruses, special ses-

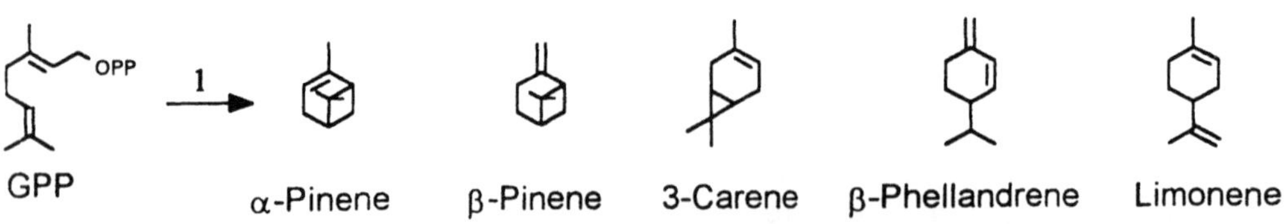

Fig. 167. Biosynthesis of monoterpenes from geranyl pyrophosphate by *Abies grandis*. *1* Monoterpene synthases; *GPP* geranyl pyrophosphate. (Steele et al. 1995)

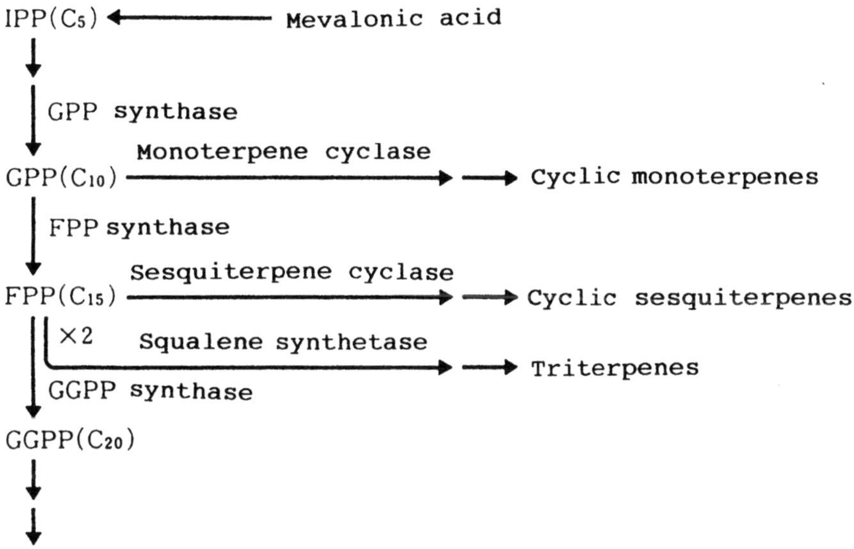

Fig. 168. Enzymes at the branching points in the mevalonate pathway. *IPP* Isopentenylpyrophosphate; *GPP* geranyl pyrophosphate; *FPP* farnesylpyrophosphate; *GGPP* geranylgeranylpyrophosphate

quiterpenes are formed as defensive substances and accumulate. In such cases, FPP is preferentially used for synthesis of sesquiterpenes.

When an elicitor from *Phytophytora parasitica* is added to a tobacco cell suspension culture, antimicrobial cyclic sesquiterpenes such as capsydol are formed and accumulate within several hours. This is followed by a decrease of sterol synthesis. In this case, a remarkable increase of the activity of sesquiterpene cyclase is continued for 30 h after addition of the elicitor, while the activity of squalene synthase is inhibited.

These cyclases in tobacco cell culture were purified as two polypeptides of 60 and 62 kDa. Western blot analyses with a specific antibody showed that the synthesis of the cyclase protein is only detected in elicitor-treated tobacco cells. This shows that the channeling of isoprenoid intermediates is conducted by the regulation of the synthesis of enzymes at the branching point (Minamikawa 1994).

Two full-length EAS genes (*EAS1* and *EAS2*) encoding 5-*epi*-aristolochene synthase that mapped 5 kb apart on one 15-kb genomic clone were isolated from an elicitor-induced tobacco. The genes consists of 1479 bp open reading frames, each containing five introns and specifying 56 828-dalton proteins. RNA blot hydridization with *cEAS1* demonstrated a mRNA induction time-course consistent with the induction of the EAS mRNA translational activity with maximum levels 4–6 h after elicitation. EAS mRNA was not detected in control cells (Facchini and Chappell 1992).

4.5.4 Biosynthesis of Carotenoids

Two molecules of GGPP are condensed to give phytoene by the mediation of phytoene synthase. Phytoene is converted to δ-carotene by dehydrogenation with phytoene desaturase. δ-Carotene, pale yellow in color, is converted to various carotenoids, such as the red lycopene, the orange β-carotene, and the yellow xanthophyll. Repeated dehydrogenation of phytoene affords carotenes, and the following stepwise oxidations of the carotene afford xanthophyll. These carotenoids play a role in attracting insects for pollination as well as birds and small animals to help in scattering seeds.

cDNA clones of phytoene synthases are obtained from soybean, tomato and pepper, and cDNA of phytoene desaturase is obtained from tomato. Since both enzymes are encoded by nuclear genes, precursors of the carotenoids are transported to chloroplasts and plastids and converted to the respective carotenoids. By illumination of tomato seedlings, the amount of carotenoid synthesized is increased, but the expression of the two enzymes was scarcely affected by the illumination. During the fruiting period of tomato, the mRNA of the synthase increased 20 times, but the mRNA of the desaturase increased only three times, suggesting that the expression of the two enzymes is regulated in different ways (Minamikawa 1994).

4.5.5 Di- and Triterpenes

Figure 169 shows typical diterpenes in conifers (Croteau and Johnson 1985). Nonvolatile acids (resin acid) in conifers, leguminous resins, phorbol irritants of Euphorbiaceae and Thymelaceae etc. are common diterpenes. When pine stems are wounded oleoresin, which contains 50–75% of diterpenes is formed. The diterpene shows a resin character and named resin acid, and resin acids tend to crystallize by oxidative polymerization.

It was found that two tricyclic diterpenes in *Pinus pinaster*, abietic acid and pimaric acid, are synthesized at the base of conifer leaves, while the dicyclic diterpene, dextropimaric acid, is synthesized by whole conifer leaves. Xylem contains monoterpenes and diterpenes such as abietic acid and pimaric acid (Croteau and Johnson 1985). The principal resin acid of grand fir, (–)-abietic acid, has been shown to originate by cyclization of the corresponding C_{20} isoprenoid precursor, geranylgeranyl pyrophosphate (GGPP), to (–)-abieta-7(8),13(14)-diene, followed by sequential oxidation of the A-ring α-methyl of the olefin to a carboxyl function involving two distinct cytochrome P450-dependent hydroxylases and an aldehyde dehydrogenase (Steele et al. 1995; Fig. 170).

Most other common resin acids represent double-bond positional isomers of abietic acid and are thought to be formed by variations on the same biogenetic theme, involving formation of different parent olefines followed by

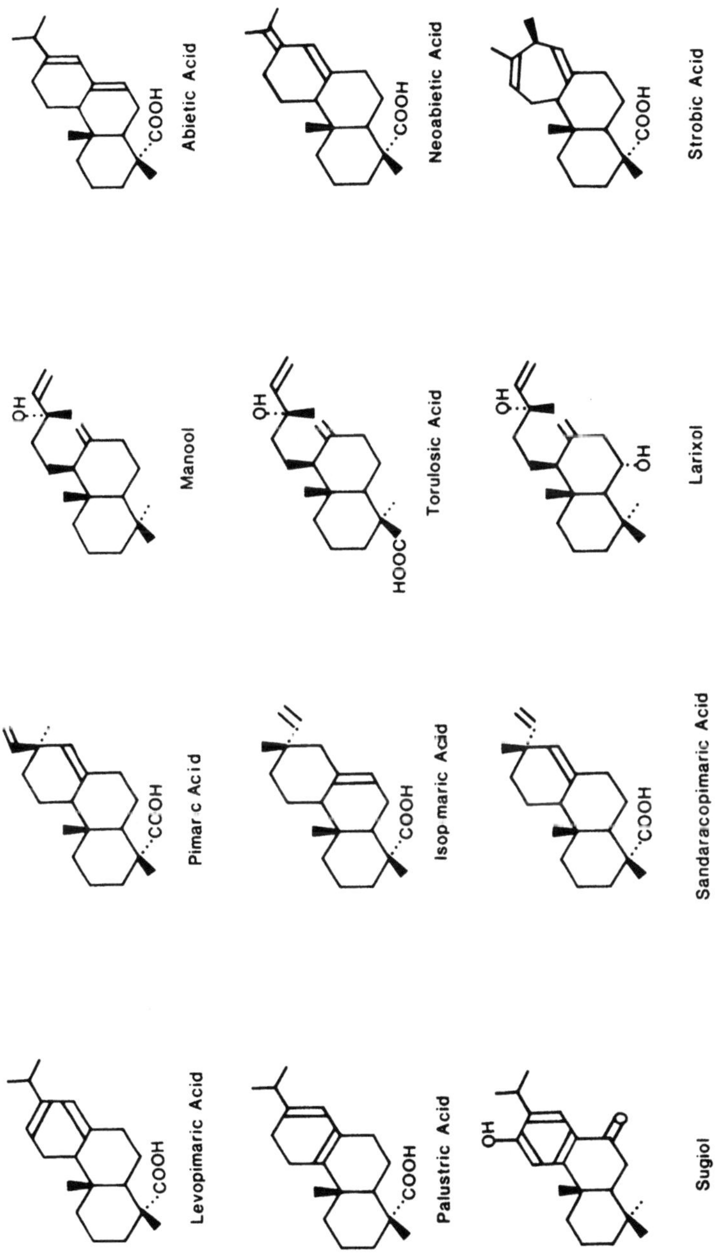

Fig. 169. Diterpenes in conifers. (Croteau and Johnson 1985)

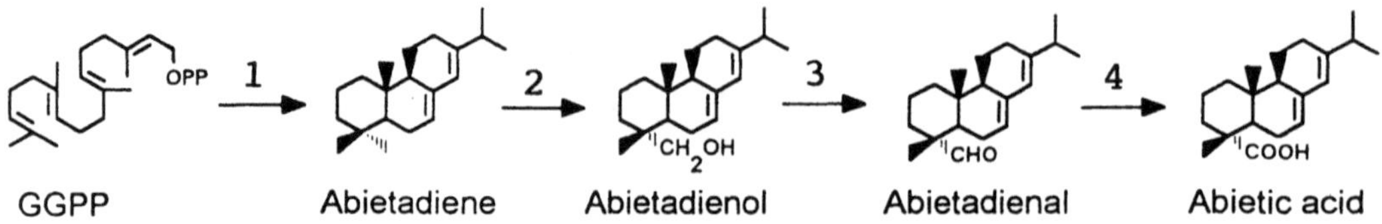

Fig. 170. Biosynthesis of abietic acid, the principal diterpenoid resin acid of *Abies grnadis* oleoresin. *GGPP* Geranylgeranyl pyrophosphate; *1* abietadiene synthase; *2* abietadiene hydroxylase; *3* abietadienol hydroxylase; *4* abietadienal dehydrogenase. (Steele et al. 1995)

similar oxidation sequences. After stem wounding in grand fir, the rate of diterpenoid resin biosynthesis increases significantly.

The first step of resin acid biosynthesis is catalyzed by abietadiene synthase. This inducible diterpene cyclase, like the monoterpene cyclases of fir, is a soluble enzyme (about 80 kDa), and has been purified and characterized. In general properties, including pH optimum, pI value, cation requirements, inhibitors, and kinetic constants, abietadiene synthase resembles other terpenoid cyclases of angiosperm and gymnosperm origin. Although the enzymatic cyclization sequence from GGPP to (–)-abieta-7(8),13(14)-diene almost certainly involves the formation of copalyl pyrophosphate and a pimaradine as stable intermediates, no evidence for separation of the corresponding partial activities has been obtained (Steele et al. 1995).

Monoterpene and diterpene biosyntheses are coordinately induced in wounded fir stems as determined by monitoring the activity of monoterpene and diterpene cyclases, and of the two P-450-dependent diterpene hydroxylases involved in the formation of (–)-abietic acid. The activity of these enzymes reaches maximum levels that are 25- to 500-fold higher than those of nonwounded control systems 10 days after wounding, and it is followed by a synchronous decline (Steele et al. 1995).

Tetracyclic diterpene, ent-kaurene, is a precursor of gibberellins in the fungus *Gibberella fujikuroi* and in higher plants (Croteau and Johnson 1985). Biosynthesis of kaurene by a partially purified enzyme preparation from *Marah macrocarpus* endosperm and other higher plants has been studied by West (Hedden et al. 1978; Fig. 171).

GGPP is converted to kaurene via two enzymatic reactions, cyclization of GGPP to copalyl pyrophosphate (CPP) and further cyclization and rearrangement of CPP to kaurene. Each enzyme has a molecular mass of 82 kDa and requires Mg^{2+} for activity. The two enzymes function as a complex, in that the CPP produced by the first enzyme is channeled to the second enzyme. Inhibitors of the first enzyme were found, and it seems that gibberellin synthesis is regulated via inhibition of the first enzyme.

A near full-length casbene synthase cDNA clone, pCS7, was isolated from elicited castor bean seedlings. The cDNA insert had a length of 1983 bases with a polyadenylate tail of 19 bases. Translation of the cDNA sequence

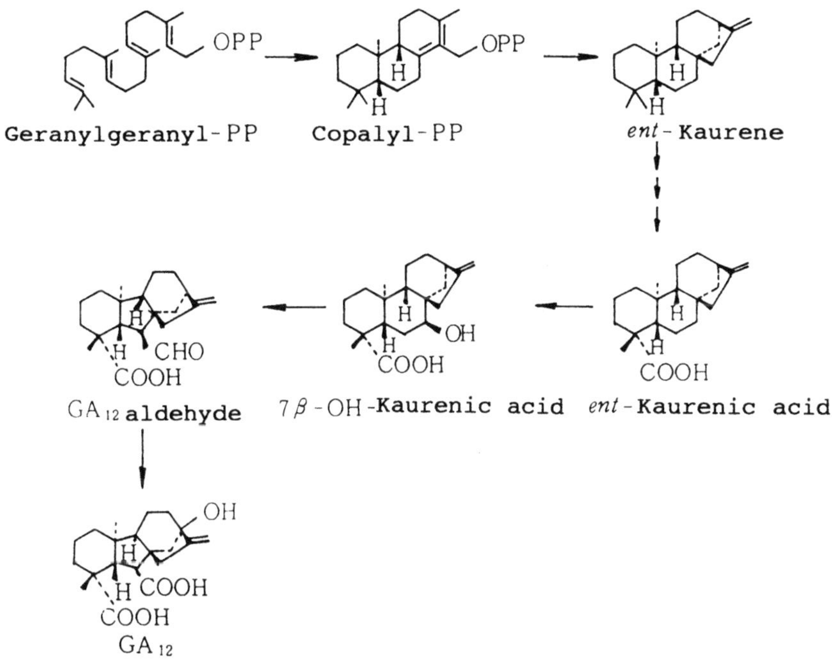

Fig. 171. Biosynthesis of kaurene and gibberellins

revealed an open reading frame, encoding a 601-amino acid protein with a predicted M_r of 68960. The deduced translation product revealed 42% identity and 65% similarity with 5-*epi*-aristolochene synthase from tobacco and 31% identity and 53% similarity with limonene synthase from spearmint. Each of the three proteins catalyzes an intramolecular cyclization of a prenyl diphosphate substrate to a specific cyclic terpenoid. Analysis of the alignment of the three proteins shows that both primary and secondary structural elements are conserved, suggesting that the genes that encode terpenoid cyclization enzymes of this type have undergone divergent evolution from an ancestral progenitor gene (Mau and West 1994).

Triterpene synthesis is initiated by a tail-to-tail condensation of two molecules of FPP to afford squalene. Other triterpenes and sterols are derived from squalene. Lanosterol and cycloartenol are suggested as essential intermediates in the biosynthesis of animal sterols and plant sterols, respectively. Squalene biosynthesis via presqualenepyrophosphate is effected by an enzyme preparation from higher plants. In addition, the conversion of ^{14}C-squalene to β-amylene has been shown to be mediated by an enzyme preparation from germinating pea seeds (Croteau and Johnson 1985).

Two overlapping cDNA clones for *Arabidopsis thaliana* squalene synthase have been isolated. Their nucleotide sequences contained an open reading frame for a 410-amino acid polypeptide (calculated M_r, 47 kDa). The deduced amino acid sequence of the polypeptide was significantly homologous (42–44% identical) to the sequences of known squalene synthases of several species, from yeast to man, but it was much less homologous to that of tomato phytoene synthase.

A cell-free extract of *E. coli* transformed with a recombinant plasmid, in the presence of NADPH and Mg^{2+}, efficiently converted [^{14}C]farnesyl diphosphate into squalene. In the absence of NADPH and the presence of Mn^{2+}, the cell-free extract formed dehydrosqualene as a secondary product. Another *E. coli* extract expressing mouse squalene synthase showed the same activity as the *Arabidopsis* enzyme, indicating that the two synthases are markedly conserved in taxonomically remote eukaryotes (Nakashima et al. 1995).

4.5.6 Biosynthesis of Polyterpenes

It seems that a continuous *cis*-condensation of isopentenylpyrophosphate (IPP) occurs by the mediation of a specific *cis*-prenyltransfease in biosynthesis of rubber (Archer et al. 1961). Lynen and Henning (1960) found that isopentenylpyrophosphate-1-^{14}C is rapidly incorporated into rubber of *Hevea brasiliensis*. Many studies have been reported on rubber biosynthesis. IPP, mevalonic acid (MVA) and hydroxymethylglutaryl-CoA (HMG-CoA) are used as substrates for rubber synthesis in *Hevea* latex. When neryl pyrophosphate (NPP, *cis*-C_{10}) is added into latex, rubber synthesis is promoted, but geranylpyrophosphate (GPP; *trans*-C_{10}) does not have this effect. The results indicate that a *cis*-specific prenyltransferase is involved in rubber synthesis. Rubber latex is composed of colloidal rubber particles, lutoids (vacuole) and cytoplasm constituents, and the transferase is active only on rubber particles.

4.5.7 Prenylation of Proteins

Recently, prenylation of the cysteine residue of a C-terminal of polypeptides was discovered as a posttranslational modification. The prenylprotein plays important roles in regulation of cell multiplication and differentiation. Guanosine triphosphate (GTP)-combined proteins with low molecular weight, which regulate cell functions in transmission of intercellular signals, are reported to be prenylated (Minamikawa 1994). In addition to ras proteins, which are involved in signal transmission in mammals, proteins responsible for the formation of cytoskeleton and intracellular transport of cellular substances are prenylated.

More than 20 species of GTP-combined proteins with low molecular weights have been identified from plants. These are mentioned as proteins involved in the transport of cell substances. It has also been shown that the γ-subunit of G proteins involved in the intercellular signal transfer, lammin as a constituent of nuclear lammina, the conjugation factor of yeast, and cyclic GMP phosphodiesterase are prenylated. These proteins can connect to the plasma membrane by prenylation and express physiological functions in the cell.

It has been shown that farnecyltransferase (FT), and geranylgeranyltransferase (GGT) I and II are involved in prenylation of proteins. By mediation of these enzymes the prenyl group is linked as a thioester to cysteine of the C-terminal of proteins. GTP-combined proteins with a cysteine residue as C-terminal have been identified from several plants. Recent investigations showed that ^{14}C-mevalonic acid is incorporated into GTP-combined protein, as in mammalian ras-protein, and into membrane-connected protein, as in nuclear lammin. It was also shown that the cell-free extract mediates the incorporation of ^{3}H-FPP and ^{3}H-GGPP into ras-protein (Minamikawa 1994).

A complete cDNA clone corresponding to the subunit of rat FT was isolated from pea seedlings and the base sequence was determined. The deduced amino acid sequence with 419 residues has a homology of 48 and 40% with those of rat and yeast, respectively (Minamikawa 1994).

Geranylgeranyl transferase activity for proteins having the C-terminal sequence Cys-Ile-Ile-Leu or Cys-Ile-Ile-Ser was identified in crude soluble yeast extracts. This activity is dependent upon the *CDC43/CAL1* gene, which is involved in budding and the control of cell polarity, but it does not require the *DPR1/RAM1* gene, which is required for the farnesylation of ras protein. The geranylgeranyl transferase activity is distinct from the protein farnesyl transferase activity and that its specificity depends in part on the extreme C-terminal leucine in the protein to be prenylated (Finegold et al. 1991).

4.5.8 Tropolones

Heartwoods of trees of the Cupressaceae and their essential oils contain strong acidic and antimicrobial C_{10} compounds. One such compound was isolated from heartwood essential oil of *Chamaecyparis taiwanensis* by Nozoe in 1949 (Nozoe et al. 1957). It is a seven-membered cyclic compound and was named hinokichiol. α, β, and γ Thujaplicins were isolated from *Thuja plicata* heartwood by Erdtman, who found that β-thujaplicin is identical with hinokichiol (Erdtman 1959). Similar compounds were identified as a fungal metabolite by Dewar and named tropolones, (2-hydroxy-2,4,6-cycloheptatorien-1-on; Bentley 1963). The heartwood of *Chamaecyparis nootkatensis* contains different species of toropolones, such as nootkatin and chanootin (Fig. 172). These tolopolones are different from fungal tolopolone, and contain isopropyl side chains. Erdtman (1959) suggested that thujaplicins are derived from carene via mevalonic acid. Cell cultures of *Thuja occidentalis* have been shown to synthesize several thujaplicin derivatives.

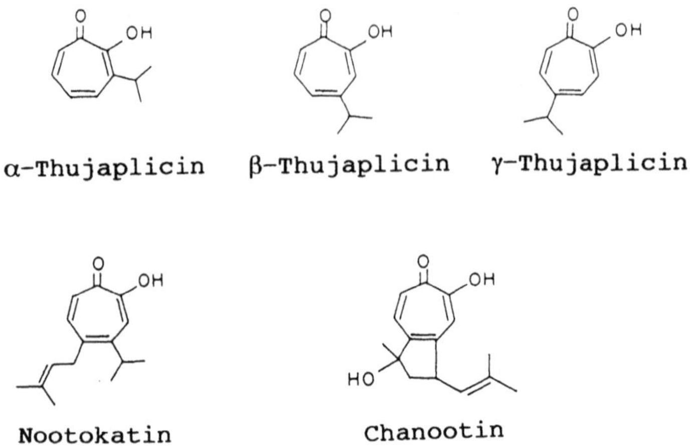

Fig. 172. Toropolones in conifers

Recently Sakai et al. (1994) found that β-thujaplicin is produced by callus cultures of *Cupressus lusitanica*, and that the thujaplicin synthesis is remarkably increased by addition of yeast extract (Inada et al. 1993) or elicitors (culture filtrates of some fungi). They found that mevalonic acid-^{14}C is incorporated into β-thujaplicin in suspension cultures of *C. lusitanica*. In addition, 80 μM compactin, an inhibitor of HMG-CoA reductase that catalyzes the reduction of 3-hydroxy-3-methylglutaryl-CoA (HMG-CoA) to mevalonic acid (see Fig. 160), completely inhibits the formation of β-thujaplicin in the suspension cultures, suggesting that β-thujaplicin is synthesized via the mevalonate pathway.

5 Formation and Development of Wood Tissues

5.1 Development of Phloem and Xylem

5.1.1 Formation of Phloem and Xylem

In the xylem, vascular bundles transport water and salts from the roots toward the tree top, while in the phloem sieve elements transport photosynthate from the foliage downward. The vascular cambium produces new xylem and phloem cells. In higher plants xylem, phloem and other tissues are formed from the apical meristem, which differentiates into primary phloem and primary xylem (see Fig. 3, Chap. 1).

5.1.1.1 Primary Xylem

The conductive elements of the primary xylem are the tracheids in the conifers and the vessels in the hardwoods. Xylem is composed of vessels, tracheids, wood fibers and xylem parenchyma. Both ends of vessels open to form perforations, and the perforated vessels are connected to form a tube. Tracheids are sharp spindle-shaped cells without perforations. Both vessels and tracheids are dead. Their cell contents are lost in differentiation. Vessels and tracheids bear characteristic patterns by partial deposition of secondary cell walls. Secondary xylem originally formed from the procambium is called protoxylem, while the later formed primary xylem is called metaxylem. Vessels of the protoxylem generally have annular or spiral cell wall thickening, while vessels of the metaxylem have reticulate or pit cell wall thickening.

5.1.1.2 Primary Phloem

Primary phloem is a complex tissue composed of sieve tubes, companion cells, phloem parenchyma, scleroids, and phloem fibers. Sieve tube elements form a sieve tube. Companion cells are formed from the same procambium cells. In the process of differentiation of sieve tube elements from procambium cells, the nucleus and vacuole membrane disappear, the mitochondria become small, and the inner membranes of the plastids disappear so that only starch granules remain. In contrast to these changes a special protein "p-protein" accumulates in the cytoplasm, forming a tube-like or fibrous mass.

The sieve elements in the phloem are of two kinds, namely sieve cells in the gymnosperms, which contain sieve areas, and sieve-tube elements in the angiosperms, which contain sieve plates. The pores in the sieve areas and sieve plates often contain callose, a β-D-1,3-glucan, which develops as a result of wounding or other environmental stresses (wound xylose). Callose is also deposited onto sieve areas and plates in dying cells (definitive callose). The pores in the sieve plates in most angiosperms are lined with P-protein, which probably serves the same function as wound callose, namely to seal the pore on wounding. The sieve elements have no nucleus or cytoplasm, and their normal functions are maintained by parenchyma cells, called companion cells, which are connected with the sieve elements via plasmodesmata. The phloem originally formed from the procambium is referred to as protophloem, and the primary phloem formed later is called metaphloem.

5.1.1.3 Vascular Cambium

The intrafascicular cambium formed from the procambium is combined with the interfascicular cambium formed from intervascular parenchyma cells to make a ring. Cells of the cambium thus formed divide and continuously produce secondary xylem on the inside, and secondary phloem on outside, resulting in thickening of roots, branches and stems (see Fig. 3, Chap. 1).

In herbs, cambial activity ceases at an early stage, whereas in some trees the cambial activity can continue for several thousand years. However, in most monocotyledons, a vascular cambium is not present.

The cambial activity of stems of woody plants changes seasonally, and when the cooler season comes, cambial activity becomes weak and ultimately ceases entirely. By such seasonal changes of cambial activity, annual rings of wood are formed.

5.1.2 Regulation of Xylem and Phloem Differentiation

As noted above, phloem and xylem are formed from cells derived from meristematic tissues. The mechanism involved in this differentiation of the cambial cells to xylem and phloem cells has been investigated by studies of whole plants, tissue cultures, and cells cultures.

5.1.2.1 Plant Hormones

Plant hormones are known to play important roles in xylem and phloem differentiation (Roberts et al. 1988). Effects of plant hormones on growth and differentiation of plant tissues at molecular levels have recently been studied.

An auxin-binding protein (ABP) was isolated and characterized from several plants such as maize, tobacco, and *Arabidopsis*. By using antiidiotypic antibodies the distribution of 65-kDa protein, which binds auxin specifically, was demonstrated in a wide range of plants. The protein is localized to the nucleus and supposed to be a candidate for an auxin receptor that mediates some nuclear process (Prasad and Jones 1991).

By photoaffinity labeling, several IAA (indole acetic acid) binding proteins with M_r of 60 kDa, 58 kDa, and 23 kDa were identified from maize (*Zea mays* L.) coleoptile tissue. Among these proteins, a 23-kDa protein shares common epitopes with an auxin-binding protein present in the plasma membrane (Feldwisch et al. 1992). The antibody of ABP inhibited the auxin-responding growth of maize coleoptiles, and the auxin response measured by polarization of membrane potential of tobacco protoplasts.

cDNAs of the ABP have been isolated from maize, tobacco and *Arabidopsis* and sequenced. ABP from maize is a glycoprotein composed of 201 amino acids, and contains a terminal signal peptide with 38 amino acid residues, indicating that ABP is synthesized in the endoplasmic reticulum. The mature ABP protein (18 kDa) is composed of 163 amino acids. The ABP derived from the ER exist as a hydrophilic soluble dimer protein (Nagata 1995, Shimomura 1995).

Recently, the mechanism of auxin response and transportation, and the auxin induction of several genes such as *parA* and *parC*, which are located in the nucleus and involved in the regulation of transcription, have been investigated. The products of *parA* and *parB* induced by auxin are involved in the induction of cell division and dedifferentiation. The auxin-responding genes *arcA*, *Aux22*, *Aux28*, and *SAUrs*, which serve to maintain cell division, have been identified in soybean hypocotyls (Nagata 1995, Shimomura 1995). Information on the molecular biology of the mechanism of auxin response is scarce and not conclusive. The response reaction of plant cells to plant hormones may proceed as follows:

> plant hormones → receptors (ABP) → message transmittance → gene expression → function (enzymes) → growth and differentiation, etc.

5.1.2.1.1 Whole Plants

Figure 173 shows how information is transmitted for induction of vascular differentiation in a plant. Vascular tissues are continuously formed from apical and root apical meristems. To effect the transport function, the vascular bundles must be continuous. Polar transport of auxin from shoot apex to root apex is considered as an important factor to realize this continuity.

When the vascular bundle of *Coleus* stem is cut, parenchyma cells are differentiated to tracheary elements. Sieve tube elements surround the cut vascular bundle, and connect to form a bypass. If leaves on the upper part of

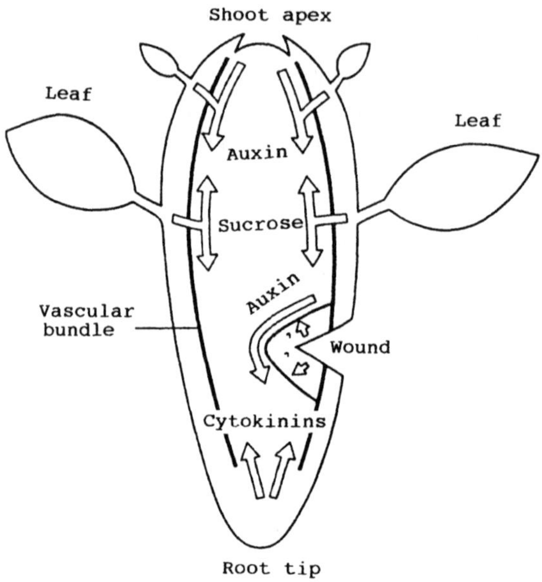

Fig. 173. Channel of transmitted information for induction of vascular differentiation in plants. (Fukuda 1994a)

a wounded tissue are removed, vascular bundle differentiation is inhibited. However, when indole acetic acid (IAA) is added to the stem, vascular bundle differentiation occurs as in a plant with leaves. It seems that polar transport of auxin from the upper leaves to the lower tissues is important for vascular bundle differentiation (Jacobs 1952).

Auxin transported to lower tissues causes differentiation of vascular bundle cells, while the differentiated cells promote auxin transport to effect vascular bundle differentiation in tissues further down. Thus, the successive differentiation of lower tissues results in a vascular bundle as a connected tube.

In vascular bundles, xylem and phloem and formed interactively, and the formation is regulated by plant hormones. High concentrations of IAA promote xylem formation, while low concentrations induce phloem formation.

In vascular bundle differentiation, cytokinins supplied by roots also play an important role. When roots of *Coleus* and sunflower are removed vascular bundle differentiation is inhibited. However, added cytokinins release the inhibition. Cytokinins added to *Coleus* stems seem to increase sensitivity to auxin and a small amount of auxin induces vessel formation.

Gibberellins are plant hormones that induce plant elongation, causing an increase in cell size in xylem and phloem differentiation. When gibberellins are added together with auxin, differentiation of phloem fiber cells is increased.

5.1.2.1.2 Tissue and Cell Cultures

Formation of xylem and phloem as in a whole plant can be reproduced in tissue and cell cultures. As shown in Fig. 174 isolated parenchyma cells from *Zinnia* leaves are differentiated to form a single cell of a tracheary element by cultivation in a suitable medium containing auxin and cytokinin. Since external conditions of the culture can be freely changed, tissue and cell cultures have been used as a model system for xylem and phloem differentiation (Fukuda 1989a, 1992).

These cultures require auxins for differentiation of xylem and phloem. When auxins are not added to the culture externally, no xylem and phloem differentiation occurs. Inconsistent results have been obtained regarding the effects of cytokinins. For example, both auxin and cytokinin are required for xylem differentiation of mesophyll cells of *Zinnia*, but for xylem differentiation of lilac callus only auxin is required. In a variety of carrot, tracheary elements are induced without addition of external cytokinin, whereas another variety requires cytokinin for xylem differentiation. The difference in cytokinin requirement in these carrot varieties was found to depend on the amount of endogenous cytokinin. Thus, the results indicate that both cytokinin and auxin are required for xylem differentiation.

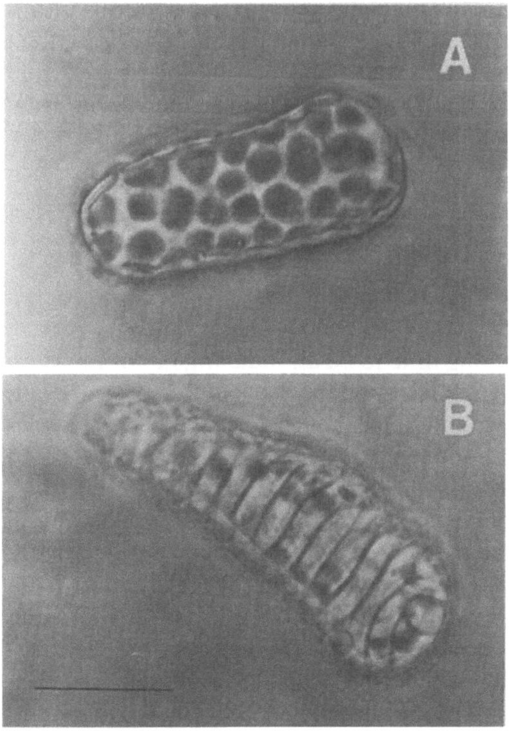

Fig. 174. A,B Differentiation of isolated cells of *Zinnia* leaf into tracheary element as single cells by cultivation. **A** Freshly isolated cell; **B** tracheary element formed at 58 h of culture without intervening cell division. (Fukuda 1989a; courtesy of Dr. H. Fukuda)

When IAA and sucrose are added to lilac callus, vascular bundles are induced. However, low concentrations (1–2%) of sucrose induced only xylem, medium concentrations (2.5–3.5%) induced both xylem and phloem, and high concentrations (4%) induced only phloem. This indicates that both plant hormones and sucrose play an important role in the induction of xylem or phloem.

Based on these facts, Northcote (1989) suggested a feedback control on differentiation of cambium, xylem and phloem in a whole plant. That is, tryptophane, a precursor of IAA, is released by autolysis of cells in xylem to cause an increased concentration of IAA. The high concentration of IAA promotes differentiation of xylem, while sucrose produced by photosynthesis promotes differentiation of phloem, which is the channel of the synthesized sucrose. Cambium is induced by cells with medium concentrations of IAA and sucrose.

In tissue and cell cultures, ethylene and brassinosteroids also play an important role in xylem and phloem differentiation. Xylem differentiation of explants of lettuce pith and soybean callus are promoted by methionine, *S*-adenosylmethionine, and 1-aminocyclopropane-1-carboxylic acid (ACC) as precursors of ethylene, whereas aminoethoxyvinylglycine, cobalt, and silver as inhibitors of ethylene inhibit xylem differentiation. The inhibition is eliminated by addition of ethylene precursors.

Uniconasole, a plant growth retardant, was found to inhibit the synthesis of gibberellins and brassinosteroids. This reagent inhibits tracheary differentiation of mesophyll cells of *Zinnia*. Interestingly, uniconasole does not affect cell division but specifically inhibits differentiation. The inhibition is restored by a trace of brassinosteroids (0.2 n*M*). The results suggest that an internal brassinosteroid is specifically involved in xylem differentiation (Fukuda 1992).

5.1.2.2 Other Factors

5.1.2.2.1 Wound Stress

Wound stress has been known for a long time to be an inducing factor in xylem and phloem differentiation. When vascular bundles of a stem or root are cut, parenchyma cells surrounding the wounded tissues are differentiated to tracheary and sieve tube elements. Wounding plays an important role also in xylem and phloem differentiation of tissues and cell cultures.

When leaf sections of *Zinnia* plants are cultured in a medium containing auxin and cytokinin, only some of the mesophyll cells near the wounded tissue are differentiated to tracheary elements, whereas when a leaf section from which one side of epidermis has been peeled off is cultured in the same medium, differentiation frequently occurs in the whole leaf section. Interestingly, not only mesophyll cells but also epidermal cells differentiate to tracheary elements under these conditions. The results suggest that the wound stress created by the removal of the epidermis is a factor inducing tracheary differentiation (Fukuda 1992).

Wounding first affects the wounded cells, and cells surrounding the wound, and only later does it affect the cells distal to the wound. Differentiated cells in wounding are not the wounded cell themselves but parenchyma cells, surrounding the wound. It seems that some signal substances are transmitted from the wounded tissue to the parenchyma. Such a signal may be an internal plant hormone such as IAA (Fig. 173).

Ethylene production is one of the fastest reactions induced by wound stress. Such ethylene induced by wound stress causes the initiation of xylem differentiation.

5.1.2.2.2 Light

Light promotes xylem and phloem differentiation in certain tissues. When phloem parenchyma of carrot is cultured in a medium with auxin and cytokinin, xylem cells are differentiated. For xylem differentiation in this case both auxin and cytokinin are required. However, xylem differentiation of a parenchyma tissue cultured without cytokinin is induced upon illumination. This indicates that the effect of cytokinin is replaced by light. The results suggest that in certain plants light is involved in xylem and phloem differentiation via formation of plant hormones (Fukuda 1992).

5.1.3 Physiological and Biochemical Processes of Xylem and Phloem Differentiation

Tracheary differentiation progresses in several steps in plants (Torrey et al. 1971, Fukuda 1992). First, cells derived from meristematic cells acquire a differentiating ability and elongate. Then, secondary cell walls consisting of cellulose, hemicelluloses and lignin are deposited in the elongated cells. Finally the cell contents, including the nucleus, disappear resulting in empty cells. Water is transported from the roots to the tree top through the empty cells. In the gymnosperms, the longitudinal tracheids serve this function. In the angiosperms, water is transported through the vessels.

5.1.3.1 Acquisition of Differentiation Ability

It is difficult to investigate the mechanism of acquisition of differentiation ability by using cells that differentiate according to the program of ontogenesis. However, the differentiation ability can be easily investigated by using tissues and cell cultures whose differentiation to tracheary elements is not predetermined. Investigations with such tissues have shown the necessity of auxins and cytokinins for acquisition of differentiation ability. However, the mechanism involved in the induction of cell differentiation by auxins and cytokinins is still not fully known.

5.1.3.2 Cell Elongation

After acquisition of differentiation ability, cells still elongate. It seems that auxins and gibberellins are generally required for this elongation. At the stage of cell elongation, synthesis of materials required for secondary wall deposition, and development of the necessary intracellular conditions take place. For example, there are changes of orientation of microtubules and actin fibers, which play important roles for morphogenesis, and there is also synthesis of tubulin, a component of microtubules. In addition, the endoplasmic reticulum, and Golgi apparatus, which are involved in polysaccharide synthesis increase.

5.1.3.3 Formation of Secondary Walls

Secondary walls with spiral and annular thickenings, anastomosis and pits are formed inside the primary walls. Secondary walls are composed of parallel oriented cellulose microfibrils. Between them lignin, hemicelluloses, and pectin are embedded, resulting in hard and strong cell walls.

5.1.3.3.1 Cellulose, Hemicelluloses and Lignin

The rate of synthesis of cellulose increases remarkably with the onset of formation of the secondary cell wall. In the gymnosperms, the rate of synthesis of the galactoglucomannans also increases, while that of xylan increases in the angiosperms. The increase of xylan synthesis is regulated by at least two processes, namely the increased synthesis of UDP-xylose, and the increase of xylan synthase which catalyzes synthesis of xylan from UDP-xylose. In contrast to the activation of xylan synthesis, the synthesis of pectin and the pectin synthase activity are inhibited.

Lignin synthesis starts after the formation of cellulose microfibrils. In cell culture experiments, the deposition of lignin stained by phloroglucinol-HCl starts several hours after the formation of the microfibrils. Many enzymes are involved in the biosynthesis of lignin, and the activities of phenylalanine ammonia-lyase (PAL), *O*-methyltransferase (OMT), 4-coumarate:CoA ligase (4CL), cinnamyl alcohol dehydrogenase (CAD) and cell wall-bound peroxidase increase on lignification. The increase has been ascribed to the synthesis of enzyme proteins. For example, increases of PAL activity and PAL mRNA occur just before lignification.

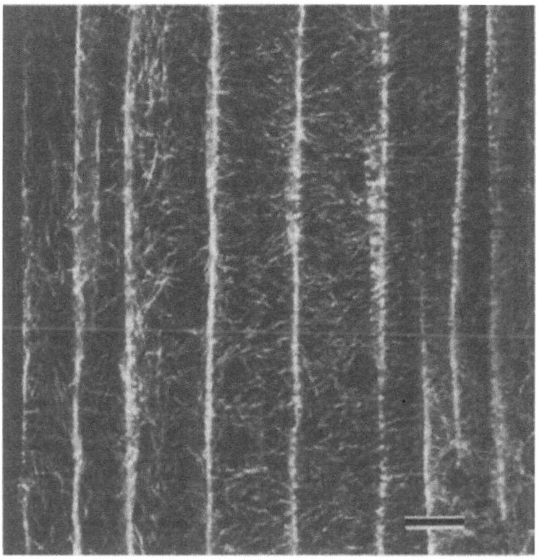

Fig. 175. Confocal laser scanning micrograph of cortical microtubules in differentiating tracheids of *Abies sacchalinensis* during cell expansion. *Scale bar* 20 μm. (Courtesy of Dr. R. Funada)

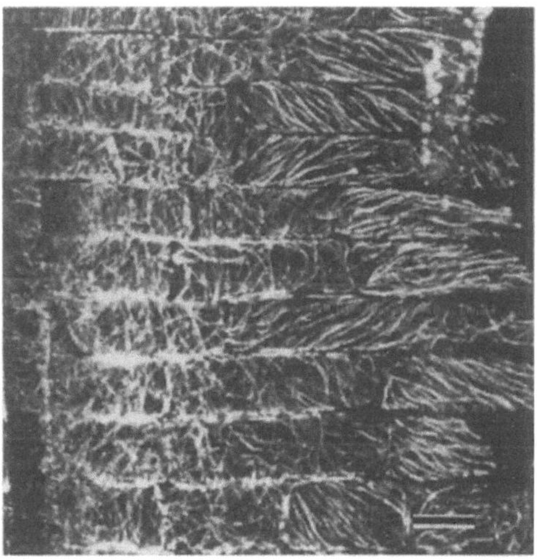

Fig. 176. Confocal laser scanning micrograph of cortical microtubules of the differentiating ray cells of *Abies sacchalinensis* during cell elongation. *Scale bar* 20 μm. (Courtesy of Dr. R. Funada)

5.1.3.3.2 Pattern of Secondary Cell Walls

The pattern of secondary cell walls depends on the orientation of the cellulose microfibrils. Electron microscope observations show that microtubules are parallel with the microfibrils immediately inside the cell walls (Helper 1981; Figs. 175 and 176).

Microtubule arrays change dramatically during the differentiation of the mesophyll cells into tracheary elements in *Zinnia*; a random, sparse network changes into lateral bundles via a random or longitudinal dense network. When the microtubule organization is disturbed by an agent such as colchicine, a microtubule-depolymerizing agent, or taxol, a microtubule-

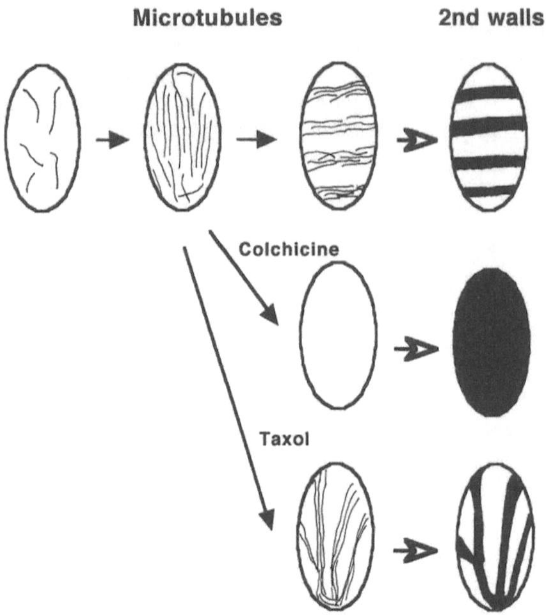

Fig. 177. Disturbance of microtubule organization and formation of unusual secondary wall thickenings by colchicine and taxol. (Fukuda 1994a)

stabilizing agent, unusual secondary wall thickenings are formed and the cellulose microfibrils lose their orientation (Fukuda and Komamine 1980, Falconer and Seagull 1985; Fig. 177). These results suggest that microtubules regulate the orientation and the regions of cellulose microfibrils through control of the movement of cellulose synthases on the surface of the plasma membrane.

Actin fibers also play an important role in the organization of secondary cell walls. Immediately before secondary cell wall formation begins, actin fibers appear inside the plasma membranes as regularly aligned short aggregations (Fig. 178). The aggregated actin fibers elongate at a right angle to the cell length and form bands of actin fibers. Microtubules are present between the actin fibers at this stage.

When the actin fibers are destroyed, microtubules are lined in parallel, and secondary walls are formed as lengthwise oriented cells walls along the microtubules. The results suggest that the patterns of secondary walls are determined through control of the orientation of microtubules by actin fibers (Fukuda et al. 1993, Fukuda 1994a).

5.1.3.4 Disappearance of Cell Content

The last stage of tracheary element differentiation is decomposition and disappearance of cell contents. The process is considered to be a programmed death process. Decomposition of the nucleus and other organelles occurs, but the decisive event is the collapse of the vacuoles, which causes the release of

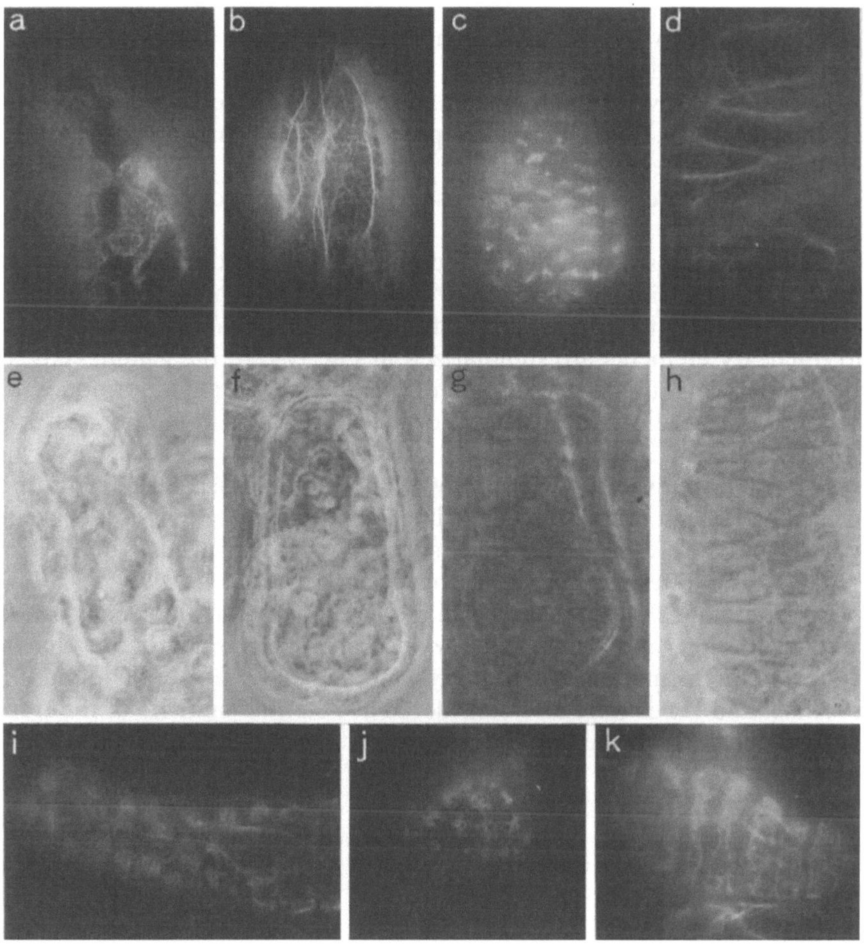

Fig. 178. a–k Sequential changes in the organization of actin filaments during differentiation of *Zinnia* mesophyll cells into tracheary elements. Actin filaments seem to change the order of **a**→**b**→**c**, and **i**→**j**→**k**→**d**. **e**, **f**, **g**, and **h** are the phase-contrast microscope figures of **a**, **b**, **c**, and **d**, respectively. (Fukuda 1994a; courtesy of Dr. H. Fukuda)

various hydrolases into the cytoplasm. As a result, various polymeric substances in the cytoplasm are hydrolyzed. Decomposition of the vacuoles occurs several hours after secondary cell wall formation in tracheary differentiation of cell cultures has taken place.

Among hydrolases effective in cell autolysis, a certain nucleic acid hydrolase and proteases appear prior to the autolysis process. These facts suggest that the expression of certain genes is activated in the autolysis process.

The loss of physiological cell activity is accompanied by decomposition of the cytoplasm, but deposition of secondary cell walls still continues. Parts of the end walls are hydrolyzed by wall hydrolases, and the formation of tracheary elements is completed.

Thus, cells derived from meristematic cells are differentiated to tracheary elements through strictly regulated processes. Tracheary differentiation is the ultimate process toward death, but the process is complicated and requires new expressions of various genes and networks of information transmittance.

Other cells constituting xylem and phloem are also differentiated to mature cells through similar complicated processes (Fukuda et al. 1993, Fukuda 1994a). The sequential events during differentiation of *Zinnia* mesophyll cells into tracheary elements are summarized in Fig. 179. In the *Zinnia* cells, the process following the initiation of differentiation by phytohormones and wounding can be divided into early, or prerestricted, and late, or restricted processes.

The late process involves a variety of differentiation-specific events, most of which have been found to be associated with the thickening of the secondary walls and autolysis. In this process, cells become engaged, probably irreversibly, in the differentiation of tracheary elements.

Investigations showed that tubulin was a relatively early molecular marker of differentiation, and that some enzymes and structural proteins that are involved in the formation of secondary walls and autolysis were late markers. Newly isolated genes were also considered as potential markers of the transition from the early to the late process (Fukuda et al. 1993).

Acquisition of differentiation activity →	**Cell elongation** →	**Formation of secondary cell wall**
Inducers; auxins, cytokinins	Preparation for secondary cell wall: orientation changes of microtubules and actin filaments Increase in ER, Golgi apparatus, and vesicles	Disappearance of cell contents Formation of secondary cell wall: synthesis of cellulose, xylan, and lignin Increase in gene expression of synthases of cellulose, hemicelluloses, and lignin Cell autolysis: disappearance of vacuole, nuclei, and other organelles Increase in nucleases, proteases, hydrolases, etc.

Fig. 179. The sequential events during differentiation of *Zinnia* mesophyll cells into tracheary elements. (Fukuda 1994a)

An extensin-like protein was isolated from metabolically active cells of differentiating and mature xylem of loblolly pine (*Pinus taeda*). The protein is present in xylem cell walls during lignification and remains as a structural component of cell walls in wood for many years after xylogenesis. It is suggested that the protein plays an important role in the differentiation of xylem (Bao et al. 1992).

5.1.4 Genes Involved in Tracheary Element Differentiation

5.1.4.1 Genes Expressed Before Vascular Bundle Differentiation

Differentiation-specific changes were observed on two-dimensional polypeptide maps of *Zinnia* cells around 12 h before secondary wall deposition begins. Two newly synthesized polypeptides appeared in the cell cultures, which induced formation of tracheary elements (TEs), and their synthesis continued at least until the secondary wall began to form. No tracheary elements were formed in the control cells.

By differential screening of a cDNA library, four clones containing cDNA inserts were isolated, and it was found that the corresponding RNAs were expressed preferentially in the differentiated cells. The cDNAs were designated tracheary element differentiation-related genes 1–4 (*TEDs 1–4*). Sequencing of the *TED* cDNAs revealed some characteristics of *TEDs* (Fukuda 1994b,c; Table 26).

Table 26. Characteristics of *TED* cDNAs. (Fukuda 1994b)

cDNA	bp	Amino acid	Characteristics	Homology	In situ expression
TED1	1700			Mitochondria F1-ATPase α-subunit	
TED2	1183	325	Hydrophobic regions	ζ-Crystallin lens of guinea pig (NADPH: quinone oxidoreductase)	Future xylem and phloem cells; procambial cells
TED3	1435	319	(NGY) motifs; repeated amino acid sequences		Differentiating or future vessel cells
TED4	535	95	Transit peptide; metal-binding "finger"	Aleurone-specific protein (B11E) of barley NS-lipid transfer protein	Differentiating or future xylem cells

TED1 cDNA corresponded to the *mt atpA* gene, (mitochondrial gene for F1-ATPase α-subunit) in *Zinnia*. A homology search revealed a significant similarity between *TED2* and the gene encoding zeta-crystalline, which was recently demonstrated to function as an NADPH: quinone oxidoreductase, from the guinea pig lens.

The polypeptide sequence deduced from *TED3* cDNA showed that *TED3* encodes a hydrophobic protein with the Asp-Gly-Tyr motif, which is repeated fifteen times. No protein was homologous to the polypeptide deduced from *TED3*. The partial sequence of the polypeptides encoded by *TED3* indicated a protein that is probably located in the secondary walls of tracheary elements (TEs). *TED3* was expressed specifically in differentiating vessel elements or in the cells that were expected to differentiate into vessel elements in the near future, in all organs tested. There was no expression of *TED3* transcripts in other types of cells, such as epidermal, mesophyll, meristematic, and phloem cells.

TED4 cDNA encoded a polypeptide of 10kDa with 11 hydrophobic amino acids at the N-terminus, which may be a transit peptide. A homology search with the nucleotide and deduced amino acid sequences of *TED4* revealed a significant similarity to those of the barley aleurone-specific clone, *B11E* (Fukuda 1994b,c). The expression of *TED4* was also restricted to vascular or future vascular cells and, in particular, to immature xylem cells. In situ hybridization with the cotyledon sections showed that *TED4* was expressed at an earlier stage of differentiation into vessel elements in the main leaf vein than *TED3*.

A cDNA clone (p-48h-10) identical to *TED4*, and other cDNA clones of which corresponding mRNAs were expressed specifically in the differentiating cultures, were also isolated from *Zinnia* cells. The expression patterns of *p-48h-10* of an auxin-induced gene were examined in *Zinnia* cells. In the isolated mesophyll cells, the mRNA accumulates in 48h of culture with 1-naphthaleneacetic acid alone. Benzyladenine does not induce expression of this gene, but the combination of 1-naphthaleneacetic acid and benzyladenine induces the mRNA accumulation about 24h earlier than does 1-naphthaleneacetic acid alone.

The mRNA is present predominantly in the cambial region in stems, leaves and roots and in the vascular bundles in flower buds, but does not occur in the apical regions of shoots or roots, suggesting that the gene is involved in the early process of vascular differentiation (Ye and Varner 1993, 1994).

In situ hybridization of various tissues of young *Zinnia* seedlings, using probes constructed from these cDNAs, indicated that the expression of these genes was restricted to the cells that were involved in vascular differentiation. In contrast to *TED3* and *TED4*, *TED2* was not expressed in xylem or in phloem of the main leaf vein; instead, it was expressed in the two small areas where parenchyma cells may have differentiated into vascular cells to form lateral veins. In root tips, *TED2* transcripts were present in restricted procambial regions that were predicted to form phloem and xylem. These results

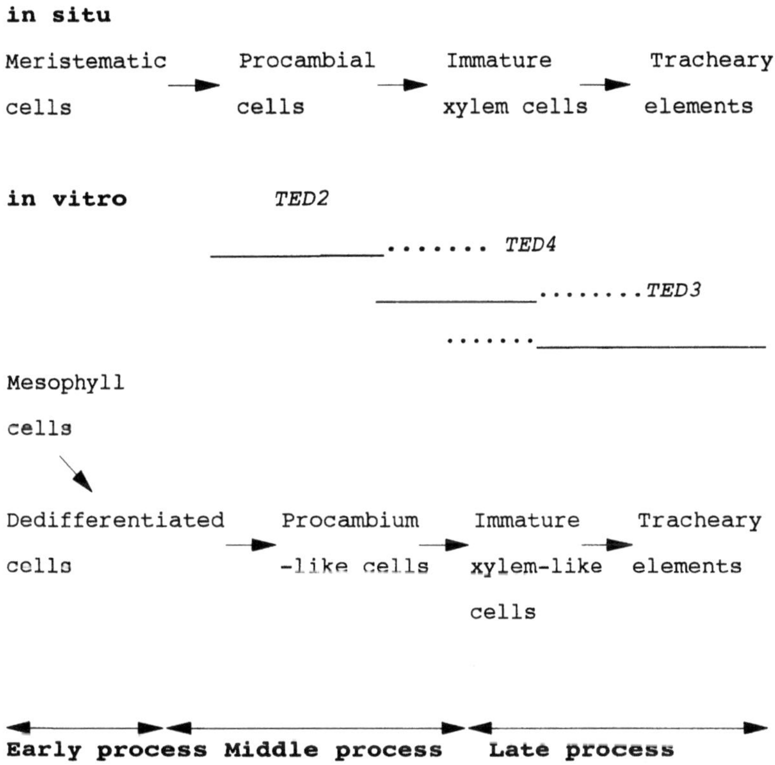

Fig. 180. Comparison of the process between redifferentiation of mesophyll cells and differentiation of meristematic cells into tracheary elements. (Fukuda 1994b)

indicate that *TED2* transcripts are accumulated in the early stage of differentiation into TEs from the precursor cells.

In situ, TEDs were expressed in the order of *TED2*, *TED4*, and *TED3*. It is suggested that cells expressing *TED2* are procambium-like cells that have the potency to become future phloem and xylem cells, while cells expressing *TED4* are differentiated into xylem cells or cells that have a potency to be xylem cells. Cells expressing *TED3* are differentiated into TEs or cells that have a potency to become TEs (Fig. 180). During differentiation from meristematic cells to vessels, potency of differentiation may be switched from pluripotency to single potency in which cells can be only vessels (Fukuda 1994b,c). *TED2*, *TED3*, and *TED4* may be markers for each stage of differentiation.

5.1.4.2 Genes Expressed in Vascular Bundle Formation

In association with the dynamic changes in the organization of microtubules during differentiation, the number of microtubules increases. Northern analysis with a fragment of the *Arabidopsis* β-1 tubulin gene has indicated that the new tubulin synthesis during TE differentiation was regulated at the transcription level. Three different cDNA clones, which share 80–81% identity at the nucleotide level and 91–96% identity at the amino acid level, were isolated from differentiating *Zinnia* cells.

Detailed analysis of the expression of the three β-tubulin genes (*ZTUB1-3*) with 3′-nontranslated regions as probes indicated that the three β-tubulin genes were expressed differentially during TE differentiation in *Zinnia* cells. The expression of *ZTUB1* and *ZTUB3* was not observed in freshly isolated *Zinnia* mesophyll cells, but the gene products increased rapidly during 24 and 48 h of culture prior to secondary wall formation. The expression of *ZTUB1* preceded that of *ZTUB3*. In contrast, *ZTUB2* showed a weak but continuous expression throughout the culture period. The rapid increase in the expression of *ZTUB1* and *ZTUB3* at the early stage would allow generation of a large amount of tubulin, which is necessary for dynamic changes in the organization of microtubules associated with secondary wall formation (Fukuda 1994b,c).

Increases in the activities of enzymes involved in secondary wall formation such as xylan synthase, cellulose synthase, PAL, OMT, coumarate: CoA ligase (4-CL), cinnamyl alcohol dehydrogenase (CAD) etc. occur during secondary wall formation. Many genes corresponding to the enzymes expressed in the late stage are being isolated and characterized. The *cis*-elements and *trans* acting factors of each gene have been compared to obtain essential information on programming of gene expression in the late process (Sect. 3.3).

In association with secondary wall thickening, autolysis of cells occurs. Protease activity increases more conspicuously in the late process of differentiation in the differentiation-inductive cultures of *Zinnia*. Partial purification of the enzyme indicated that it is a cysteine protease with a molecular mass of ca. 30 kDa.

Ye and Varner (1993) have isolated a cDNA clone for a protease similar to papain as a differentiation-specific clone from *Zinnia* cells. In addition, Thelen and Northcote (1989) found that several nucleases appeared in the autolysis process of tracheary elements of *Zinnia* cells, and that an RNase of 17 kDa is produced transiently during the initial induction of other nucleases in differentiating *Zinnia* cells. The analysis of the mechanism of autolysis is very important for elucidating the secondary differentiation of plant cells.

5.1.4.3 Regulation of Gene Expression in Vascular Bundle Differentiation

By studies of transgenic plants with several genes combined with promoters and reporters, a unique pattern of gene expression in vascular cells has been found. Various regulation regions are located in the upstream region of genes, and the gene expression is strictly regulated according to the developmental program and environmental conditions. In particular, there are factors which inhibit gene expression within and outside the vascular tissues. This results in a characteristic gene expression in the specific vascular tissue.

A glycine-rich protein (GRP), a hydroxyproline-rich glycoprotein (HRGP), and a proline-rich protein are located in vascular tissues. Two *GRP* genes (*GRP1.8* and *GRP1.0*) were isolated from *Phaseolus vulgaris*, and it was found that *GRP1.8* was expressed in xylem cells (Keller and Baumgartner 1991). Four different regulation regions (*SE1*, *SE2*, *NRE*, and *RSE*) are located in the promoter region of *GRP1.8. SE1 and SE2* accelerate specific expression in the vascular tissue, while *RSE* accelerates specific expression in the root. In contrast, *NRE* inhibits gene expression in tissues other than the vascular bundles, such as the cortex.

In *PAL2* from *Phaseolus vulgaris* a region that accelerates gene expression in the xylem but inhibits it in the phloem is located between –119 and –135, which also contains a promoter region. This region contains a TCCACCAACCCC sequence which commonly occurs in *PAL* as the *pal 1* box (CCA(A/C)C(A/T)AAC(C/T)CC) from parsley and *Arabidopsis*. The sequence was found to be a *cis* element of chalcone synthase activated by stress, and of other stress-inducing genes (Leyva et al. 1992). Analysis of the promoter of 4-coumarate:CoA ligase 1 which catalyzes the formation of 4-coumaroyl CoA (*4CL-1*) revealed that three regions (*B1*, *C*, and *D*) composed of 40 nucleotides, which directly regulate the expression of *4CL-1* in vascular tissues, are located up to –244 (Hauffe et al. 1991). The *B1* and *D* are the regions that accelerate the gene expression both in xylem and phloem. The *C* region was found to be involved in acceleration of the gene expression in the xylem but in inhibition of expression in the phloem. A motif, the sequence CTCACCAACCCC, was also found in the *C* region.

Recently, a xylem-specific *myb* gene has been isolated. It is suggested that a certain *myb* protein (a *trans*-acting factor) is combined with a specific *cis* element of genes encoding enzymes in the phenylpropanoid pathway. The analysis of *myb* as a candidate for the transcription factor is in progress as part of a program to elucidate the coordinated expression of the characteristic enzymes involved in vessel formation (Fukuda 1994c).

5.1.5 Deposition of Polysaccharides and Lignin in Xylem Cell Walls

Tree stems grow by repeated cell divisions of the vascular cambium to produce xylem inside and phloem outside. A cell plate is formed between two daughter nuclei after nuclear division of cambial cells. Cell plates are formed by fusion of many vesicles derived from the Golgi apparatus and the endoplasmic reticulum (Fig. 181).

Investigations with tobacco cell cultures showed that the cell plate contains β-1, 3-glucan and xyloglucan, and that the former is synthesized by the enzymes located on the plasma membrane, while the latter is formed by the

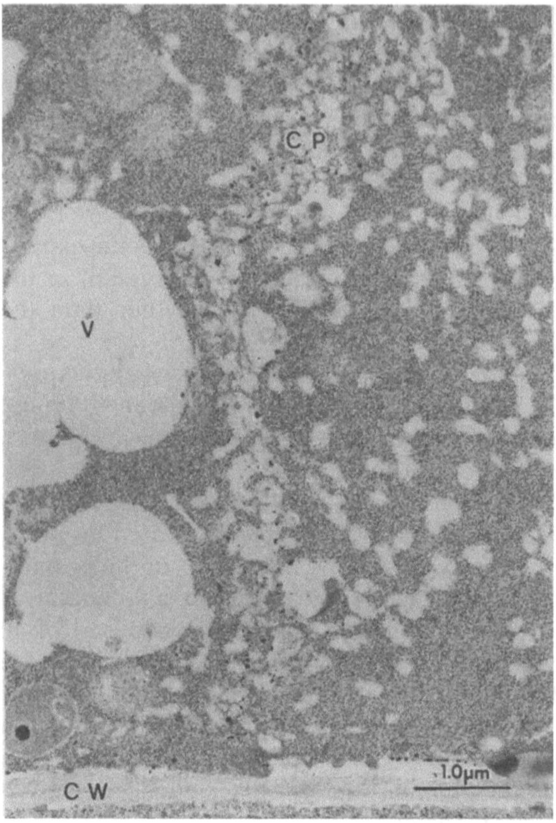

Fig. 181. Tobacco cell cultures forming a cell plate. After nuclear division many Golgi and ER vecicles aggregate to form cell plate. *CP* Cell plate; *CW* cell wall; *V* vacuole. (Courtesy of Y. Kashiwagawa)

Golgi vesicles (Kakimoto and Shibaoka 1992). The cell plate also contains pectin mainly composed of polygalacturonan, which forms a gel in the presence of Ca^{2+}.

After completion of cell division the xylem cells enlarge, and this is followed by an increase in diameter and length. Cell walls formed at this stage are composed of a primary wall only. The primary wall is mostly composed of cellulose, hemicelluloses, such as xylan, xyloglucan and rhamnogalacturonan (pectin), and protein, but the cellulose content is rather low compared with that of the secondary walls (25%).

Cytochemical observations and biochemical analyses of the cambial zone of *Acer pseudoplatanus* showed that young tangential walls (T walls) consist of a loose microfibrillar skeleton and a matrix rich in methylated pectins. No recognizable middle lamella is present. In contrast, in radial walls (R walls) a middle lamella, which contains a large amount of acidic pectins, is present. Cytochemical and immunological studies of cell wall formation in primary meristems demonstrated that nascent cell walls contained esterified pectins and fibrillar polysaccharides (cellulose and xylan) as their main components. During the ensuing maturation of the cell wall within the meristem, newly synthesized polysaccharides are added and a spatial redistribution of polysaccharides occurs, leading to the formation of a proper middle lamella (Catesson 1994).

An important step in primary wall evolution is the deesterification of pectins through pectin methylesterase activity, which releases protons and acidic polygalacturonans (Carpita and Gibeaut 1993). The resulting drop in pH allows cell wall loosening and auxin-induced growth.

Biochemical studies on the cambial zone in poplar (Baïer et al. 1994) showed that the total amount of pectins per gram of cell wall was rather stable throughout the annual cycle, but that the composition of the pectin changed dramatically. The overall decrease in uronic acid content occurring at the end of the growing season concealed an increase of nine times in esterified pectins concomitant with a drop in pectin methylesterase activity. Acidic polygalacturonans accumulate into cell wall, stiffening either by cross linking with polypeptides or by gel formation in the presence of Ca^{2+} (Catesson 1994).

Electron micrography of cell walls by rapid freeze etching showed that primary walls consist of lamellae of cellulose microfibrils. Xyloglucan bridges the microfibrils and keeps the spaces between them (Maccann et al. 1990).

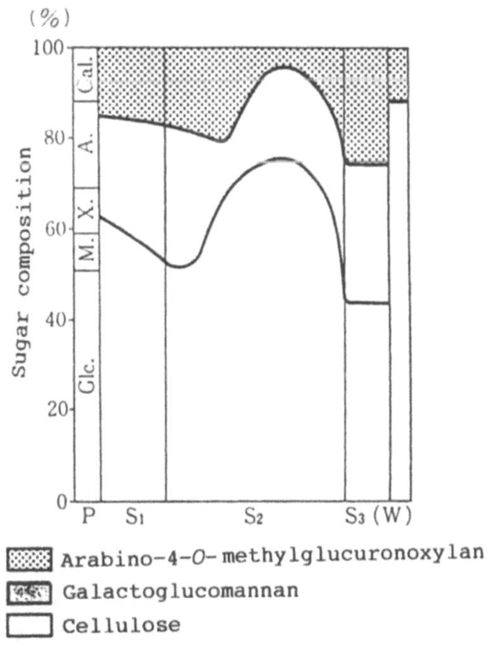

Fig. 182. Composition of polysaccharides in the secondary wall of a tracheid of *Cryptomeria japonica*. Hemicelluloses are abundant in the S_1, S_2, and S_3 layers. *P* Primary wall; S_1 outer layer of secondary wall; S_2 middle layer; S_3 inner layer; *W* warty layer; *Glc* glucose; *M* mannose; *X* xylose; *A* arabinose; *Gal* galactose. (Takabe 1994)

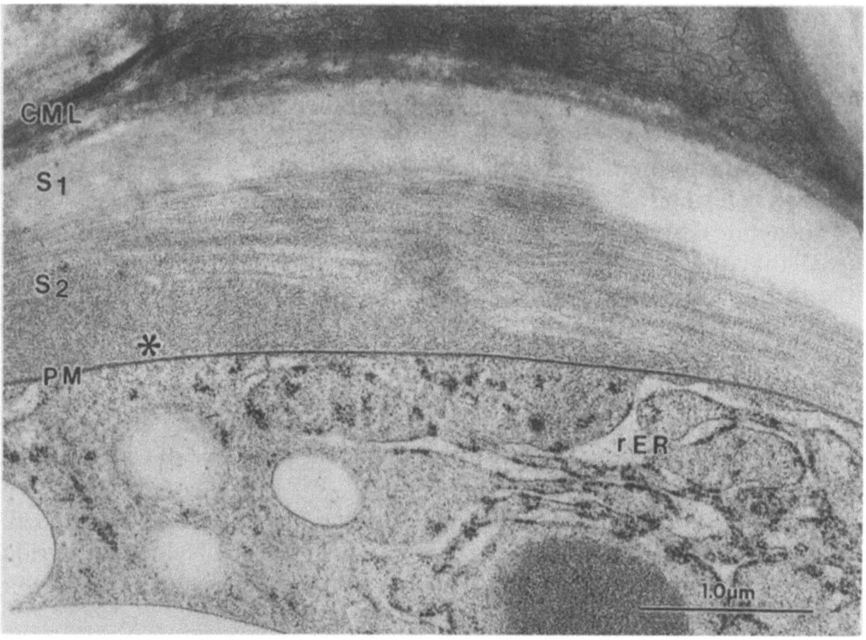

Fig. 183. Tracheid of *Cryptomeria japonica* forming a secondary wall. The innermost layer (*) of the cell wall, which is composed of hemicelluloses and rich in water, appears to be homogeneous but at the outer side fibrils can be observed. *CML* Compound middle lamella; S_1 outer layer of secondary wall; S_2 middle layer; *PM* cell membrane; *rER* rough endoplasmic reticulum. (Courtesy of K. Sano)

During formation of secondary cell walls in softwoods, hemicellulose deposition is active at the stages of outer and inner layer (S_1 and S_3) formation, while cellulose deposition is active at the stage of middle layer (S_2) formation (Takabe et al. 1983). As a result the S_1 and S_3 layers of softwood secondary walls are rich in hemicelluloses, while the S_2 layer is rich in cellulose (Fig. 182). These results are supported by cytochemical and immunocytochemical investigations (Vian et al. 1986).

Electron microscope–autoradiography of growing xylem tissues indicated that cellulose is accumulated by appositional deposition, while hemicelluloses are accumulated into cellulose networks by intussusceptional deposition (Ray 1967). However, recent elctronmicroscope studies with rapidly frozen wood samples suggest that some of the hemicelluloses previously transported from the Golgi apparatus absorb water to make a sol that covers the inner surfaces of the cell walls as a thin layer (Inomata et al. 1992; Fig. 183). Cellulose synthesized by TCs is subsequently released into the hemicellulose sol and deposited. Cellulose microfibrils deposited in this way form bands of lamellae composed of several microfibrils together with hemicelluloses. The orientation of the lamellae controlled by the microtubules is changed 15–25° on deposition of respective lamellae (Kataoka et al. 1992; Fig. 184). The thickness of the

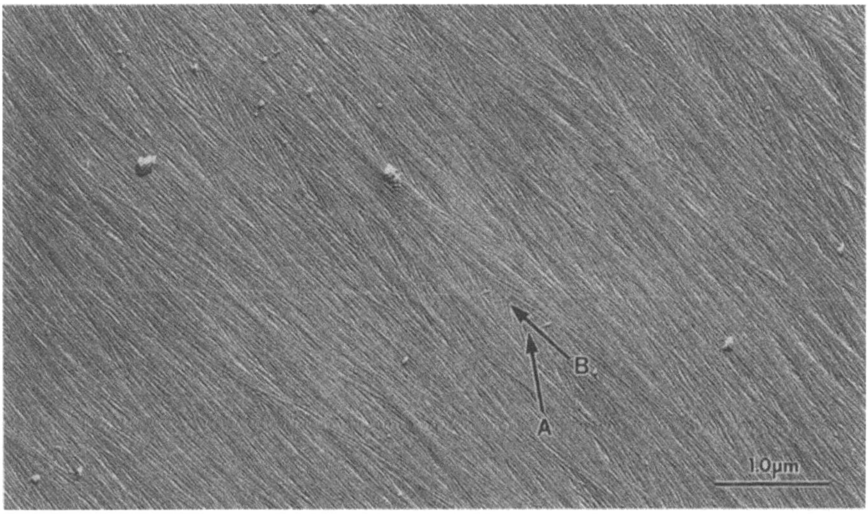

Fig. 184. Replica of the inner surface of a tracheid of *Chamaecyparis obtusa*. On the microfibrils (*A*), microfibrils (*B*) are piled up at 15°–25°. (Courtesy of S. Ohta)

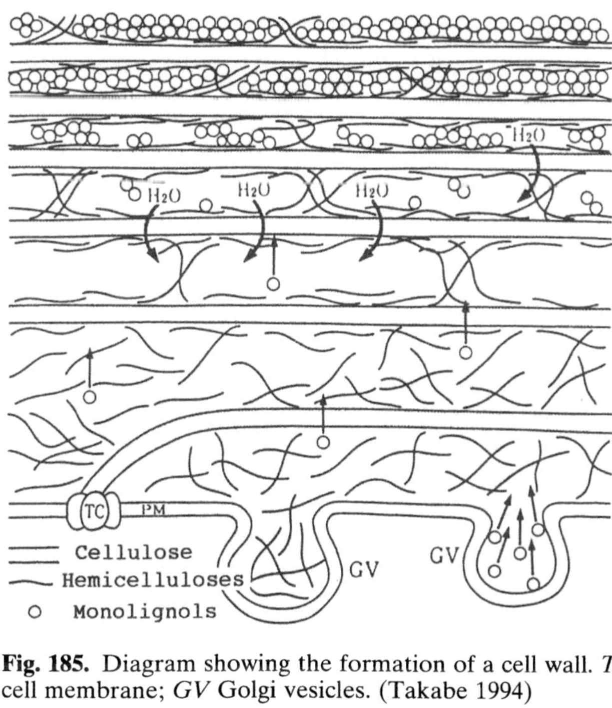

Fig. 185. Diagram showing the formation of a cell wall. *TC* Terminal complexes; *PM* cell membrane; *GV* Golgi vesicles. (Takabe 1994)

cell walls increases by changing the microfibril orientation continually. The deposited cellulose and hemicelluloses still absorb sufficient water to maintain cell activity, but the water is gradually released from the inner to the outer side as a result of dehydrogenative polymerization of monolignols (lignification), leading to a relatively hydrophobic cell wall (Fig. 185).

5.1.5.1 Deposition of Lignin in Xylem Cell Wall

Lignin, like cellulose and hemicelluloses, is synthesized by individual xylem cells. The synthesis of lignin is initiated after cell elongation has ceased. Microradioautographic observations with labeled phenylalanine and cinnamic acid suggest that the ER and the Golgi apparatus are involved in the synthesis and transportation of the lignin precursors (Fujita and Harada 1978, Takabe et al. 1985; Fig. 25).

Monolignols supplied to the cell walls are transported to the hemicellulose gel surrounding the cellulose microfibrils and dehydrogenatively polymerized by the mediation of cell wall-bound peroxidases and laccases. In this process lignins become chemically linked to hemicelluloses via quinonemethide intermediates (Fig. 112), filling the microspaces between the microfibrils in the cell wall. By deposition of the lignin, the cell walls become rigid and more hydrophobic. The tracheids in the gymnosperms and the fibers in the angiosperms thus become able to offer necessary physical support, and transportation of water becomes more smooth in the tracheids and vessels, respectively. The formation and structure of lignin are different in the primary xylem, which is derived from the apical meristem, and in the secondary xylem, derived from the vascular cambium. It is the latter tissue that is referred to here.

5.1.5.2 Formation and Structure of Conifer Lignin

The process of lignification of cell walls can be observed microscopically by taking advantage of the UV absorption of lignins around 270–280 nm (Wardrop and Bland 1959), and by microautoradiography of lignin labeled with radioactive tracers (Saleh et al. 1967, Fujita et al. 1983).

Lignification of conifer tracheids takes place in three stages. First, at the same time as initiation of the S_1 layer occurs, active deposition of lignin begins at the cell corners in the primary wall and middle lamella. Gradual deposition of lignin then continues in these regions and in the S_1 layer at the same time as the S_2 and S_3 layers are being formed. The active lignification occurs in the secondary wall during the formation of the S_3 layer. Finally, lignification is completed in the middle lamella and the entire secondary wall (Takabe et al. 1981, Terashima and Fukushima 1988, Terashima et al. 1988, Fukushima and Terashima 1991a). These processes have been confirmed by interference microscopy (Donaldson 1991).

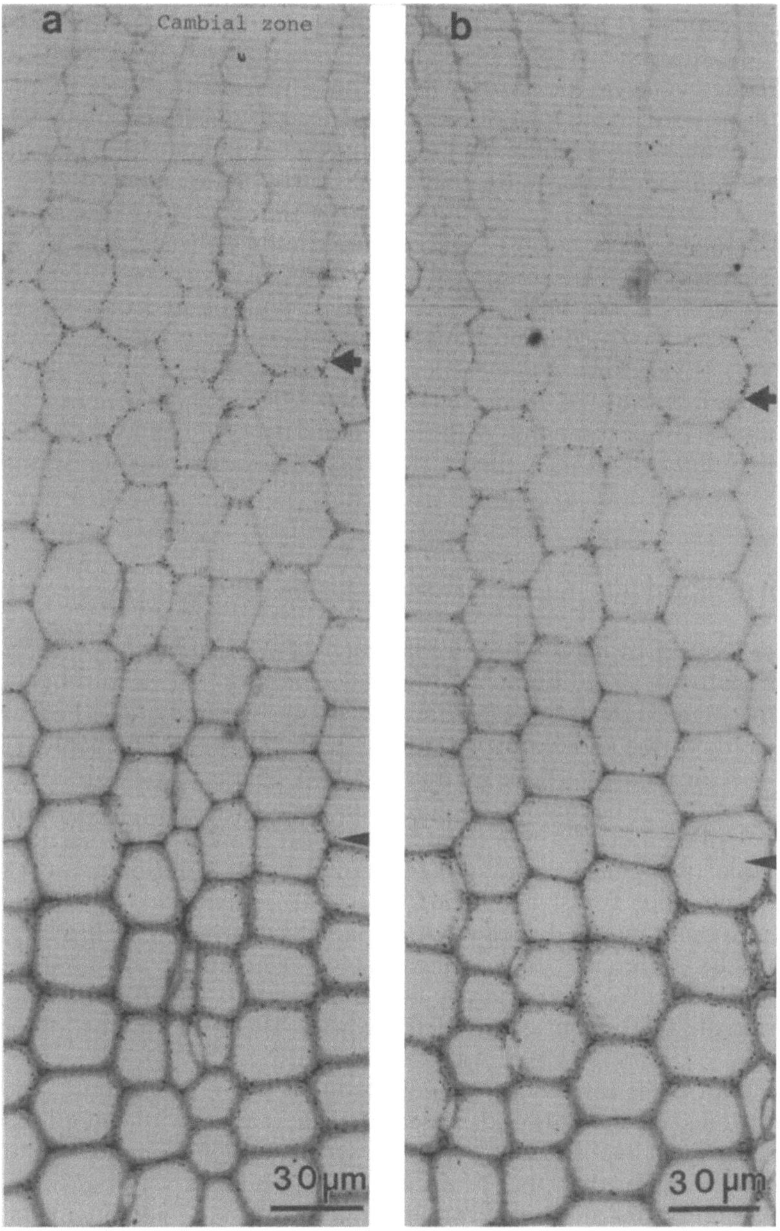

Fig. 186. a,b Autoradiography of differentiating wood of *Pinus thunbergii* fed with **a** coniferin [2-^{3}H of aromatic ring] or **b** [5-^{3}H of aromatic ring]. Incorporation into compound middle lamella (*arrow*) in **b** is lower than in **a**, suggesting that 5-^{3}H is released in the process of condensation. (Terashima et al. 1988; courtesy of Dr. N. Terashima)

Several investigators have found that the lignin in the middle lamella of conifers is a *p*-hydroxyphenyl-guaiacyl lignin, whereas the lignin in the secondary wall is of the guaiacyl type. As a result of this situation, the lignin in the middle lamella is more highly condensed and less reactive than the lignin in the wall (Whiting et al. 1981, Whiting and Goring 1982a,b,c, Fukuda and Terashima 1988, Terashima and Fukushima 1988, Fukushima and Terashima 1991a,b; Fig. 186).

Others, however, have failed to corroborate these claims (Sorvari et al. 1986, Westermark et al. 1988).

Lignification of the tracheids in compression wood differs from that in normal wood. The lignin concentration in the outer part of the S_2 Layer (the S_2-L layer) is very high (50–60%) and still relatively high in the inner portion (35–40%). Because of the presence of intercellular spaces in compression wood, almost all of the lignin is located inside the cell wall (Wood and Goring 1971, Fukazawa 1974, Fujita et al. 1978, Fukushima and Terashima 1991a,b).

5.1.5.3 Formation and Structure of Hardwood Lignin

The secondary xylem of hardwoods is composed of vessels, libriform fibers and ray parenchyma. The formation and structure of hardwood lignin have been investigated by various methods, such as UV microscopy (Fergus and Goring 1970, Musha and Goring 1975, Takabe et al. 1992), histochemistry using lignin-specific color reactions (Wu et al. 1990), energy dispersive X-ray analysis (EDXA) of brominated or mercurated wood sections (Saka and Goring 1985, Eriksson et al. 1988), microautoradiography of wood sections fed with radioactive guaiacyl or syringyl monolignols (Terashima et al. 1986, Fukushima and Terashima 1990), chemical analyses of secondary xylem and compound middle lamella: middle lamella + primary wall (CML) separated by the difference of specific gravity (Hardell et al. 1980a,b, Sorvari et al. 1983, 1986), and analysis of lignified CML in differentiating xylem (Cho et al. 1980, Eom et al. 1987, Kim et al. 1993), among other methods.

The results of these investigations indicated that lignin deposited in the initial lignification stage in hardwoods is rich in guaiacyl lignin with condensed structures, while lignin deposited later is rich in syringyl lignin. Lignification of vessels progresses earlier than that of wood fibers. In both vessels and wood fibers, lignification begins at the cell corners, followed by the middle lamella, the primary wall and the secondary wall in this order. The secondary wall in fibers and ray cells of *Betula papyrifera* contain a syringyl lignin. In the vessels, in contrast, both middle lamella and secondary wall contain a guaiacyl-rich lignin (Fergus and Goring 1970, Musha and Goring 1975).

These results which demonstrate such heterogeneity of lignin in cell walls are partly supported by enzymatic analysis of phloem and xylem of poplar wood. Sarni et al. (1984) and Grand et al. (1985) assayed the activities of hydroxycinnamate: CoA ligase and cinnamyl alcohol dehydrogenase involved in lignin synthesis in the cambial zone and in newly formed xylem and phloem tissues. They found that the isozymes of these enzymes are localized in certain tissues, and that their substrate specificities for syringyl and guaiacyl precursors are different.

5.1.5.4 Formation and Structure of Grass Lignin

Enzymes involved in the shikimate and cinnamate pathways were isolated from bamboo shoots and characterized, using feeding experiments with ^{14}C-labeled lignin precursors (Higuchi 1969b). The activities of 3-dehydroquinate hydrolyase, 3-dehydroshikimate reductase, and aromatic amino acid transaminases are greatest at the top of the bamboo shoots, and decrease gradually toward the lower parts of the shoots. The amounts of shikimic acid, L-phenylalanine and L-tyrosine in bamboo shoots seem to follow the pattern of the enzyme activity. Meanwhile, the activities of PAL, TAL, and OMT when forming lignin precursors increase progressively from the top to the lower parts, but these activities decrease again at the base of the bamboo shoots, where the vessels and parenchyma tissues are already lignified.

These results suggest that in the early stages of bamboo growth, the enzymes in the shikimate pathway and aromatic amino acid synthesis are dominantly active, but with onset of lignification the metabolic systems are transformed to the synthesis of lignin precursors.

Bamboo lignin is a typical grass lignin and gives a large amount of *p*-hydroxybenzalbehyde as well as vanillin and syringaldehyde on alkaline nitrobenzene oxidation (Higuchi 1958a, Higuchi et al. 1967e). The *p*-hydroxybenzaldehyde produced is mainly due to the esterification of *p*-coumaric acid with the lignin polymer, and therefore, the aldehyde amount is reduced to about one third by previous alkaline hydrolysis.

Alkaline hydrolysis and acidolysis of model compounds such as veratryl *p*-coumarate, 3-(3,4-dimethoxyphenyl)propyl *p*-coumarate, and bamboo milled wood lignin (MWL) revealed that at least 80% of the *p*-coumaric acid in bamboo lignin is esterified with primary hydroxyl groups (γ position) in the propyl side chain; the content of benzyl esters of the acid in the side chain was estimated to be less than 20%. Dehydrogenative polymer (DHP) of coniferyl *p*-coumarate, as a possible precursor of *p*-coumaroyl ester of lignin, with coniferyl alcohol by horseradish peroxidase showed that the *p*-coumaric acid moiety of the ester is dehydrogenated to give a DHP connected not only by original γ-ester linkage but also partly by ether or C-C bonds at C5(3) or the aromatic rings, or Cβ of the side chains of the *p*-coumaric acid. The content of γ-ester of the acid in the DHP determined by alkaline hydrolysis was 5–16%.

The results indicated that most of the *p*-coumaric acid in bamboo lignin is linked to primary hydroxyl groups in the lignin side chains, and that relatively few ester and ether linkages are formed at the benzyl carbon via quinone methide intermediates (Higuchi 1990).

Recently Ralph et al. (1994a) found that maize lignin from rind tissue of stem internodes contains large amounts of esterified *p*-coumaric acid. ^{13}C-^{1}H correlative NMR experiments of the lignin, α-and γ-*p*-coumaroylated lignin models compounds, and a specifically labeled synthetic coniferyl alcohol/coniferyl-*p*-coumarate dehydrogenation polymer revealed that the *p*-coumaric acid is attached exclusively at the γ-position as in bamboo lignin. Since the *p*-coumaroyl moieties retain their phenolic group the investigators suggested that *p*-coumarate ester monomers are incorporated into the lignin late in the process and subsequent to incorporation of substantial amounts of syringyl lignin.

Ethanolysis of bamboo lignin gives moderate amounts of vanilloyl and syringoyl methyl ketones with a small quantity of *p*-hydroxybenzoyl methyl ketone, indicating the occurrence of a small amount of *p*-hydroxyphenylglycerol-β-aryl ether substructure in bamboo lignin (Higuchi and Kawamura 1966).

Permanganate oxidation of methylated bamboo MWL gives moderate amounts of 4-methoxyisophthalic, isohemipinic, *p*-anisic and veratric acids, which are derived from the condensed and uncondensed substructures of bamboo lignin, respectively (Yamasaki and Higuchi 1971). The results indicate that, except for the esterified *p*-coumaric acid, the *p*-hydroxyphenyl moieties of bamboo lignin mainly exists as condensed substructures.

Scalbert (1984) showed that wheat straw lignin gives a very small amount of *p*-hydroxyphenylglycerol derivatives with large amounts of guaiacyl- and syringylglycerol derivatives on thioacidolysis, suggesting that most of the *p*-hydroxyphenyl moiety of grass lignin occurs as a condensed substructure.

An excellent review article on grass lignin by Ralph and Helm (1993) should be consulted for details of grass lignin.

5.1.5.5 Deposition of Lignin at the Molecular Level

Lignin is chemically connected to hemicelluloses embedded between the cellulose microfibrils (Watanabe et al. 1986, 1987). Ruel et al. (1978) proposed a lamella model of the distribution of cell wall components in secondary cell wall (Fig. 14, Chap. 1). They suggested that the spaces between microfibrils are about 7 nm (for the S_2 layer of spruce wood; Goring 1977) and the space for lignin deposition in the matrices is about 3.5 nm in width (Westermark 1982). In accordance with such suggestions, studies by Raman microspectroscopy showed that aromatic rings of lignin in spruce secondary walls are oriented in parallel with the microfibril orientation in the intercellulose microfibril space (Atalla and Agarwal 1985, Agarwal and Atalla 1986). A model of the localization of lignin and hemicelluloses between the cellulose microfibrils was proposed based on these observations by Terashima et al. (1993; Fig. 187).

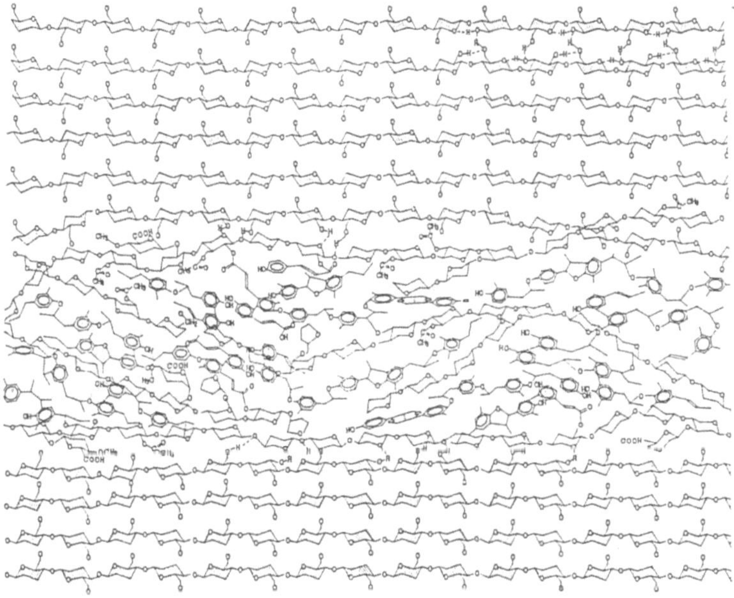

Fig. 187. Model structure of lignified secondary cell wall. (Terashima et al. 1993)

5.1.5.6 Transportation and Polymerization of Monolignols in Cell Walls

In conifers and several hardwoods coniferin and syringin are present, respectively (Freudenberg and Harkin 1963, Terazawa et al. 1984). Freudenberg (1959), Kratzl et al. (1957) and Billek (1959) found that radioactive coniferin fed to spruce is hydrolyzed by β-glucosidase to glucose and coniferyl alcohol which is incorporated into lignin.

The distribution of β-glucosidase in the cambial zone was confirmed by indigo formation from indican fed to cross-sections of spruce (Freudenberg 1959). Indican is hydrolyzed by β-glucosidase to glucose and indoxyl which is oxidized to a blue pigment, indigo. The blue color of indigo was observed between the cambium and the lignified xylem.

Terashima and his coworkers (Terashima et al. 1988, Terashima and Fukushima 1988, Fukushima and Terashima 1990, 1991a,b, Matsui et al. 1994a,b) recently found that radioactive monolignol glucosides are incorporated into lignifying woods of ginkgo, black pine, poplar, beech, lilac and *Magnolia*. These precursors were selectively incorporated into certain cells at different stages of cell wall differentiation. β-Glucosidase activity was detected in the inner layer of secondary walls of differentiating wood. Terashima proposed that monolignol glucosides are essential lignin precursors and serve as the chemical entities transported into the cell wall.

Coniferin- and syringin-specific β-glucosidases associated with tracheids and other xylem tissues were found in spruce (*Picea abies*; Marcinowski and

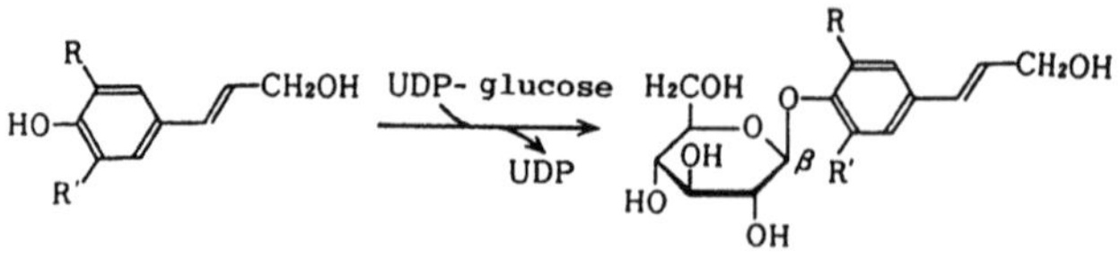

Fig. 188. Formation of monolignol glucoside. R-H, R′-OCH_3 = coniferyl alcohol and coniferin. R,R′-OCH_3 = sinapyl alcohol and syringin

Grisebach 1978, Marcinowski et al. 1979). In addition, a glucosyltransferase specific for monolignols was identified (Ibrahim 1977, Schmid and Grisebach 1982, Yamamoto et al. 1990, Whetten and Sederoff 1995; Fig. 188). The transferase, which uses UDP-glucose as glucose donor, was found to be located in the cytoplasm of epidermal and vascular bundle cells of spruce seedlings by immunochemical analysis using FITC-labeled antibodies (Schmid et al. 1982). The enzyme is also present in beech bark (Yamamoto et al. 1990) and in the cambial zone of loblloly pine (*Pinus taeda*; Davin and Lewis 1992).

These results seem to lend support to Terashima's ideas, but attempts to correlate β-glucosidase activity with presumed lignification in a variety of cell cultures and young trees (Fukushima et al. 1994) have not been successful.

There is no direct correlation between coniferin or syringin turnover and lignification (Marcinowski and Grisebach 1977, Fukushima et al. 1994), suggesting that monolignol glucosides are not the preferred chemical entities as precursors for lignification.

As noted in Section 4.3.4, the formation of monolignols from phenylalanine in conifers and hardwoods and the quantities produced, are regulated by substrate specificity, allosteric regulation, and channeling etc. of specific enzymes such as OMT, ferulate-5-hydroxylase, cinnamoyl-CoA ligase, and cinnamyl alcohol dehydrogenase. The formation of sinapic acid is especially restricted in hardwoods, and sinapic acid cannot be reduced to sinapyl alcohol in conifers.

The addition of exogeneous syringin to conifers, which cannot synthesize it, does not lead to a normal lignification, but implies a metabolic ability of conifers to modify syringin as a xenobiotic. It seems that monolignol glucosides are storage products and only used as needed when additional and rapid lignification is required.

A recent immunochemical investigation (Awano et al. 1994) showed that cinnamyl alcohol dehydrogenase is located in the plasma membrane of lignifying tracheary elements and vacuole of *Zinnia elegans* and *Aralia cordata*. It seems that monolignols formed in the plasma membrane of lignifying tissues are immediately converted to lignin by peroxidase bound to the cell wall, while monolignols formed in the vacuole are glucosylated to their glucosides and stored before incorporation into lignin.

6 Formation of Earlywood, Latewood, and Heartwood

6.1 Regulation of Formation of Earlywood and Latewood

A clear periodicity in light and dark color is observed on the cross surface of trees grown in temperate and warm-temperate zones, especially conifers. Each period corresponds to a growing ring (an annual ring in temperate and warm-temperate zones). The light and dark colors are ascribed to the different cells formed in the earlier and the later parts of the growing season. Wood formed in the first half of growing season is composed of cells with large diameters in the radial direction and thin cell walls, while wood formed in the latter half is composed of cells with small diameters and thick cell walls. The former wood is called earlywood, while the latter is called latewood, as noted before (Chap. 1). The proportion of latewood in an annual ring is characteristic of each species (Fig. 189).

This variation is mainly ascribed to either the shift of cell wall thickness from earlywood to latewood or decrease of cell diameter of the less thick cells (Saiki 1965, Kubo 1983). Even in the same species, the ratio of latewood changes with differences of thickening growth and the position of the wood in the tree stem (Larson 1960, Funada et al. 1989; Fig. 190). Such differences depend on the formation period of earlywood and latewood, the rate of formation of wood, etc. (Denne 1979, Funada et al. 1987, 1990). The process of earlywood and latewood formation, that is the frequency of division of cambial cells, the enlargement of newly formed xylem cells, and cell wall thickening are controlled by plant hormones (Savidge and Wareing 1981, Little and Savidge 1987). Auxins, abscisic acid, cytokinins, gibberellins, ethylene, brassinosteroids, and jasmonic acid are known to be involved in earlywood and latewood formation (Fig. 191).

6.1.1 Auxins

6.1.1.1 Formation and Distribution of Auxins in Conifers

Indole acetic acid (IAA) is largely synthesized by buds in the tree crown, developing needles, and shoots. In the dormant period in winter IAA is synthesized by dormant buds and old leaves (Savidge and Wareing 1982). The occurrence of small amounts of endogenous IAA in the cambial zone suggests the formation of IAA by cambial cells (Little and Wareing 1981, Wodzicki et

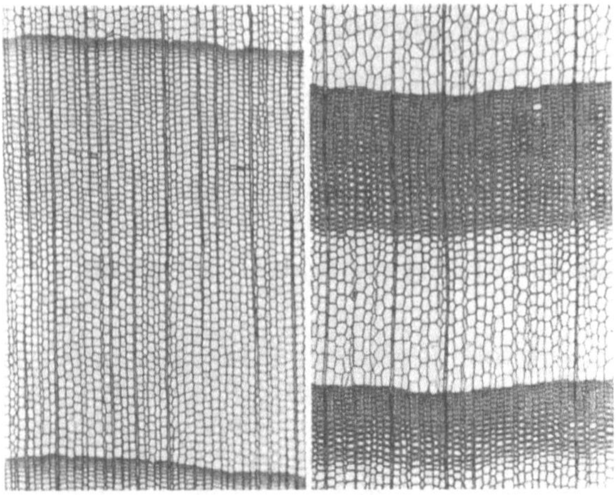

Fig. 189. Transition from earlywood to latewood. *Left* Gradual transition (*Thujopsis dolabrata*); *right* abrupt transition (*Larix leptolepis*). (Courtesy of Dr. R. Funada)

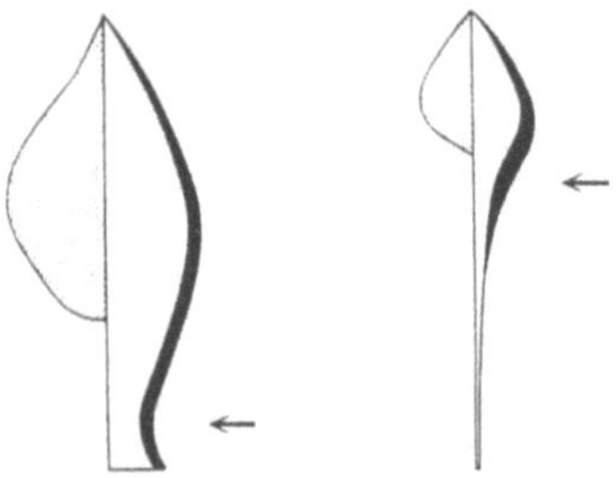

Fig. 190. Diagrammatic representation of the distribution of annual ring characteristics in relation to crown profiles. *Dotted area* Weight of green foliage; *white area* width of earlywood; *black* width of latewood; *arrows* stem positions with the maximum percentage of latewood. (Funada et al. 1989)

al. 1987, Sundberg et al. 1990). IAA synthesized by the tree crown is basipetally transported mainly to cambial cells and their differentiating cells (Little 1981, Odani 1985, Wodzicki et al. 1987).

When radioactive IAA was fed to *Pinus echinata* a high radioactivity was found in cambial and differentiating cells (Nix and Wodzicki 1974). The transport rate of IAA was estimated to be about 5–7 mm/h in red pine (*Pinus densiflora*) and balsam fir (*Abies balsamea*; Little 1981, Odani 1985). The transport rate did not change with seasonal changes or with active and resting periods of cambial activity.

The occurrence of IAA in the cambial zone (cambium, differentiating phloem and xylem) of conifers has been confirmed in many species (Little et al. 1978, Funada et al. 1987, 1990, Savidge 1990). In *Pinus sylvestris*, IAA is

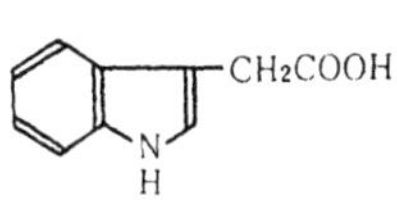

Indole-acetic acid

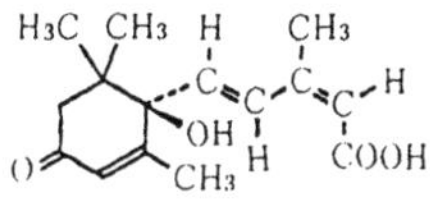

Abscisic acid

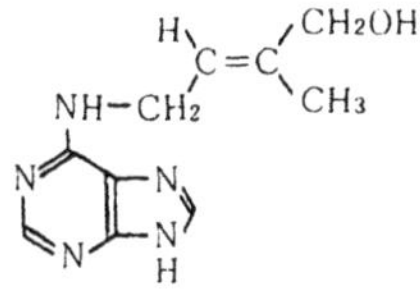

trans-Zeatin *trans*-

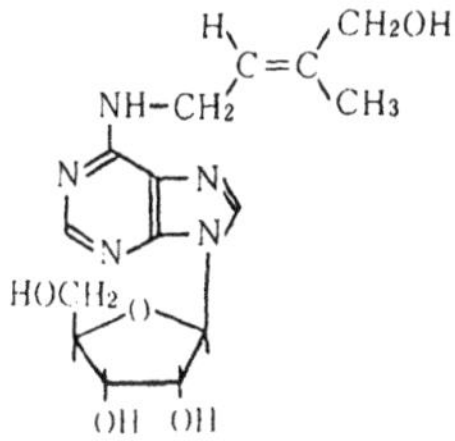

trans- **Ribosylzeatin**

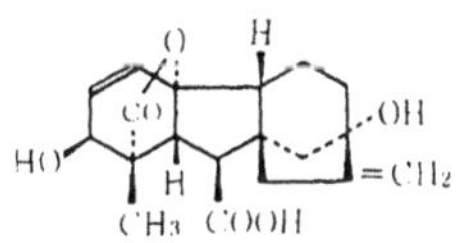

Geberellin A3

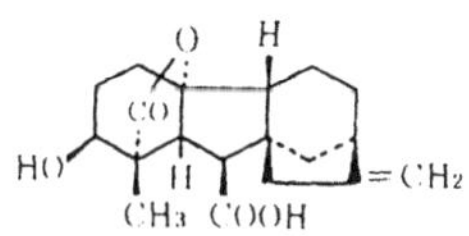

Gibberellin A4

$CH_2=CH_2$

Ethylene

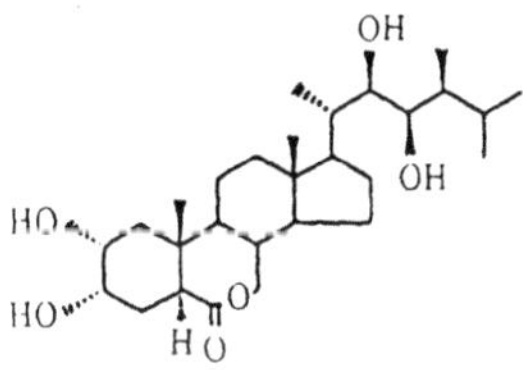

Brassinolide

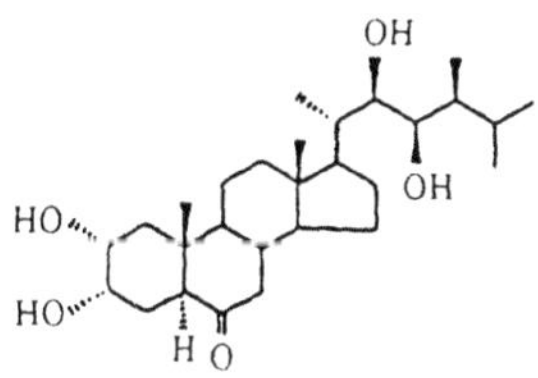

Castasterone

Fig. 191. Chemical structures of plant hormones

mostly present in the cambial zone (Sundberg et al. 1990), while in *Pinus contora* IAA occurs in differentiating xylem (Savidge et al. 1982). The amount of IAA in the cambial zone differs depending on the weight of the tree crown and the part of the stem where it is located (Funada et al. 1987; Fig. 192).

6.1.1.2 Effect of Indole Acetic Acid on Formation of Tree Stems

When disbudding or defoliation was effected before initiation of cambial activity in early spring cambial activity was inhibited, except when IAA was administered (Onaka 1950, Odani 1975, Denne and Wilson 1977). The amount of xylem formed increased in accordance with the amount of IAA added (Sundberg and Little 1990). A continuous supply of IAA to cambial cells was required to maintain fusiform initials, and when IAA was lacking these cells

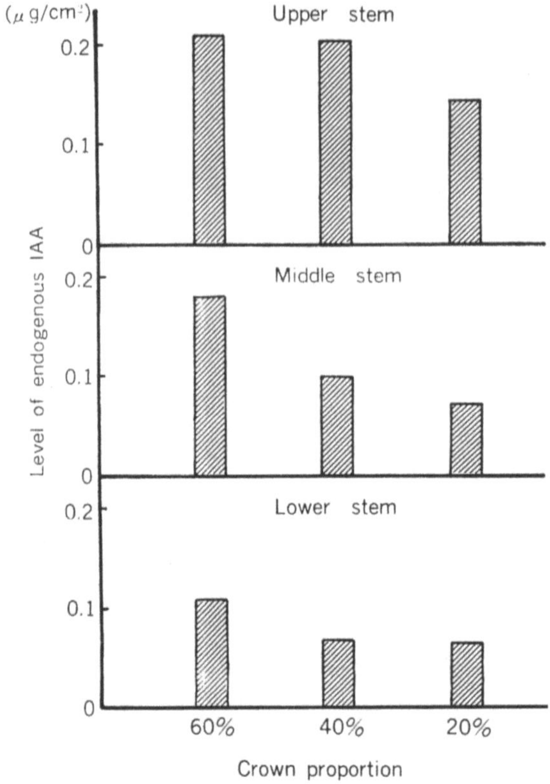

Fig. 192. Level of endogenous IAA in September at different stem heights of *Pinus densiflora* having different crown proportions. (Funada et al. 1987)

were differentiated into axial parenchyma cells and not to tracheids (Savidge 1983).

When *Pinus sylvestris* (Wareing 1958) and *Pinus resinosa* (Larson 1964) were subjected to a long-day treatment, continuous shoot elongation and formation of earlywood cells occurred, whereas on short-day treatment shoot elongation ceased and latewood cells were induced. When IAA was supplied externally to the latewood-forming saplings in the short-day treatments, a band of earlywood cells was formed (Larson 1960).

These results indicate that changes in cell diameter are regulated by IAA. This is consistent with the fact that the decrease in diameter of conifer tracheids in the summer is reversed by administration of IAA under natural growing conditions (Balatinecz and Kennedy 1968, Odani 1975, Denne and Wilson 1977). When shoot growth ceased, the productive ability of IAA decreased acropetally from the lower part of the stem, inducing latewood formation from the lower part. In contrast, the thickness of cell walls has been shown to be controlled by the amount of photosynthate available (Larson 1960).

Concentration of endogenous IAA in the cambial zone of a tree stem exhibits a seasonal change (Little and Wareing 1981, Savidge et al. 1982, Savidge and Wareing 1984, Sandberg and Ericsson 1987, Funada et al. 1990). The concentration of IAA increased from spring to early summer, and then decreased toward the autumn to the level found in the spring. In winter, IAA was present at a low level. The stage of rapid decrease of IAA coincides well with the change from earlywood to latewood formation (Sundberg et al. 1987). When the concentration of endogenous IAA decreased earlier than usual as a result of a change in growing conditions, the transfer of earlywood to latewood occurred earlier. These findings indicate that IAA is a major factor in the control of elongation of xylem-differentiating cells, the formation of earlywood and latewood, and the transition of earlywood to latewood.

Since a supply of IAA is required to maintain cambial activity, it has been suggested that IAA controls the termination of cambial activity (Larson 1960). However, the termination of cambial activity occurs at a stage of relatively high endogenous IAA concentration, and the final decrease of IAA concentration takes place after all cambial activity has ceased (Little and Wareing 1981, Sundberg et al. 1987, Funada 1994). External addition of IAA at the termination stage of cambial activity did not stop the termination (Denne and Wilson 1977, Little and Wareing 1981). These results suggest that even if IAA is supplied, cambial activity ceases at a certain time of the season. For termination of cambial activity, that is, induction of dormancy, other factors may be involved. A change in the sensitivity of cambial cells to IAA has been suggested to be one of the factors involved (Zajaczkowski 1973, Denne and Wilson 1977).

When IAA was supplied to tree stems cambial activity increased, but the effect changed seasonally, and after summer the effect was very small (Zajaczkowski 1973, Denne and Wilson 1977, Riding and Little 1984, Sundberg et al. 1987).

The exact nature of any cambial sensitivity to IAA is scarcely known, but it seems that the change is induced by a change of temperature and day length, and is related to structural and histochemical changes in the cambial cells (Riding and Little 1984, Mellerowicz et al. 1992b).

It was found that a reconstruction of cell membranes occurred before dormancy. If IAA receptor or carrier proteins are present on the plasma membranes these proteins may tentatively be transported or decreased, thus inducing a low sensitivity for IAA (Catesson and Lachaud 1993).

When IAA was supplied to *Picea abies*, both the thickening period and the cell wall thickness increased (Denne and Wilson 1977). IAA is involved in enlargement of cells and promotion of cell division. However, it is suggested that in differentiation of cambial cells, a tracheid-differentiation factor (TDF) is essential, and that earlywood formation requires high concentrations of IAA and TDF, while latewood formation is induced by low concentration of IAA and high concentrations of TDF (Savidge and Wareing 1981).

6.1.2 Abscisic Acid

The formation of latewood of *Larix occidentalis* is induced by growth-inhibiting substances (Balatinecz and Kennedy 1968). The amount of these substances increases at the stage of latewood formation (Wodzicki and Wodzicki 1980). Thus, it has been suggested that abscisic acid (ABA), which is a dormancy-inducing substance, might be involved in the formation of latewood.

ABA has been found in the cambial zone of conifers (Little et al. 1978, Savidge and Wareing 1984, Funada et al. 1988). When exogenous ABA was administered to stems, it inhibited division and enlargement of the cambial cells (Little and Eidt 1968, Jenkins and Shepherd 1974). In addition, a high concentration of ABA was found in dormant shoots of *Pseudotsuga menziesii* (Webber et al. 1979). These results suggest that ABA acts as a growth-inhibiting substance and regulates cambial activity.

However, recent investigations on the concentrations of ABA in the cambial zones of *Picea sitchensis* (Little and Wareing 1981), *Pinus contorta* (Savidge and Wareing 1984), and *Pinus densiflora* (Funada et al. 1988) have shown that the concentration of ABA did not change seasonally and that it did not change when earlywood was transformed into latewood or at the termination of cambial activity. In addition, a negative result on growth inhibition by external addition of ABA has been reported (Pharis et al. 1981). Accordingly, any direct control of cambial activity by ABA seems doubtful.

6.1.3 Cytokinins

Cytokinins are substances promoting cell division, and many physiological effects of cytokinins are known. When cytokinins are added externally to tree stems, cambial activity is stimulated, and the stimulation is promoted by IAA (Philipson and Coutts 1980). Cytokinins are essential for tissue and cell cultures (Kuraishi 1988).

Native cytokinins are all adenine derivatives, and classified into free base, riboside, ribotide, and glucoside types. The free base and the riboside types are active cytokinins for biological assay (Matsubara 1980). The biosynthetic pathway of cytokinins is not fully elucidated, but it has been suggested that they are synthesized by the root apical meristem, and transported to the stem through vascular bundles. Several kinds of cytokinins were found in the roots of *Pseudotsuga menziesii* (Doumas et al. 1989), while in the cambial zones of *Abies balsamea* (Little et al. 1979) *trans*-zeatin, and *trans*-ribosylzeatin were identified. In *Cryptomeria japonica* (Funada et al. 1992a) and *Larix leptolepis* (Funada et al. 1992b) zeatin, *trans*-ribosylzeatin, *cis*-ribosylzeatin, isopentenyl-adenine, and isopentenyladenosine are present.

Analyses of endogenous cytokinins in trees have not been successful so far. The reason for this is that there are many cytokinin derivatives with different polarities, they are present in very low quantities, and their extraction and purification from plant materials is difficult.

Enzyme-immunoassay showed that the ribosylzeatin concentration in the cambial zone did not change significantly at the initiation stage of cambial activity, and that it was not correlated with resumption of cambial activity in *Picea abies* (Eklund 1991). Analysis of cytokinins in the cambial zone of *Larix leptolepis* by the isotope-dilution method also showed that the concentration of cytokinins was low in comparison with those of IAA and ABA, and that it did not change on reactivation of the cambium in the spring (Funada et al. 1992b).

6.1.4 Gibberellins

Gibberellins are synthesized via ent-kaurene, a diterpene, derived from mevalonic acid (Fig. 171). Gibberellins contain an ent-gibberellan skeleton (C_{20} gibberellin) or ent-norgibberellan skeleton (C_{19} gibberellin) and are numbered in the order of their identification from plants. More than 70 gibberellins have been isolated, and their glucosides have been identified. Among these gibberellins, A_1, A_3, A_4 and A_7 are highly active (Kuraishi 1988). Gibberellins promote elongation growth, floral bud formation and fructification, and are involved in dwarf expression.

In the shoots of *Picea abies*, gibberellins A_1, A_3, and A_9 were found (Oden et al. 1982, 1987). It is known that the concentration of endogenous gibberellins A_1, A_3, and A_9 increases coincidentally with the growth of *Picea sitchensis* shoots (Moritz et al. 1990).

When a small amount of gibberellin was added to a stem of *Pinus radiata*, trunk xylem formation and tracheid diameter increased (Pharis et al. 1981). However, in *Pinus densiflora* and *Picea sitchensis* administration of gibberellin alone was not effective, but when gibberellin was added together with IAA, xylem differentiation was promoted synergistically (Odani 1975, Denne and Wilson 1977).

In the cambial zone of *Pinus sylvestris* gibberellins A_4, A_7, and A_9 were identified, and when A_4 and A_7 were administered, xylem formation was promoted and the concentration of endogenous gibberellins and IAA increased (Wang et al. 1992). All these results suggest that the effect of gibberellins is expressed in coordination with IAA.

6.1.5 Ethylene

Ethylene inhibits elongation growth of stems but promotes thickening growth, defoliation, and fruit and leaf abscission (promotion of absciss layer formation). It has been suggested that ethylene is an information transmitter among plants (Kuraishi 1988).

When ethrel, an ethylene-forming substance, was added to stems of *Pinus radiata* (Barker 1979), *P. strobus* (Brown and Leopold 1973), *P. halepensis* and *P. densiflora*, diameter growth of the trunks increased (Yamamoto and Kozlowski 1987). This is ascribed to an increase of tracheid formation caused by ethylene. An increase of ethylene concentration in the cambial zone of *Picea abies* has been observed at the stage of active differentiation (Eklund 1991).

It is well known that ethylene formation is induced by IAA (Yang and Hoffman 1984). The concentration of ethylene changes with seasonal changes in the concentration of endogenous IAA, and the cooperation of both hormones has therefore been suggested for promotion of cambial activity.

It was further found that ethylene formation was promoted on the lower side of bent branches (Blake et al. 1980), and that 1-aminocyclopropane-1-carboxylic acid (ACC) was detected only at the lower side of such branches (Savidge et al. 1983). These results suggest that the increased concentration of ethylene is intimately related to gravitropism and compression wood formation.

6.1.6 Brassinosteroids

In 1979, Grove et al. identified a new plant hormone, brassinolide, from pollen of *Brassica campestris*. Subsequently, about 30 derivatives of brassinolide have been identified in plants. These compounds are generically named brassinosteroids. Brassinosteroids promote elongation and thickening growth. The physiological effect is expressed at remarkably low concentrations compared with other plant hormones (Kuraishi 1988).

In shoots of *Picea sitchensis*, castasterone, which is a precursor of brassinolide, and typhasterol were found (Yokota et al. 1985), while in the cambial zone of *Pinus sylvestris* castasterone and brassinolide are present (Kim et al. 1990).

It is known that brassinosteroids act on grasses synergistically with IAA (Katsumi 1985), but the role of brassinosteroids in regulation of cambial activity in conifers is not known. Synergism with IAA was observed when both hormones were added to grasses at the same time, or when IAA was added after treatment with brassinosteroid, but in treatment of grasses in the reverse order no synergism was found. These results suggest that brassinosteroids increase the susceptibility of cells to IAA. The same synergism is conceivable in regulation of cambial activity of trees.

These investigations suggest that plant hormones are involved in control of cambial activity of trees. The hormone may induce the expression of certain genes as noted in Section 5.1.2.1. although the mechanism of gene expression by hormones remains to be established.

6.2 Regulation of Heartwood Formation

6.2.1 Histological Characteristics of Heartwood

In 1923 Craib found a narrow band of pale wood around the heartwood of *Taxus* sp. and that the water content of the pale wood was lower than that of sapwood and heartwood. He suggested that the pale wood represented an initiation stage of heartwood formation. Since then similar regions have been found in many species and named intermediate or transition wood (Hillis 1968, Higuchi 1970, 1976; Fig. 7, Chap. 1).

It is known that when cambium is differentiated to sapwood, prosenchyma cells are dead, but ray and axial parenchyma cells survive in the sapwood. Frey-Wyssling and Bosshard (1959) found that living parenchyma cells of sapwood of several European conifers and hardwoods contain oval nuclei at the initial stage, but the shape of the nuclei changes to round in aged sapwood. The nuclei later lose chromatin and disappear. Further studies (Higuchi et al. 1964, 1967a,b,c, Fukazawa and Higuchi 1965, 1966) showed that the volume of nuclei, the amount of RNA, and the numbers of mitochondria, Golgi apparatus, endoplasmic reticulum, plastids and ribosomes decrease during the differentiation of cambium to sapwood. In contrast, the volume of vacuoles which store heartwood phenolics increases in accordance with the gradual decrease in physiological activity of the ray parenchyma cells.

In contrast, the histological observations of Chattaway (1952) showed that the physiological activity of intermediate wood increases considerably when tyloses are formed. Hugentobler (1965) found that the volume of nuclei and nucleoli in ray parenchyma cells of intermediate wood is largest in particular trees that form a very dark heartwood. He suggested that the increase in the volume of the nucleoli is related to an increase in RNA production.

These results suggest that the physiological activity of the organelles in ray parenchyma cells is remarkably different in the cambial zone and the sapwood including intermediate wood. This could be related to the formation of heartwood extractives.

6.2.2 Heartwood and Extractives

Various wood extractives such as terpenes, flavonoids, lignans, stilbenes, tropolones, among others, are found in heartwood (Hillis 1987). The content of polyphenols is generally higher in the heartwood of old trees than in that of young trees, and decreases with increasing stem height and from the periphery toward the central portion of the stem.

Fig. 193 shows the distributions of dihydroquercetin in wood disks obtained from different tree heights of *Pseudotsuga menziesii* and *Larix occiden-*

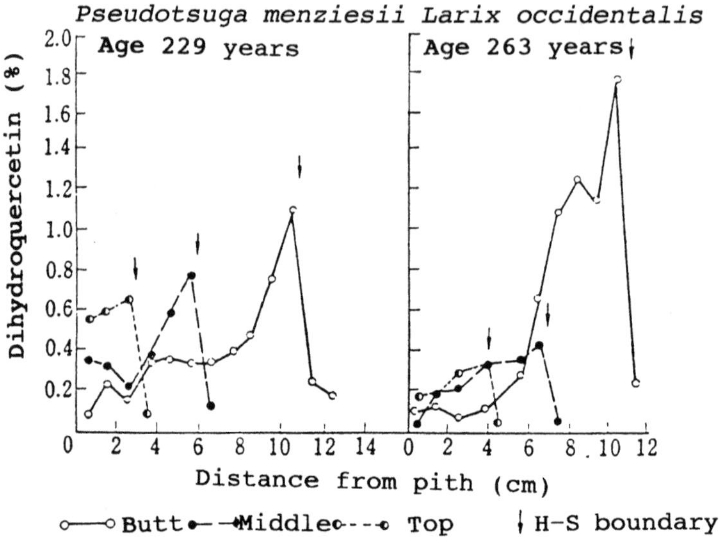

Fig. 193. Distribution of dihydroquercetin (taxifolin) in the wood disks obtained from different tree heights of *Pseudotsuga menziesii* (*left*), and *Larix occidentalis* (*right*). ○—○ Butt; •---• middle; ◐---◐ top; *arrow* heartwood-sapwood boundary. (Gardner and Barton 1960)

talis (Gardner and Barton 1960). A small amount of dihydroquercetin was found in the outer sapwood of these species. The amount increased remarkably at the boundary between the sapwood and heartwood, and in the outer heartwood but it decreased in the inner heartwood. The concentration of dihydroquercetin in the heartwood was highest at the base of the tree and decreased with increasing stem height. It also decreased from the periphery to the pith. This flavonoid was located only in the ray parenchyma.

Similar distribution patterns were found for arabinogalactan of *Larix* (Chochieva et al. 1959, Côté et al. 1966, Côté and Timell 1967a,b), for proanthocyanidins of *Eucalyptus* (Hillis 1966), *P. menziesii* (Hancock 1957) and *Acacia* (Hillis 1960, Roux and Evelyn 1960) and for resins of *Abies amabilis* (Isenberg 1936) and *Picea abies* (Engström and Back 1959).

The content of extractive was different in different specimens of the same species and in different parts of the stem. The heartwood of slow-growing large trees was generally darker and contained larger amounts of extractive than did young fast-growing trees (Hillis 1987). The heartwood of branches and roots of *Picea abies* contained more lignans than the stem heartwood (Ekman 1979).

Figure 194 shows the distribution of arabinogalactan from the center to the periphery of the heartwood of an old (aged 300 years) tree of *Larix occidentalis* (Hillis 1987). A reverse relation was found between the content of arabinogalactan and average width of the annual ring. The content of ara-

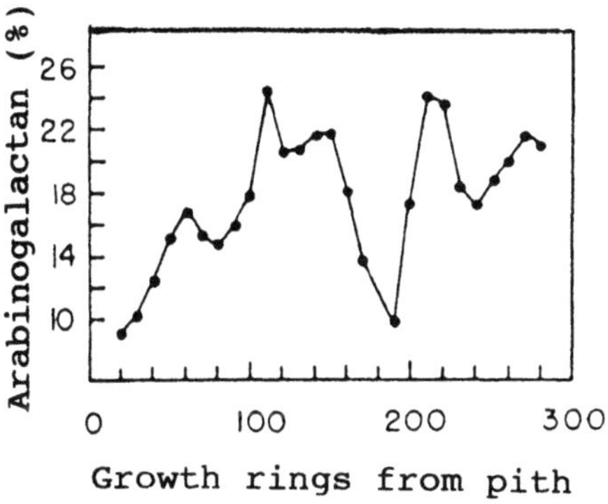

Fig. 194. Variation in arabinogalactan content with distance from the center of a *Larix occidentalis*, and width of the ten growth rings present in each of the sections used. (Côté et al. 1966)

binogalactan was only 0.3% in the sapwood but 24.5% in the heartwood. It was also found that a 26-year old fast-growing *L. decidua* contained 0.60% arabinogalactan in the sapwood and that the content increased to only 7.68% in the heartwood. The arabinogalactan occurs largely in the lumen of the heartwood tracheids (Côté et al. 1966, Côté and Timell 1967).

In *Pinus sylvestris* and *P. radiata*, triglycerides decreased rapidly and free fatty acids increased in the narrow intermediate wood (Pensar 1967, 1968, 1969). Saturated fatty acids were more common in the heartwood than unsaturated acids, while in the sapwood the content of unsaturated fatty acids was higher than that of saturated acids.

Various lignans and toropolones were present mostly at the inner 50–100 annual rings from the dark heartwood boundary of *Thuja plicata* (Swan et al. 1969, Swan and Jiang 1970), and these toropolones were not synthesized by very young trees.

Red heartwood of *Cryptomeria japonica* contains significant amounts of several norlignans such as sequirin C, agatharesinol, sugiresinol, hydroxysugiresinol, yateresinol and hinokiresinol, whereas black heartwood has fewer norlignans together with hot-water soluble dark-colored substances (M_r higher than 10 kDa) composed of phenolic components, arabinogalactan, and arabinoglucuronoxylan. Chromatographic, IR and NMR analyses suggested that the phenolic components of the dark-colored substances consisted of oxidatively polymerized norlignans (Hayashi et al. 1994) Extractives of *Cryptomeria japonica* attained a maximum concentration on the inner inside of three annual rings from the boundary of the heartwood (Nobuchi and Harada 1985).

6.2.3 Metabolism of Ray Parenchyma Cells

Generally, starch and fat are present in parenchyma cells of the sapwood. These are used with sugars (sucrose) transported via vascular bundles from the leaves, for wood and phloem growth and synthesis of heartwood extractives.

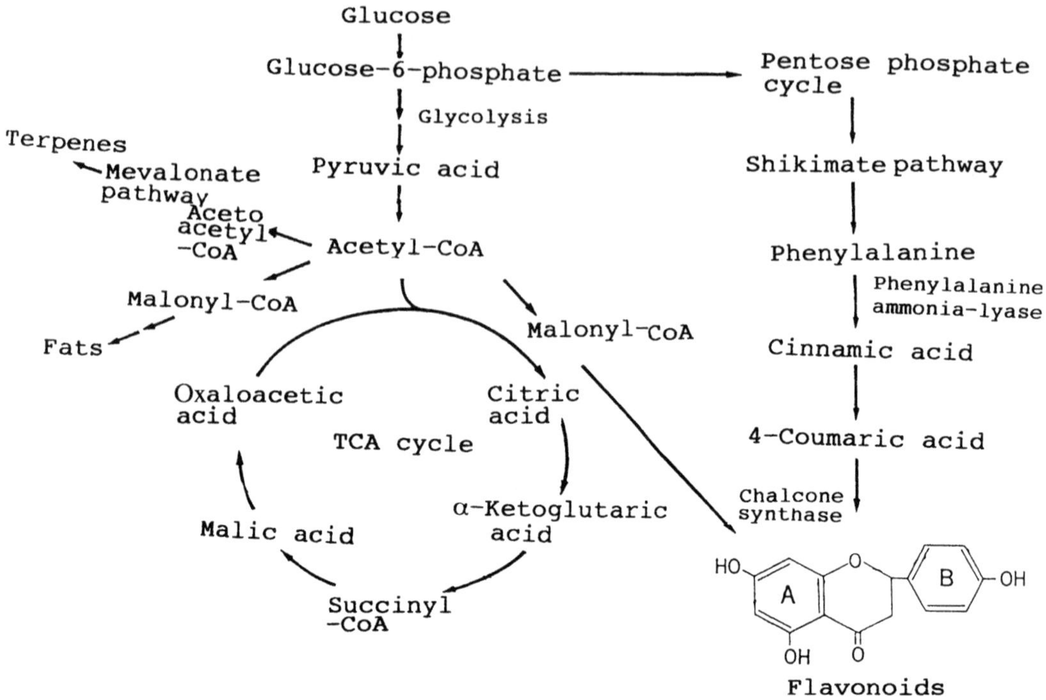

Fig. 195. Biosynthetic pathways of heartwood extractives

Sugars transported from leaves to ray parenchyma cells of intermediate wood could be converted to acetyl-CoA via the glycolysis–TCA cycle, and then used for the synthesis of A rings of flavonoids and stilbenes via the malonyl-CoA pathway, or terpenes via the mevalonic acid pathway. On the other hand, sugars could be converted to pentosephosphates via the pentose phosphate cycle, leading to B rings of flavonoids and stilbenes, and hydrolyzable tannins via the shikimate–cinnamate pathway. By cooperation of the biosynthetic reactions in these pathways, heartwood extractives could be synthesized in parenchyma cells of the intermediate wood (Hillis 1968, Higuchi 1970, 1992; Fig. 195).

Under what physiological conditions are the genes which encode the enzymes involved in the synthesis of heartwood extractives in ray parenchyma cells induced? How do these enzymes cooperate with the enzymes of primary metabolism for the synthesis of these extractives?

Xylem cells derived by the division of cambial cells initiate lignin synthesis during secondary cell wall formation as noted in Sections 4.3.4 and 5.1.4. Precursors of the main heartwood extractives, such as flavonoids, stilbenes, tannins and lignan are formed from cinnamic acid derived from L-phenylalanine (see Sect. 4.3).

It seems that during heartwood formation genes encoding new enzyme (isozyme) systems (such as chalcone synthase, terpenoid synthases, etc.) that are involved in the synthesis of heartwood extractives are induced in ray parenchyma cells. Elucidation of the metabolic pattern of ray parenchyma cells in intermediate woods may give a clue to this problem.

Previous investigations (Higuchi et al. 1967c) showed that the apparent metabolic activity per gram of wood sections generally decreased from the cambium to the intermediate wood. Respiration and activity of phenylalanine ammonia-lyase also decreased in this way. The results are in accordance with that nitrogen content of sections located between the cambium and the intermediate wood decreased considerably (Figs. 196a, 196b and 197).

a
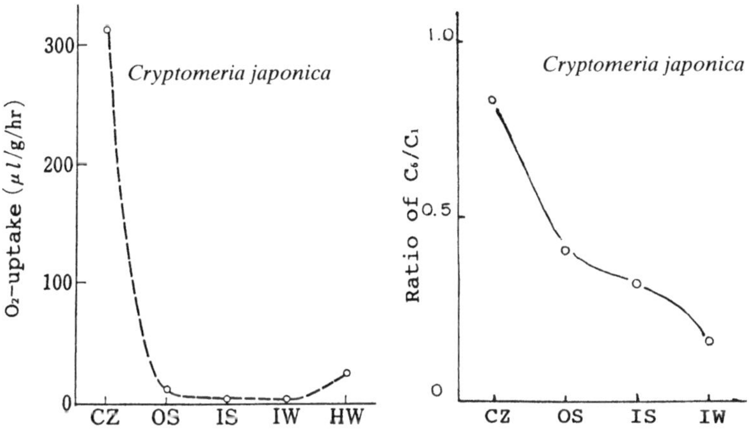

b

Fig. 196. **a** O_2 uptake (O_2 uptake by fresh wood, μl g^{-1} h^{-1}); **b** the pattern of respiratory breakdown of glucose-1-^{14}C and glucose-6-^{14}C degradation by *Cryptomeria japonica* wood. *CZ* Cambial zone; *OS* outer sapwood; *IS* inner sapwood; *IW* intermediate wood; *HW* heartwood. (Higuchi et al. 1967c)

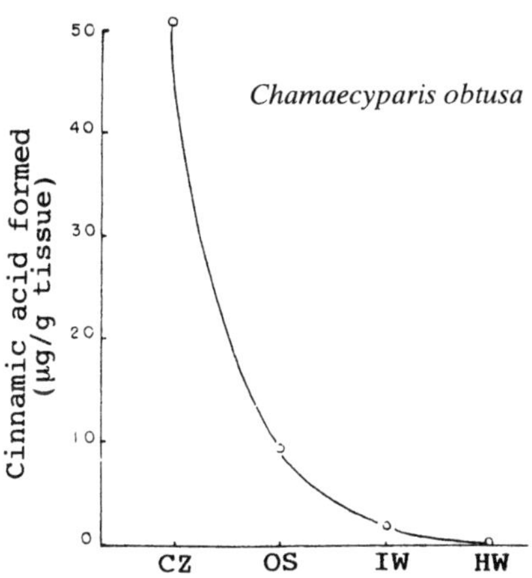

Fig. 197. Activity of phenylalanine ammonia-lyase in xylem tissues of *Cryptomeria japonica*. (Higuchi et al. 1967c)

However, Ziegler (1968) found that both protein nitrogen and amino acid nitrogen in the inner sapwood of *Pinus strobus*, *Larix decidua* and *Fagus sylvatica* significantly increased compared to the outer sapwood, and that water-soluble vitamins such as thiamine, riboflavine, nicotinamide, pyridoxine, biotin etc. also increased in concentration near the intermediate wood. They suggested that the increase of these vitamins, which are coenzymes of carboxylases, dehydrogenases, transaminases and fatty acid synthetases, is related to the increase of physiological activity when sapwood is transformed into intermediate wood. Furthermore, Magel and Höll (1993) found that starch, sugars, and adenine nucleotides in *Fagus sylvatica* wood were highest in the youngest sapwood, but negligible in the discolored wood. ATP/ADP ratios at the sapwood-discolored wood boundary were fairly high, showing an enhanced metabolic activity during the process of transformation of sapwood into discoloration wood. Shain and Mackay (1973) measured respiration and activities of malate dehydrogenase and glucose-6-phosphate dehydrogenase in cambium, sapwood, intermediate wood and heartwood, and found that both respiration and enzyme activities were higher in the intermediate wood than in the inner sapwood, especially in the winter.

In addition, Hillis (1975, 1987) found that when sapwood of *Rhus succdanea*, *Azadirachta indica* and *Samanea saman* was treated with 2-chloroethylphosphonic acid (an ethylene-forming substance), starch granules in the parenchyma cells of the treated sapwood disappeared, while lipid and polyphenols increased, and heartwood was formed. The activities of lipase, acid phosphatase, succinate dehydrogenase, ATPase, and peroxidase in axial and ray parenchyma cells also increased. Hillis (1975) also detected ethylene in the intermediate wood of *Eucalyptus tereticornus* and found that the ethylene formation was enhanced in autumn and winter. Ethylene formation in intermediate wood of *Pinus radiata* was also observed in the winter (Shain and Hillis 1973). On the other hand, Jorgensen (1961) found that pinosylvin, a heartwood phenol, was induced when sapwood of *Pinus radiata* was kept in an atmosphere of air, carbon dioxide and ethylene. From these results Hillis (1987) suggested that the enzymes involved in the formation of heartwood extractives are induced by ethylene in both conifers and hardwoods.

Recently Magel et al. (1991) investigated the seasonal activities of phenylalanine ammonia-lyase (PAL; Fig. 198), and chalcone synthase (CHS; Fig. 199) in xylem of *Robinia pseudoacacia*. They found that PAL was active in the youngest wood near the cambium in April and September but in intermediate wood in all seasons, while CHS was active only in the intermediate wood. From these results they suggested that PAL in the youngest wood is involved in lignin biosynthesis but in the intermediate wood in flavonoid synthesis, while CHS is only involved in flavonoid synthesis. In the intermediate wood high activities of PAL and CHS were found in November suggesting that the heartwood formation occurs in this season. In addition, specific flavonoids of this tree increased in intermediate wood in September and November. The

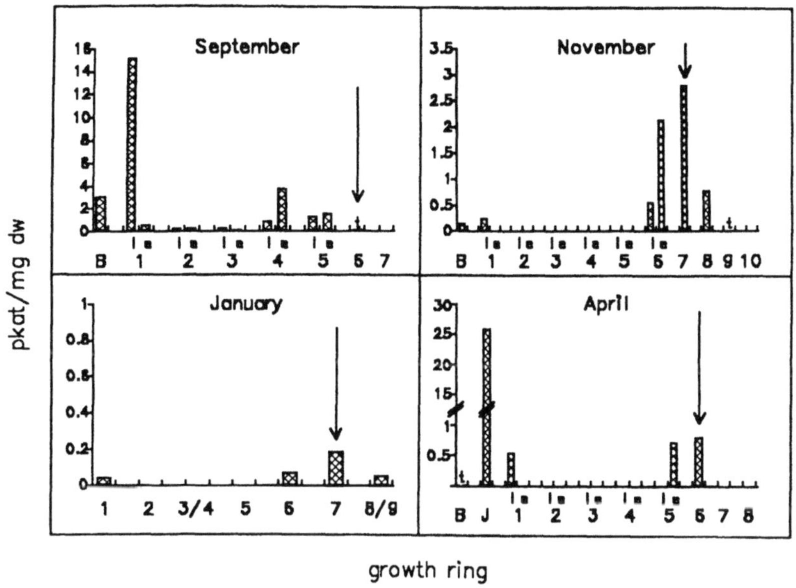

Fig. 198. PAL activity in individual annual growth ring tissues of the trunks of *Robinia pseudoacacia* L. cut at different times of the year (September, November, January, and April). *B* Bark; *1,2,3, . . . n* growth rings of increasing age. *1* Youngest, outermost growth ring; *3/4* or *8/9*, the tissues of two very small growth rings were not separated. *l* Latewood; *e* earlywood; *J* cell layers of the earlywood of the youngest growth ring which had been formed up to the harvest date; *t* trace amount of enzyme activity; *arrows* heartwood-sapwood border. (Magel et al. 1991)

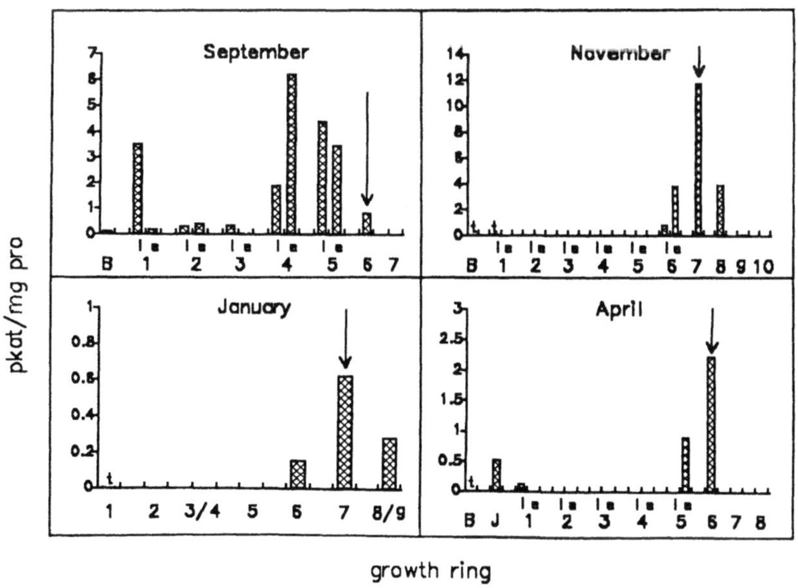

Fig. 199. Chalcone synthase activity in individual annual growth ring tissues of the trunks of *Robinia pseudoacacia* L. (Mager et al. 1991)

Sapwood ray parenchyma →	**Intermediate wood ray parenchyma** →	**Heartwood ray parenchyma**
Inducer: ethylene, wounding, etc.	Activation of genes encoding enzymes of general phenylpropanoids, metabolism, and synthesis of heartwood extractives such as PAL, 4-CL, CHS, CHI, etc.	Accumulation of heartwood extractives Autolysis

Fig. 200. The process of heartwood formation

activities of PAL and CHS were higher in earlywood than latewood. The results suggest that heartwood formation occurs in earlywood.

6.2.4 Gene Expression in Heartwood Formation

It has been shown that PAL, a key enzyme leading to phenylpropanoid metabolism (see Sects. 4.3.3, 4.3.4 and 4.3.6) is encoded by a small family of genes in bean (Cramer et al. 1989, Liang et al. 1989a), parsley (Lois et al. 1989), rice (Minami et al. 1989), tobacco and a poplar (Osakabe et al. 1994). In a conifer, PAL is encoded by a single gene, which is exceptional (Whetten and Sederoff 1991). The isoforms of PAL genes are expressed differently in different organs and in response to different environmental stimuli, which include wounding, infection, light, and other factors.

Chalcone synthase (CHS), which regulates the first step of flavonoid synthesis, is also encoded by a gene family comprising several members. This gene family is founds in such plants as soybean and chick pea (Barz and Roland 1992), *Petunia* (Gerats and Martin 1992), and a French bean (Ryder et al. 1987). The various isoforms are differently regulated by various physiological factors and environmental stimuli.

Enzymes involved in the general phenylpropanoid metabolism are mostly encoded by several genes or gene families. The expression of genes in the syntheses of heartwood extractives (flavonoids and anthocyanins etc.) should therefore be regulated by various biotic and abiotic factors.

The characterization of genes with promoters in flavonoid synthesis in intermediate wood should lead to better information regarding the mechanism of heartwood formation. The process and mechanism of heartwood formation can be summarized as shown in Fig. 200.

The concept of "physiological activation" in intermediate wood proposed by Chattaway (1952) could be understood as the physiological differentiation of sapwood ray parenchyma cells during heartwood formation: first, alteration of the metabolism of sapwood ray parenchyma cells by ethylene, such as induction and activation of PAL, 4-CL, CHS, CHI, the pentose phosphate cycle, malonyl-CoA pathways, synthesis of lipids and terpenes, and then activation of hydrolases and oxidases by destruction of vacuoles of ray parenchyma cells in the intermediate wood to give heartwood.

References

Abe H, Funada R, Ohtani J, Fukazawa K 1995 Changes in the arrangement of microtubules and microfibrils in differentiating conifer tracheids during the expansion of cells. Ann Bot 75:305–310

Adler E 1977 Lignin chemistry-past, present and future. Wood Sci Technol 11:169–218

Agarwal UP, Atalla RH 1986 In-situ Raman microprobe studies of plant cell walls: macromolecular organization and compositional variability in the secondary wall of Picea mariana (Mill.)B.S.P. Planta 169:325–332

Agarwal PK, Rastogi RP 1982 Two lignans from Cedrus deodara. Phytochemistry 21:1459–1461

Aguamah GE, Langcake P, Leworthy DP, Page JA, Pryce RJ, Strange RN 1981 Two novel stilbene phytoalexins from Arachis hypogaea. Phytochemistry 20:1381–1383

Albersheim P 1975 The walls of growing plants. Sci Am 232:80–95

Alberts B, Bray D, Lewis J, Raff M, Robert K, Watson JD 1994 Molecular biology of the cell, 3rd edn. Garland, New York

Amor Y, Haigler CH, Johnson S, Wainscott M, Delmer DP 1995 A membrane-associated form of sucrose synthase and its potential role in synthesis of cellulose and callose in plants. Proc Natl Acad Sci USA 92:9353–9357

Amor Y, Mayer R, Benziman M, Delmer DP 1991 Evidence for a cyclic diguanylic acid-dependent cellulose synthase in plants. Plant Cell 3:989–995

Amrhein N 1991 Inhibitors for pathways of biosynthesis in plants. Schweiz Lab Z 48:384–388

Amrhein N, Gödeke KH 1977 α-Amminoxy-β-phenylpropionic acid – a potent inhibitor of L-phenylalanine ammonia-lyase in vitro and in vivo. Plant Sci Lett 8:313–317

Aoyama M, Sakakibara A 1978 Isolation of two new dilignols from hydrolysis products of hardwood lignin. Mokuzai Gakkaishi 24:422–423

Archer BL, Ayrey G, Cockbain EG, McSweeney GP 1961 Incorporation of [1-^{14}C]-isopentenyl pyrophosphate into polyisoprene. Nature 189:663–664

Atalla RH, Agarwal UP 1985 Raman microprobe evidence for lignin orientation in native state of woody tissue. Science 227:636–638

Atalla RH, Vander Hart DL 1984 Native cellulose: a composite of two distinct crystalline forms. Science 223:283–285

Awano T, Takabe K, Fujita M, Saiki H, Hibino T, Shibata D 1994 Distribution of CAD in lignifying cells. Abstr 44th Annu Meet Jpn Wood Res Soc, p 178

Ayres DC 1969 Incorporation of L-[U-^{14}C]-β-phenylalanine into the lignan podophylotoxin. Tetrahedron Lett (11):883–886

Ayres DC, Loike JD 1990 Lignans, chemical, biological and clinical properties. Cambridge Univ Press, Cambridge, 402 pp

Baba K, Sone Y, Misaki A, Hayashi T 1994 Localization of xyloglucan in the macromolecular complex composed of xyloglucan and cellulose in pea stems. Plant Cell Physiol 35:439–444

Bach TJ, Boronat A, Caelles C, Ferrer A, Weber T, Wettstein A 1991 Aspects related to mevalonate biosynthesis in plants. Lipid 26:637–684

Bacic A, Delmer DP 1981 Stimulation of membrane-associated polysaccharide syntheses by a membrane potential in developing cotton fibers. Planta 152:346–351

Baïer M, Goldberg R, Catesson AM, Liberman M, Bouchemal N, Michon V, Hervé du Penhoat C 1994 Pectin changes in samples containing poplar cambium and inner bark in relation to the seasonal cycle. Planta 193:446–454

Bailey IW 1923 The cambium and its derivative tissues. IV. The increase in girth of the cambium. Am J Bot 10:499–509

Bailey RW, Hassid WZ 1966 Xylan synthesis from uridine-diphosphate-D-xylose by particulate preparations from corncobs. Proc Natl Acad Sci USA 56:1586–1593

Balatinecz JJ, Kennedy RW 1968 Mechanism of earlywood-latewood differentiation in Larix decidua. Tappi 51:414–422

Bamber RK 1976 Heartwood, its function and formation. Wood Sci Technol 10:1–8

Bao W, O'Malley DM, Sederoff RR 1992 Wood contains a cell-wall structural protein. Proc Natl Acad Sci USA 89:6604–6608

Bao W, O'Malley DM, Whetten R, Sederoff RR 1993 A laccase associated with lignification in loblolly pine xylem. Science 260:672–674

Barber GA, Elbein AD, Hassid WZ 1964 The synthesis of cellulose by enzyme systems from higher plants. J Biol Chem 239:4056–4061

Barker EJ 1979 Growth and wood properties of Pinus radiata in relation to applied ethylene. N Z J For Sci 9:15–19

Barnes RL 1962 Glutamine synthesis and translocation in pine. Plant Physiol 37:323–326

Barnes RL 1963a Organic nitrogen compounds in tree xylem sap. For Sci 9:98–102

Barnes RL 1963b Nitrogen transport in the xylem of trees. J For 61:50–51

Barz W, Roland W 1992 Biosynthesis and metabolism of isoflavones and pterocarpan phytoalexins in chickpea, soybean and phytopathogenic fungi. In: Stafford HA, Ibrahim RK (eds) Recent advances in phytochemistry, vol 26. Phenolic metabolism in plants. Plenum Press, New York, 139–164

Bate NJ, Orr J, Ni W, Meromi A, Nadler-Hassar T, Doerner PW, Dixon RA, Lamb CJ, Elkind Y 1994 Quantitative relationship between phenylalanine ammonia-lyase levels and phenylpropanoid accumulation in transgenic tobacco identifies a rate-determining step in natural product synthesis. Proc Natl Acad Sci USA 91:7608–7612

Baudracco S, Grima-Pettanati J, Boudet AM, Gahan PB 1993 Quantitative cytochemical localization of cinnamyl alcohol dehydrogenase activity in plant tissues. Phytochem Analysis 4:205–209

Becker-André M, Schulze-Lefert P, Hahlbrock K 1991 Structural comparison, modes of expression, and putative *cis*-acting elements of the two 4-coumarate: CoA ligase genes in potato. J Biol Chem 266:8551–8559

Beld MGHM, Martin C, Huits H, Stuitje AR, Gerats AGM 1989 Flavonoid synthesis in Petunia hybrida: partial characterisation of dihydroflavonol 4-reductase genes. Plant Mol Biol 13:491–502

Bentley R 1963 Biosynthesis of tropolones in Penicillium stipitatum. J Biol Chem 238:1895–1902

Bentley R 1990 The shikimate pathway. A metabolic tree with many branches. Crit Rev Biochem Mol Biol 25:307–384

Benverniste I, Lesot A, Hasenfratz M-P, Kochs G, Drust F 1991 Multiple forms of NADPH-cytochrome P450 reductase in higher plants. Biochem Biophys Res Commun 177:105–112

Billek G 1959 Zur Isolierung und Bestimmung aktiver Bruchstücke (Phenylpropan) des Lignins nach Aufname markierter Verbindungen. In: Kratzl K, Billek G (eds) Biochemistry of wood. Pergamon Press, London, 207–214

Billek G, Kindl H 1962 Phenolic content of the Saxifragaceae family. Monatsh Chem 93:85–98

Billek G, Ziegler W 1962 The biosynthesis of plant stilbenes III. The biosynthesis of pinosylvin monomethyl ether. Monatsh Chem 93:1430–1440

Birch AJ, Donovan FW 1953 Studies in relation to biosynthesis I. Some possible routes to derivatives of orcinol and phloroglucinol. Aust J Chem 6:360–368
Björkman A, Person B 1957 Studies on finely divided wood. Svensk Papperstidn 60:158–169
Blake TJ, Pharis RP, Reid DM 1980 Ethylene, giberellins, auxin, and the apical control of branch angle in a conifer, Cupressum arizonica. Planta 148:64–68
Boland MJ, Wong E 1975 Purification and kinetic properties of chalcone-flavanone isomerase from soy bean. Eur J Biochem 50:383–389
Bollard EG 1958 Nitrogenous compounds in tree xylem sap. In: Thiman KV (ed) The physiology of forest trees. Ronald Press, New York, 83–93
Bolwell GP, Dalessandro G, Northcote DH 1985 Decrease of polygalacturonic acid synthase during xylem differentiation in sycamore. Phytochemistry 24:699–702
Boudet AM, Lapierre C, Grima-Pettenati J 1995 Biochemistry and molecular biology of lignification. New Phytol 129:203–236
Brauns FE 1952 The chemistry of lignin. Academic Press, New York, 215–216
Brett C, Waldron K 1990 Physiology and biochemistry of plant cell walls. Unwin Hyman, London, 194pp
Britsch L, Heller W, Grisebach H 1981 Conversion of flavanone to flavone, dihydroflavonol and flavonol with an enzyme system from cell cultures of parsley. Z Naturforsch 36C:742–750
Broomhead AJ, Rahman MMA, Dewick PM, Jackson DE, Lucas JA 1991 Matairesinol as precursor of Podophyllum lignans. Phytochemistry 30:1489–1492
Brown KM, Leopold AC 1973 Ethylene and the regulation of growth in pine. Can J For Res 3:143–145
Brown RM Jr, Franke WW, Kleing H, Falk H, Sitte P 1970 Scale formation in chrysophycean algae I. Cellulose and noncellulosic wall components made by the Golgi apparatus. J Cell Biol 45:246–271
Brown RM Jr, Montezinos D 1976 Cellulose microfibrils: visualization of biosynthetic and orienting complexes in association with plasma membrane. Proc Natl Acad Sci USA 73:143–147
Brown RM Jr, Saxena IM, Kudlicka K 1996 Cellulose biosynthesis in higher plants. Trends Plant Sci 1:149–156
Brown SA 1961 Chemistry of lignification. Science 134:305–313
Brown SA, Neish AC 1955a Shikimic acid as a precursor in lignin biosynthesis. Nature 175:688–690
Brown SA, Neish AC 1955b Studies on lignin biosynthesis using isotopic carbon IV. Formation from some aromatic monomers. Can J Biochem Physiol 33:948–962
Brown SA, Neish AC 1956 Studies on lignin biosynthesis using isotopic carbon V. Comparative studies on different plant species. Can J Biochem Physiol 34:767–778
Brown SA, Neish AC 1959 Studies on lignin biosynthesis using isotopic carbon VIII. Isolation of radioactive hydrogenolysis products of lignins. J Am Chem Soc 81:2419–2424
Brucker W, Drehmann U 1958 Formation of phenolic carboxylic acid from ^{14}C-labeled substrates by Phycomyces. Arch Microbiol 30:396–408
Brummell DA, Maclachlan GA 1989 Formation and functions of xyloglucan and derivatives. Plant cell wall polymers, biogenesis and biodegradation. ACS Symp Ser 399:18–32
Bruno G, Sipilä J, Mäkelä T 1989 On the mechanism of formation of non-cyclic benzyl ethers during lignin biosynthesis. I. The reactivity of β-O-4 quinone methides with phenols and alcohols. Holzforschung 43:55–59
Bugos RC, Chiang VL, Campbell WH 1989 Isolation of *O*-methyltransferase associated with lignin biosynthesis in aspen. In: Tappi Proc, Int Symp Wood and Pulping Chemistry. Tappi Press, Atlanta, 345–347

Bugos RC, Chiang VL, Campbell WH 1991 cDNA cloning, sequence analysis and seasonal expression of lignin bi-specific caffeic acid/5-hydroxyferulic acid *O*-methyltransferase of aspen. Plant Mol Biol 17:1203–1215
Bugos RC, Chiang VL, Campbell WH 1992 Characterization of bi-specific caffeic acid/5-hydroxyferulic acid *O*-methyltransferase from aspen. Phytochemistry 31:1495–1498
Bureau TE, Brown RM Jr 1987 In vitro synthesis of cellulose II from a cytoplasmic membrane fraction of Acetobacter xylinum. Proc Natl Acad Sci USA 84:6985–6989
Callaghan T, Benziman M 1984 High rates of in vitro synthesis of 1,4-β-D-glucan in cell free preparations from Phaseolus aureus. Nature 311:165–167
Calvin M, Benson AA 1948 The pathway of carbon in photosynthesis. Science 107:476–479
Calvin M, Benson AA 1949 The pathway of carbon in photosynthesis IV. The identity and sequence of the intermediates in sucrose synthesis. Science 109:140–142
Camirand C, Maclachlan GA 1986 Biosynthesis of fucose-containing xyloglucan nonasaccharide by pea microsomal membranes. Plant Physiol 82:379–383
Camm EL, Towers GHN 1973 Phenylalanine ammonia lyase. Phytochemistry 12:961–973
Carpita NC, Delmer DP 1981 Concentration and metabolic turnover of UDP-glucose in developing cotton fibers. J Biol Chem 256:308–331
Carpita NC, Gibeaut M 1993 Structural models of primary cell walls in flowering plants: consistency of molecular structure with the physical properties of the walls during growth. Plant J 3:1–30
Catesson AM 1994 Cambial ultrastructure and biochemistry: changes in relation to vascular tissue differentiation and the seasonal cycle. Int J Plant Sci 155:251–261
Catesson AM, Funada R, Robert-Baby D, Quinet-Szély M, Chu-Bâ J 1994 Biochemical and cytochemical cell wall changes across the cambial zone. IAWA J 15:91–101
Catesson AM, Lachaud S 1993 Le cambium, structure, fonctionnement et contrôle de l'activité saisonnière. Acta Bot Gallica 140:337–350
Chapple C 1994 Genetic characterization of secondary metabolism in Arabidopsis. In: Ellis BE, Kuroki GW, Stafford HA (eds) Genetic engineering of plant secondary metabolism. Plenum Press, New York, 251–273
Chapple C, Vogt T, Ellis BE, Somerville CR 1992 An Arabidopsis mutant defective in the general phenylpropanoid pathway. Plant Cell 4:1413–1424
Chattaway MM 1952 The sapwood-heartwood transformation. Aust For 16:25–34
Chiang VL, Podia GK, Tsai CJ, Kumar AS, Yu J 1994 Regulation of bi-functional *O*-methyltransferase gene expression in transgenic quaking aspen (Populus tremuloides). Proc Int Wood Biotech Symp, Tokyo, 15–22
Cho NS, Lee LY, Meshitsuka G, Nakano J 1980 On the characteristics of hardwood compound middle lamella lignin. Mokuzai Gakkaishi 26:527–533
Chochieva MM, Tsvetaeva IP, Yur'eva MK, Zaitseva AF, Petropavlovskii GA, Nikitin NI 1959 Distribution of arabinogalactan in Larix dahurica. Chem Abstr 53:4439
Chopin J, Hauteville M, Joshi BS, Gaward DH 1978 A novel example of a natural 2,5-dihydroxyflavanone from Unona lawii. Phytochemistry 17:332–334
Chourey PS 1981 Genetic control of sucrose synthetase in maize endosperm. Mol Gen Genet 184:372–376
Chourey PS, Latham MD, Still PE 1986 Expression of two sucrose synthase genes in endosperm and seedling cells of maize: evidence of tissue-specific polymerization of protomers. Mol Gen Genet 203:251–255
Chu A, Dinkova A, Davin LB, Bedgar DL, Lewis NG 1993 Stereospecificity of (+)-pinoresinol and (+)-lariciresinol reductases from Forsythia intermedia. J Biol Chem 268:27026–27033
Clegg MT 1993 Chloroplast gene sequences and the study of plant evolution. Proc Natl Acad Sci USA 90:363–367

Collazo P, Montoliu L, Puigdomenech P, Rigau J 1992 Structure and expression of the lignin *O*-methyltransferase gene from Zea mays L. Plant Mol Biol 20:857–867

Conn EE (ed) 1986 The shikimic acid pathway. Plenum Press, New York, 356pp

Conn EE, Stumpf PK 1972 Outlines of biochemistry, 3rd edn. Translated into Japanese by Tamiya N and Yagi T, Tokyo Kagaku Dojin, Tokyo, 226–312

Conn EE, Swain T 1961 Biosynthesis of gallic acid in higher plants. Chem Ind:592–593

Connelly JA, Conn EE 1986 Tyrosine biosynthesis in Sorghum bicolor: isolation and regulatory properties of arogenate dehydrogenase. Z Naturforsch 41C:69–78

Cornthwait DC, Haslam E 1965 Gallotannins part IX. The biosynthesis of gallic acid in Rhus typhina. J Chem Soc:3008–3011

Côté WA Jr 1967 Wood ultrastructure – an atlas of electron micrographs. Univ Washington Press, Seattle, 62pp

Côté WA Jr, Day AC, Simson BW, Timell TE 1966 Studies on larch arabinogalactan I. The distribution of arabinogalactan in larch wood. Holzforschung 20:178–192

Côté WA Jr, Kutscha NP, Simson BW, Timell TE 1968 Studies on compression wood VI. Distribution of polysaccharides in the cell wall of tracheids from compression wood of balsam fir (Abies balsamea (L.) Mill.). Tappi 51:33–40

Côté WA Jr, Timell TE 1967a Studies on larch arabinogalactan in larch wood. Holzforschung 20:178–192

Côté WA Jr, Timell TE 1967b Studies on larch arabinogalactan III. Distribution of arabinogalactan in tamarack. Tappi 50:285–289

Cottignies A 1983 Dormance totale et distribution normale des teneurs en ADN dans le point végétatif du frêne. CR Acad Sci Ser III Sci Vie 296:829–832

Cottle W, Kolattukudy PE 1982 Biosynthesis, deposition, and partial characterizarion of potato suberin phenolics. Plant Physiol 69:393–399

Craib WG 1923 Regional spread of moisture in the wood of trees III. Notes R Bot Gardon Edinb XIV(LXVI) (Jan), 1–11

Cramer CL, Edwards K, Dron M, Liang X, Dildine SL, Bolwell GP, Dixon RA, Lamb CJ, Schuch W 1989 Phenylalanine ammonia-lyase gene organization and structure. Plant Mol Biol 12:367–383

Cronshaw J 1965 Cytoplasmic fine structure and cell wall development in differentiating xylem elements. In: Côté WA Jr (ed) Cellular ultrastructure of woody plants. Syracuse Univ Press, Syracuse, 99–124

Croteau R, Johnson MA 1985 Biosynthesis of terpenoid wood extractives. In: Higuchi T (ed) Biosynthesis and biodegradation of wood components. Academic Press, Orlando, 379–439

Czaninski Y, Sachot RM, Catesson AM 1993 Cytochemical localization of hydrogen peroxide in lignifying cell walls. Ann Bot 72:547–550

Czichi V, Kindl H 1977 Phenylalanine ammonia-lyase and cinnamic acid hydroxylase as assembled consecutive enzymes on microsomal membranes of cucumber cotyledons: cooperation and subcellular distribution. Planta 134:133–143

Dalessandro G, Northcote DH 1977a Possible control site of polysaccharides synthesis during growth and wall expansion of pea seedlings (Pisum sativum L.). Planta 134:39–44

Dalessandro G, Northcote DH 1977b Changes in enzymic activity of nucleoside diphosphate sugar interconversions during differentiation of cambium to xylem in sycamore and poplar. Biochem J 162:267–279

Dalessandro G, Northcote DH 1981 Xylan synthetase activity in differentiated xylem cells of sycamore trees (Acer pseudoplatanus). Planta 151:53–60

Davin LB, Bedgar DL, Katayama T, Lewis NG 1992 On the stereospecific synthesis of (+)-pinoresinol in Forsythia suspensa from its achiral precursor, coniferyl alcohol. Phytochemistry 31:3869–3874

Davin LB, Lewis NG 1992 Phenylpropanoid metabolism: Biosynthesis of monolignols, lignans and neolignans, lignins, and suberins. In: Stafford HA, Ibrahim RK (eds) Recent advances in phytochemistry. Phenolic metabolism in plants. Plenum Press, New York, 325–375

Davin LB, Umezawa T, Lewis NG 1991 Enantioselective separations in phytochemistry. In: Fischer NH, Isman MB, Stafford HA (eds) Recent advances in Phytochemistry 25. Modern phytochemical methods. Plenum Press, New York, 75–112

Davis BD 1955 Intermediates in amino acid biosynthesis. Adv Enzymol 16:247–312

Dean JFD, Eriksson K-EL 1994 Laccase and the deposition of lignin in vascular plants. Holzforschung 48:21–33

De-Eknamkul W, Ellis BE 1988 Purification and characterization of prephenate aminotransferase from Anchusa officinalis cell cultures. Arch Biochem Biophys 267:87–94

Della-Cioppa G, Bauer C, Klein BK, Sha DM, Fraley RT, Kishore GM 1986 Translocation of the precursor of 5-enolpyruvylshikimate 3-phosphate synthase into chloroplast of higher plants in vitro. Proc Natl Acad Sci USA 83:6873–6877

Delmer DP 1972 The purification and properties of sucrose synthetase from etiolated Phaseolus aureus seedlings. J Biol Chem 247:3822–3828

Delmer DP 1983 Biosynthesis of cellulose. Adv Carbohyd Chem Biochem 41:105–153

Delmer DP 1987 Cellulose biosynthesis. Annu Rev Plant Physiol 38:259–290

Delmer DP, Amor Y 1995 Cellulose biosynthesis. Plant Cell 7:987–1000

Delmer DP, Thelen M, Marsden MPF 1984 Regulatory mechanisms for the synthesis of β-glucans in plants. In: Dugger WM, Bartnicki-Garcia S (eds) Structure, function, and biosynthesis of plant cell walls. Am Soc Plant Physiol, Rockville, Maryland, 133–149

Delmer DP, Solomon M, Read SM 1991 Direct photolabelling with [^{32}P]UDP-glucose for identification of a cotton fiber callose synthase. Plant Physiol 95:556–563

Denne MP 1979 Wood structure and production within the trunk and branches of Picea sitchensis in relation to canopy formation. Can J For Res 9:406–427

Denne MP, Wilson JE 1977 Some quantitative effects of indole-acetic acid on the wood production and tracheid dimensions of Picea. Planta 134:223–228

Dennis DT, Turpin DH 1992 Plant physiology, biochemistry and molecular biology. Longman, Singapore, 529 pp

Denzel K, Gross GG 1991 Biosynthesis of gallotannins. Enzymatic disproportionation of 1,6-digalloylglucose to 1,2,6-trigalloylglucose and 6-galloylglucose by an acyltransferase from leaves of Rhus typhina L. Planta 184:285–289

De Silva J, Jarman CD, Arrowsmith DA, Stronach MS, Chengappa S, Sidebottom C, Rid JSG 1993 Molecular characterization of a xyloglucan-specific endo 1,4-β-D-glucanase (xyloglucan endo-transglycosylase) from nasturtium seeds. Plant J 3:701–711

De Vlaming P, Kho KFF 1976 4,2′,4′,6′-Tetrahydroxychalcone in pollen of Petunia hybrida. Phytochemistry 15:348–349

Dev Sukh 1989 Terpenoids. In: Rowe JW (ed) Products of woody plants II. Springer, Berlin Heidelberg New York, 691–807

Dewick PM 1989 Biosynthesis of lignans. In: Atta-ur-Rahman (ed) Studies on natural products chemistry. 5. Structure elucidation (Part B). Elsevier, Amsterdam, 459–503

Dewick PM, Haslam E 1969 Phenol biosynthesis in higher plants. Gallic acid. Biochem J 113:537–542

Dewick PM, Martin M 1979a Biosynthesis of pterocarpan and isoflavan phytoalexins in Medicago sativa: the biochemical interconversion of pterocarpans and 2′-hydroxyisoflavans. Phytochemistry 18:591–596

Dewick PM, Martin M 1979b Biosynthesis of pterocarpan, isoflavan and coumestan metabolites of Medicago sativa: chalcone, isoflavone and isoflavanone precursors. Phytochemistry 18:597–602

Dhugga KS, Gallagher SR, Ray PM 1991 Plant polypeptides reversibly glycosylated by UDP-glucose: possible components of Golgi β-glucan synthase in pea cells. J Biol Chem 266:21977–21984

Dhugga KS, Ray PM 1991a Isoelectric focusing of plant plasma membrane proteins. Further evidence that a 55 kDa polypeptide is associated with β-1,3-glucan synthase activity from pea. Plant Physiol 95:1302–1305

Dhugga KS, Ray PM 1991b A 55 kDa plasma membrane-associated polypeptide is involved in β-1,3-glucan synthase activity in pea tissue. FEBS Lett 278:283–286

Diesperger H, Sandermann H Jr 1978 Evidence for an electrophilic intermediate in the microsomal hydroxylation of cinnamic acid in plants. FEBS Lett 85:333–336

Dixon RA, Choudhary AD, Dalkin K, Edwards R, Fahrendolf T, Gowri G, Harrison MJ, Lamb CJ, Loake GL, Maxwell CA, Orr J, Paiva NL 1992 Molecular biology of stress-induced phenylpropanoid and isoflavonoid biosynthesis in alfalfa. In: Stafford HA, Ibrahim RK (eds) Recent advances in phytochemistry. Phenolic metabolism in Plants. Plenum Press, New York, 91–138

Donaldson LA 1991 Seasonal changes in lignin distribution during tracheid development in Pinus radiata D. Don. Wood Sci Technol 25:15–24

Doodeman M, Tabak AJT, Schram A, Bennink GHI 1982 Hydroxylation of cinnamic acids and flavonoids during biosynthesis of anthocyanins in Petunia hybrida. Planta 154:546–549

Dooner HK, Nelson OE 1977 Genetic control of UDP-glucose: flavonol 3-*O*-glucosyltransferase in the endosperm of maize. Biochem Genet 15:509–519

Dore WH 1919 The proximate analysis of wood. Ind Eng Chem II:556–563

Douglas CJ, Hauffe KD, Ites-Morales M-E, Ellard M, Paszkowski U, Hahlbrock K, Dangl JL 1991 Exonic sequences are required for elicitor and light activation of a plant defence gene, but promoter sequences are sufficient for tissue specific expression. EMBO J 10:1767–1775

Douglas CJ, Hoffmann H, Schulz W, Hahlbrock K 1987 Structure and elicitor or U.V. light-stimulated expression of two 4-coumarate: CoA ligase genes in parsley. EMBO J 6:1189–1195

Doumas P, Bonnet-Masimbert M, Zaerr JB 1989 Evidence of cytokinin bases, ribosides and glucosides in roots of Douglas fir, Pseudotsuga menziesii. Tree Physiol 5:63–72

Driouich A, Lainé AC, Vian B, Faye L 1992 Characterization and localization of laccase forms in stem and cell cultures of sycamore. Plant J 2:13–24

Dugger WM, Bartinicki-Garcia S (eds) 1984 Structure, function and biosynthesis of plant cell walls. Proc 7th Annu Symp Bot, Univ California, Riverside. Am Soc Plant Physiol, Rockville, Maryland, 507 pp

Du Luca V, Ibrahim RK 1985 Enzymatic synthesis of polymethylated flavonols in Chrrsosplenium americanum 1. Partial purification and some properties of *S*-adenosyl-*L*-methionine: flavonol 3-,6-,7-, and 4′-*O*-methyltransferases. Arch Biochem Biophys 238:596–605

Dumas B, Van Doorsselaere J, Gielen J, Legrand M, Fritig B, Van Montagu M, Inzé D 1992 Nucleotide sequence of a complementary DNA encoding *O*-methyltransferase from poplar. Plant Physiol 98:796–797

Dyer WE, Henstrand JM, Handa AK, Herrmann KM 1989 Wounding induces the first enzyme of the shikimate pathway in Solanaceae. Proc Natl Acad Sci USA 86:7370–7373

Dyer WE, Weaver LM, Zhao J, Kuhn DN, Weller SC, Herrmann KM 1990 A cDNA encoding 3-deoxy-D-arabino-heptulosonate-7-phosphate synthase from Solanum tuberosum L. J Biol Chem 265:1608–1614

Ebel J, Grisebach H 1973 Reduction of cinnamic acids to cinnamyl alcohols with an enzyme preparation from cell suspension cultures of soy bean (Glycine max.). FEBS Lett 30:141–143

Ebel J, Hahlbrock K 1982 Biosynthesis. In: Harborne JB, Mabry TJ (eds) The flavonoids: advances in research. Chapman and Hall, London, 641–679

Eijkman JF 1885 Sur les principes constituants de l'Illicium religiosum (sieb.) (Shikimi-no-Ki en Japonais). Rec Trav Chim Pays-Bas 4:32–54

Eklund L 1991 Hormone levels in the cambial region of intact Picea abies during the onset of cambial activity. Physiol Plant 82:385–388

Ekman R 1979 Distribution of lignans in Norway spruce. Acta Acad Åbo 39:B(3)6

El-Basyouni SZ, Neish AC, Towers GHN 1964 The phenolic acids in wheat III. Insoluble derivatives of phenolic cinnamic acids as natural intermediates in lignin biosynthesis. Phytochemistry 3:627–639

El-Basyouni SZ, Chen D, Ibrahim RK, Neish AC, Towers GHN 1964 The biosynthesis of hydroxybenzoic acids in higher plants. Phytochemistry 3:485–492

Elbein AD 1969 Biosynthesis of a cell wall glucomannan in mung bean seedlings. J Biol Chem 244:1608–1616

Elkind Y, Edwards R, Mavandad M, Hedrick SA, Ribak O, Dixon RA, Lamb CJ 1990 Abnormal plant development and down-regulation of phenylpropanoid biosynthesis in transgenic tobacco containing a heterologous phenylalanine ammonia-lyase gene. Proc Natl Acad Sci USA 87:9057–9061

Elstner EF, Heupel A 1976 Formation of hydrogen peroxide by isolated cell walls from horseradish (Aromoracia lapathifolia Gilib.). Planta 130:175–180

Engelsma G, Van Bruggen JMH 1971 Ethylene production and enzyme induction in excised plant tissues. Plant Physiol 41:94–96

Engström P, Back E 1959 The proportion of canal resin in Picea abies pulpwood. The resin in parenchymatous cells and resin canals of conifers VIII. Svensk Papperstidn 62:545–553

Eom TJ, Meshitsuka G, Nakano J 1987 Chemical characteristics of lignin in the differentiating xylem of a hardwood II. Mokuzai Gakkaishi 33:576–581

Erdtman H 1933 Dehydrierungen in der Coniferylreihe. (I). Dehydrodieugenol und dehydrodiisoeugenol. Biochem Z 258:172–180

Erdtman H 1959 Conifer chemistry and taxonomy of conifers. In: Kratzl K, Billek G (eds) Biochemistry of wood. Proc 4th Int Congr Biochem. Pergamon Press, London, 1–27

Eriksson I, Libdrandt O, Westermark U 1988 Lignin distribution in birch (Betula verrucosa) as determined by mercurization with SEM- and TEM-EDXA. Wood Sci Technol 22:251–257

Facchini PJ, Chappell J 1992 Gene family for an elicitor-induced sesquiterpene cyclase in tobacco. Proc Natl Acad Sci USA 89:11088–11092

Falconer MM, Seagull RW 1985 Xylogenesis in tissue culture: Taxol effects on microtubule reorientation and lateral association in differentiation cells. Protoplasma 128:157–166

Feldwisch J, Zettl R, Hesse F, Schell J, Palme K 1992 An auxin-binding protein is localized to the plasma membrane of maize coleoptile cells: identification by photoaffinity labeling and purification of a 23-kDa polypeptide. Proc Natl Acad Sci USA 89:475–479

Fergus BJ, Goring DAI 1970 The distribution of lignin in birchwood as determined by ultraviolet microscopy. Holzforschung 24:118–124

Fergus BJ, Procter AR, Scott JAN, Goring DAI 1969 The distribution of lignin in spruce wood determined by ultraviolet microscopy. Wood Sci Technol 3:117–138

Feuillet C, Boudet AM, Grima-Pettenati J 1993 Nucleotide sequence of a cDNA encoding cinnamyl alcohol dehydrogenase from Eucalyptus. Plant Physiol 103:1447

Fickenscher K, Scheibe R 1986 Purification and properties of the cytoplasmic glucose-6-phosphate dehydrogenase from pea leaves. Arch Biochem Biophys 247:393–402

Fillatti JJ, Sellmer J, McCown B, Haissig B, Comai L 1987 Agrobacterium mediated transformation and regeneration of Populus. Mol Gen Genet 206:192–199

Finegold AA, Johnson DI, Farnsworth CC, Gelb MH, Judd SR, Glomset JA, Tamanoi F 1991 Protein geranylgeranyl transferase of Saccharomyces cerevisiae is specific for Cys-Xaa-Xaa-Leu motif proteins and requires the *CDC43* gene product but not the *DPR1* gene product. Proc Natl Acad Sci USA 88:4448–4452

Fink J, Jeblick W, Kass H 1990 Partial purification and immunological characterization of 1,3-β glucan synthase from suspension cell of Glycine max. Planta 181:343–348

Finkle BJ, Masri MS 1964 Methylation of polyhydroxyaromatic compounds by pampas grass *O*-methyltransferase. Biochim Biophys Acta 85:167–169

Finkle BJ, Nelson RF 1963 Enzymic reactions with phenolic compounds: a meta-*O*-methyltransferase in plants. Biochim Biophys Acta 76:747–749

Fliegmann J, Schröder G, Schanz S, Britsch L, Schröder J 1992 Molecular analysis of chalcone and dihydropinosylvin synthase from Scots pine (Pinus sylvestris), and differential regulation of these and related enzyme activities in stressed plants. Plant Mol Biol 18:489–503

Forkmann G 1977 Precursors and genetic control of anthocyanin synthesis in Matthiola incana R. Br. Planta 137:159–163

Forkmann G 1980 The B-ring hydroxylation pattern of intermediates of anthocyanin synthesis in pelargonidin- and cyanidin-producing lines of Matthiola incana. Planta 148:157–161

Forkmann G, Dangelmayer B 1980 Genetic control of chalcone isomerase activity in flowers of Dianthus caryophyllus. Biochem Genet 18:519–527

Forkmann G, Heller W, Grisebach H 1980 Anthocyanin biosynthesis in flowers of Matthiola incana. Flavanone 3- and flavonoid 3′-hydroxylases. Z Naturforsch 35C:691–695

Franck B, Stange A 1981 Nachweis einer biocyclischen Zwischenstufe der Anthrachinon-Biosynthese. Liebigs Ann Chem (12):2106–2116

Freudenberg K 1959 Biochemische Vorgänge bei der Holzbildung. In: Kratzl K, Billek G (eds) Biochemistry of wood. Pergamon Press, London, 121–136

Freudenberg K 1965 Lignin: its constitution and formation from *p*-hydroxycinnamyl alcohols. Science 148:595–600

Freudenberg K 1968 The constitution and biosynthesis of Lignin. In: Freudenberg K, Neish AC (eds) Constitution and biosynthesis of lignin. Molecular biology, biochemistry and biophysics. Springer, Berlin Heidelberg New York, 47–122

Freudenberg K, Grion G 1959 Beitrag zum Bildungsmechanismus des Lignins und der Lignin-Kohlenhydratbindung. Chem Ber 92:1355–1363

Freudenberg K, Harkin JM 1963 The glucosides of cambial sap of spruce. Phytochemistry 2:189–219

Freudenberg K, Kraft R, Heimberger W 1951 Über den Sinapin-alkohol, den Coniferylalkohol und ihre Dehydrierungspolymerisate. Chem Ber 84:472–476

Freudenberg K, Niedercorn F 1958 Anwendung radioaktiver Isotope bei der Erforschung des Lignins VIII. Umwandlung des Phenylalanins in Coniferin und Fichtenlignin. Chem Ber 91:591–597

Frey-Wyssling A 1954 The fine structure of cellulose microfibrils. Science 119:80–82

Frey-Wyssling A, Bosshard HH 1959 Cytology of the ray cells in sapwood and heartwood. Holzforschung 13:129–137

Fritsch H, Grisebach H 1975 Biosynthesis of cyanidin in cell cultures of Haplopappas gracilis. Phytochemistry 14:2437–2442

Fritzemeier K-H, Kindl H 1981 Coordinate induction by UV light of stilbene synthase, phenylalanine ammonia-lyase and cinnamate-4-hydroxylase in leaves of Vitaceae. Planta 151:48–52

Fritzemeier K-H, Kindl H 1983 9,10-Dehydrophenanthrenes as phytoalexins of Orchidaceae. Biosynthetic studies in vitro and in vivo proving the route from L-phenylalanine to dihydro-*m*-cinnamic acid, dihydrostilbene and dihydrophenanthrens. Eur J Biochem 133:543–550

Frost DJ, Read SM, Drake RR, Haley BE, Wasserman BP 1990 Identification of UDP-glucose-binding polypeptide of callose synthase from Beta vulgaris L. by photoaffinity labeling with 5-azido-UDP-glucose. J Biol Chem 265:2162–2167

Fry SC, Northcote DH 1983 Sugar-nucleotide precursors of arabinopyranosyl, arabinofuranosyl, and xylopyranosyl residues in spinach polysaccharides. Plant Physiol 73:1055–1061

Fry SC, Smith RC, Renwick KF, Martin DJ, Hodge SK, Matthews KJ 1992 Xyloglucan endotransglycosylase, a new wall-loosening enzyme activity from plants. Biochem J 282:821–828

Fujimoto H, Higuchi T 1977 Biosynthesis of liriodendrin by Liriodendron tulipifera. Wood Res (Kyoto) 62:1–10

Fujita M, Harada H 1978 Autoradiographic investigations of cell wall development I. Tritiated glucose assimilation in relation to cellulose and hemicellulose deposition. Mokuzai Gakkaishi 24:435–440

Fujita M, Saiki H, Harada H 1978 The secondary wall formation of compression wood tracheids II. Cell wall thickening and lignification. Mokuzai Gakkaishi 24:158–163

Fujita M, Saiki H, Harada H 1983 Deposition of cellulose, hemicelluloses and lignin in the differentiating tracheids. Proc 1983 Int Symp Wood Pulping Chem 1:14–19

Fukazawa K 1974 The distribution of lignin in compression- and lateral-wood of Abies sachalinensis using ultraviolet microscopy. Res Bull Coll Exp For Hokkaido Univ 31:87–114

Fukazawa K, Higuchi T 1965 Studies on the mechanism of heartwood formation II. Some observations on nucleus and DNA content in the ray parenchyma cell. Mokuzai Gakkaishi 11:196–201

Fukazawa K, Higuchi T 1966 Studies on the mechanism of heartwood formation IV. RNA content in the ray parenchyma cell. Mokuzai Gakkaishi 12:221–225

Fukuda H 1989a Cytodifferentiation in isolated single cells. Bot Mag Tokyo 192:491–501

Fukuda H 1989b Regulation of tubulin degradation in isolated Zinnia mesophyll cells in culture. Plant Cell Physiol 30:243–252

Fukuda H 1992 Tracheary element differentiation as a model system of cell differentiation. Int Rev Cytol 136:289–332

Fukuda H 1994a Differentiation of xylem and phloem. In: Higuchi T (ed) Wood molecular biology. Buneido Publ, Tokyo, 56–70

Fukuda H 1994b Redifferentiation of single mesophyll cells into tracheary elements. Int J Plant Sci 155:262–271

Fukuda H 1994c Molecular biology of vascular bundle formation. Saibo Kogaku, Plant Cell Technol Ser 1:62–72

Fukuda H, Komamine A 1980 Direct evidence for cytodifferentiation to tracheary elements without intervening mitosis in a culture of single cells isolated from the mesophyll of Zinnia elegans. Plant Physiol 65:61–64

Fukuda H, Sato Y, Yoshimura T, Demura T 1993 Molecular mechanisms of xylem differentiation. J Plant Res Spec Issue 3:97–107

Fukuda T, Terashima N 1988 Heterogeneity in formation of lignin XII. Deposition of chemical components during the formation of cell walls of black pine and poplar. Mokuzai Gakkaishi 34:604–608

Fukushima K 1994 Deposition of lignin in xylem cell walls. In: Higuchi T (ed) Wood molecular biology. Buneido Publ, Tokyo, 74–83

Fukushima K, Taguchi S, Matsui N, Yasuda S 1994 Distribution of monolignol glucosides in trees. Proc 39th Lignin Symp, Fukuoka, Faculty of Agric Kyushu Univ, Japan, 37–40

Fukushima K, Terashima N 1990 Heterogeneity in formation of lignin XIII. Formation of *p*-hydroxyphenyl lignin in various hardwoods visualized by microautoradiography. J Wood Chem Technol 10:413–433

Fukushima K, Terashima N 1991a Heterogeneity in formation of lignin XIV. Formation and structure of lignin in differentiating xylem of Ginkgo biloba. Holzforschung 45:87–94

Fukushima K, Terashima N 1991b Heterogeneity in formation of lignin. Wood Sci Technol 25:371–381

Fukushima K, Yamada H, Yasuda S, Takabe K, Terashima N 1993 Studies on the formation and structure of lignin in the primary xylem of angiospermous tree (Magnolia kobus) by microautoradiography and UV microscopy. Proc 1993 Int Symp Wood Pulp Chem 1:101–110

Funada R 1994 Regulation of the formation of early- and late woods by plant hormones. In: Higuchi T (ed) Wood molecular biology. Buneido Publ, Tokyo, 187–197

Funada R, Abe H, Imaizumi H, Ohtani J, Fukazawa K 1995 Changes in orientation of cortical microtubules in the differentiating tracheids of Abies sacchalinensis. Abstr 40th Annu Conf Jpn Wood Res Soc, p 26

Funada R, Kubo T, Fushitani M 1987 Relationship between patterns of latewood formation and levels of endogenous auxin in the trunk of Sugi (Cryptomeria japonica). Mokuzai Gakkaishi 33:253–260

Funada R, Kubo T, Fushitani M 1989 Vertical distribution of annual ring characteristics in relation to the crown profile in Akamatsu (Pinus densiflora). Mokuzai Gakkaishi 35:944–947

Funada R, Kubo T, Fushitani M 1990 Early- and late wood formation in Pinus densiflora trees with different amounts of crowns. IAWA Bull 11:281–288

Funada R, Kubo T, Sugiyama T, Fushitani M, Fukazawa K 1992b Levels in endogenous plant hormones in the cambial region of Larix leptolepis stem during the onset of cambial activity. IUFRO All-Division 5 Conf, Nancy 1:231

Funada R, Mizukami E, Kubo T, Fushitani M, Sugiyama T 1990 Distribution of indole-3-acetic acid and compression wood formation in the stems of inclined Cryptomeria japonica. Holzforschung 44:331–334

Funada R, Sugiyama T, Kubo T, Fushitani M 1987 Determination of indole-3-acetic acid levels in Pinus densiflora using the isotope dilution method. Mokuzai Gakkaishi 33:83–87

Funada R, Sugiyama T, Kubo T, Fushitani M 1988 Determination of abscisic acid in Pinus densiflora by selected ion monitoring. Plant Physiol 88:525–527

Funada R, Sugiyama T, Kubo T, Fushitani M 1992a Identification of endogenous cytokinins in the cambial region of Cryptomeria japonica stem. Mokuzai Gakkaishi 38:317–320

Gabriac B, Reichhart DW, Teutsch H, Durst F 1991 Purification and immunocharacterization of a plant cytochrome P450: the cinnamic acid 4-hydroxylase. Arch Biochem Biophys 288:302–309

Gahan PB 1988 Xylem and phloem differentiation in perspective. In: Roberts LW, Gahan PB, Aloni R (eds) Vascular differentiation and plant growth regulators. Springer, Berlin Heidelberg New York, 1–21

Gaines CG, Byng GS, Whitaker RJ, Jensen RA 1982 L-tyrosine regulation and biosynthesis via arogenate dehydrogenase in suspension cultured cells of Nicotiana silvestris Speg. et Comes. Planta 156:233–240

Gamborg OL 1966 Aromatic metabolism in plants II. Enzymes of the shikimate pathway in suspension cultures of plant cells. Can J Biochem 44:791–799

Gardner IC, Leaf G 1960 Translocation of citrulline in Alunus glutinosa. Plant Physiol 35:948–950
Gardner JAF, Barton GM 1960 The distribution of dihydroquercetin in Douglas-fir and western larch. For Prod J 10:171–173
Gardner KH, Blackwell J 1974 The structure of native cellulose. Biopolymers 13:1975–2001
Garner CC, Herrmann KM 1985 Operator mutations of the Escherichia coli aroF gene. J Biol Chem 260:3820–3825
Gasser CS, Winter JA, Hironaka CM, Shah DM 1989 Structure, expression, and evolution of the 5-enolpyruvyl shikimate-3-phosphate synthase genes of Petunia and tomato. J Biol Chem 263:4280–4289
Gaudillere M, Monties B 1989 Biochemical and biosynthetic studies on lignification of Gramineae. Plant cell wall polymers, biogenesis and biodegradation. ACS Symp Ser 399:182–192
Geissman TA, Swain T 1957 Biosynthesis of flavonoid compounds in higher plants. Chem Ind:984
GenBank Accession Z17369, L07634, L11046
Gerats AG, Martin C 1992 Flavonoid synthesis in Petunia hybrida; genetics and molecular biology of flower color. In: Stafford HA, Ibrahim PK (eds) Phenolic metabolism in plants. Plenum Press, New York, 165–199
Giddings TH Jr, Brown DL, Staehelin LA 1980 Visualization of particle complexes in the plasma membrane of Micrasterias denticulata associated with the formation of cellulose fibrils in primary and secondary walls. J Cell Biol 84:327–339
Glaser L 1958 The synthesis of cellulose in cell free extracts of Acetobacter xylinum. J Biol Chem 232:627–636
Goers SK, Jensen RA 1984 The differential allosteric regulation of two chorismate-mutase isozymes of Nicotiana silvestris. Planta 162:117–124
Goffner D, Joffroy I, Grima-Pettenati J, Halpin C, Knight ME, Schuch W, Boudet AM 1992 Purification and characterization of isoforms of cinnamyl alcohol dehydrogenase from Eucalyptus xylem. Planta 188:48–53
Gordon R, Maclachlan G 1989 Incorporation of UDP-[^{14}C]glucose into xyloglucan by pea membranes. Plant Physiol 91:373–378
Goring DAI 1977 A speculative picture of the delignification process. In: Arthur JC Jr (ed) Cellulose chemistry and technology. ACS Symp Series 48:273–274
Goring DAI, Timell TE 1962 Molecular weight of native cellulose. Tappi 40:454–460
Görlach J, Beck A, Henstrand JM, Handa AK, Hermann KM, Schmid J, Amrhein N 1993 Differential expression of tomato (Lycopersicon esculentum L.) gene encoding shikimate pathway isoenzymes. I. 3-deoxy-D-arabino-heptulosonate 7-phosphate synthase. Plant Mol Biol 23:697–706
Gornicki P, Podkowinski J, Scappino LA, DiMaio J, Ward E, Haselkorn R 1994 Wheat acetyl-coenzyme A carboxylase: cDNA and protein structure. Proc Natl Acad Sci USA 91:6860–6864
Gotanda H, Yamauchi S, Ishibashi F, Taniguchi E 1993 Chromanoanalogs of insecticidal neolignans of the 1,4-benzodioxane type. Biosci Biotech Biochem 57:884–889
Goto T, Harada H, Saiki H 1973 Cross-sectional view of microfibrils in valonia (Valonia macrophysa). Mokuzai Gakkaishi 19:463–468
Goto T, Harada H, Saiki H 1978 Fine structure of cellulose microfibrils in poplar gelatinous layer and valonia. Wood Sci Technol 12:223–231
Gottlieb OR, Yoshida M 1989 Lignans. In: Rowe JW (ed) Natural products of woody plants. Springer, Berlin Heidelberg New York, 439–511
Gowri G, Bugos RC, Campbell WH, Maxwell CA, Dixon RA 1991 Stress responses in alfalfa (Medicago sativa L.). X. Molecular cloning and expression of *S*-adenosyl-L-methionine: caffeic acid 3-*O*-methyltransferase, a key enzyme of lignin biosynthesis. Plant Physiol 97:7–14

Gowri G, Paiva NL, Dixon RA 1991 Stress responses in alfalfa (Medicago sativa L.) 12. Sequence analysis of phenylalnine ammonia-lyase (PAL) cDNA clones and appearance of PAL transcripts in elicitor-treated cell cultures and developing plants. Plant Mol Biol 17:415–429

Graham TL, Kim JE, Graham MY 1990 Role of constitutive isoflavone conjugates in the accumulation of glyceollin in soybean infected with Phytophthora megasperma. Mol Plant-Microb Interact 3:157–166

Grand C 1984 Ferulic acid 5-hydroxylase: a new cytochrome P-450 dependent enzyme from higher plant microsomes involved in lignin synthesis. FEBS Lett 169:7–11

Grand C, Boudet A, Boudet AM 1983 Isoenzymes of hydroxycinnamate: CoA ligase from poplar stems. Properties and tissue distribution. Planta 158:225–229

Grand C, Sarni F, Boudet AM 1985 Inhibition of cinnamyl alcohol dehydrogenase activity and lignin synthesis in poplar (Populus × euramericana Dobe) tissues by two organic compounds. Planta 163:232–237

Greathouse GA 1959 On the enzymic polysaccharide synthesis In: Kratzl K, Billek G (eds) Biochemistry of wood. Pergamon Press, London, 76–81

Grima-Pettenati J, Campargue C, Boudet A, Boudet AM 1994 Purification and characterization of cinnamyl alcohol dehydrogenase isoforms from Phaseolus vulgaris. Phytochemistry 37:941–947

Grima-Pettenati J, Feuillet C, Goffner D, Borderies G, Boudet AM 1993 Molecular cloning and expression of a Eucalyptus gunii cDNA clone encoding cinnamyl alcohol dehydrogenase. Plant Mol Biol 21:1085–1095

Grisebach H 1957 Zur Biogenese des Cyanidins I. Versuche mit Acetat –[1 – ^{14}C] und Acetat –[2 – ^{14}C]. Z Naturforsch 12b:227–231

Grisebach H 1959 Zur Biogenese der Isoflavone. Z Naturforsch 14b:802–809

Grisebach H 1985 Biosynthesis of flavonoids. In: Higuchi T (ed) Biosynthesis and biodegradation of wood components. Academic Press, Orlando, 291–324

Grisebach H, Doerr N 1960 Zur Biogenese der Isoflavone II. Über den Mechanismus der Umlagerung. Z Naturforsch 15B:284–286

Grisebach H, Kellner S 1965 Untersuchungen zum Mechanismus der Umwanderung von 4,2′,4′,6′-Tetrahydroxychalkon in Taxifolin in Chamaecyparis obtusa. Z Naturforsch 20B:446–450

Gröbe A 1989 Properties of cellulose materials. In: Brandup J, Immergut EH (eds) Polymer handbook, 3rd edn. Wiley Interscience, New York, V117–V119

Gross GG 1977 Biosynthesis of lignin and related monomers. Rec Adv Phytochem 11:141–184

Gross GG 1983a Synthesis of mono-, di- and trigalloyl-β-D-glucose by β-glucogallin dependent galloyltransferase from oak leaves. Z Naturforsch 38C:519–523

Gross GG 1983b Partial purification and properties of UDP-glucose: vanillate 1-*O*-glucosyl transferase from oak leaves. Phytochemistry 22:2179–2182

Gross GG 1985 Biosynthesis and metabolism of phenolic acids and monolignols. In: Higuchi T (ed) Biosynthesis and biodegradation of wood components. Academic Press, Orlando, 229–271

Gross GG 1992 Enzymatic synthesis of gallotannins and related compounds. In: Stafford HA, Ibrahim RK (eds) Phenolic metabolism in plants. Plenum Press, New York, 297–324

Gross GG, Janse C, Elstner EF 1977 Involvement of malate, monophenols, and superoxide radical in hydrogen peroxide formation by isolated cell walls from horseradish (Aromoracia lapathifolia Gilib). Planta 136:271–276

Grove MD, Spencer GF, Rohwedder WK, Mandava N, Worley JF, Warthen JD, Steffens GL, Flippen-Anderson JL, Cook JC 1979 Brassinolide, a plant growth-promoting steroid isolated from Brassica napus pollen. Nature 281:216–217

Gupta RB, Khanna RN, Manchanda VP 1979 Synthesis of 3,3′-bis (2-methyl-1,4-naphthoquinone), a naturally occurring binaphthoquinone isolated from Asplenium laciniatum. Ind J Chem 18B:217–218

Haddock EA, Gupta RK, Al-Shafi SRK, Layden K, Haslam E, Magnolato D 1982 The metabolism of gallic acid and hexahydroxydiphenic acid in plants. Biogenetic and molecular taxonomic considerations. Phytochemistry 21:1049–1062

Hagmann M, Grisebach H 1984 Enzymatic rearrangement of flavanone to isoflavone. FEBS Lett 175:199–202

Hagmann M, Heller W, Grisebach H 1983 Induction and characterization of a microsomal flavonoid 3′-hydroxylase from parsley cell cultures. Eur J Biochem 134:547–554

Hahlbrock K, Wong E, Schill L, Grisebach H 1970 Comparison of chalcone-flavanone isomerase heteroenzymes and isoenzymes. Phytochemistry 9:949–958

Hahlbrock K, Grisebach H 1975 Biosynthesis of flavonoids. In: Harborne JB, Marbry TJ, Marbry H (eds) The flavonoids. Chapman and Hall, London, 866–915

Hahlbrock K, Grisebach H 1979 Enzymic controls in the biosynthesis of lignin and flavonoids. Annu Rev Plant Physiol 30:105–130

Hahlbrock K, Scheel D 1989 Physiology and molecular biology of phenylpropanoid metabolism. Annu Rev Plant Physiol Plant Mol Biol 40:347–369

Haigler CH, Brown RM Jr 1986 Transport of rosettes from the Golgi apparatus to the plasama membrane in isolated mesophyll cells of Zinnia elegans during differentiation to tracheary elements in suspension culture. Protoplasma 134:111–120

Hain R, Reif HJ, Krause E, Langebartels R, Kindl H, Vornam B, Wiese W, Schmelzer E, Schreier PH, Stöcker RH, Stenzel K 1993 Disease resistance results from foreign phytoalexin expression in a novel plant. Nature 361:153–156

Hakamatsuka T, Ebizuka Y, Sankawa U 1991 Induced isoflavonoids from copper chloride-treated stems of Pueraria lobata. Phytochemistry 30:1481–1482

Hakamatsuka T, Noguchi H, Ebizuka Y, Sankawa U 1990 Isoflavone synthase from cell suspension cultures of Pueraria lobata. Chem Pharm Bull 38:1942–1945

Hakamatsuka T, Shinkai K, Noguchi H, Ebinuma Y, Sankawa U 1992 Isoflavone dimers from yeast extract-treated cell suspension cultures of Pueraria lobata. Z Naturforsch 47C:177–182

Halpin C, Knight ME, Grima-Petternati J, Goffner D, Boudet A, Schuch W 1992 Purification and characterization of cinnamyl alcohol dehydrogenase from tobacco stems. Plant Physiol 98:12–16

Hancock MV 1957 The distribution of dihydroquercetin and leucocyanidin in a Douglas-fir tree. For Prod J 7:335–338

Hanson KR, Havir EA 1981 Phenylalanine ammonia-lyase. In: Stumpf PK, Conn EE (eds) The biochemistry of plants, vol 7. Academic Press, New York, 577–625

Harada H 1968 Structure of wood tissue. In: Migita N, Yonezawa Y, Kondo T (eds) Wood chemistry. Kyouritsu Publ, Tokyo, 1–25

Harada H, 1976a Cell wall. In: Shimaji K, Sudo S, Harada H (eds) Wood tissues. Morikita Publ, Tokyo, 47–101

Harada H 1976b Reaction wood. In: Shimaji K, Sudo S, Harada H (eds) Wood tissues. Morikita Publ, Tokyo, 216–223

Harada H, Côté WA Jr 1985 Structure of wood In: Higuchi T (ed) Biosynthesis and biodegradation of wood components. Academic Press, Orlando, 1–42

Harada H, Goto T 1982 The structure of cellulose microfibrils in valonia. In: Brown RM Jr (ed) Cellulose and other natural polymer system. Plenum Press, New York, 383–401

Harazdina G, Kreuzaler F, Hahlbrock K, Grisebach H 1976 Substrate specificity of flavanone synthase from cell suspension cultures of parsley and structure of release products in vitro. Arch Biochem Biophys 175:392–399

Harborne JB 1980 Plant phenolics. In: Bell EA (ed) Encyclopedia of plant physiology. New Series vol 8. Secondary plant products. Springer, Berlin Heidelberg New York, 329–402

Harborne JB, Mabry TJ (eds) 1982 The flavonoids: advances in research. Chapman & Hall, London, 744pp

Hardell H, Leary GJ, Stoll M, Westermark U 1980a Variations in lignin structure in defined morphological parts of spruce. Svensk Papperstidn 83:44–49

Hardell H, Leary GJ, Stoll M, Westermark U 1980b Variations of lignin structure in defined morphological parts of birch. Svensk Papperstidn 83:71–74

Harkin JM, Obst TR 1973 Lignification in trees: indication of exclusive peroxidase participation. Science 180:296–297

Harrison BJ, Stickland RG 1974 Precursors and genetic control of pigmentation. 2. Genotype analysis of pigment controlling genes in acyanic phenotypes in Antirrhinum majous. Heredity 33:112–115

Harrison K 1960 A guide book to biochemistry. Translated by Iwai K, Tokyo Kagaku Dojin, Tokyo, 91–102

Hashem M, Brucker W 1962 The formation of phenylcarboxylic acid from shikimic acid by Phycomyces. Flora 152:57–67

Haslam E 1974 The shikimate pathway. Butterworths, London, 316pp

Haslam E 1989 Gallic acid derivatives and hydrolyzable tannins. In: Rowe JW (ed) Natural products of woody plants I. Chemicals extraneous to the lignocellulosic cell wall. Springer, Berlin Heidelberg New York, 399–438

Hauffe KD, Lee SP, Subramaniam R, Douglas CJ 1993 Combinatorial interractions between positive and negative *cis*-acting elements control spatial patterns of *4CL-1* expression in transgenic tobacco. Plant J 4:235–253

Hauffe KD, Paszkowski U, Schlze-Lefert P, Hahlbrock K, Dangl J, Douglas CJ 1991 A parsley 4CL-1 promotor fragment specifies complex expression patterns in transgenic tobacco. Plant Cell 3:435–443

Hawkins AR 1987 The complex Arom locus of Aspergillus nidulans. Curr Genet 11:491–498

Haworth RD 1936 Natural resins. Annu Rep Prog Chem 33:266–279

Hayashi T 1989 Xyloglucan in the primary cell wall. Annu Rev Plant Physiol Plant Mol Biol 40:139–168

Hayashi T 1994 Polysaccharides. In: Higuchi T (ed) Wood molecular biology. Buneido Publ, Tokyo, 87–101

Hayashi T, Koyama T, Matsuda K 1988 Formation of UDP-xylose and xyloglucan in soybean Golgi membranes. Plant Physiol 87:341–345

Hayashi T, Maclachlan G 1984 Pea xyloglucan and cellulose. I. Macromolecular organization. Plant Physiol 75:596–604

Hayashi T, Matsuda K 1981a Biosynthesis of xyloglucan in suspension-cultured soybean cells. Occurrence and some properties of xyloglucan 4-β-D-glucosyltransferase and 6-α-D-xylosyltransferase. J Biol Chem 256:11117–11122

Hayashi T, Matsuda K 1981b Biosynthesis of xyloglucan from UDP-glucose and UDP-xylose in the cell free system. Plant Cell Physiol 22:517–523

Hayashi T, Nakajima T, Matsuda K 1984 Biosynthesis of xyloglucan in suspension-cultured soybean cells processing of the oligosaccharide building blocks. Agric Biol Chem 48:1023–1027

Hayashi T, Read SM, Bussell J, Lin FC, Brown RM Jr, Delmer DP 1987 UDP-glucose: (1–3)β-glucan synthases from mung bean and cotton: differential effects of Ca and Mg on enzyme properties and on macromolecular structure of the glucan product. Plant Physiol 83:1054–1062

Hayashi Y 1990 Chemical studies on the wood extractives of Zelkova serrata (Thunb). Makino and related species in Ulmaceae. PhD Thesis, Kyoto Univ, 1–142

Hayashi Y, Kato A, Ito Y 1994 Chemical studies on black heartwood of Cryptomeria japonica (2). Abstr 44th Annu Meet Jpn Wood Res Soc, Kinki Univ Nara, Japan, p 491

Hedden P, MacMillan J, Phinney BO 1978 The metabolism of the gibberellins. Annu Rev Plant Physiol 29:149–192

Heide L, Leistner E 1982 Synthesis of naturally occurring phenylated naphthalene derivatives. Isolation of a new prenylnaphthoquinone from Galium mollugo. Z Naturforsch C: Biosci 37C:354–362

Heide L, Leistner E 1983 Enzyme activities in extracts of anthraquinone-containing cells of Galium mollugo. Phytochemistry 22:659–662

Heide L, Nishioka N, Fukui H, Tabata M 1989 Enzymatic regulation of shikonin biosynthesis in Lithospermum erythrorhizon cell cultures. Phytochemistry 28:1873–1877

Heide L, Tabata M 1987 Geranylpyrophosphate: *p*-hydroxybenzoate geranyltransferase activity in extracts of Lithospermum erythrorhizon cell cultures. Phytochemistry 26:1651–1655

Heiniger U, Delmer DP 1977 UDP-glucose: glucan synthetase in developing cotton fibers II. Structure of the reaction product. Plant Physiol 59:719–723

Heller W 1986 Flavonoid biosynthesis. An overview. In: Cody V, Middleton E Jr, Harborne JB (eds) Progress in clinical and biological research, vol 213. Plant flavonoids in biology and medicine, biochemical pharmacological, and structure-activity relationships. Alan R Liss, New York, 25–42

Heller W, Forkmann 1988 Biosynthesis. In: Harborne JB (ed) The flavonoids. Chapman & Hall, London, 399–425

Heller W, Forkmann G, Britsch L, Grisebach H 1985 Enzymatic reduction of the (+)-dihydroflavonols to flavan-3,4-cis-diols with flower extracts from Matthiola incana and its role in anthocyanin biosynthesis. Planta 165:284–287

Heller W, Hahlbrock K 1980 Highly purified flavanone synthase from parsley catalyzes the formation of naringenin chalcone. Arch Biochem Biophys 200:617–619

Helper PK 1981 Morphogenesis of tracheary elements and guard cells. In: Kiermayer O (ed) Cytomorphogenesis in plants. Springer, Wien Berlin Heidelberg, 327–347

Helper PK, Rice RM, Terranova WA 1972 Cytochemical localization of peroxidase activity in wound vessel members of Coleus. Can J Bot 50:977–983

Herrmann KM 1995 The shikimate pathway as an entry to aromatic secondary metabolism. Plant Physiol 107:7–12

Heyn ANJ 1966 The microcrystalline structure of cellulose in cell walls of cotton, ramie, and jute fibers as revealed by negative staining of sections. J Cell Biol 29:181–197

Hibino T, Chem J-Q, Shibata D, Higuchi T 1994 Nucleotide sequence of a Eucalyptus botryoides gene encoding cinnamyl alcohol dehydrogenase. Plant Physiol 104:305–306

Hibino T, Shibata D, Chen J-Q, Higuchi T 1993a Cinnamyl alcohol dehydrogenase from Aralia cordata: cloning of the cDNA and expression of the gene in lignified tissue. Plant Cell Physiol 34:659–665

Hibino T, Shibata D, Umezawa T, Higuchi T 1993b Purification and partial sequences of Aralia cordata cinnamyl alcohol dehydrogenase. Phytochemistry 32:565–567

Higuchi T 1958a Studies on the chemical properties of the lignin of bamboo stalk. J Biochem (Jpn) 45:675–685

Higuchi T 1958b Further studies on phenol oxidase related to the lignin biosynthesis. J Biochem (Jpn) 45:515–528

Higuchi T 1959 Studies on the biosynthesis of lignin. In: Kratzl K, Billek G (eds) Biochemistry of wood. Pergamon Press, London, 161–188

Higuchi T 1962 Studies of lignin biosynthesis using isotopic carbon X. Formation of lignin from phenylpropanoids in tissue culture of white pine. Can J Biochem Physiol 40:31–34

Higuchi T 1966 Role of phenylalanine deaminase and tyrase in the lignification of bamboo. Agric Biol Chem 30:667–673
Higuchi T 1969a Tree biochemistry. Kyoritsu Publ, Tokyo, 133 pp
Higuchi T 1969b Bamboo lignin and its biosynthesis. Wood Res (Kyoto) 48:1–14
Higuchi T 1970 Mechanism of heartwood formation. Mokuzai Gakkaishi 16:99–103
Higuchi T 1976 Biochemical aspects of lignification and heartwood formation. Wood Res (Kyoto) 59/60:180–199
Higuchi T 1983 Biosynthesis and microbial degradation of lignin. In: Akazawa T, Asahi T, Imaseki H (eds) The new frontiers in plant biochemistry. Jpn Sci Soc Press, Tokyo, 23–46
Higuchi T 1985 Biosynthesis of lignin. In: Higuchi T (ed) Biosynthesis and biodegradation of wood components. Academic Press, Orlando, 141–160
Higuchi T 1987 Chemistry and biochemistry of bamboo. Bamboo J 4:132–145
Higuchi T 1989 Bamboo. In: Schniewind AP (ed) Concise encyclopedia of wood and wood-based materials. Pergamon Press, London, 19–26
Higuchi T 1990 Lignin biochemistry: biosynthesis and biodegradation. Wood Sci Technol 24:23–63
Higuchi T 1992 Wood biochemistry. Buneido Publ, Tokyo, 246 pp
Higuchi T, Barnoud F 1964 Les lignines dans les tissus végétaux cultivés in vitro. Chimie Biochimie de la Lignin, Cellulose, Hémicelluloses. Actes du Symp Int de Grenoble Juillet 1964. Les Imprimeries Réunies de Chambéry, France, 255–274
Higuchi T, Brown SA 1963a Studies of lignin biosynthesis using isotopic carbon XII. The biosynthesis and metabolism of sinapic acid. Can J Biochem Physiol 41:613–620
Higuchi T, Brown SA 1963b Studies of lignin biosynthesis using isotopic carbon XIII. The phenylpropanoid system in lignification. Can J Biochem Physiol 41:621–628
Higuchi T, Fukazawa K, Nakashima S 1964 Study on the mechanism of heartwood formation 1. Histochemistry of the wood tissue. Mokuzai Gakkaishi 10:235–241
Higuchi T, Fukazawa K, Shimada M 1967c Biochemical studies on the heartwood formation. Res Bull Coll Exp For Hokkaido Univ 25:167–194
Higuchi T, Ito Y 1958 Dehydrogenation products of coniferyl alcohol formed by the action of mushroom phenol oxidase, rhus-laccase and radish peroxidase. J Biochem (Jpn) 45:575–579
Higuchi T, Ito Y, Kawamura I 1967d *p*-Hydroxyphenylpropane component of grass lignin and role of tyrosine ammonia-lyase in its formation. Phytochemistry 6:875–881
Higuchi T, Ito Y, Shimada M, Kawamura I 1967e Chemical properties of milled wood lignin of grasses. Phytochemistry 6:1551–1556
Higuchi T, Kawamura I 1966 Occurrence of *p*-hydroxyphenylglycerol-β-aryl ether structure in lignins. Holzforschung 20:17–21
Higuchi T, Kawamura I, Morimoto I, Kanda T 1954 The formation of lignin in bamboo shoot II. J Jpn For Soc 36:101–106
Higuchi T, Onda Y, Fujimoto Y 1969 Biochemical aspects of heartwood formation with special reference to the site of biogenesis of heartwood compounds. Wood Res (Kyoto) 48:15–30
Higuchi T, Shimada M, Nakatsubo F, Tanahashi M 1977 Differences in biosynthesis of guaiacyl and syringyl lignins in woods. Wood Sci Technol 11:153–167
Higuchi T, Shimada M, Ohashi H 1967f Role of *O*-methyltransferase in the lignification of bamboo. Agric Biol Chem 31:1459–1465
Higuchi T, Shimada M, Watanabe K 1967a Studies on the mechanism of heartwood formation V. Changes in the pattern of glucose metabolism in heartwood formation. Mokuzai Gakkaishi 13:269–273
Higuchi T, Shimada M, Watanabe K 1967b Studies on the mechanism of heartwood formation VI. On the artificial heartwood of Cryptomeria japonica and Pinus densiflora. Mokuzai Gakkaishi 13:274–279

Hillis WE 1960 Factors influencing the formation of phloem and heartwood polyphenols. Holzforschung 14:105–110
Hillis WE 1966 Variation in polyphenol composition within species of Eucalyptus. Phytochemistry 5:541–556
Hillis WE 1968 Chemical aspects of heartwood formation. Wood Sci Technol 2:241–259
Hillis WE 1975 Etylene and extraneous material formation in woody tissues. Phytochemistry 14:2559–2562
Hillis WE 1985 Biosynthesis of tannins. In: Higuchi T (ed) Biosynthesis and biodegradation of wood components. Academic Press, Orlando, 325–347
Hillis WE 1987 Heartwood and tree exudates. Springer, Berlin Heidelberg New York, 100–101
Hoffmann GC, Timell TE 1970 Isolation of a β-1,3-glucan (laricinan) from compression wood of Larix larcina. Wood Sci Technol 4:159–162
Hoffmann GC, Timell TE 1972 Polysaccharides in compression wood of tamarack (Larix laricina). I. Isolation and characterization of laricinan, an acidic glucan. Svensk Papperstidn 75:135–142
Hofmann AS, Gross GG 1990 Biosynthesis of gallotannins: formation of polygalloylglucoses by enzymatic acylation of 1,2,3,4,6-penta-*O*-galloylglucose. Arch Biochem Biophys 283:530–532
Hogetsu T 1989 Plant microtubules. Saibou 21:3–7
Holloway PJ 1983 Some variations in the composition of suberin from the cork layers of higher plants. Phytochemistry 22:495–502
Holton TA, Brugliera F, Lester DR, Tanaka Y, Hyland CD, Menting JGT, Lu C-Y, Farcy E, Stavenson TW, Cornish EC 1993 Cloning and expression of cytochrome P450 genes controlling flower colour. Nature 366:276–279
Hori H, Elbein AD 1985 Biosynthesis of plant cell wall polysaccharides. In: Higuchi T (ed) Biosynthesis and biodegradation of wood components. Academic Press, Orlando, 109–139
Horii F, Hirai A, Kitamaru R 1987b CP/MAS ^{13}C-NMR spectra of the crystalline components of native celluloses. Macromolecules 20:2117–2120
Horii F, Yamamoto A, Kitamaru R, Tanahashi M, Higuchi T 1987a Transformation of native cellulose crystals induced by saturated steam at high temperatures. Macromolecules 20:2946–2949
Hrazdina G, Kreuzaler F, Hahlbrock K, Grisebach H 1976 Substrate specificity of flavanone synthase from cell suspension cultures of parsley and structure of release products in vitro. Arch Biochem Biophys 175:392–399
Hugentobler U 1965 Zur Cytologie der Kernholzbildung. Vierteljahrschr Naturforsch Ges Zürich 110(2):321–342
Hutchinson A, Taper CD, Towers GHN 1959 Phlordzin in Malus. Can J Biochem Physiol 37:901–910
Hwang BH, Sakakibara A 1981 Hydrogenolysis of protolignin XVIII. Isolation of a new dimeric compound with a heterocycle involving α-, β-diether. Holzforschung 35:297–300
Hyodo H, Kuroda H, Yang SF 1978 Induction of phenylalanine ammonia-lyase and increase in phenolics in lettuce leaves in relation to the development of russet spotting caused by ethylene. Plant Physiol 62:31–35
Ibrahim RK 1977 Glucosylation of lignin precursors by uridine diphosphate glucose: coniferyl alcohol glucosyltransferase in higher plants. Z Pflanzenphysiol 85:253–262
Ibrahim RK, Towers GHN 1962 Studies in hydrangenol in Hydrangea macrophylla II. Biosynthesis of hydrangenol from ^{14}C-labeled compounds. Can J Biochem Physiol 40:449–453
Iiyama K, Lam TBT, Stone BA 1990 Phenolic acid bridges between polysaccharides and lignin in wheat internodes. Phytochemistry 29:733–737

Imahori K, Yamakawa T 1990 Dictionary of biochemistry, 2nd edn. Tokyo Kagaku Dojin, Tokyo, 461 pp

Indada S, Tsutsumi Y, Sakai K 1993 Elicitor of the β-thujaplicin accumulation in callus cultures of Cupressus lusitanica. J Fac Agr Kyushu Univ 38:119–126

Inagaki Y, Hisatomi Y, Suzuki T, Kasahara K, Iida 1994 Isolation of a *suppressor-mutator/enhancer*-like transposable element, Tpn1, from Japanese morning glory bearing variegated flowers. Plant Cell 6:375–383

Inomata F, Takabe K, Saiki H 1992 Cell wall formation of conifer tracheid as revealed by rapid-freeze and substitution method. J Electron Microsc 41:369–374

Inoue H, Ueda S, Inoue K, Matsumura H 1979 Biosynthesis of shikonin in callus cultures of Lithospermum erythrorhizon. Phytochemistry 18:1301–1308

Isenberg IH 1936 Age and the chemical composition of white fir wood. J Am Chem Soc 58:2231–2233

Ishii H, Ishikawa T, Mihara M, Akaike M 1983 Studies on the chemical constituents of Rutaceous plants. XLVIII. The chemical constituents of Xanthoxylum ailanthoides Sieb. et Zucc. [Fagara ailanthoides (Sieb. et Engl.]. (3) Isolation of the chemical consituents of the bark. Yakugaku Zasshi 103:279–292

Ishii H, Ohida H, Haginiwa J 1972 Studies on the alkaloids of Rutaceous plants. XIX. The chemical constituents of Xanthoxylum inerme Koidz. (Fagara boninensis Koidz.). (I). isolation of the chemical constituents from bark and wood. Yakugaku Zasshi 92:118–128

Itoh T 1985 Recent progress in cellulose biosynthesis. Wood Res Tech Notes (Kyoto) 21:1–22

Itoh T 1989 Biogenesis of cellulose microfibrils and the role of microtubules in green algae. In: Lewis NG, Paice MG (eds) Plant cell wall polymers: biogenesis and biodegradation. ACS Symp Series 399:257–277

Itoh T 1994a Cellulose microfibril. In: Higuchi T (ed) Wood molecular biology. Buneido Publ, Tokyo, 102–108

Itoh T 1994b Cytoskeleton. In: Higuchi T (ed) Wood molecular biology. Buneido Publ, Tokyo, 17–27

Jackson DE, Dewick PM 1984a Biosynthesis of Podophyllum lignans. I. Cinnamic acid precursors of podophyllotoxin in Podophyllum hexandrum. Phytochemistry 23: 1029–1035

Jackson DE, Dewick PM 1984b Biosynthesis of Podophyllum lignans. II. Interconversions of aryltetralins lignans in Podophyllum hexandrum. Phytochemistry 23:1037–1042

Jacobs WP 1952 The role of auxin in differentiation of xylem around a wound. Am J Bot 39:301–309

Jaeck E, Dumas B, Geoffroy P, Favet N, Inzé D, Van Montagu M, Fritig B, Legrand M 1992 Regulation of enzymes involved in lignin biosynthesis: induction of *O*-methyltransferase mRNAs during the hypersensitive reaction of tobacco to tobacco mosaic virus. Mol Plant-Microbe Interact 5:294–300

Jenkins PA, Shepherd KR 1974 Seasonal changes in levels of indole-acetic acid and abscisic acid in stem tissues of Pinus radiata. NZ J For Sci 4:511–519

Jensen RA, Morris P, Bonner C, Zamir LO 1989 Biochemical interface between aromatic amino acid biosynthesis and secondary metabolism. Plant cell wall polymers, biogenesis and biodegradation. ACS Symp Ser 399:89–107

Jensen RG 1990 Ribulose 1,5-bisphosphate carboxylase/oxygenase: mechanisms, activation, and regulation. In: Dennis DT, Turpin DH (eds) Plant physiology, biochemistry and molecular biology. Longman Scientific & Technical, Singapore, 224–238

John ME, Crow LJ 1992 Gene expression in cotton (Gossypium hirsutum L.) fiber: cloning of the mRNAs. Proc Natl Acad Sci USA 89:5769–5773

Jonsson LMV, Aarsman MEG, Schram AW, Bennink GJH 1982 Methylation of anthocyanins by cell-free extracts of flower buds of Petunia hybrida. Phytochemistry 21:2457–2459

Jorgensen E 1961 The formation of pinosylvin and its monometyl ether in the sapwood of Pinus resinosa. Can J Bot 39:1765–1772

Jung E, Zamir LO, Jensen RA 1986 Chloroplasts of higher plants synthesize L-phenylalanine via arogenate. Proc Natl Acad Sci USA 83:7231–7235

Jung HG, Deetz DA 1993 Cell wall lignification and degradability. In: Jung HG, Buxton DR, Hatfield RD, Ralph J (eds) Forage cell wall structure and digestibility. Am Soc Agronomy, Madison, 315–346

Kakimoto T, Shibaoka H 1992 Synthesis of polysaccharides in phragmoplasts isolated from tobacco BY-2 cells. Plant Cell Physiol 33:353–361

Kamen M 1963 Primary processes in photosynthesis. Academic Press, New York, 182pp

Kamil WM, Dewick PM 1986a Biosynthesis of the lignans α- and β-peltatin. Phytochemistry 25:2089–2092

Kamil WM, Dewick PM 1986b Biosynthetic relationship of aryltetralin lactone lignans to dibenzylbutyrolactone lignans. Phytochemistry 25:2093–2102

Kamsteeg J, Van Brederode J, Van Nigtevecht G 1978 Identification, properties and genetic control of UDP-glucose: cyanidin 3-rhamnosyl(1–6)-glucoside 5-*O*-glucosyltransferase isolated from petals of the red campion (Silene dioica). Biochem Genet 16:1059–1071

Kamsteeg J, Van Brederode J, Van Nigtevecht G 1979 Genetics of anthocyanin formation in petals of red campion (Silene dioica). Genetica (The Hague) 51:5–14

Kamsteeg J, Van Brederode J, Van Nigtevecht G 1980a Genetical and biochemical evidence that the hydroxylation pattern of the anthocyanin B-ring in Silene dioica is determined at the *p*-coumaroyl coenzyme A stage. Phytochemistry 19:1459–1462

Kamsteeg J, Van Brederode J, Van Nigtevecht G 1980b Identification, properties and genetic control of UDP-L-rhamnose: anthocyanidin 3-*O*-glucoside, 6″-*O*-rhamnosyl transferase isolated from petals of the red campion (Silene dioita). Z Naturforsch 35C:249–257

Kamsteeg J, Van Brederode J, Verschuren PM, Van Nigtevecht G 1981 Identification, properties and genetic control of *p*-coumaroyl coenzyme A 3-hydroxylase isolated from petals of Silene dioica. Z Pflanzenphysiol 102:435–442

Karr AL Jr 1972 Isolation of an enzyme system which will catalyze the glycosylation of extensin. Plant Physiol 50:275–282

Kashikawa Y, Takabe K, Fujita M, Saiki H 1995 Observation of cortical microtubules in xylem cells of Eucalyptus globulus by immunostaining. Abstr 40th Annu Conf Jpn Wood Res Soc, Tsukuba, Japan, p 25

Kataoka Y, Saiki H, Fujita M 1992 Arrangement and superimposition of cellulose microfibrils in the secondary walls of coniferous tracheids. Mokuzai Gakkaishi 38:327–335

Katayama T, Davin LB, Chu A, Lewis NG 1993 Novel benzylic ether reductions in lignan biogenesis in Forsythia intermedia. Phytochemistry 33:581–591

Katayama T, Davin LB, Lewis NG 1992 An extraordinary accumulation of (−)-pinoresinol in cell-free extracts of Forsythia intermedia: evidence for enantiospecific reduction of (+)-pinoresinol. Phytochemistry 31:3875–3881

Kato Y, Munakata K 1978 Dibenzylbutyrolactones. In: Rao CBS (ed) Chemistry of lignans. Andhra Univ Press, Andhra Pradesh, 95–122

Katsumi M 1985 Interaction of a brassinosteroid with IAA and GA_3 in the elongation of cucumber hypocotyl sections. Plant Cell Physiol 26:615–625

Kaufman S 1987 Section II. Biosynthesis of the aromatic amino acids. Methods in enzymology, vol 142. Academic Press, Orlando, 294–518

Kauss H, Hassid WZ 1967 Biosynthesis of the 4-*O*-methyl-D-glucuronic acid unit of hemicellulose B by transmethylation from *S*-adenosyl-L-methionine. J Biol Chem 242:1680–1684

Kawalleck P, Plesch G, Hahlbrock K, Somssich IE 1992 Induction by fungal elicitor of *S*-adenosyl-L-methionine synthase and *S*-adenosyl-L-homocysteine hydrolase mRNAs in cultured cells and leaves of Petroselinum crispum. Proc Natl Acad Sci USA 89:4713–4717

Kawamura I, Higuchi T 1964 Comparative studies of milled wood lignins from different taxonomical origins by IR spectrometry. Chimie Biochimie de la Lignin, Cellulose, Hémicelluloses. Actes du Symp Int de Grenoble Juillet 1964. Les Imprimeries Réunies de Chambéry, France, 439–456

Keith B, Dong X, Frederick M, Ausubel FM, Fink GR 1991 Differential induction of 3-deoxy-D-arabino-heptulosonate 7-phosphate synthase genes in Arabidopsis thaliana by wounding and pathogenic attack. Proc Natl Acad Sci USA 88:8821–8825

Keller B, Baumgartner C 1991 Vascular-specific expression of the bean GRP 1.8 gene is negatively regulated. Plant Cell 3:1051–1061

Kelley PM, Tolan DR 1986 The complete amino acid sequence for the anaerobically induced aldolase from maize derived from cDNA clones. Plant Physol 82:1076–1080

Khan AW, Colvin JR 1961 Synthesis of bacterial cellulose from labeled precursor. Science 133:2014–2015

Kho KFF, Bennink GJH, Wiering H 1975 Anthocyanin synthesis in a white flowering mutant of Petunia hybrida by a complementation technique. Planta 127:271–279

Kho KFF, Bolsman-Louwen AC, Vuik JC, Bennink GJH 1977 Anthocyanin synthesis in a white flowering mutant of Petunia hybrida II. Accumulation of dihydroflavonol intermediates in white-flowering mutant; uptake of intermediates in isolated corollas and conversion into anthocyanins. Planta 135:109–118

Kim SK, Abe H, Little CHA, Pharis RP 1990 Identification of two brassinosteroids from the cambial region of Scots pine (Pinus silvestris) by gas chromatography-mass spectrometry after detection using a dwarf rice lamina inclination bioassay. Plant Physiol 94:1709–1713

Kim YS, Meshitsuka G, Ishizu A 1993 Structural inhomogeneity of lignin. Proc 1993 Int Symp Wood Pulp Chem 3:15

Kimmerer TW, Stringer MA 1988 Alcohol dehydrogenase and ethanol in the stems of trees. Evidence for anaerobic metabolism in the vascular cambium. Plant Physiol 87:693–697

Kimura M, Osawa S (eds) 1989 History of life. Iwanami Shoten, Tokyo, 25–60

Kindel PK 1973 Occurrence and metabolism of D-apiose in Lemna minor. In: Loewus F (ed) Biogenesis of plant cell wall polysaccharides. Academic Press, New York, 85–94

Kindl H 1985 Biosynthesis of stilbenes. In: Higuchi T (ed) Biosynthesis and biodegradation of wood components. Academic Press, Orlando, 349–377

Knight ME, Halpin C, Schuch W 1992 Identification and characterization of cDNA clones encoding cinnamyl alcohol dehydrogenase from tobacco. Plant Mol Biol 19:793–801

Knobloch K-H, Hahlbrock K 1975 Isoenzymes of *p*-coumarate: CoA ligase from cell suspension cultures of Glycine max. Eur J Biochem 52:311–320

Kochs G, Grisebach H 1986 Enzymic synthesis of isoflavones. Eur J Biochem 155:311–318

Kojima M, Takeuchi W 1989 Detection and characterization of *p*-coumaric acid hydroxylase in mung bean (Vigna mungo) seedlings. J Biochem (Jpn) 105:265–270

Kolattukudy PE, Espelie KE 1985 Biosynthesis of waxes. In: Higuchi T (ed) Bio-

synthesis and biodegradation of wood components. Academic Press, Orlando, 197–202
Kolattukudy PE, Espelie KE 1989 Chemistry, biochemistry, and function of suberin and associated waxes. In: Rowe JW (ed) Natural products of woody plants I. Springer, Berlin Heidelberg New York, 304–367
Kolpak FJ, Blackwell J 1976 Determination of the structure of cellulose II. J Macromol 9:273–278
Köster J, Strack D, Barz W 1983 High performance liquid chromatographic separation and structural elucidation of isoflavone 7-*O*-glucoside 6″-malonate from Cicer arietinum. Planta Med 48:131–135
Koukol J, Conn EE 1961 The metabolism of aromatic compounds in higher plants IV. Purification and properties of the phenylalanine deaminase of Hordeum vulgare. J Biol Chem 236:2692–2698
Kratzl K, Billek G, Klein E, Buchtera K 1957 Über das Verhalten von markiertem Coniferin in der verholzenden Pflanze. Monatsh Chem 88:721–723
Kreuzaler F, Hahlbrock K 1975a Enzymic synthesis of an aromatic ring from acetate units. Eur J Biochem 56:205–213
Kreuzaler F, Hahlbrock K 1975b Partial purification and some properties of flavanone synthase from cell suspension cultures of Petroselinum hortense. Arch Biochem Biophys 169:84–90
Kreuzaler F, Light RJ, Hahlbrock K 1978 Flavanone synthase catalyzes CO_2 exchange and decarboxylation of malonyl-CoA. FEBS Lett 94:175–178
Kreuzaler F, Ragg H, Fautz E, Kuhn DH, Hahlbrock K 1983 UV-induction of chalcone synthase mRNA in cell suspension cultures of Petroselinum hortense. Proc Natl Acad Sci USA 80:2591–2593
Kubo T 1983 Characteristics of latewood formation in Japanese coniferous woods. Mokuzai Gakkaishi 29:635–640
Kuhn DN, Chappel J, Boudet A, Hahlbrock K 1984 Induction of phenylalanine ammonia-lyase and 4-coumarate: CoA ligase mRNAs in cultured plant cells by UV light or fungal elicitor. Proc Natl Acad Sci USA 81:1102–1106
Kuraishi S 1988 Plant hormones. Tokyo Univ Press, Tokyo, 142 pp
Kuroda H 1983 Comparative studies on *O*-methyltransferases involved in lignin biosynthesis. Wood Res (Kyoto) 69:91–135
Kuroda H, Higuchi T 1982 *O*-Methyltransferase as a tool to evaluate the lignin evolution. Wood Res (Kyoto) 68:1–7
Kuroda H, Higuchi T 1994 Phenylpropanoid derivatives. In: Higuchi T (ed) Wood molecular biology. Buneido Publ, Tokyo, 109–140
Kuroda H, Shimada M, Higuchi T 1975 Purification and properties of *O*-methyltransferase involved in the biosynthesis of gymnosperm lignin. Phytochemistry 14:1759–1763
Kuroda H, Shimada M, Higuchi T 1981 Characterization of a lignin-specific *O*-methyltransferase in aspen. Phytochemistry 20:2635–2639
Kuroki GW, Conn EE 1988 Increased chorismate mutase levels as a response to wounding in Solanum tuberosum L. tubers. Plant Physiol 86:895–898
Kutsuki H 1981 Biochemical differences in the formation of angiosperm and gymnosperm lignins. PhD Thesis, Kyoto Univ, 123 pp
Kutsuki H, Higuchi T 1981 Activities of some enzymes of lignin formation in reaction wood of Thuja orientalis, Metasequoia glyptostroboides and Robinia pseudoacacia. Planta 152:365–368
Kutsuki H, Shimada M, Higuchi T 1981 The role of *O*-methyltransferase in the formation of lignin and lignans in the genus Thuja. Mokuzai Gakkaishi 27:39–48
Kutsuki H, Shimada M, Higuchi T 1982a Distribution and role of *p*-hydroxycinnamate: CoA ligase in lignin biosynthesis. Phytochemistry 21:267–271

Kutsuki H, Shimada M, Higuchi T 1982b Regulatory role of cinnamyl alcohol dehydrogenase in the formation of guaiacyl and syringyl lignins. Phytochemistry 21:19–23

Lacombe E, Poeydomenge O, Van Dooresselaere J, Piquemal J, Goffner D, Boudet AM, Grima-Pettenati J 1994 Cloning and characterization of an Eucalyptus gunnii cDNA encoding cinnamoyl CoA reductase, a key enzyme in lignification. 22nd Aharon Katzir-Katchalsky Conf on Plant molecular biology, biotechnology and environment, Cologne, Germany

Lagrimini LM, Burkhart W, Moyer M, Rothstein S 1987 Molecular cloning of complementary DNA encoding the lignin-forming peroxidase from tobacco: molecular analysis and tissue specific expression. Proc Natl Acad Sci USA 84:7542–7546

Lagrimini LM, Rothstein S 1987 Tissue specificity of tobacco peroxidase isozymes and their induction by wounding and tobacco mosaic virus infection. Plant Physiol 84:438–442

Lam TBT, Iiyama K, Stone B 1990 Distribution of free and combined phenolic acids in wheat internodes. Phytochemistry 29:429–433

Lam TBT, Iiyama K, Stone B 1991 Quantitative determination of lignin-polysaccharide associations in wheat internode cell walls: ferulic acid bridges. Proc 6th Int Symp Wood Pulping Chem, Melbourne, Australia, 29–33

Lam TBT, Iiyama K, Stone B 1992 Cinnamic acid bridges between cell wall polymers in wheat and phalaris internodes. Phytochemistry 31:1177–1183

Lanz T, Schröder G, Schröder J 1990 Differential regulation of genes for resveratrol synthase in cell cultures of Arachis hypogaea L. Planta 181:169–175

Larson PR 1960 A physiological consideration of the springwood summerwood transition in red pine. For Sci 6:110–122

Larson PR 1964 Some indirect effects of environment on wood formation. In: Zimmermann MH (ed) The formation of wood in forest trees. Academic Press, New York, 345–365

Last RL, Bissinger PH, Mahoney D, Radwanski ER, Fink GR 1991 Tryptophan mutants in Arabidopsis: the consequences of duplicated tryptophan synthase β gene. Plant Cell 3:345–358

Lea PJ 1993 Nitrogen metabolism. In: Lea PJ, Leegood RC (eds) Plant biochemistry and molecular biology. Wiley, New York, 155–180

Lea PJ, Leegood RC 1993 Plant biochemistry and molecular biology. Wiley, New York, 312 pp

Ledbetter MC, Porter KR 1963 A "microtubule" in plant cell fine structure. J Cell Biol 19:239–250

Leegood RC 1993 The calvin cycle and photorespiration. In: Lea PJ, Leegood RC (eds) Plant biochemistry and molecular biology. Wiley, New York, 27–45

Legrand M, Fritig B, Hirth L 1978 *o*-Diphenol *O*-methyltransferases of healty and tobacco-mosaic-virus-infected hypersensitive tobacco. Planta 144:101–108

Leistner E 1985 Occurrence and biosynthesis of quinones in woody plants. In: Higuchi T (ed) Biosynthesis and biodegradation of wood components. Academic Press, Orlando, 273–290

Letcher RM, Nhamo LRM 1972 Chemical constituents of the Combretaceae. Part III. Substituted phenanthrens, 9,10-dihydrophenanthrenes, and bibenzyls from heartwood of Combretum psidioides. J Chem Soc Perkin Trans I:2941–2946

Letcher RM, Nhamo LRM, Gumiro T 1972 Chemical constituents of the Combretaceae. Part II. Substituted phenanthrenes and 9,10-dihydrophenanthrenes and a substituted bibenzyl from the heartwood of Combretum molle. J Chem Soc Perkin Trans I:206–210

Lewis NG, Yamamoto E 1990 Lignins: occurrence biosynthesis and biodegradation. Annu Rev Plant Physiol Plant Mol Biol 41:455–496

Lewis NG, Yamamoto E, Wooten JB, Just G, Ohashi H, Towers GHN 1987 Monitoring biosynthesis of wheat cell wall phenylpropanoids in situ. Science 237:1344–1346

Lewinsohn E, Gijzen M, Croteau RB 1991 Regulation of monoterpene biosynthesis in conifer defence. In: Parish NWD, Trzaskos JM (eds) Regulation of isoterpenoid metabolism. ACS Symp Ser 497:8

Leyva A, Liang X, Pintor-Toro JA, Dixon RA, Lamb CJ 1992 *cis*-Element combinations determine phenylalanin ammonia-lyase gene tissue-specific expression patterns. Plant Cell 4:263–271

Li L, Brown RM Jr 1993 β-Glucan synthesis in the cotton fiber. II. Regulation and kinetic properties of β-glucan synthase. Plant Physiol 101:1143–1148

Li L, Drake RR Jr, Clement S, Brown RM Jr 1993 β-Glucan synthesis in the cotton fiber. III. Identification of UDP-glucose-binding subunits of β-glucan synthases by photoaffinity labeling with [β-^{32}P]5′-N_3-UDP-glucose. Plant Physiol 101:1149–1156

Liang X, Dron M, Cramer C, Dixon RA, Lamb CJ 1989a Differential regulation of phenylalanine ammonia-lyase genes during development and by environmental cues. J Biol Chem 264:14486–14492

Liang X, Dron M, Schmid J, Dixon RA, Lamb CJ 1989b Developmental and environmental regulation of a phenylalanine ammonia-lyase-β-glucuronidase gene fusion in transgenic tobacco plants. Proc Natl Acad Sci USA 86:9284–9288

Light RJ, Hahlbrock K 1980 Randomization of the flavonoid A ring during biosynthesis of kaempferol from [1,2-$^{13}C_2$] acetate in cell suspension cultures of parsley. Z Naturforsch 35C:717–772

Lin FC, Brown RM Jr, Cooper JB, Delmer DP 1985 Synthesis of fibrils in vitro by a solubilized cellulose synthase from Acetobacter xylinum. Science 230:822–825

Lin FC, Brown RM Jr, Drake RR, Haley BE 1990 Identification of the uridine-5′-diphosphoglucose(UDP-Glc) binding subunit of cellulose synthase in Acetobacter xylinum using the photoaffinity probe-5-azido-UDP-Glc. J Biol Chem 265:4782–4784

Little CHA 1981 Effect of cambial dormancy state on the transport of [1-^{14}C]indol-3-ylacetic acid in Abies balsamea shoots. Can J Bot 59:342–348

Little CHA, Andrew DM, Silk PJ, Strunz GM 1979 Identification of cytokinins zeatin and zeatin riboside in Abies balsamea. Phytochemistry 18:1219–1220

Little CHA, Eidt DC 1968 Effect of abscisic acid on budbreak and transpiration in woody species. Nature 220:498–499

Little CHA, Heald JK, Browning G 1978 Identification and measurement of indolacetic and abscisic acids in the cambial region of Picea sitchensis (Borg.) Carr. by combined gas chromatography-mass spectrometry. Planta 139:133–138

Little CHA, Wareing PF 1981 Control of cambial activity and dormancy in Picea sitchensis by indol-3-ylacetic and abscisic acid. Can J Bot 59:1480–1493

Little CHA, Savidge RA 1987 The role of plant growth regulators in forest tree cambial growth. Plant Growth Reg 6:137–169

Little EL 1993 The Audubon Society Field Guide to North American trees. Knopf, New York, 25–26

Liu L, Dean JFD, Friedman WE, Eriksson K-EL 1994 A laccase-like phenoloxidase is correlated with lignin biosynthesis in Zinnia elegans stem tissues. Plant J 6:213–224

Loake GJ, Faktor O, Lamb CJ, Dixon RA 1992 Combination of H-box [CCTACC(N)$_7$CT] and G-box (CACGTG) *cis* elements is necessary for feed-forward stimulation of a chalcone synthase promoter by the phenylpropanoid-pathway intermediate *p*-coumaric acid. Proc Natl Acad Sci USA 89:9230–9234

Loewus F, Chen MS, Loewus MW 1973 The myo-inositol oxidation pathway to cell wall polysaccharides. In: Loewus F (ed) Biogenesis of plant cell wall polysaccharides. Academic Press, New York, 1–27

Logemann E, Parniske M, Hahlbrock K 1995 Modes of expression and common structural features of the complete phenylalanine ammonia-lyase gene family in parsley. Proc Natl Acad Sci USA 92:5905–5909

Lois R, Dietrich A, Hahlbrock K, Schulz W 1989 A phenylalanine ammonia-lyase gene from parsley: structure, regulation and identification of elicitor and light responsive *cis*-acting elements. EMBO J 8:1641–1648
Lüderitz T, Grisebach H 1981 Enzymatic synthesis of lignin precursors. Comparison of cinnamoyl-CoA reductase and cinnamyl alcohol: $NADP^+$ dehydrogenase from spruce (Picea abies L.) and soybean (Glycine max L.). Eur J Biochem 119:115–124
Lynen F, Eggerer H, Henning U, Kessel I 1958 Farnesyl-pyrophosphat und 3-Methyl-Δ^3-butenyl-1-pyrophosphat, die biologischen Vorstufen des Squalens. Angew Chem 70:738–742
Lynen F, Henning U 1960 Über den biologischen Weg zum Naturkautschuk. Angew Chem 72:820–828
Maccan MC, Wells B, Roberts K 1990 Direct visualization of cross-links in the primary plant cell wall. J Cell Sci 96:323–334
MacKay JJ, Liu W, Whetten R, Sederoff RR, O'Malley DM 1995 Genetic analysis of cinnamyl alcohol dehydrogenase in loblolly pine: single gene inheritance, molecular characterization and evolution. Mol Gen Genet 247:537–545
MacRae WD, Towers GHN 1984 Biological activities of lignans. Phytochemistry 23: 1207–1220
Mäder M, Nessel A, Bopp M 1977 Über die physiologische Bedeutung der Peroxidase-Isoenzym-Gruppen des Tabaks anhand einiger biochemischer Eigenschaften II. pH-Optima, Michaelis-Konstanten, Maximale Oxidationsraten. Z Pflanzenphysiol 82:247–260
Mäder M, Ungemach J, Schloss P 1980 The role of peroxidase isoenzyme groups of Nicotiana tabacum in hydrogen peroxide formation. Planta 147:467–470
Maekawa E 1974 Studies on the hemicelluloses and related polysaccharides of bamboo and bamboo shoot. PhD Thesis, Kyoto Univ, 177pp
Magel EA, Höll W 1993 Storage carbohydrates and ademine nucleotides in tranks of Fagus sylvatica L. in relation to discoloned wood. Holzfonschung 47:19–24
Magel EA, Drouet A, Claudot AC, Ziegler H 1991 Formation of heartwood substances in the stem of Robinia pseudoacacia L. Trees 5:203–207
Manners GD, Jurd L 1977 New natural products from marine borer resistant woods. J Agric Food Chem 25:726–730
Mansell RL, Gross GG, Stöckigt J, Frank H, Zenk MH 1974 Purification and properties of cinnamyl alcohol dehydrogenae from higher plants involved in lignin biosynthesis. Phytochemistry 13:2427–2435
Mansell RL, Stöckigt J, Zenk MH 1972 Reduction of ferulic acid to coniferyl alcohol in a cell free system from a higher plant. Z Pflanzenphysiol 68:286–288
Marcinowski S, Falk H, Hammer DK, Hoyer B, Grisebach H 1979 Appearance and localization of a β-glucosidase hydrolyzing coniferin in spruce (Picea abies) seedlings. Planta 152:578–586
Marcinowski S, Grisebach H 1977 Turnover of coniferin in pine seedlings. Phytochemistry 16:1665–166
Marcinowski S, Grisebach H 1978 Enzymology of lignification. Cell wall-bound β-glucosidases for coniferin from spruce (Picea abies) seedlings. Eur J Biochem 87:37–44
Martin C, Carpenter R, Sommer H, Saedler H, Coen ES 1985 Molecular analysis of instability in flower pigmentation in Antirrhinum majus, following the isolation of the pallida locus by transposon tagging. EMBO J 4:1625–1630
Mascara T, Fincher GB 1982 Biosynthesis of arabinogalactan-protein in Lolium multiflorum (rye grass) endosperm cells. II. In vitro incorporation of galactosyl residues from UDP-galactose into polymeric products. Aust J Plant Physiol 9:31–45
Matern U, Potts JRM, Hahlbrock K 1981 Two flavonoid-specific malonyltransferase from cell suspension cultures of Petroselinum hortense: partial purification and some properties of malonyl-CoA: flavone/flavonol-7-*O*-glucoside malonyltransferase and

malonyl-coenzyme A: flavonol-3-*O*-glucoside malonyltransferase. Arch Biochem Biophys 208:233–241

Matern U, Heller W, Himmelspach K 1983 Conformational changes of apigenin 7-*O*-(6-*O*-malonylglucoside), a vacuolar pigment from parsley, with solvent composition and proton concentration. Eur J Biochem 133:439–448

Matsubara S 1980 Review: structure-activity relationships of cytokinins. Phytochemistry 19:2239–2252

Matsui N, Fukushima K, Kamada K, Nishikawa Y, Yasuda S, Terashima N 1994a On the behavior of monolignol glucosides in lignin biosynthesis. I. Synthesis of monolignol glucosides labeled with ^2H at the hydroxymethyl group of side chain, and polymerization of the labeled monolignols in vitro. Holzforschung 48:215–221

Matsui N, Fukushima K, Yasuda S, Terashima N 1994b On the behavior of monolignol glucosides in lignin biosynthesis. II. Synthesis of monolignol glucosides labeled with ^3H at the hydroxymethyl group of side chain, and incorporation of the label into Magnolia and Ginkgo lignin. Holzforschung 48:375–380

Matsumura S, Stumpf PK 1968 Fat metabolism in higher plants XXXV. Partial primary structure of spinach acyl carrier protein. Arch Biochem Biophys 125:932–941

Mau CJD, West CA 1994 Cloning of casbene synthase cDNA: evidence for conserved structural features among terpenoid cyclases in plants. Proc Natl Acad Sci USA 91:8497–8501

Mavandad M, Edwards R, Liang X, Lamb CJ, Dixon RA 1990 Effects of trans-cinnamic acid on expression of the bean phenylalanine ammonia-lyase gene family. Plant Physiol 94:671–680

McCarty DR, Shaw JR, Hannah LC 1986 The cloning, genetic mapping, and expression of the constitutive sucrose synthase locus of maize. Proc Natl Acad Sci USA 83:9099–9103

McCormick S 1978 Pigment synthesis in maize aleurone from precursors fed to anthocyanin mutants. Biochem Genet 16:777–786

Mehdy MC, Lamb CJ 1987 Chalcon isomerase cDNA cloning and mRNA induction by fungal elicitor, wounding and infection. EMBO J 6:1527–1533

Meier H 1964 General chemistry of cell walls and distribution of the chemical constituents across the walls. In: Zimmerman MH (ed) The formation of wood in forest trees. Academic Press, New York, 137–151

Meier H 1985 Localization of polysaccharides in wood cell walls. In: Higuchi T (ed) Biosynthesis and biodegradation of wood components. Academic Press, Orlando, 43–62

Meier H, Buchs L, Buchala SL, Homewood T 1981 (1→)-β-D-glucan (callose) is a probable intermediate in biosynthesis of cellulose of cotton fibers. Nature 289:821–822

Meier H, Wilkie KCB 1959 The distribution of polysaccharides in the cell-wall of tracheids of pine (Pinus silvestris L.) Holzforschung 13:177–182

Meijer AH, Souer E, Verpoorte R, Hogej HC 1993 Isolation of cytochrome P-450 cDNA clones from the higher plant Catharanthus roseus by a PCR strategy. Plant Mol Biol 22:379–383

Meikle PJ, Ng KF, Johnson E, Hoogenraad NJ, Stone BA 1991 The β-glucan synthase from Lolium multiflorum. Detergent solubilization, purification using monoclonal antibodies, and photoaffinity labeling with a novel photoreactive pyrimidine analog of uridine 5′-diphosphoglucose. J Biol Chem 266:22569–22579

Melchior F, Kindl H 1991 Coordinate- and elicitor-dependent expression of stilbene synthase and phenylalanine ammonia- lyase genes in Vitis cv. optima. Arch Biochem Biophys 288:552–557

Mellerowicz EJ, Coleman WK, Riding T, Little CHA 1992a Periodicity of cambial activity in Abies balsamea II. Effects of temperature and photoperiod on the size of the nuclear genome in fusiform cambial cells. Physiol Plant 85:526–530

Mellerowicz EJ, Coleman WK, Riding T, Little CHA 1992b Periodicity of cambial activity in Abies balsamea. I. Effects of temperature and photoperiod on cambial dormancy and frost hardiness. Physiol Plant 85:515–525
Mellerowicz EJ, Riding T, Little CHA 1989 Genomic variability in the vascular cambium of Abies balsamea. Can J Bot 67:990–996
Miernyk JA 1990 Glycolysis, the oxidative pentose phosphate pathway and anaerobic respiration. In: Dennis DT, Turpin DH (eds) Plant physiology, biochemistry and molecular biology. Longman Scientific & Technical, Singapore, 77–100
Migita N 1968 Chemical composition of woods. In: Migita N, Yonezawa Y, Kondo T (eds) Wood chemistry. Kyoritsu Publ, Tokyo, 65–80
Minami E, Ozeki Y, Matsuoka M, Koizuka N, Tanaka Y 1989 Structure and some characterization of the gene for phenylalanine ammonia-lyase from rice plants. Eur J Biochem 185:19–25
Minamikawa T 1992 Secondary metabolism. In: Miyaji S (ed) Metabolism. Asakura Publ, Tokyo, 172–193
Minamikawa T 1994 Isoprenoids. In: Higuchi T (ed) Wood molecular biology. Buneido Publ, Tokyo, 166–177
Mizutani M, Ward E, DiMaio J, Ohta D, Ryals J, Sato R 1993 Molecular cloning and sequencing of a cDNA encoding mung bean cytochrome P450 (P450C4H) possessing cinnamate 4-hydroxylase activity. Biochem Biophys Res Commun 190:875–880
Monties B 1985 Recent advances on lignin inhomogeneity. In: Van Sumere CF, Lea PJ (eds) The biochemistry of plant phenolics. Clarendon Press, Oxford, 161–181
Moore PJ, Staehelin LA 1988 Immunogold localization of the cell-wall-matrix polysaccharides rhamnogalacturonan I and xyloglucan during cell expansion and cytokinesis in Trifolium pratense L.; implication for secretory pathways. Planta 174:433–445
Mori T 1977 Amino acid metabolism in saplings of Cryptomeria japonica. The role of citrulline. PhD Theis, Kyoto Univ, 85 pp
Moritz T, Philipson JJ, Oden PC 1990 Quantitation of gibberellins A_1, A_3, A_4, A_9 and a putative A_9-conjugate in grafts of sitka spruce (Picea sitchensis) during the period of shoot elongation. Plant Physiol 93:1476–1481
Mors WB, Magelhaes MT, Gottlieb OR 1960 The chemistry of the genus Aniba fragrans, a valid species. Andis Assoc Brasil Quim 19:193–197
Mors WB, Magalhaes MT, Gottlieb OR 1961 Naturally occurring anomatic derivatives of monocyclic α-pyrones. Fortsch Chem Org Naturstoffe 20:131–164
Mühlethaler K 1965 The fine structure of the cellulose microfibril. In: Côté WA (ed) Cellular ultrastructure of woody plants. Syracuse Univ Press, Syracuse, 191–198
Müller W-U, Leistner E 1978a Metabolic relation between naphthalene derivatives in Juglans. Phytochemistry 17:1735–1738
Müller W-U, Leistner E 1978b Aglycones and glycosides of oxygenated naphthalenes and a glycosyltransferase from Juglans. Phytochemistry 17:1739–1742
Murphy DJ 1993 Plant lipid: their metabolism, function and utilization In: Lea PJ, Leegood RC (eds) Plant biochemistry and molecular biology. Wiley, New York, 113–128
Musha Y, Goring ADI 1975 Distribution of syringyl and guaiacyl moieties in hardwoods as indicated by ultraviolet microscopy. Wood Sci Technol 9:45–58
Nabeta K, Nakahara K, Yonekubo J, Okuyama H, Sasaya T 1991 lignan biosynthesis in Larix leptolepis callus. Phytochemistry 30:3591–3593
Nagata T 1995 Multiform of auxin responding genes. In: Nagata T, Uchimiya H (eds) Gene expression in plants. Kodansha Scientific, Tokyo, 69–80
Nakamura W 1967 Studies on the biosynthesis of lignin. I. Disproof against the catalytic activity of laccase in the oxidation of coniferyl alcohol. J Biochem (Jpn) 62:54–60
Nakamura Y, Fushiki H, Higuchi T 1974 Metabolic differences between gymnosperms and angiosperms in the formation of syringyl lignin. Phytochemistry 13:1777–1784

Nakamura Y, Higuchi T 1976 Ester linkage of *p*-coumaric acid in bamboo lignin. Holzforschung 30:187–191

Nakamura Y, Higuchi T 1978a Ester linkage of *p*-coumaric acid in bamboo lignin. II. Syntheses of coniferyl *p*-hydroxybenzoate and coniferyl *p*-coumarate as possible precursors of aromatic esters in lignin. Cell Chem Technol 12:199–208

Nakamura Y, Higuchi T 1978b Ester linkage of *p*-coumaric acid in bamboo lignin III. Dehydrogenative polymerization of coniferyl *p*-benzoate and coniferyl *p*-coumarate. Cell Chem Technol 12:209–221

Nakashima T, Inoue T, Oka A, Nishino T, Osumi T, Hata S 1995 Cloning, expression, and characterization of cDNAs encoding Arabidopsis thaliana squalene synthase. Proc Natl Acad Sci USA 92:2328–2332

Neish AC 1961 The formation of *m*- and *p*-coumaric acids by enzymatic deamination of the corresponding isomers of tyrosine. Phytochemistry 1:1–24

Neish AC, Towers GHN, Chen D, El-Basyouni SZ, Ibrahim RH 1964 The biosynthesis of hydroxybenzoic acids in higher plants. Phytochemistry 3:485–492

Nelson ND 1978 Xylem ethylene, phenol-oxidizing enzymes, and nitrogen, and heartwood formation in walnut and cherry. Can J Bot 56:626–634

Nelson ND, Rietveld WJ, Isebrands JG 1981 Xylem ethylene production in five black walnut families in the early stages of heartwood formation. For Sci 27:537–543

Nimz H, Gaber H 1965 Isolierung von DL-Syringaresinol aus Buchenholz. Chem Ber 98:538–539

Nimz HH, Robert D, Faix O, Nemer M 1981 Carbon-13 NMR spectra of lignins. 8. Structural differences between lignins of hardwoods, softwoods, grass and compression wood. Holzforschung 35:16–26

Nishibe S 1989 Structure elucidation and biological activities of phenylpropanoids, coumarins and lignans from medicinal plants. In: Atta-ur-Rahman (ed) Studies in natural products chemistry. 5. Structure elucidation (Part B). Elsevier, Amsterdam, 505–548

Nishitani K, Okazawa K, Asada K, Kato I 1993 European patent application No. 93302230.0

Nix LE, Wodzicki TJ 1974 The radial distribution and metabolism of IAA-^{14}C in Pinus echinata stems in relation to wood formation. Can J Bot 52:1349–1355

Nobuchi T, Harada H 1985 Ultrastructural changes in parenchyma cells of sugi (Cryptomeria japonica) associated with heartwood formation. Mokuzai Gakkaishi 31:965–973

Noltmann EA 1972 Aldose-ketose isomerases. In: Boyer PD (ed) The enzymes, vol 6, 3rd edn. Academic Press, New York, 271–354

Norberg PH, Meier H 1966 Physical and chemical properties of the gelatinous layer in tension wood fibres of aspen (Populus tremula L). Holzforschung 20:174–178

Northcote DH 1968 The organization of the endoplasmic reticulum, the Golgi bodies and microtubules during cell division and subsequent growth. In: Pridham JB (ed) Plant cell organelles. Academic Press, New York, 179–197

Northcote DH 1989 Control of plant cell wall biogenesis: an overview. In: Lewis NG, Paice MG (eds) Plant cell wall polymers. ACS Symp Ser 399:1–15

Northcote DH, Davey R, Lay J 1989 Use of antisera to localize callose, xylan and arabinogalactan in the cell plate, primary and secondary walls of plant cells. Planta 178:353–366

Northcote DH, Pickett-Heaps JB 1966 A function of the Golgi apparatus in polysaccharide synthesis and transport in the root-cap cells of wheat. Biochem J 98:159–167

Nozoe T, Takase K, Ogata M 1957 β-Dolabrin: a new natural tropolone. Chem Ind (Lond) (31):1070

Odani K 1975 The effect of indole acetic acid and chilling on cambial activity of Pinus densiflora. J Jpn For Soc 57:112–116

Odani K 1985 Indole-3-acetic acid transport in pine shoots under the stage of true dormancy. J Jpn For Soc 67:332–334

Oden PC, Andersson B, Gref R 1982 Identification of gibberellin A_9 in extracts of Norway spruce (Picea abies L. Karst) by combined gas chromatography-mass spectrometry. J Chromatogr 247:133–140

Oden PC, Schwenen L, Graebe JE 1987 Identification of gibberellins in Norway spruce (Picea abies L. Karst) by combined gas chromatography-mass spectrometry. Plant Physiol 84:516–519

Ogura K 1993 Mechanism of chain elongation in isoprenoid biosynthesis. Proc 29th Symp Phytochemistry, Kyoto, Japan, 42–51

Ohana P, Delmer DP, Volman G, Steffens JC, Matthews DE, Benziman M 1991 β-Furfuryl-β-glucoside. An endogenous activator of higher plant UDP-glucose: (1→3)-β-glucan synthase. J Biol Chem 266:13742–13745

Ohashi H 1994 Flavonoid. In: Higuchi T (ed) Wood molecular biology. Buneido Publ, Tokyo, 145–155

Ohashi H, Yamamoto E, Lewis NG, Towers GHN 1987 5-Hydroxyferulic acid in Zea mays and Hordeum vulgare cell walls. Phytochemistry 26:1915–1916

Ohta A, Shimada M, Higuchi T, Takahashi M 1991 A new type of *O*-methyltransferase involved in biosynthesis of secondary metabolites of a brown rot fungus Lentinus lepideus. Mokuzai Gakkaishi 37:275–280

Okamoto T, Yazaki K, Tabata M 1995 Biosynthesis of shikonin derivatives from L-phenylalanine via deoxyshikonin in Lithospermum cell cultures and cell-free extracts. Phytochemistry 38:83–88

Okano T, Koyanagi A 1986 Structural variation of native cellulose related to its source. Biopolymers 25:851–861

Okano T, Koyanagi A, Kondo Y, Sarko A 1989 Structural variation of native cellulose related to its source. In: Schurch C (ed) Cellulose and wood. Chemistry and technology. Wiley, New York, 53–65

Okuda K, Li L, Kudlicka K, Kuga S, Brown RM Jr 1993 β-Glucan synthesis in the cotton fiber I. Identification of β-1,4- and β-1,3-glucans synthesized in vitro. Plant Physiol 101:1131–1142

O'Malley DM, Porter S, Sederoff RR 1992 Purification, characterization, and cloning of cinnamyl alcohol dehydrogenase in loblolly pine (Pinus taeda L.). Plant Physiol 98:1364–1371

O'Malley DM, Whetten R, Bao W, Chen C-L, Sederoff RR 1993 The role of laccase in lignification. Plant J 4:887–892

Omori S, Sakakibara A 1971 Hydrolysis of lignin with dioxane and water. IX. Isolation of D,L-syringaresinol from hydrolysis products of yachidamo (Fraxinus mandshurica). Mokuzai Gakkaishi 17:464–467

Onaka F 1950 Studies on compression and tension wood. Wood Res (Kyoto) 1:1–88

Orme-Johnson WH 1977 Biochemistry of nitrogenase. Basic Life Sci 9 (Genet Eng Nitrogen Fixation): 317–332

Osakabe K, Koyama H, Kawai S, Katayama Y, Morohoshi N 1994 Molecular cloning and the nucleotide sequences of two novel cDNA that encode anionic peroxidase of Populus kitakamiensis. Plant Sci 103:167–175

Osakabe Y, Ohtsubo Y, Kawai S, Katayama Y, Morohoshi N 1995 Structure and tissue-specific expression of genes for phenylalanine ammonia-lyase from a hybrid aspen, Populus kitakamiensis. Plant Sci 105:217–226

Ozawa S, Davin IB, Lewis NG 1993 Formation of (−)-arctigenin in Forsythia intermedia. Phytochemistry 32:643–652

Pakusch A-E, Kneusel RE, Matern U 1989 *S*-Adenosyl-L-methionine: trans-caffeoyl-coenzyme *A* 3-*O*-methyl transferase from elicitor-treated parsley cell suspension cultures. Arch Biochem Biophys 271:488–494

Panayotato N, Villemez CL 1973 The formation of a β-(1→4)-D-galactan chain catalyzed by a Phaseolus aureus enzyme. Bicochem J 133:263–271

Paoletti F, Aldinucci D 1986 Immunoaffinity purification of rat liver transketolase: evidence for multiple forms of the enzyme. Arch Biochem Biophys 245:212–219

Parameswaran N, Liese W 1981 The fine structure of bamboo. In: Higuchi (ed) Bamboo production and utilization. Wood Res Inst, Kyoto Univ, 178–183

Paré PW, Wang H-B, Davin LB, Lewis NG 1994 (+)-Pinoresinol synthase: a stereoselective oxidase catalysing 8,8′-lignan formation in Forsythia intermedia. Tetrahedron Lett 35:4731–4734

Patschke L, Grisebach H 1965 Zur Biogenese der Flavonoid X. Über den Zeitpunkt der Hydroxylierung im Ring B. Z Naturforsch 20B:1039–1042

Pelman J, Boerjan W, Engler G, Seurinck J, Botterman J, Alliotte T, Van Montagu M, Inzé D 1989 Strong cellular preference in the expression of a housekeeping gene of Arabidopsis thaliana encoding *S*-adenosyl-L-methionine synthase. Plant Cell 1:81–93

Pensar G 1967 The distribution and composition of extractives in wood. 1. Earlywood and latewood in spruce. Inst Träkemi Åbo Akad, Åbo, 31 pp

Pensar G 1968 The distribution and composition of extractives in wood. 2. Earlywood and latewood in pine. Inst Träkemi Åbo Akad, Åbo, 33 pp

Pensar G 1969 Studies on the radial distribution and composition of wood rosin in cross sections of Scandinavian spruce and pine. Inst Träkemi Åbo Akad, Åbo, 35 pp

Perrin PW, Towers GHN 1973 Hispidin biosynthesis in cultures of Polyporus hispidus. Phytochemistry 12:589–592

Pharis RP, Jenkins PA, Aoki H, Sasaa T 1981 Hormonal physiology of wood growth in Pinus radiata D. Don: effects of gibberellin A_4 and the influence of abscisic acid upon [^{3}H] gibberellin A_4 metabolism. Aust J Plant Physiol 8:559–57

Phelps JE, McGinnes EA Jr, Smolinski M, Saniewski M, Pieniqzek J 1977 A note on the formation of compression wood induced by morphactin IT 3456 in Thuja shoots. Wood Fiber 8:223–227

Phelps JE, Saniewski M, Pieniqzek J, McGinnes EA Jr 1974 A note on the structure of morphactin-induced wood in two coniferous species. Wood Fiber 6:13–17

Philipson JJ, Coutts MP 1980 Effects of growth hormone application on the secondary growth of roots and stems in Picea sitchensis (Bong.) Carr. Ann Bot 46:747–755

Picket-Heaps JD 1968 Xylem wall deposition. Autoradiographic investigations using lignin precursors. Protoplasma 65:181–205

Pinto JEBP, Suzich JA, Herrmann KM 1986 3-Deoxy-D-arabino-heptulosonate 7-phosphate synthase from potato tuber (Solanum tuberosum L.). Plant Physiol 82:1040–1044

Popjak G 1959 The biosynthesis of allylic alcohols from [2-^{14}C] mevalonate in liver enzyme preparations and their relation to synthesis of squalene. Tetrahedron Lett 19:19–23

Porter KR, Claude A, Fullam EF 1945 A study of tissue culture cells by electron microscopy. J Exp Med 81:233–246

Porter KR, Machado RD 1969 Studies on the endoplasmic reticulum IV. Its form and distribution during mitosis in cells of onion root tip. J Biophys Biochem Cytol 7:167–180

Porter LJ 1989 Condensed tannins. In: Rowe JW (ed) Natural products of woody plants I. Springer, Berlin Heidelberg New York, 651–690

Potts JR, Weklych R, Conn EE 1974 Metabolism of aromatic compounds in higher plants XI. 4-Hydroxylation of cinnamic acid by sorghum microsomes and the requirement for cytochrome P-450. J Biol Chem 249:5019–5026

Poulton JE 1981 Transmethylation and demetylation reactions in the metabolism of secondary plant products. In: Conn EE (ed) The biochemistry of plants, vol 7. Secondary plant products. Academic Press, Orlando, 667–723

Prasad PV, Jones AM 1991 Putative receptor for the plant growth hormone auxin identified and characterized by anti-idiotypic antibodies. Proc Natl Acad Sci USA 88:5479–5483

Preston RD 1964 Structural and mechanical aspects of plant cell walls with particular reference to synthesis and growth. In: Zimmermann MH (ed) The formation of wood in forest trees. Academic Press, New York, 169–188

Preston RD 1971 Negative staining and cellulose microfibril size. J Microsc 93:7–13

Preston RD 1974 The physical biology of plant cell walls. Chapman & Halls, London, 491 pp

Preston RD, Cronshaw J 1958 Constitution of the fibrillar and non-fibrillar components of the walls of Valonia venticosa. Nature 181:248–250

Rahman MMA, Dewick PM, Jackson DE, Lucas JA 1990 Biosynthesis of lignans in Forsythia intermedia. Phytochemistry 29:1841–1846

Ralph J, Helm RF, Quideau S 1992b Lignin-feruloyl ester cross-links in grasses. Part 2. Model compound syntheses. J Chem Soc Perkin Trans 1:2971–2980

Ralph J, Helm RF, Quideau S, Hatfield RD 1992a Lignin-feruloyl ester cross-links in grasses. Part 1. Incorporation of feruloyl esters into conferyl alcohol dehydrogenation polymers. J Chem Soc Perkin Trans 1:2961–2967

Ralph J, Helm RF 1993 Lignin/hydroxycinnamic acid/polysaccharide complexes: synthetic models for regiochemical characterization. In: Forage cell wall structure and digestibility. Am Soc Agronomy, Madison, 201–246

Ralph J, Hatfield RD, Quideau S, Helm RF, Grabber JH, Jung H-JG 1994a Pathway of *p*-coumaric acid incorporation into maize lignin as revealed by NMR. J Am Chem Soc 116:9448–9456

Ralph J, Quideau S, Grabber JH, Hatfield RD 1994b Identification and synthesis of new ferulic acid dehydrodimers present in grass cell walls. J Chem Soc Perkin Trans 1:3485–3498

Rao CBS 1978 Chemistry of lignans. Andhra Univ Press, Andhra Pradesh, 377 pp

Ray PM 1967 Radioautographic study of cell wall deposition in growing plant cells. J Cell Biol 35:659–674

Ray PM 1980 Cooperative action of β-glucan synthase and UDP-xylose xylosyl transferase of Golgi membrane in the synthesis of xyloglucan-like polysaccharide. Biochim Biophys Acta 629:431–444

Rees T 1990 Carbon metabolism in mitochondria. In: Dennis DT, Turpin DH (eds) Plant physiology, biochemistry and molecular biology. Longman Scientific & Technical, Singapore, 106–123

Reimold U, Kröger M, Kreuzaler F, Hahlbrock K 1983 Cloning and 3′ non-coding nucleotide sequence of chalcone synthase mRNA and assignment of amino acid sequence of the enzyme. EMBO J 2:1801–1806

Reinold S, Hauffe KD, Douglas CJ 1993 Tobacco and parsley 4-coumarate: coenzyme A ligase genes are temporally and spatially regulated in a cell type-specific manner during tobacco flower development. Plant Physiol 101:373–383

Revol JF 1985 Changes of the d spacing in cellulose crystals during lattice imaging. J Mat Sci Lett 4:1347–1349

Rhodes MJC, Wooltorton LSC 1974 Reduction of the CoA thioesters of *p*-coumaric and ferulic acids by extract of aged Brassica napo-brassica root tissue. Phytochemistry 13:107–110

Rhodes MJC, Wooltorton LSN 1975 Enzymes involved in the reduction of ferulic acid to coniferyl alcohol during the aging of disks of swede root tissues. Phytochemistry 14:1235–1240

Riding RT, Little CHA 1984 Anatomy and histochemistry of Abies balsamea cambial zone cells during the onset and breaking of dormancy. Can J Bot 62:2570–2579

Roberts LW, Gahan PB, Aloni R 1988 Vascular differentiation and plant growth regulators. Springer, Berlin Heidelberg New York, 154 pp

Rolfs C-H, Fritzemeier K-H, Kindl H 1981 Cultured cells of Arachis hypogaea susceptible to induction of stilbene synthase (resveratrol-forming). Plant Cell Rep 1:83–85

Rolfs C-H, Kindl H 1984 Stilbene synthase and chalcone synthase. Plant Physiol 75:489–492

Ross P, Aloni Y, Weinhouse H, Michaeli D, Weinberger-Ohama P, Mayer R, Benziman M 1986 Control of cellulose synthesis in Acetobacter xylinum. A unique guanyl oligonucleotide is the immediate activator of the cellulose synthase. Carbohydr Res 149:101–117

Ross P, Weinhouse H, Aloni Y, Michaeli D, Weinberger-Ohana P, Mayer R, Braun S, De Vroom E, Der Vroom GA, Van der Marel GA, Van Broom JH, Benziman M 1987 Regulation of cellulose synthesis in Acetobacter xylinum by cyclic diguanylic acid. Nature 325:279–281

Roux DG 1958 Biogenesis of condensed tannins from leucoanthocyanins. Nature 181:1454–1456

Roux DG, Evelyn SR 1958 Condensed tannins 3. Biogenesis of condensed tannins based on leucoanthocyanins. Biochem J 70:344–349

Roux DG, Evelyn SR 1960 Condensed tannins. 4. The distribution of tannins in heartwoods of Acacia mollissima and Schinopsis spp. Biochem J 76:17–23

Rowe JW, Seikel MK, Roy ND, Jorgensen E 1972 Chemotaxonomy of Ulmus. Phytochemistry 11:2513–2517

Rubin JL, Jensen RA 1985 Differentially regulated isozymes of 3-deoxy-D-arabino-heptulosonate-7-phosphate synthase from seedlings of Vigna radiata L. Wilczek. Plant Physiol 79:711–718

Ruel K, Barnoud F, Goring DAI 1978 Lamellation in S_2 layer of softwood tracheids as demonstrated by scanning transmission electron microscopy. Wood Sci Technol 12:287–291

Rupprich N, Hildebrand H, Kindl H 1980 Substrate specificity in vivo and in vitro in the formation of stilbenes. Biosynthesis of rhaponticin. Arch Biochem Biophys 200:72–78

Rupprich N, Kindl H 1978 Stilbene synthases and stilbene carboxylate synthase I. Enzymic synthesis of 3,5,4′-trihydroxystilbene from *p*-coumaroyl coenzyme A and malonyl coenzyme A. Hoppe-Seyler Z Physiol Chem 359:165–175

Russel DW 1971 The metabolism of aromatic compounds in higher plants. J Biol Chem 246:3870–3878

Ryder TB, Hedrick SA, Bell JN, Liang X, Clouse SD, Lamb CJ 1987 Organization and differential activation of a gene family encoding the plant defence enzyme chalcone synthase in Phaseolus vulgaris. Mol Gen Genet 210:219–233

Sackett DL 1993 Podophyllotoxin, steganacin and combretastatin: Natural products that bind at the colchicine site of tubulin. Pharmacol Ther 59:163–228

Saiki H 1965 Studies on annual ring structure of coniferous wood. IV. Effects of diameter decrease and cell wall thickening of tracheid on late wood formation. Mokuzai Gakkaishi 11:1–6

Saiki H 1970 Proportion of component layers in tracheid wall of early wood and late wood of some conifers. Mokuzai Gakkaishi 16:244–249

Saiki H 1982 The structure of domestic and important woods in Japan. Jpn For Technol Assoc, Tokyo, 218 pp

Saka S, Goring DAI 1985 Localization of lignins in wood cell walls. In: Higuchi T (ed) Biosynthesis and biodegradation of wood components. Academic Press, Orlando, 51–62

Sakai K, Kusaba K, Tsutsumi Y, Shiraishi T 1994 Secondary metabolites in cell culture of woody plants III. Formation of β-thujaplicin in Cupressus lusitanica callus cultures treated with fungal elicitors. Mokuzai Gakkaishi 40:1–5

Sakakibara A 1980 A structural model of softwood lignin. Wood Sci Technol 14:89–100

Sakakibara A 1983 Chemical structure of lignin related mainly to degradation products. In: Higuchi T, Chang H-M, Kirk TK (eds) Recent advances in lignin biodegradation research. UNI Publ, Tokyo, 12–33
Saleh NAM, Fritsch H, Kreuzaler F, Grisebach H 1976a UDP-glucose: cyanidin 3-*O*-glucosyltransferase from cell cultures of Haplopappus gracilis. Planta 133:41–45
Saleh NAM, Fritsch H, Kreuzaler F, Grisebach H 1978 Flavanone synthase from cell suspension cultures of Haplopappus gracilis and comparison with the synthase from parsley. Phytochemistry 17:183–186
Saleh NAM, Poulton JE, Grisebach H 1976b UDP-glucose: cyanidin 3-*O*-glucosyltransferase from red cabbage seedlings. Phytochemistry 15:1865–1868
Saleh TM, Leney L, Sarkanen KV 1967 Radioautographic studies of cottonwood, Douglas fir and wheat plants. Holzforschung 21:116–120
Sandermann W 1966 Wood constituents, their chemistry and biochemistry. Naturwissenschaften 53:513–524
Sandermann WS, Dietrichs HH 1959 Chemische Untersuchungen an Teakholz. Holzforschung 13:137–148
Sandberg G, Ericsson A 1987 IAA concentration in the leading shoot and living stem bark of Scotch pine: seasonal variation and effects of pruning. Tree Physiol 3:173–184
Sarkanen KV, Ludwig CH 1971 Lignins: occurrence, formation, structure and reactions. Wiley, New York, 59–60
Sarko A, Muggli R 1974 Packing analysis of carbohydrates and polysaccharides. III. Valonia cellulose and cellulose II. Macromolecules 7:486–494
Sarni F, Grand C, Boudet AM 1984 Purification and properties of cinnamoyl-CoA reductase and cinnamyl alcohol dehydrogenase from poplar stem (Populus × euramericana). Eur J Biochem 139:259–265
Sato Ka 1992 Energy metabolism. In: Miyaji S (ed) Photosynthesis. Asakura Publ, Tokyo, 18–36
Sato Ki 1992 Molecular structure and function of photosynthesizing organ. In: Miyaji S (ed) Photosynthesis. Asakura Publ, Tokyo, 1–18
Sato Y, Sugiyama M, Gorecki RJ, Fukuda H, Komamine A 1993 Interrelationship between lignin deposition and the activities of peroxidase isoenzymes in differentiating tracheary elements of Zinnia: analysis using L-α-aminooxy-β-phenylpropionic acid and 2-aminoindan-2-phosphoric acid. Planta 189:584–589
Sato Y, Sugiyama M, Komamine A, Fukuda H 1995 Separation and characterization of the isoenzymes of wall-bound peroxidase from cultured Zinnia cells during tracheary element differentiation. Planta 196:141–147
Savidge RA 1983 The role of plant hormones in higher plant cellular differentiation. 2. Experiments with the vascular cambium, and sclereid and tracheid differentiation in the pine, Pinus contorta. Histochem J 15:447–466
Savidge RA 1990 Characterization of indole-3-ylacetic acid in developing secondary xylem of 26 Canadian species by combined gas chromatography-mass spectrometry. Can J Bot 68:521–523
Savidge RA, Heald JK, Wareing PF 1982 Non-uniform distribution and seasonal variation of endogenous indol-3-ylacetic acid in the cambial region of Pinus contorta Dougl. Planta 155:89–92
Savidge RA, Mutumba GMC, Heald JK, Wareing PF 1983 Gas chromatography-mass spectroscopy identification of 1-aminocyclopropane-1-carboxylic acid in compression wood vascular cambium of Pinus contorta Dougl. Plant Physiol 71:434–436
Savidge RA, Udagama-Randeniya P 1992 Cell wall-bound coniferyl alcohol oxidase associated with lignification in conifers. Phytochemistry 31:2959–2966
Savidge RA, Wareing PF 1981 Plant growth regulators and the differentiation of vascular elements. In: Barnett JR (ed) Xylem cell development. Castle House, Tunbridge Wells, 192–235

Savidge RA, Wareing PF 1982 Apparent auxin production and transport during winter in the nongrowing pine tree. Can J Bot 60:681–691

Savidge RA, Wareing PF 1984 Seasonal cambial activity and xylem development in Pinus contorta in relation to endogenous IAA and (S)-abscisic acid levels. Can J For Res 14:676–682

Saxena IM, Kudlicka K, Okuda K, Brown RM Jr 1994 Characterization of genes in the cellulose-synthesizing operon (*acs* operon) of Acetobacter xylinum: implications for cellulose crystallization. J Bacteriol 176:5735–5752

Saxena IM, Lin FC, Brown RM Jr 1990 Cloning and sequencing of the cellulose synthase catalytic subunit gene of Acetobacter xylinum. Plant Mol Biol 15:673–683

Saxena IM, Lin FC, Brown RM Jr 1991 Identification of a new gene in an operon for cellulose biosynthesis in Acetobacter xylinum. Plant Mol Biol 16:947–954

Scalbert AB 1984 Caracterisation des lignins de paille de blé: fractionnement, association avec les oses et les acides phénoliques. Thèse de Docteur-Ingénieur. Inst Natl Agron Paris-Grignon, 58pp

Scalbert AB, Monties JY, Lallemand E, Guittet E, Roland C 1985 Ether linkage between phenolic acids and lignin fractions from wheat straw. Phytochemistry 24:1359–1362

Scalbert AB, Monties JY, Rolando C, Sierr-Escudero A 1986 Formation of ether linkage between phenolic acids and Gramineae lignin: a possible mechanism involving quinone methides. Holzforschung 40:191–195

Schaller A, Schmid J, Leibinger U, Amrhein N 1991 Molecular cloning and analysis of a cDNA coding for chorismate synthase from the higher plant Corydalis sempervirens Pers. J Biol Chem 266:21434–21438

Schanz S, Schröder G, Schröder J 1992 Stilbene synthase from Scots pine (Pinus sylvestris). FEBS Lett 313:71–74

Schmid G, Grisebach H 1982 Enzymatic synthesis of lignin precursors: Purification and properties of UDP-glucose: coniferyl alcohol glucosyltransferase from cambial sap of spruce (Picea abies L.). Eur J Biochem 123:363–370

Schmid G, Hammer DK, Ritterbusch A, Grisebach H 1982 Appearance and immunohistochemical localization of UDP-glucose: coniferyl alcohol glucosyltransferase in spruce (Picea abies (L.) Karst. seedlings. Planta 156:207–212

Schmid J, Schaller A, Leibinger U, Boll W, Amrhein N 1992 The in vitro synthesized tomato shikimate kinase precursor is enzymatically active and is imported and processed to the mature enzyme by chloroplasts. Plant J 2:375–383

Schmidt CL, Danneel H-J, Schulz B, Buchanan BB 1990 Shikimate kinase from spinach chloroplasts. Plant Physiol 93:758–766

Schmitt D, Pakusch A-E, Matern U 1991 Molecular cloning, induction, and taxonomic distribution of caffeoyl-CoA 3-*O*-methyltransferase, an enzyme involved in disease resistance. J Biol Chem 266:17416–17423

Schöppner A, Kindl H 1979 Stilbene synthase (pinosylvin synthase) and its induction by ultraviolet light. FEBS Lett 108:349–352

Schöppner A, Kindl H 1984 Purification and properties of a stilbene synthase from induced cell suspension cultures of peanut. J Biol Chem 259:6806–6811

Schorger AW 1917 A study of the chemistry of wood I. Methods and results of analysis of some American species. II. Discussion of methods and results. Ind Eng Chem 9:556–561, 561–566

Schröder G, Schröder J 1992 A single change of histidine to glutamin alters the substrate preference of a stilbene synthase. J Biol Chem 267:20558–20560

Schubert KR 1982 Symbiotic nitrogen fixation in nodulated leguminous and nonleguminous angiosperms. In: Mitsui A, Black CC (eds) Handbook of biosolar resources. CRC Press, Orlando, 265–282

Schwekendiek A, Pfeffer G, Kindl H 1992 Pine stilbene synthase cDNA, a tool for probing environmental stress. FEBS Lett 301:41–44

Scott JAN, Procter AR, Fergus BJ, Goring DAI 1969 The application of ultraviolet microscopy to the distribution of lignin in wood. Description and validity of the technique. Wood Sci Technol 3:73–92

Shain L, Hillis WE 1973 Etylene production in xylem of Pinus radiata in relation to heartwood formation. Can J Bot 51:1331–1335

Shain L, Mackay JFG 1973 Seasonal fluctuations in respiration of aging xylem in relation to heartwood formation in Pinus radiata. Can J Bot 51:737–741

Shanklin J, Somerville C 1991 Stearoyl-acyl-carrier-protein desaturase from higher plants is structurally unrelated to the animal and fungal homologs. Proc Natl Acad Sci USA 88:2510–2514

Shibata D 1994 Gene expression and its control. In: Higuchi T (ed) Wood molecular biology. Buneido Publ, Tokyo, 29–56

Shibata S, Yamazaki M 1957 The biogenesis of rutin. Pharm Bull Tokyo 5:501–502

Shigematsu M, Goto A, Yoshida S, Tanahashi M, Shinoda Y 1994 Hydrophobic regions of hemicelluloses estimated by fluorescent probe method. Mokuzai Gakkaishi 40:1214–1218

Shimada M, Fushiki H, Higuchi T 1973a Metabolism of biochemical formation of the methoxyl groups in soft and hardwood lignins. Mokuzai Gakkaishi 19:13–21

Shimada M, Kuroda H, Higuchi T 1973b Evidence for the formation of methoxyl groups of ferulic and sinapic acids in Bambusa by the same *O*-methyltransferase. Phytochemistry 12:2873–2875

Shimada M, Ohashi H, Higuchi T 1970 *O*-Methyltransferases involved in the biosynthesis of lignins. Phytochemistry 9:2463–2470

Shimaji K 1976a Growth of trees. In: Shimaji K, Sudo S, Harada H (eds) Wood tissues. Morikita Publ, Tokyo, 17–27

Shimaji K 1976b Cambium. In: Shimaji K, Sudo S, Harada H (eds) Wood tissues. Morikita Publ, Tokyo, 28–45

Shimomura S 1995 What is auxin-binding protein? In: Nagata T, Uchimiya H (eds) Gene expression in plants. Kodansha Scientific, Tokyo, 58–69

Shorrosh BS, Dixon RA, Ohlrogge JB 1994 Molecular cloning, characterization, and elicitation of acetyl-CoA carboxylase from alfalfa. Proc Natl Acad Sci USA 91:4323–4327

Siegel SM 1969 Evidence for the presence of lignin in moss gametophytes. Am J Bot 56:175–179

Siehl DL, Conn EE 1988 Kinetic and regulatory properties of arogenate dehydratase in seedlings of Sorghum bicolor (L.). Arch Biochem Biophys 260:822–829

Simon P 1993 Diversity and conservation of plant peroxidases. Plant Peroxidase Newsl 1:4–7

Simson BW, Timell TE 1978 Polysaccharides in cambial tissues of Populus tremuloides and Tilia americana I. Isolation, fractionation, and chemical composition of the cambial tissue. Cell Chem Technol 12:39–50

Singh SV, Guha SRD 1981 Indian experience of paper-making from bamboo. In: Higuchi (ed) Bamboo production and utilization. XVII IUFRO World Congr, Kyoto, 33–38

Sipilä J, Bruno G 1991a On the mechanism of formation of non-cyclic benzyl ethers during lignin biosynthesis. Part 2. The effect of pH on the reaction between a β-O-4 type quinone methide with vanillyl alcohol in water-dioxane solutions. The stability of non-cyclic benzyl aryl ethers during lignin biosynthesis. Holzforschung 45:275–278

Sipilä J, Bruno G 1991b On the mechanism of formation of non-cyclic benzyl ethers during lignin biosynthesis. Part 3. The reactivity of a β-O-4 type quinone methide with methyl-α-D-glucopyranoside in competition with vanillyl alcohol. The formation and stability of benzyl ethers between lignin and carbohydrates. Holzforschung 45:3–7

Sipilä J, Bruno G 1991c On the mechanism of formation of non-cyclic benzyl ethers during lignin biosynthesis. Part 4. The reactions of a β-O-4 type quinone methide with carboxylic acids in the presence of phenols. The formation and stability of benzyl esters between lignin and carbohydrates. Holzforschung 45:9–11

Smith CJ 1993 Carbohydrate chemistry. In: Lea PJ, Leagood RC (eds) Plant biochemistry and molecular biology. Wiley, New York, 73–111

Smith RJ, Gallon JR 1993 Nitrogen fixation. In: Lea PJ, Leegood RC (eds) Plant biochemistry and molecular biology. Wiley, New York, 129–153

Smolinski M, Saniewski M, Pieniqzek J 1972 The effect of morphactin IT3456 on cambial activity and wood differentiation in Picea excelsa. Bull Acad Pol Sci 20:431–435

Soliday CL, Kolattukudy PE 1977 Biosynthesis of cutin. ω-Hydroxylation of fatty acids by a microsomal preparation from germinating Vicia faba. Plant Physiol 59:1116–1121

Soliday CL, Kolattukudy PE 1978 Midchain hydroxylation of 16-hydroxypalmitic acid by the endoplasmic reticulum fraction from germinating Vicia faba. Arch Biochem Biophys 188:338–347

Song L, Poulter CD 1994 Yeast farnesyl-diphosphate synthase: site-directed mutagenesis of residues in highly conserved prenyltransferase domain I and II. Proc Natl Acal Sci USA 91:3044–3048

Sonobe S 1989 Actin fiber. Cell 21:8–11

Sorvari J, Pietarila V, Nygren-Konttinen A, Klemola A, Laine JE, Sjöström E 1983 Attempts at isolating and characterizing secondary wall and middle lamella from spruce wood (Picea abies). Paperi Puu 65:117–121

Sorvari J, Sjöström E, Klemola A, Laine JE 1986 Chemical characterization of wood constituents, especially lignin, in fraction separated from middle lamella and secondary wall of Norway spruce (Picea abies). Wood Sci Technol 20:35–52

Specht JE, Gorz HJ, Haskins FA 1976 Genetic regulation of flavonoid content in seeds and seedlings of Melilotus alba. Phytochemistry 15:133–134

Stafford HA, Ibrahim RK (eds) 1992 Phenolic metabolism in plants. Plenum Press, New York, 411 pp

Stafford HA, Lester HH 1985 Flavan-3-ol biosynthesis. The conversion of (+)-dihydroquercetin and flavan-3,4-*cis*-diol(leucoanthocyanidin) to (+)-catechin by reductases extracted from cell suspension cultures of Douglas fir. Plant Physiol 76:184–186

Stafford HA, Lester HH 1984 Flavan-3-ol biosynthesis. The conversion of (+)-dihydroquercetin and its flavan-3,4-*cis*-diol(leucodelphinidin) to (+)-gallocatechin by reductases extracted from tissue cultures of Ginkgo biloba and Psudotsuga menziesii. Plant Physiol 78:791–794

Steele CL, Lewinsohn E, Croteau R 1995 Induced oleoresin biosynthesis in grand fir as a defense against bark beetles. Proc Natl Acad Sci USA 92:4164–4168

Steinrücken HC, Amrhein N 1980 The herbicide glyphosate is a potent inhibitor of 5-enolpyruvylshikimic acid-3-phosphate synthase. Biochem Biophys Res Commun 94:1207–1212

Stenmark SL, Pierson DL, Glover GI, Jensen RA 1974 Blue-green bacteria synthesize L-tyrosin by the pretyrosin pathway. Nature 247:290–292

Sterijades R, Dean JFD, Eriksson K-EL 1992 Laccase from sycamore maple (Acer pseudoplatanus) polymerizes monolignols. Plant Physiol 99:1162–1168

Sterjiades R, Dean JFD, Gamble G, Himmelsbach DS, Eriksson K-E 1993 Extracellular laccases and peroxidases from sycamore maple (Acer pseudoplatanus) cell suspension cultures. Reactions with monolignols and lignin model compounds. Planta 190:75–87

Stevenson R 1978 Diarylbutanes. In: Rao CBS (ed) Chemistry of lignans. Andhra Univ Press, Andhra Pradesh, 65–94

Stewart CM 1966 The chemistry of secondary growth in trees. CSIRO, Australia, 20:107

Stickland RG, Harrison BJ 1974a Precursors and genetic control of pigmentation. 1. Induced biosynthesis of pelargonidin, cyanidin and delphinidin in *Antirrhinum majus*. Heredity 33:108–112

Stickland RG, Harrison BJ 1974b Precursors and genetic control of pigmentation. 2. Genotype analysis of pigment controlling genes in acyanic phenotypes in *Antirrhinum majus*. Heredety 33:112–115

Stöckigt J, Klischies M 1977 Biosynthesis of lignans Part I. Biosynthesis of arctiin and Phillyrin. Holzforschung 31:41–44

Stöckigt J, Mansell RL, Gross GG, Zenk MH 1973 Enzymic reduction of *p*-coumaric acid via *p*-coumaroyl-CoA to *p*-coumaryl alcohol by a cell free system from Forsythia sp. Z Pflanzenphysiol 70:305–307

Stotz G, De Vlaming P, Wiering H, Schram AW, Forkman G 1985 Genetic and biochemical studies on flavonoid 3′-hydroxylation in flower of *Petunia hybrida*. Theor Appl Genet 70:300–305

Strickland KP 1967 The biogenesis of the lipids In: Bernfeld P (ed) Biogenesis of natural compounds. Pergamon Press, London, 103–205

Stumpf PK, Conn EE (eds) 1981 Secondary products. The biochemistry of plants, vol 7. Academic Press, San Diego, 798 pp

Su JC, Preiss J 1978 Purification and properties of sucrose synthase from maize kernels. Plant Physiol 61:389–393

Sudo S 1976 Cellular constitutions of wood. In: Shimaji K, Sudo S, Harada H (eds) Wood tissues. Morikita Publ, Tokyo, 111–199

Sugiura M 1986 *Nicotiana tabacum* chloroplast DNA 155844 bp. Bulletin, Center for Gene Research, Nagoya Univ

Sugiura M 1989a *Oryza sativa* chloroplast DNA 134525 bp. Bulletin, Center for Gene Research, Nagoya Univ

Sugiura M 1989b The chloroplast genome. In: Marcus A (ed) The biochemistry of plants, vol 15. Academic Press, San Diego, 133–150

Sugiura M 1992 Plant molecular biology, vol 37. Tanpakushitsu, Kakusan, Kouso, 1368 pp

Sugiura M 1993 *Pinus thunbergii* chloroplast DNA 119707 bp. Bulletin, Center for Gene Research, Nagoya Univ

Sugiyama J 1994 Multiformity of cellulose microfibrils. Cell Commun (Jpn)1:6–12

Sugiyama J, Harada H, Fujiyoshi Y, Uyeda N 1985a Lattice images from ultrathin sections of cellulose microfibrils in the cell wall of *Valonia macrophysa* Kutz. Planta 166:161–168

Sugiyama J, Harada H, Fujiyoshi Y, Uyeda N 1985b Observations of cellulose microfibrils in *Valonia macrophysa* by high resolution electron microscopy. Mokuzai Gakkaishi 31:61–67

Sugiyama J, Vuong R, Chanzy H 1991 Electron diffraction study of the two crystalline phases occurring in native cellulose from an algal cell wall. Macromolecules 24:4168–4175

Sundberg B, Little CHA 1990 Tracheid production in response to changes in the internal level of indole-3-acetic acid in 1-year-old shoots of Scots pine. Plant Physiol 94:1721–1727

Sundberg B, Little CHA, Riding RT, Sandberg G 1987 Levels of endogenous indole-3-acetic acid in the vascular cambium region of Abies balsmea trees during the activity-rest-quiescence transition. Physiol Plant 71:163–170

Sundberg B, Little CHA, Cui K 1990 Distribution of indole-3-acetic acid and the occurrence of its alkali-labile conjugates in the extraxylary region of *Pinus sylvestris* stems. Plant Physiol 93:1295–1302

Süss K-H, Arkona C, Manteuffel R, Adler K 1993 Calvin cycle multienzyme complexes are bound to chloroplast thylakoid membranes of higher plants in situ. Proc Natl Acad Sci USA 90:5514–5518

Sütfeld R, Wiermann R 1978 The occurrence of two different SAM: 3,4-dihydric phenol 3-*O*-methyltransferases in tulip anthers. Biochem Physiol Pflanz 172:111–124

Sütfeld R, Wiermann R 1981 Purification of chalcone synthase from tulip anthers and comparison with the synthase from Cosmos petals. Z Naturforsch 36C:30–34

Sutter A, Grisebach H 1973 UDP-glucose: flavonol 3-*O*-glucosyl transferase from cell suspension cultures of parsley. Biochim Biophys Acta 309:289–295

Sutter A, Ortmann R, Grisebach H 1972 Purification and properties of an enzyme from cell suspension cultures of parsley catalyzing the transfer of D-glucose from UDP-D-glucose to flavonoids. Biochim Biophys Acta 258:71–87

Swan EP, Jiang KS 1970 Formation of heartwood extractives in western red ceder. Tappi 53:844–846

Swan EP, Jiang KS, Gardner JFA 1969 The lignans of Thuja plicata and sapwood-heartwood transformation. Phytochemistry 8:345–351

Tabak AJH, Meyer H, Bennink GJH 1978 Modification of the B-ring during flavonoid biosynthesis in Petunia hybrida: introduction of the 3′hydroxyl group is regulated by the gene Ht1. Planta 139:67–71

Takabe K 1994 Ultrastructure and function of the cell. In: Higuchi T (ed) Wood molecular biology. Buneido Publ, Tokyo, 1–17

Takabe K, Fujita M, Harada H, Saiki H 1981 Lignification process of Japanese black pine (*Pinus thunbergii Parl*) tracheids. Mokuzai Gakkaishi 27:813–820

Takabe K, Fujita M, Harada H, Saiki H 1983 Changes in the composition and the absolute amount of sugars with the development of Cryptomeria tracheids. Mokuzai Gakkaishi 29:183–189

Takabe K, Fujita M, Harada H, Saiki H 1985 Autoradiographic investigation of lignification in the cell walls of Cryptomeria (Cryptomeria japonice D. Don). Mokuzai Gakkaishi 31:613–619

Takabe K, Harada H, Fukazawa K 1989 Deposition of cell wall components in conifer tracheids. In: Lewis NG, Paice MG (eds) Plant cell wall polymers. ACS Symp Ser 399:47–66

Takabe K, Miyauchi S, Tsunoda R, Fukazawa K 1992 Distribution of guaiacyl and syringyl lignins in Japanese beech (Fagus crenata): variation within an annual ring. IAWA Bull 13:105–112

Takabe K, Saiki H, Shibata D, Hibino T 1993 Localization of cinnamyl alcohol dehydrogenase in Aralia cordata cells. Immunochemical detection. Abstr 43rd Annu Meet Jpn Wood Res Soc, Morioka, Japan, p 33

Tamolang FN, Lopez FR, Semana JA, Casin RF, Espiloy ZB 1980 Properties and utilization of Philippine erect bamboos. In: Lessard G, Chouinard A (eds) Bamboo research in Asia. IDRC, Ottawa, Canada

Tamura G 1957 Further studies on the chemical structure of "Hiochi acid". Bull Agric Chem Soc Jpn 21:202–203

Tanahashi M, Higuchi T 1981 Dehydrogenative polymerization of monolignols by peroxidase and H_2O_2 in a dialysis tube I. preparation of highly polymerized DHPs. Wood Res (Kyoto) 67:29–42

Tanahashi M, Higuchi T 1990 Effect of the hydrophobic regions of hemicelluloses on dehydrogenative polymerization of sinapyl alcohol. Mokuzai Gakkaishi 36:424–428

Tanahashi M, Takeuchi H, Higuchi T 1976 Dehydrogenative polymerization of 3,5-disubstituted *p*-coumaryl alcohols. Wood Res (Kyoto) 61:44–53

Tanaka Y, Matsuoka M, Yamamoto N, Ohashi Y, Kano-Murakami Y, Ozeki Y 1989 Structure and characterization of a cDNA clone for phenylalanine ammonia-lyase from cut-injured roots of sweet potato. Plant Physiol 90:1403–1407

Tepfer SS, Nougarède A, Rondet P 1981 Seasonal studies of vegetative buds of Helianthus tuberosus: concentration of nuclei in phase G1 during winter dormancy. Can J Bot 59:1918–1927

Terashima N, Fukushima K 1988 Heterogeneity in formation of lignin XI. An autoradiographic study of the heterogeneous formation and structure of pine lignin. Wood Sci Technol 22:259–270

Terashima N, Fukushima K 1989 Biogenesis and structure of macromolecular lignin in the cell wall of tree xylem as studied by microautography. In: Lewis NG, Paice MG (eds) Plant cell wall polymers. ACS Symp Ser 399:160–168

Terashima N, Fukushima K, Sano Y, Takabe K 1988 Heterogeneity in formation of lignin X. Visualization of lignification process in differentiating xylem of pine by microautoradiography. Holzforschung 42:347–350

Terashima N, Fukushima K, Takabe K 1986 Heterogeneity in formation of lignin VIII. An autoradiographic study on the formation of guaiacyl and syringyl lignin in Magnolia kobus DC. Holzforschung 40(Suppl):101–105

Terashima N, Fukushima K, He L-F, Takabe K 1993 Comprehensive model of the lignified plant cell wall. In: Forage cell wall structure and digestibility. American Soc Agronomy, Madison U.S.A., 247–270

Terazawa M, Okuyama H, Miyake M 1984 Isolation of coniferin and syringin from the cambial tissue and inner-bark of some angiospermous woods. Mokuzai Gakkaishi 30:409–412

Teutsch HG, Hasenfratz MP, Lesot A, Stoltz C, Garnier J-M, Jeltsch J-M, Durst F, Werck-Reichhart D 1993 Isolation and sequence of a cDNA encoding the Jerusalem artichoke cinnamate 4-hydroxylase, a major plant cytochrome P450, involved in the general phenylpropanoid pathway. Proc Natl Acad Sci USA 90:4102–4106

Thelen MP, Northcote DH 1989 Identification and purification of a nuclease from Zinnia elegans L.: a potential molecular marker for xylogenesis. Planta 179:181–195

Thompson GA, Scherer DE, Van Aken SF, Kenny JW, Young HL, Shintani DK, Kridl JC, Knauf VC 1991 Primary structures of the precursor and mature forms of stearoyl-acyl carrier protein desaturase from safflower embryos and requirement of feredoxin for enzyme activity. Proc Natl Acal Sci USA 88:2578–2582

Thomson RH 1979 Recent advances in the chemistry and biochemistry of quinone pigments. In: Swain T, Harborne JB, Van Sumere CF (eds) Recent advances in phytochemistry, vol 12. Plenum Press, New York, 287–312

Timell TE 1969 The chemical composition of tension wood. Svensk Papperstidn 72:173–181

Timell TE 1978 Ultrastructure of compression wood in Ginkgo biloba. Wood Sci Technol 12:89–103

Timell TE 1982 Recent progress in the chemistry and topochemistry of compression wood. Wood Sci Technol 16:83–122

Timell TE 1983 Origin and evolution of compression wood. Holzforschung 37:1–10

Timell TE 1986 Compression wood in gymnosperms. Springer, Berlin Heidelberg New York, 2150 pp

Torrey JG, Fosket DE, Helper PK 1971 Xylem formation: a paradigm of cytodifferentiation in higher plants. Am Sci 59:338–352

Towers GHN, Stafford HA (eds) 1990 Biochemistry of the mevalonic acid pathway to terpenoids. Plenum Press, New York, 352 pp

Tsukada M, Tabata M 1984 Intercellular localization and secretion of naphthoquinone pigments in cell cultures of Lithospermum erythrorhizon. Planta Med 51:338–340

Tsutsumi Y, Nishida T, Sakai K 1994 Lignin biosynthesis in woody angiosperm tissues III. Isolation of substrate-specific peroxidases related to the dehydrogenative polymerization of sinapyl and coniferyl alcohols from Populus callus cultures. Mokuzai Gakkaishi 40:1348–1354

Tsutsumi Y, Sakai K 1994 Lignin biosynthesis in woody angiosperm tissues II. Peroxidase related to syringyl and guaiacyl lignin biosynthesis in Populus callus cultures. Mokuzai Gakkaishi 40:744–750

Umezawa T 1994 Lignan. In: Higuchi T (ed) Wood molecalar biology. Buneido Publ, Tokyo, 140–145

Umezawa T, Davin LB, Lewis NG 1990a Formation of the lignan, (–)-secoisolariciresinol, by cell free extracts of Forsythia intermedia. Biochem Biophys Res Commun 171:1008–1014

Umezawa T, Davin LB, Lewis NG 1991 Formation of lignans (–)-secoisolariciresinol and (–)-matairesinol with Forsythia intermedia cell-free extracts. J Biol Chem 266:10210–10217

Umezawa T, Davin LB, Yamamoto E, Kingston DGI, Lewis NG 1990b Lignan biosynthesis in Forsythia species. J Chem Soc Chem Commun:1405–1408

Umezawa T, Isohata T, Kuroda H, Higuchi T, Shimada M 1992 Chiral HPLC and LC-MS analysis of several lignans. In: Kuwahara M, Shimada M (eds) Biotechnology in pulp and paper industry. Uni Publishers, Tokyo, 507–512

Umezawa T, Kuroda H, Isohata T, Higuchi T, Shimada M 1994 Enantioselective lignan synthesis by cell-free extracts of Forsythia koreana. Biosci Biotech Biochem 58:230–234

Umezawa T, Shimada M 1996a Formation of the lignan (+)-secoisolariciresinol by cell free extracts of *Anctium lappa*. Biosci Biotech Biochem 60:736–737

Umezawa T, Shimada M 1996b Enantiomeric composition of (–)-pinoresinol, (+)-matairesinol and (+)-wikstromol isolated from Wikstroemia sikokiana. Mokuzai Gakkaishi 42:180–185

Underhill EW, Watkin JE, Neish AC 1957 Biosynthesis of quercetin in buck wheat I. Can J Biochem Physiol 35:219–228

Updegraff DM 1969 Semimicro determination of cellulose in biological materials. Anal Biochem 32:420–424

Van Doorsselaere J, Cabrillac D, Grima-Pettenati J, Boudet AM 1994 Isolation of Eucalyptus gunnii clones coding for cinnamoyl CoA reductase, a key enzyme in lignin synthesis. 22nd Aharon Katzir-Katchalsky Conf on Plant molecular biology, biotechnology and environment, Cologne, Germany

Van Doorsselaere J, Dumas B, Baucher M, Fritig B, Legrand M, Van Montagu M, Inzé D 1993b One-step purification and characterization of a lignin-specific O-methyltransferase from poplar. Gene 133:213–217

Van Doorsselaere J, Gielen J, Van Montagu M, Inzé D 1993a A cDNA encoding *S*-adenosyl-L-methionine synthetase from poplar. Plant Physiol 102:1365–1366

Van Tunen AJ, Hartman SA, Mur LA, Mol JNM 1989 Regulation of chalcon flavanone isomerase (CHI) gene expression in Petunia hybrida: the use of alternative promoters in corolla, anthers, and pollen. Plant Mol Biol 12:539–551

Van Tunen AJ, Koes RE, Spelt CE, Van Der Krol AR, Stuitje AR, Mol JNM 1988 Cloning of the two chalcone flavanone isomerase genes from Petunia hybrida: coordinate, light regulated and differential expression of flavonoid genes. EMBO J 7:1257–1263

Vaughan PFT, Butt VS 1970 The action of *o*-dihydric phenols in the hydroxylation of *p*-coumaric acid by a phenolase from leaves of spinach beet (Beta vulgaris L). Biochem J 119:89–94

Vaughan PFT, Eason R, Paton JY, Ritchie GA 1975 Molecular weight and amino acid composition of purified spinach leaf phenolase. Phytochemistry 14:2383–2386

Vian B, Reis D, Mosiniak M, Roland JC 1986 The glucuronoxylans and the helicoidal shift in cellulose microfibrils in linden (Tilia platyphyllos) wood. Cytochemistry in vitro and isolated molecules. Protoplasma 131:185–199

Villemez CL, Lin T-Y, Hassid WZ 1965 Biosynthesis of the polygalacturonic acid chain of pectin by a particulate enzyme preparation from Phaseolus aureus seedlings. Proc Natl Acad Sci USA 54:1626–1632

Volman G, Ohana P, Benziman M 1995 Biochemistry and molecular biology of cellulose biosynthesis. Carbohydrates in Europe. pp 20–27

Voo KS, Whetten RW, O'Malley DM, Sederoff RR 1995 4-Coumarate:coenzyme A ligase from loblolly pine xylem. 1. Isolation, characterization, and complementary DNA cloning. Plant Physiol 108:85–97

Von Rudloff E, Jorgensen E 1963 Biosynthesis of pinosylvin in the sapwood of Pinus resinosa. Phytochemistry 2:297–304

Wada M, Sugiyama J, Okano T 1995 Two crystalline phase (Iα/Iβ) system of native celluloses in relation to plant phylogenesis. Mokuzai Gakkaishi 41:186–192

Waldron KW, Brett CT 1983 A glucuronyltransferase involved in glucuronoxylan synthesis in pea (Pisum sativum) epicotyls. Biochem J 213:115–122

Wahlberg I, Karlsson K, Enzell CR 1972 Non-volatile constituents of deertongue leaf. Acta Chem Scand 26:1383–1388

Wang Q, Little CHA, Sheng C, Oden PC, Pharis RP 1992 Effects of exogeneous gibberellin A4/7 on tracheid production, longitudinal growth and the levels of indole-3-acetic acid and gibberellins A4, A7 and A9 in the terminal shoot of Pinus sylvestris seedlings. Physiol Plant 86:202–208

Wang Y, Herrmann KM, Weller SC, Goldsbrough PB 1991 Cloning and nucleotide sequence of a complementary DNA encoding 3-deoxy-D-arabino-heptulosonate 7-phosphate synthase from tobacco. Plant Physiol 97:847–848

Ward RS 1993 Lignans, neolignans, and related compounds. Nat Prod Rep 10:1–28

Ward RS 1995 Lignans, neolignans, and related compounds. Nat Prod Rep 12:183–204

Wardrop AB 1965 Cellular differentiation in xylem. In: Côté WA (ed) Cellular ultrastructure of woody plants. Syracuse Univ Press, Syracuse, 61–97

Wardrop AB, Bland DE 1959 The process of lignification in woody plants. In: Kratzl K, Billek G (eds) Biochemistry of wood. Pergamon Press, London, 92–116

Wareing PF 1958 The physiology of cambial activity. J Inst Wood Sci 1:34–42

Watanabe I (ed) 1990a Dictionary of biotechnology. Kodansha Publ, Tokyo, 448

Watanabe I (ed) 1990b Dictionary of biotechnology. Kodansha Publ, Tokyo, 415

Watanabe I (ed) 1990c Dictionary of biotechnology. Kodansha Publ, Tokyo, 438

Watanabe T, Imamura T, Koshijima T, Kuwahara M 1992 Quantitative analysis of ester bonds between lignin and glucuronoxylan in lignin carbohydrate complexes. In: Kuwahara M, Shimada M (eds) Biotechnology in pulp and paper industry. UNI Publ, Tokyo, 279–283

Watanabe T, Imamura T, Koshijima T, Kuwahara M 1993 Determination of ester bonds between lignin and glucuronoxylan in Fagus crenata wood. Proc 7th Int Symp Wood Pulp Chem 1:82–86

Watanabe T, Kaizu S, Koshijima T 1986 Binding sites of carbohydrate moieties toward lignin in lignn carbohydrate complexes from Pinus densiflora wood. Chem Lett:1871–1874

Watanabe T, Koshijima T, Azuma J 1987 Selective cleavage of carbohydrate lignin bonds in lignin-carbohydrate complex from Pinus densiflora wood. Proc 4th Int Symp Wood Pulp Chem 2:45–48

Watkin JE, Underhill EW, Neish AC 1957 Biosynthesis of quercetin in buckwheat II. Can J Biochem 35:229–237

Watson JD, Hopkins NH, Roberts JW, Seitz JA, Weiner AM 1977 Molecular biology of the genes, 4th edn. (Translated into Japanese by Matsubara K, Nakamura K, Miura K). Kagaku Dojin Publ, Tokyo, 1163 pp

Watson JD, Michael G, Witkowski J, Zoller M 1992 Recombinant DNA, 2nd edn. Sci Am Books, Freeman, New York, 626pp
Webber JE, Laver ML, Zaerr JB, Lavender DP 1979 Seasonal variation of abscisic acid in the dormant shoots of Douglas-fir. Can J Bot 57:534–538
Weinges K, Spänig R 1967 Lignans and cyclolignans. In: Taylor WI, Battersby AR (eds) Oxidative coupling of phenols. Dekker, New York, 323–355
Welle R, Grisebach H 1988 Isolation of a novel NADPH-dependent reductase which coacts with chalcone synthase in the biosynthesis of 6′-deoxychalcone. FEBS Lett 236:221–225
Welle R, Schröder G, Schiltz E, Grisebach H, Schröder J 1991 Induced plant responces to pathogen attack. Analysis and heterologous expression of the key enzyme in the biosynthesis of phytoalexin in soybean (Glycine max Merr. cv. Harosoy 63). Eur J Biochem 196:423–430
Wellard HJ 1954 Variation in the lattice spacing of cellulose. J Polym Sci 13:471–476
Wengenmayer H, Ebel J, Grisebach H 1974 Purification and properties of a S-adenosylmethionine: isoflavone 4′-*O*-methyl transferase from cell suspension cultures of Cicer arietimum L. Eur J Biochem 50:135–143
Wengenmayer H, Ebel J, Grisebach H 1976 Enzymatic synthesis of lignin precursors. Purification and properties of a cinnamoyl-CoA: NADPH reductase from cell suspension cultures of soy bean (Glycine max). Eur J Biochem 65:529–536
Westermark U 1982 Calcium promoted phenolic coupling by superoxide radical- a possible lignification reaction in wood. Wood Sci Technol 16:71–78
Westermark U 1985 The occurrence of *p*-hydroxyphenylpropane units in the middle-lamella lignin of spruce (Picea abies). Wood Sci Technol 19:223–232
Westermark U, Lidbrandt O, Eriksson I 1988 Lignin distribution in spruce (Picea abies) determined by mercurization with SEM-EDXA technique. Wood Sci Technol 22:243–250
Whetten R, Sederoff R 1991 Genetic engineering of wood. For Ecol Managem 43:301–316
Whetten R, Sederoff R 1995 Lignin biosynthesis. Plant Cell 7:1001–1013
Whiting DA 1985 Lignans and neolignans. Nat Prod Rep 2:191–211
Whiting DA 1987 Lignans, neolignans, and related compounds. Nat Prod Rep 4:499–525
Whiting DA 1990 Lignans, neolignans, and related compounds. Nat Prod Rep 7:349–364
Whiting P, Favis BD, St-Germain FGT, Goring DAI 1981 Fractional separation of middle lamella and secondary wall tissue from spruce wood. Wood Chem Technol 1:29–42
Whiting P, Goring DAI 1982a Relative reactivities of middle lamella and secondary wall lignin of black spruce wood. Holzforschung 36:303–306
Whiting P, Goring DAI 1982b Chemical characterization of tissue fractions from the middle lamella and secondary wall of black spruce tracheids. Wood Sci Technol 16:261–267
Whiting P, Goring DAI 1982c Phenolic hydroxyl analysis of lignin by pyrolytic gas chromatography. Pap Puu 65:592–594, 599
Wiering H 1974 Genetics of flower colour in Petunia hybrida hort. Gen Phaenen 17:117–134
Wiese W, Vornam B, Krause E, Kindl H 1994 Structural organization and differential expression of three stilbene synthase genes located on a 13kb grapevine DNA fragment. Plant Mol Biol 26:667–677
Winter HC, Burris RH 1976 Nitrogenase. Annu Rev Biochem 45:409–426
Wodzicki TJ, Abe H, Wodzicki AB, Pharis RP, Cohen JD 1987 Investigations on the nature of the auxin-wave in the cambial region of pine stems. Plant Physiol 84:135–143

Wodzicki TJ, Wodzicki AB 1980 Seasonal abscisic acid accumulation in stem cambial region of Pinus silvestris, and its contribution to the hypothesis of a late-wood control system in conifers. Physiol Plant 48:443–447

Wolter KE, Harkin JM, Kirk TK 1974 Guaiacyl lignin associated with vessels in aspen callus cultures. Physiol Plant 31:140–143

Wong HC, Fear AL, Calhoon RD, Eichinger GH, Mayer R, Amikam D, Benziman M, Gelfand DH, Meade JH, Emerick AW, Bruner R, Ben-Bassat A, Tal R 1990 Genetic organization of the cellulose synthase operon in Acetobacter xylinum. Proc Natl Acad Sci USA 87:8130–8134

Wood JR, Goring DAI 1971 Distribution of lignin in stem wood and branch wood of Douglasfir. Pulp Pap Mag Can 72:T95–T102

Wright LD, Cressen EL, Skeggs HR, MacRae GED, Hoffmann CH, Walf DE, Folkers K 1956 Isolation of a new acetate-replacing factor. J Am Chem Soc 78:5273–5275

Wu J, Fukazawa K, Ohtani J 1990 Lignin analysis in some tropical hardwoods using ultraviolet microscopy. Res Bull College Exp For Hokkaido Univ 47:353–366

Wyrambik D, Grisebach H 1975 Purification and properties of isoenzymes of cinnamyl alcohol dehydrogenase from soybean-cell-suspension cultures. Eur J Biochem 59:9–15

Yamada A 1981 Synthesis of glycerides. In: Asahi T (ed) Metabolism I. Plant physiology. Asakura Publ, Tokyo, 113–130

Yamaguchi H, Nakatsubo F, Katsura Y, Murakami K 1990 Characterization of (+)- and (−)-syringaresinol di-β-D-glucosides. Holzforschung 44:381–385

Yamaguchi K, Itoh T, Shimaji K 1980 Compression wood induced by 1-N-naphthylphthalamic acid (NPA), an IAA transport inhibitor. Wood Sci Technol 14:181–185

Yamaguchi K, Shimaji K, Itoh T 1983 Simultaneous inhibition and induction of compression wood formation by morphactin in artificially inclined stems of Japanese larch (Larix leptolepis Gordon). Wood Sci Technol 17:81–89

Yamakado M, Kawai S, Katayama Y, Morohoshi N 1991 Cloning and analysis of genomic sequences amplified enzymatically for phenylalanine ammonia-lyases from poplar and tobacco. Mokuzai Gakkaishi 37:761–765

Yamamoto E, Inciong Ma EJ, Davin LB, Lewis NG 1990 Formation of *cis*-coniferin in cell-free extracts of Fagus grandifolia Ehrh. bark. Plant Physiol 94:209–213

Yamamoto F, Kozlowski TT 1987 Effects of flooding, tilting of stems, and ethrel application on growth, stem anatomy and ethylene production of Pinus densiflora seedlings. J Exp Bot 38:293–310

Yamasaki T, Higuchi T 1971 *p*-Hydroxyphenyl component of grass lignin. Mokuzai Gakkaishi 17:117–121

Yang SF, Hoffman NE 1984 Ethylene biosynthesis and its regulation in higher plants. Annu Rev Plant Physiol 35:155–189

Yasuda T, Sakakibara A 1981 Hydrogenolysis of protolignin in compression wood V. Isolation of two trimeric compounds with γ-lactone ring. Holzforschung 35:183–187

Ye Z-H, Varner JE 1993 Gene expression pattern associated with in vitro tracheary element formation from isolated mesophyll cells of Zinnia elegans. Plant Physiol 103:805–813

Ye Z-H, Varner JE 1994 Expression of an auxin- and cytokinin-regulated gene in cambial regions in Zinnia. Proc Natl Acad Sci USA 91:6539–6543

Yokota T, Arima M, Takahashi N, Crozier A 1985 Steroidal plant growth regulators, castasterone and typhasterol (2-deoxycastasterone) from the shoots of sitka spruce (Picea sitchensis). Phytochemistry 24:1333–1335

Yoshizawa N, Satoh M, Yokota S, Ide T 1993 Formation and structure of reaction wood in Buxus microphylla var. insularis Nakai. Wood Sci Technol 27:1–10

Zajaczkowski S 1973 Auxin stimulation of cambial activity in Pinus silvestris. I. The differential cambial response. Physiol Plant 29:281–287

Zaprometov MN 1962 The mechanism of catechol biosynthesis. Biokhimia 27:366–376
Zenk MH 1964 Zur Frage der Biosynthese von Gallussäure. Z Naturforsch 19B:83–84
Zenk MH, El-Shagi H, Schulte U 1975 Anthraquinone production by cell suspension cultures of Morinda citrifolia. Planta Med (Suppl):79–101
Zettl R, Feldwisch J, Boland W, Schell J, Palme K 1992 5′-azido-[3,6-$^3H_2$]-1-naphthylphthalamic acid, a photoactivatable probe for naphthylphthalamic acid receptor proteins from higher plants: identification of a 23-kDa protein from maize coleoptile plasma membranes. Proc Natl Acad Sci USA 89:480–484
Zhang GF, Staehelin LA 1992 Functional compartmentation of the Golgi apparatus of plant cells. Plant Physiol 99:1070–1083
Zhao J, Herrmann KM 1992 Cloning and sequencing of a second cDNA encoding 3-deoxy-D-arabino-heptulosonate 7-phosphate synthase from Solanum tuberosum L. Plant Physiol 100:1075–1076
Ziegler H 1968 Biological aspects of heartwood formation. Holz Roh-Werkst 26:61–68

Subject Index

Springer and the environment

At Springer we firmly believe that an international science publisher has a special obligation to the environment, and our corporate policies consistently reflect this conviction.

We also expect our business partners – paper mills, printers, packaging manufacturers, etc. – to commit themselves to using materials and production processes that do not harm the environment. The paper in this book is made from low- or no-chlorine pulp and is acid free, in conformance with international standards for paper permanency.

Springer

MIX
Papier aus verantwortungsvollen Quellen
Paper from responsible sources
FSC® C105338

If you have any concerns about our products,
you can contact us on
ProductSafety@springernature.com

In case Publisher is established outside the EU,
the EU authorized representative is:
Springer Nature Customer Service Center GmbH
Europaplatz 3, 69115 Heidelberg, Germany

Printed by Libri Plureos GmbH
in Hamburg, Germany